“The rapidly growing climate movement will benefit from this compendium of timely introspection-on-the-go, as we try to figure out how to do what we do more effectively!”

(Bill McKibben, President and Co-Founder of 350.org, USA)

“This handbook is an extremely valuable overview and analysis of the global climate change movement. It shows the range and complexity of this movement, and provides unique insights into its structure, goals, and future prospects. It sets a high standard for future scholarship to meet.”

(Robert J. Brulle, Professor of Sociology and Environmental Science, Drexel University, USA)

ROUTLEDGE HANDBOOK OF THE CLIMATE CHANGE MOVEMENT

This handbook provides a comprehensive overview of the growing transnational climate movement. A dual focus on climate politics and civil society provides a hitherto unavailable broad and systematic analysis of the current global movement, highlighting how its dynamic and diverse character can play an important role in environmental politics and climate protection.

The range of contributors, from well-known academics to activist-scholars, look at climate movements in the developed and developing world, north and south, small and large, central and marginal. The movement is examined as a whole and as single actors, thereby capturing its scope, structure, development, activities, and influence. The book thoroughly addresses theoretical approaches, from classic social movement theory to the influence of environmental justice frames, and follows this with a systematic focus on regions, specific NGOs and activists, cases and strategies, as well as relations with peripheral groups.

In its breadth, balance, and depth, this accessible volume offers a fresh and important take on the question of social mobilization around climate change, making it an essential text for advanced undergraduates, postgraduate students, and researchers in the social sciences.

Matthias Dietz is a political scientist at the Centre for Social Policy Research at the University of Bremen, Germany.

Heiko Garrelts is a political scientist at the Research Centre for Sustainability Studies at the University of Bremen, Germany.

ROUTLEDGE HANDBOOK OF THE CLIMATE CHANGE MOVEMENT

Edited by Matthias Dietz and Heiko Garrelts

Routledge
Taylor & Francis Group
LONDON AND NEW YORK

earthscan
from Routledge

First published 2014
by Routledge
2 Park Square, Milton Park, Abingdon, Oxfordshire OX14 4RN

and by Routledge
711 Third Avenue, New York, NY 10017
First issued in paperback 2016

Routledge is an imprint of the Taylor & Francis Group, an informa business

British Library Cataloguing in Publication Data
A catalogue record for this book is available from the British Library

Library of Congress Cataloging-in-Publication Data
Routledge handbook of the climate change movement / edited by Matthias Dietz and Heiko Garrelts.
pages cm. – (Routledge international handbooks)
Includes bibliographical references and index.
1. Climatic changes–Political aspects. 2. Global warming–Political aspects. 3. Climate change mitigation–Political aspects. 4. Green movement. 5. Environmentalism. 6. Civil society. 7. Climatic changes–Social aspects. 8. Global warming–Social aspects. 9. Climate change mitigation–Social aspects. I. Dietz, Matthias. II. Garrelts, Heiko. III. Title: Handbook of the climate change movement.
QC903.R69 2014
363.738'74–dc23

2013024886

ISBN 13: 978-1-138-23223-5(pbk)
ISBN 13: 978-0-415-83925-9(hbk)

Typeset in Bembo
by Wearset Ltd, Boldon, Tyne and Wear

CONTENTS

ILLUSTRATIONS

Figures

Tables

CONTRIBUTORS

Hans Baer is Associate Professor in the School of Social and Political Sciences at the University of Melbourne. He has published three climate change-related books: *Global Warming and the Political Ecology of Health* (2009) with Merrill Singer, *Climate Politics and the Climate Movement in Australia* (2012) with Verity Burgmann, and *Global Capitalism and Climate Change: The Need for an Alternative World System* (2012).

Philip Bedall is an environmental scientist pursuing a PhD in Political Science at Kassel University in Germany. His research is concerned with climate discourses in transnational NGOs and social movements, with a focus on climate justice and the (re)production of neoliberalism. In addition to his academic work he is involved in the global justice movement.

Patrick Bond is a political economist with longstanding research interests and NGO work in urban communities and with global justice movements in several countries. He teaches political economy and eco-social policy, directs the Centre for Civil Society (Howard College, University of KwaZulu-Natal, Durban, South Africa), and is involved in research on economic justice, geopolitics, climate, energy, and water.

Achim Brunnengräber is Associate Professor at the Department of Political and Social Sciences of the Free University Berlin. His main areas of research are international political economy and global governance as well as environmental, climate, and energy policy.

Donatella della Porta is Professor of Sociology at the European University Institute in Florence, Italy. Her research focuses on social movements, political violence, and terrorism as well as on corruption.

Vito De Lucia is Doctoral Research Fellow at the Faculty of Law of the University of Tromsø, Norway. His research interests include international environmental law, climate justice, critical ecological approaches as well as legal theory and philosophy.

Matthias Dietz is a political scientist and works at the Centre for Social Policy Research of the University of Bremen, Germany. The emphasis of his research lies on environmental and energy policy, social movements and participation.

Heiko Garrelts is a political scientist working at the Research Center for Sustainability Studies, University of Bremen (Germany) and at the Center for Transnational Studies (Working Group Climate Change and Transnational Policy, Universities of Bremen and Oldenburg). Fields of his research cover issues of environmental governance and urban and regional development. He is especially interested in the role of NGOs in political processes.

Christoph Görg studied political sciences, sociology, and philosophy in Frankfurt/Main. Since 2008 he has been Professor for Environmental Governance at the University of Kassel, Germany. Also, he is Head of the Department of Environmental Politics at the Helmholtz Center for Environmental Research in Leipzig. His main areas of research are multi-level governance, biodiversity policy, and societal relationships with nature.

Elaine Graham-Leigh is a member of the socialist organization Counterfire and a former member of the steering committee of the Campaign against Climate Change. She lives in London and is currently working on issues of food and climate change.

Arne Harms is a PhD candidate at the Free University Berlin. He also gives lectures at the University of Cologne. He is an ethnologist and carries out research on environmental relations and migration in South Asia and the Anglo-Saxon Caribbean.

Jonas Hein is a qualified geographer and works at the German Development Institute in Bonn, Department of Environmental Policy and Management of Natural Resources. His fields of research are international forest and climate policies, REDD (Reducing Emissions from Deforestation and Degradation), and environmental change and migration.

Johannes Herbeck is a qualified geographer. Since 2008 he has worked at the Research Centre for Sustainability Studies of the University of Bremen, Germany. His PhD thesis focuses on responses of humanitarian organizations to climate change, taking West Africa as an example. Furthermore, he works as a junior researcher in the BMBF project 'nordwest2050 – Prospects for Climate-Adapted Innovation Processes in North Western Germany.'

Silja Klepp is a social anthropologist and works at the Research Centre for Sustainability Studies at the University of Bremen, Germany. She has done field research in Kiribati, Vanuatu, New Zealand, Italy, Libya, Malta, and Zambia. She is a member of the German Young Academy of Scientists (Die Junge Akademie). Her research focuses on migration and refugees, European migration and border policies, cultural aspects of climate change, legal anthropology, and political ecology.

Claudia König is a PhD candidate at the Free University Berlin. Currently she is a visiting research associate at Universidade Federal do Pará in Brazil. She works on issues of indigenous people, climate protection as well as on economic instruments in climate policy.

Georg Kössler has studied political science in Germany and Sweden. He is involved with the NGO gegestromberlin. Before he started working for the fraction of the Green Party in the German Parliament, he was engaged in the Heinrich Böll Foundation and was a member of the board of the youth organization of the Green Party.

Johannes Kruse is a PhD fellow at the Bremen International Graduate School of Social Sciences (BIGSSS), Germany. His research interests include international environmental and climate policy as well as the role of transnational actors from civil society in international politics.

Melanie Müller is a PhD candidate at the Free University Berlin. Her dissertation topic deals with the development of the South African environmental movement. She receives a scholarship from the German Federal Environmental Foundation. Her areas of research are international environmental and climate politics, development policy, and social movements.

Andreas Niederberger is Associate Professor for Philosophy at the Goethe University Frankfurt/Main. Currently he is Visiting Professor at the University of Duisburg-Essen. His main areas of research are political theory, democratic theory, and climate justice.

Louisa Parks is Lecturer at the School of Social and Political Sciences of the University of Lincoln in the UK. She carries out research on the subjects of social movements, civil society, and the European Union.

Oliver Powalla is a PhD candidate at the University of Marburg, Germany. He works on sociology of knowledge, environmental sociology as well as on political economy.

Patrick Schroeder works in Beijing as International Advisor to the China Association for NGO Cooperation (CANGO) to support the coordination and development of the China Civil Climate Action Network (CCAN). This position is supported by the GIZ/Center for International Cooperation and Development (CIM). He also works as a consultant for the Collaborating Centre on Sustainable Consumption and Production in Wuppertal, Germany. He holds a Masters in International Relations and a PhD in Environmental Studies.

Dörte Segebart is Assistant Professor for Development and Gender at the Institute of Geographical Science of the Free University Berlin. The emphasis of her work lies in the field of humans and the environment. She conducts research on governance, participation, and sustainable management of natural resources. Thereby she focuses on Latin America, especially Brazil.

Brian Tokar is an activist and author, Director of the Vermont-based Institute for Social Ecology, and Lecturer in Environmental Studies at the University of Vermont. He is the author of *The Green Alternative* (1987), *Earth for Sale* (1997), and *Toward Climate Justice* (2010).

Heike Walk is Project Director at the Center of Technology and Society (Technical Universität Berlin) and at the WZB Berlin Social Research Center. Her research interests are new social movements, participation, multi-level governance, sustainable development as well as environmental and climate policy.

ABBREVIATIONS

ACF	Australian Conservation Foundation
AGW-LCA	Ad hoc Working Group on Long-term Cooperative Action
ALBA	Bolivarian Alliance for the Americas
AOSIS	Alliance of Small Island States
AYCC	Australian Youth Climate Coalition
BUKO	Bundeskonferenz Internationalismus (Federal Conference Internationalism)
BUND	Bund für Umwelt und Naturschutz (Friends of the Earth Germany)
CAG	Local Climate Action Group
CAN	Climate Action Network
CBD	Convention on Biological Diversity
CCAN	China Civil Climate Action Network
CCB	Climate Community and Biodiversity Standard
CCBA	Climate Community and Biodiversity Standard Alliance
CCS	carbon capture and storage
CDM	Clean Development Mechanism
CI	Conservation International
CJA	Climate Justice Action
CJN!	Climate Justice Now!
CO_2	carbon dioxide
COP	Conference of Parties
CPRS	Carbon Pollution Reduction Scheme
CSO	civil society organizations
CTW	Carbon Trade Watch
DKKV	Deutsches Komitee für Katastrophenvorsorge (German Committee for Disaster Prevention)
DRR	disaster risk reduction
EAC	Energy Action Coalition
ECOSOC	United Nations Economic and Social Council
EDF	Environmental Defense Fund
EEZ	Exclusive Economic Zone
EPA	Environmental Protection Agency

FoEI	Friends of the Earth International
FSC	Forest Stewardship Council
G77	Group of 77
G8	Group of 8
GCCA	Global Campaign for Climate Action
GDP	gross domestic product
GEF	Global Environment Facility
GGCA	Gender and Climate Alliance
GHA	Global Humanitarian Assistance
IASC	Inter-Agency Standing Committee
ICHRP	International Council on Human Rights Policy
ICRC	International Committee of the Red Cross
IDMC	Internal Displacement Monitoring Center
IED	Institute for Environment and Development
IEN	Indigenous Environmental Network
IETA	International Emission Trading Association
IFRC	International Federation of Red Cross and Red Crescent Societies
ILO	International Labour Organization
IPCC	Intergovernmental Panel on Climate Change
IPRs	intellectual property rights
IMF	International Monetary Fund
JI	Joint Implementation
K!BN	Klima!Bewegungsnetzwerk (Climate Movement Network)
LDC	least developed countries
MCJ	Mobilization for Climate Justice
MSF	Médecins Sans Frontières
MSV	Many Strong Voices
NABU	Naturschutzbund Deutschland (Nature and Biodiversity Conservation Union)
NAPA	National Adaptation Program of Action
NASA	National Aeronautics and Space Administration
NGO	nongovernmental organization
NTAC	Never Trust A COP
OCHA	Office for the Coordination of Humanitarian Affairs
OECD	Organisation for Economic Co-operation and Development
ppm	parts per million
REDD	Reducing Emission from Deforestation and Forest Degradation
RTNA	Rising Tide North America
RTUK	Rising Tide United Kingdom
SEE	Society for Ecology and Entrepreneurs
SIDS	Small Island Developing States
SSN	SouthSouthNorth
TCP	The Climate Project
TEEB	The Economics of Ecosystems and Biodiversity
TEK	traditional ecological knowledge
TMN	transnational municipal network
TNC	The Nature Conservancy
TRIPS	Trade-Related Aspects of Intellectual Property Rights
TWN	Third World Network

UN	United Nations
UNCED	United Nations Conference on Environment and Development
UNFCCC	United Nations Framework Convention on Climate Change
UNHCR	United Nations High Commissioner for Refugees
VCS	Verified Carbon Standard
WBGU	Wissenschaftlicher Beitrat der Bundesregierung Globale Umweltfragen (German Advisory Council on Global Change)
WEACT	West Harlem Environmental Action
WHH	Welthungerhilfe (German World Hunger Aid)
WTO	World Trade Organization
WWF	World Wildlife Fund for Nature

1

INTRODUCTION

Contours of the transnational climate movement – conception and contents of the handbook[1]

Heiko Garrelts and Matthias Dietz

1 Introduction

This handbook is one of the first of its kind to attempt an extensive documentation and analysis of the multiple facets of a social movement which emerged only a few years ago and which is focused on climate change. Activists, nongovernmental organizations (NGOs) and numerous other actors from many different sections of society are engaged – at local, regional, national, and transnational levels – in struggles against the causes and perpetrators of global warming and for a socially just approach to climate change mitigation. The climate movement formulates demands for more decisive action, in particular at the international policy level. It has grown in parallel with the increasing public salience of anthropogenic climate change which assumed relevant proportions with the publication of the Intergovernmental Panel on Climate Change's Fourth Assessment Report in 2007 and reached a peak during the 2009 United Nations Climate Change Conference in Copenhagen. The climate movement continues to be most visible, both internationally and in the media, during the annual UN conferences on climate change. In late 2011, for example, several thousand members of the movement demonstrated in Durban, South Africa and elsewhere in support of the inhabitants of Pacific islands threatened by rising sea-levels (Henn 2011). A protest march through Doha of 800 people to coincide with the Conference of the Parties (COP) which was held on December 1, 2012 was in fact the first political demonstration ever to take place in the host country of Qatar. Some observers have even spoken since of an incipient Arab Youth Climate Movement (Vidal 2012). Other protests and campaigns which continue throughout the year find less resonance in the media. Climate camps, a specific form of protest which goes back several years, were held in countries such as Germany and Italy in the summer of 2012. Numerous campaigns and social actions of civil disobedience against the construction of oil pipelines took place in the US and Canada in 2011 and 2012.

2 Lines of differentiation in the movement

Actors in the climate movement regard global warming as an existential threat against which extensive and decisive action must be taken. Protestors insist that emissions must be radically reduced, energy obtained from entirely new sources, and lifestyles and economic systems altered if we are to avoid catastrophic climate change. For many activists, the irreversible tipping point after which events and their impact can no longer be controlled or limited has almost been reached. For many activists climate change also raises issues of social justice, as global warming affects some regions and sections of the population more strongly than others, unequal resources and capabilities for adaptation are available, and the volume of emissions and therefore the share of responsibility for climate change varies from country to country. These and similar positions are advocated on all the continents – albeit more pointedly in Europe, North and South America, and Australia. There are at least some civil society actors in many countries who are actively engaged in climate politics and who to some extent also are connected to the structures of the transnational climate movement. This is the case in both India as well as the Middle East and North Africa region where activists called for climate action in the guise, for example, of the Arab IndyACT organization or in the framework of a workshop run by the globally active TckTckTck organization (Rigg 2012).

The climate movement has considerable, albeit limited, potential for mobilization. The movement is made up of hundreds of organizations, several thousand activists, and large numbers of sympathizers and supporters. Millions of signatures for its causes have already been collected on the Internet, around 50,000 people demonstrated in Copenhagen and London in 2009, and thousands of people regularly take part in global days of action organized by the movement. Despite its limited size and visibility the climate movement has already had a distinct impact. A good example is the planning moratorium imposed by President Obama on the Keystone XL pipeline, which was to transport oil taken from Canadian tar sands to refineries in the US, following protests by the climate movement. In Germany, protests by anti-aviation expansion activists initially thwarted plans for the construction of a third runway at Munich Airport. The impact of activities by the climate movement are also apparent at the individual level: their members install solar panels, plant trees, or, as with the 10:10 campaign, make voluntary commitments to cut their own carbon emissions, find out about environmentally friendly products on portals such as Utopia, and offset carbon emissions by donating to organizations, such as MyClimate, which provide financial support to projects that reduce the emission of greenhouse gases. Carbon offset providers such as these, which operate on a noncommercial basis, are also peripheral members of the climate movement.

Although demands for reductions in carbon emissions are common ground for all the actors involved, the climate movement is nonetheless anything but a homogeneous phenomenon. It is not only divided along lines of nationality, there are also at least three other lines of differentiation. Demands for climate justice are central for one section of the climate movement which identifies capitalist structures as the cause of climate change and diagnoses a deep-reaching crisis of the current political and economic system. Key importance is attached to demands for *distributive* and *procedural justice*. This part of the climate movement is referred to occasionally in the book as radical, in the strict sense of the word. A second group of actors is committed to a paradigm of *ecological modernization* according to which environmental problems can be resolved politically, economically, and technologically within the context of real existing institutions and power structures. Importantly, adherents of the modernization camp also believe that their aims are compatible with continuing economic growth. Their belief in science and technology, in the power of governments and markets opens up space for *processes of institutional learning* and political *reforms* (Bailey et al. 2011). This group believes that the dynamics of capitalist growth should therefore be used to develop and disseminate even more advanced environmentally friendly technologies and to create a large number of green jobs in the process. This then is the dividing line along which the various groups and wings of the climate movement move away from each other at tangents. The Climate Justice Now! coalition, the La Via Campesina network, which acts as a voice for indigenous and peasant groups, as well as small and inchoately formalized NGOs such as Plane Stupid, are good examples of actors in the first, more radical group. Established environmental organizations, such as Friends of the Earth, and internet-based campaign groups, like 350.org, are more typical of the second group, which is committed to nurturing solutions which are immanently available within the existing system.

A second line of differentiation refers to the *type of actors* in various groups. 'Traditional' nongovernmental organizations of the ilk of Greenpeace or the WWF have long since ceased to be the main actors and driving forces of the climate movement. Totally informal or scantly organized actors have joined the scene – in some cases at the international level, in others locally. State officials from countries such as Bolivia, Venezuela, or the small Pacific island states have also begun to make common cause with climate activists both during and outside the confines of UN conferences. Institutions and individuals in the world of science are of central importance for the climate movement. Climate protection activities are also now engaged in by other organizations and actors which had long ignored or even denied the existence of climate change. Of particular importance in this connection are market-based actors, several of which have renounced their previous fundamental opposition and have now integrated climate protection concerns into their corporate strategies. The remarkable story of renewable energies and the economic instruments deployed as part of prevailing climate policy (see below) have incentivized many market-based players to take action (Bulkeley and Newell 2010; Paterson 2011). Other actors involved in combating climate change and who have established relationships with the climate movement in the process include cities, municipalities, and local authorities. These players – which have founded their own transnational municipal networks – are both the causal agents of climate change and are in turn affected by it (Kern and Bulkeley 2009; Bulkeley 2011). One such network is the Climate Alliance of European Cities with Indigenous Rainforest Peoples. Other actors, such as humanitarian organizations, are engaging with climate change issues, which for a very long time made up no part of their agenda, by supporting adaptation projects. Humanitarian organizations maintain contacts with the climate movement via links such as the Climate Action Network (CAN). Finally, consumers in the Northern hemisphere are becoming increasingly aware of climate change issues and their moral implications for consumption behavior. All in all, private actors are now significant agents of action to combat climate change, including in a transnational context (Bulkeley and Newell 2010).

The third line of differentiation follows age and gender. In this context the conflict about climate change is *simultaneously* an intergenerational clash, or is at least framed in this way (see Boese 2011). This is a perspective in which antique politicians and managers are the adversaries of young "avant-garde climate savers." Daniel Boese, who counts himself as part of this avant-garde, believes he has seen the "birth of the biggest youth movement of all times, and one which is more political and international than the movement of 1968 ever was" (Boese 2011: 26, our translation). For Boese "Facebook, Wikis and Google.docs have made a reality of the slogan of international solidarity" (Boese 2011: 26, our translation). Lipschutz and McKendry (2011) also broach the same issue – from an analytical perspective. They explain growing public interest in climate change in terms of a combination of moral panic[2] and self-interest or at least worry about the negative impact (physical or economic) of climate change on future conditions of life (Lipschutz and McKendry 2011: 373). Gender-specific organizations are relative newcomers to the field of climate change and civil society. In recent years transnational networks with names such as 'Gender CC – Women for Climate Justice' have been founded alongside existing organizations. This trend primarily reflects the unequal impact which climate change, and the extreme weather events it brings in its wake, has on men and women. At the same time, these groups highlight the proactive role of women (as agents of change) in adjusting to climate change and in reducing emissions (Röhr et al. 2008; Schalatek 2011; for an overview see Skinner 2011).

3 Climate change as a field of policy

With regard to the *field of policy* in which the actors and organizations which are examined in this book 'move,' it is important above all to underline the sheer scope of the 'climate change challenge.' This is due to the factor of energy consumption, on the one hand, and the special significance of fossil fuels on the other. According to Rosenbaum and Mautz the

> large-scale fossil fuel technologies which have emerged since the mid-19th century are the central motor driving the modern industrialised countries and represent a system which is firmly entrenched in developed societies and which has been followed and adopted by newly industrialised and emerging countries in the last third of the 20th century.
>
> *(Rosenbaum and Mautz 2011: 405, our translation)*

The authors refer in their essay to the physicist Amory Lovins. The latter's concept of soft energy paths describes complex socio-technical systems consisting of interlinked and interdependent elements. These elements are both technological (e.g., centralized production plants geared towards aggregate growth), and social (e.g., energy-intensive lifestyles and consumption patterns, close links with the world of politics, public administrations, and trade unions, imperialistic energy foreign policy) (Rosenbaum and Mautz 2011: 405). The goal today is to surmount these socio-technical systems, or at the very least their negative implications. The economist Elmar Altvater elucidated the scale of imminent social transformation with reference, among other factors, to the essential properties of fossil fuels: the ability to put them to use anywhere and anytime, the advantages of fossil mobility, and the flexible use of this source of energy in all walks of life and areas of work (Altvater 2006: 172). The congruence between social formation, technology, markets, and fossil fuels is of central importance for Altvater. When "the age of fossil capitalism comes to a close" and it is time to devise and implement a renewable energy regime, the "social formation of capitalism will also need to be adjusted: a transition more profound and all encompassing than the French or Russian revolutions ever were" (Altvater 2006: 182, our translation; see also Altvater 2005). The last Flagship Report produced by the German Federal Government's Advisory Council on Global Change (WBGU) lends further weight to the urgent challenge of climate change (WBGU 2011). While it approaches the same issues from a very different angle than Altvater, the council also describes "how the requisite transformation encompasses profound changes to infrastructures, production processes, regulation systems and lifestyles and extends to a new kind of interaction between politics, society, science and the economy" (WBGU 2011: 1, our translation). The causes of climate change can be found in almost all fields of contemporary social development. Climate change goes far beyond the issue of pollution (Glover 2006). The crucial factor now is that the climate movement is up against hugely powerful economic actors in all these different fields (Newell and Paterson 2010: 78). The power of these actors and their political allies is the main reason why, within the current political process, energy and climate policy continue to fall within the remit of separate institutional bodies:

> When it comes to climate change, all eyes turn to carbon emissions and the global impact they have. However, emissions have only to do with the *output* side of the fossil

> energy system. As a result, climate change is described, defined and politically constructed in the dominant discourse as a global environmental problem. If, on the other hand, the focus is shifted to the consumption of coal, gas and oil, in other words to the *input* side of fossilism, then account must also be taken of regional and national interests.
>
> *(Brunnengräber 2008: 138, our translation, emphasis in the original)*

One section of the climate movement also pursues and reinforces the output-oriented approach and aims to 'cap' total global emissions. Another part of the movement begins precisely with the input side of the fossilist equation referred to by Brunnengräber.

According to Rosenbaum and Mautz (2011: 406), the energy system of the developed industrial societies was supported by a broad social consensus until the end of the 1960s. This consensus dissolved in subsequent decades under the pressure of diverse debates and very specific events – the projected exhaustion of fossil energy sources, the entire 'energy security' complex (including in the light of OPEC's 1973 oil embargo), the threats to the continuing existence of humankind itself posed by environmental destruction and its climatic effects – and the players responsible for fossil energy systems were forced to respond to controversial public discussions. This was thanks in no small part to the "interaction of critical scientists and civil society" (Rosenbaum and Mautz 2011: 406, our translation). These controversial public debates – or more precisely, the anything but self-explanatory *politicization of climate change* (Viehöver 2011) – gave birth around two decades ago to international climate politics. The two major milestones in international climate politics have been the signing of the United Nations Framework Convention on Climate Change (UNFCCC) in 1992 and the establishment of the Intergovernmental Panel on Climate Change (IPCC) four years earlier. This means that there now exists an international convention which defines its objective as preventing dangerous anthropogenic interference with the climate system as well as slowing down global warming and mitigating its effects (UNFCCC, Article 2). A decisive issue in the political process is to what extent and in what way industrialized countries limit and reduce their carbon emissions. The background is that while these countries account for around 75 percent of all carbon emissions, they are at the same time home to a mere 25 percent of the global population. Of key importance now is that there has been a shift in recent years in the global distribution of energy consumption which has stabilized in the industrialized countries and grown rapidly in the countries of the Global South (particularly in newly industrialized countries). At the same time the issue of how to manage the impact of climate change (droughts, flooding, etc.) has assumed increasing urgency. According to Goodman (2009: 510) the balance of power in the climate crisis is shifting *in favor* of the Global South.[3] In this respect "climate in North–South relationships" (Dietz and Brunnengräber 2008, our translation) is a key source of dynamism in international climate politics (Paterson 2011: 612; Rosenbaum and Mautz 2011: 400). The climate movement has set its sights firmly on the North–South dynamic. At the same time, the movement itself shapes this dynamic – particularly in terms of the idea of justice which it advocates.

According to Paterson (2011: 611–612), another highly dynamic aspect of climate change politics is the very close relationship which exists between science and policymakers. Scientists pointed out the causes and possible consequences of anthropogenic climate change – and gave the discussion a transnational and international dimension – as early as the 1970s (Viehöver 2011: 683). In today's politics of climate change *policymaking and politics* are regarded now more than ever as being *beholden to science* because "in view of completely new kinds of global risk, politicians are increasingly forced to rely on the problem-solving and legitimation resources of experts" (Viehöver 2011: 680, our translation; see also Weingart et al. 2008). There is thus a close empirical connection between the expertise of the Intergovernmental Panel on Climate Change and climate policy negotiations on the international stage. The reports published by the IPCC represent something resembling the "limits of the possible" (Paterson 2011: 611; see also Goodman 2009: 510) and have played a key role in generating the current dominant perspective in the field with all that this entails for climate policy, i.e., that, for the climate, it is irrelevant where exactly harmful emissions such as CO_2 are produced and where they are reduced (Paterson 2011: 61). However, what this discourse often misses is the fact that emissions are caused in some regions by people who are trying to meet their basic needs and in others by people who are consuming luxury goods (Agarwal 2000, cited in Bailey et al. 2011: 693). Yet it is only against the background of a perspective in which the omnipresence of greenhouse gases, on the one hand, and *efficiency* requirements, on the other, are central that it is possible to explain why instruments such as emissions trading have become established. The knowledge we have gained about climate change, however, is based on computer-generated simulations which work with imprecise input data and which are used to develop models. Climate researchers have now moved on from calculating linear relationships and now favor complex processes which are coupled with uncertain knowledge which has "no direct correspondence to reality" (Grittmann 2012: 178, our translation; see also Beck 2009). This is another reason why climate change science consistently attracts opposition, in many cases in the form of contradicting second expert opinions. The challenges which make up elements of a "*politicization of the natural and technological sciences*" (Rosenbaum and Mautz 2011: 406, authors' note) are intended to prevent or at the very least delay effective climate policy regulation (North 2011: 1586–1587; Paterson 2011: 612). At the same time, some critical voices express the fear that a diagnosis of climate change by natural scientists could reinforce the illusion that the use of technology would provide an adequate "solution to the problem" of the climate crisis (Goodman 2009: 510). In addition, climate science can assume the features of an "elitist mode of knowledge" (Goodman 2009: 510). All in all, the process of a politicization of supposedly "hard facts" is more complex. It needs only be added at this point that the high degree of relevance the problem of climate change possesses is itself a discourse effect for which one prerequisite – the politicization of the issue – has already been met (Viehöver 2011: 676; see also Weingart et al. 2008). Actors in the climate movement are embedded at various sites in this dynamic – they produce their own knowledge, feed knowledge into political processes, or communicate knowledge by means of visualizations (e.g., using the image of polar bears threatened by climate change) and use it in political campaigns – in many cases by harnessing the mass media (Hulme 2009; Manzo 2009; Grittmann 2012).

4 Preliminary notes regarding the term social movement

The chapters and title of the handbook refer to the climate movement as the object of investigation. We have not used the term 'movement' in an unconsidered fashion, nor indeed have we overused the term in the way which draws the venom of many researchers (e.g., see Batliwala 2002: 398). The climate movement meets many of the standard definitions of social movements as

> an action system, formed for a certain period of time and based on collective identity, of mobilized networks of groups and organizations which aim to bring about, prevent, or reverse social change by means of protest – if necessary, violent protest.
>
> *(Rucht 1994: 76–77, our translation)*[4]

The climate movement endeavors to realize its (i) shared, higher-order objective of averting climate change primarily by (ii) means of various forms of protest which have assumed *renewed priority* over and above the lobbying and advocacy activities of NGOs over the last three to four years. The climate movement has already existed for several years and thus fulfills the third criteria (iii) of remaining in action for a relatively prolonged period of time. However, it is doubtful, bearing in mind the lines of differentiation drawn above, whether the climate movement's diverse actors meet the criterion (iv) of a shared identity. Conflicts between different groups and actors – who are keen to put a perceptible distance between each other – have at times assumed unmistakably antagonistic proportions (e.g., see Dietz 2010). This is also the consequence of several formalized actors taking part in protests while at the same time being 'compromised' by their active role in the concrete implementation of existing climate policy instruments. This co-opted involvement is regarded highly critically by parts of the climate movement and has led on several occasions to conflicts on the ground with other movement actors (see also Unmüßig 2011). The element of protest is thus now a highly complex phenomenon. Protest is not merely directed against the proponents of fossilism, but also against climate policy actors in civil society themselves. Despite these conflicts there is still much to be said for the use of the term 'movement' in this book. First, conflicts and attempts by social movements to disassociate themselves from formalized actors are not in themselves unusual[5] (Brunnengräber 2005: 334–335; Schultze 2008: 428). Second, certain tentative rapprochements between climate groups became apparent in 2010 (Dietz 2012). And third, the frames and positions of climate groups which were previously especially critical and keen to fuel conflicts now appear to enjoy majority support.

'Climate movement' is used in this book as a generic and collective term which covers actors of various kinds with very different backgrounds in national and issue-oriented submovements. Members of the climate movement often have closer and longer lasting relationships with each other than with the collective movement itself. These relationships become apparent and activated in contexts such as the UN Climate Change Conferences, in large networks and campaigns, or in shared demands, such as for climate justice. References in this book to climate movements in the plural therefore designate the wings or submovements which, at the middle level, form the climate movement as a whole together with organizations and activists at the grassroots level. In some respects the climate movement may be described as a loose, but nonetheless highly active umbrella structure which is supported, shaped, and used by a multiplicity of civil society actors who are active in climate politics.

It is important in any attempt at description and definition to remember that social movements, and thereby the climate movement as well, do not refer to just any possibly insignificant actors chosen at random. Social movements are more than single NGOs, citizens' initiatives, or activists. They are complex, in many cases decentralized networks, whose members act in solidarity with each other, have very strong convictions, and who, in many cases, are normatively or ideologically motivated in their struggle for shared aims. Social movements can produce a powerful dynamic, generate considerable political and media pressure, and make a decisive contribution to social change. Germany's antinuclear movement is a case in point, managing as it did to create a broad skepticism in wider society about nuclear power – a technology which was still greeted euphorically in the 1970s – and to make a decisive push towards its ultimate phasing out (Rucht 2008: 264–265). The global justice movement successfully engendered more critical perceptions of neoliberalism (Rucht and Roth 2008: 510) and delivered ideas and templates for political responses to the financial crisis, such as the imposition of taxes on financial markets. Social movements in Eastern Europe and, most recently, in the Arab world have even contributed to "revolutionary upheavals" and the transformation of entire systems (Roth and Rucht 2008: 31, our translation).

5 Publications on the subject

Since 2000 a small branch of international research has grown up around the subject of the climate movement, and has become even stronger since 2005. The paucity of books written to date is compensated for by some dozens of essays published in this field. Recent publications on the subject are, among other things, *Politics of Climate Justice: Paralysis Above, Movement Below* (monography, Bond 2012) and "Articulating Climate Justice in Copenhagen: Antagonism, the Commons and Solidarity" (journal article, Chatterton et al. 2012), one compendium on the "Civilising of the Climate Regime" (Brunnengräber 2011, our translation), and, last but not least, another monography on the engagement of the "Generation Facebook" in the climate movement (Boese 2011). As is the case with most of the other publications, these books address certain selected aspects or sections of the climate movement. Brunnengräber has collected texts on various aspects of current climate politics which not only shed critical light on the role of NGOs but also investigate the activities of other actors active in the climate policy arena (such as the media). This compendium is referred to repeatedly in this book. Boese deals with the youth climate movement and Bond as well as Chatterton elaborate on the climate justice movement. The issue of climate justice and the actors struggling for it are frequent objects of investigation in publications on the climate movement (see Dorsey 2007; Brand et al. 2009; De Lucia 2009; Goodman 2009; Carrington 2010; Evans 2010; Tokar 2010; Bond 2012; Russell et al. 2012). The titles are, for example, "Towards Climate Justice: Perspectives on the Climate Crisis and Social Change" (Tokar 2010) or "Climate Justice: The Emerging Movement against Green Capitalism" (Dawson 2010). Publications which are not limited to climate justice issues and which take a broader look at the climate movement are relatively few and far apart; North's work (2011) is a good example; the same is the case with Hans Baer (Baer and Reuter 2011), Philip Bedall (2011), or Matthias Dietz (2010). The literature which deals specifically with the climate movement contains a considerable number of texts which focus on individual countries. Several papers have been written on the climate movement in Australia (Baer 2009; Diesendorf 2010; Baer and Burgmann 2012), the US (Moser 2007; Dayaneni 2009), and Great Britain (Skrimshire 2007; Pusey and Russell 2010; North 2011) for example, as well as in a few other countries (for Russia, see Neuner 2011). The perspective of most of the publications which appeared in the early years of the climate movement tends not to have been on climate justice. Some of the first texts to appear on the climate movement were *Climate for Change: Non-state Actors and the Global Politics of the Greenhouse* (Newell 2000) and "Building Networks from the Outside: International Movements, Japanese NGOs and the Kyoto Climate Change Conference" (Reinmann 2002). One common feature of these early publications is also that they handle the term 'movement' circumspectly and are strongly focused on climate negotiations at the UN.

There are also publications which tackle the subjects of climate change and society from a much broader angle. The climate movement is just one of a number of topics addressed by these publications which also cover such issues as the role of science, hermeneutic conflicts regarding the existence of climate change, or the economic and social consequences of climate change (e.g., in the form of climate refugees). Among these publications is the *Routledge Handbook of Climate Change and Society* (Lever-Tracy 2010) and the *Oxford Handbook of Climate Change and Society* (Dryzek et al. 2011). Elements and actors of the climate movement are referred to in these books; however, the term climate movement is not used, with exceptions of Lipschutz and McKendry (2011) and Routledge (2011).

After tailing off in the mid-1990s, research on social movements in general appears to have received a new lease of life recently. This is due to the numerous protests which have taken place against the global financial crisis, in the Arab world or against large-scale construction projects. Several new social movements have arisen as part of a new "movement cycle" (Brunnengräber 2012, our translation) which are now being studied by scholars (e.g., see Beinin 2012; Castaeda 2012; Kraushaar 2012; Rucht 2012). These new social movements include the Occupy movement, which is active in a number of different countries, the anti-austerity or anticuts movement in the United Kingdom, the 15-M Movement (the Indignants) in Spain, or the protest movement against the Stuttgart21 railway station in Germany. Even if the climate movement is somewhat older than these new movements, it is nonetheless considered to be part of this current movement cycle (Brunnengräber 2012: 43) and has numerous links with movements which have emerged more recently. Diverse connections also exist between the likewise older and frequently studied global justice movement and the climate movement (see Goodman 2009). The criticism of capitalism and demands for climate justice expressed by the radical wing of the climate movement, for example, places the movement in fairly close proximity to the goals and activists of the global justice movement (see recently, e.g., Banse and Habermann 2012 or della Porta and Rucht 2012). This wing of the climate movement is even regarded by some actors and authors as being the successor to the global justice movement (Müller 2008; North 2011: 1583).

Finally, the climate movement is also relevant in the research which has been undertaken on a considerable scale in recent years on the transnationalization of protests and movements (e.g., della Porta 2006b; della Porta and Diani 2006; Byrd 2010; Cottle 2011; Herkenrath 2011) and the political use made of the Internet (e.g., Mattoni 2009; Roberts 2009; Mercea 2012). Civil society groups from a wide variety of countries work as part of transnational networks – such as CAN – in the climate movement, for example. They also make intensive and innovative use of the Internet to communicate, to mobilize, to engage in online protest, and to document protest action.

6 Conception and contents of the handbook

This publication takes up where existing research leaves off. Adopting the handbook format, its purpose is to depict and analyze the climate movement in a systematic fashion. We have chosen the broad and fundamental approach of a handbook because, given that the climate movement is still in its formative years, relatively small and still attracts only limited media attention, it has not as yet been the subject of in-depth scholarly research. The overall aim of the handbook is to collate, examine, and expand on current knowledge about the climate movement. The handbook endeavors to describe the climate movement, to consider the movement in the context of previous research on social movements, and finally to weigh up its influence and potential for bringing about change. The handbook considers some fundamental issues: Who belongs to the climate movement? What does the movement do? What influence, characteristics, and development is the movement considered to have?

With this goal in mind, the climate movement is regarded as a whole and 'in its constituent parts.' The theory-led chapters in the first part of the book take an overarching look at the movement and larger topics and issues. Framing approaches are applied in this part alongside democracy, justice, or critical theories. The purpose of these chapters is to assess which theoretical approaches and methodological tools are appropriate for studying and understanding the climate movement and showing in what areas research could be profitably undertaken in the future. The handbook character is particularly apparent in the second part of the book which analyzes particular examples on a chapter by chapter basis of national and sectoral submovements, different types of actors and arenas, and important organizations and activists.

A portrait of the climate movement would not be complete without some account being taken of the time dimension. This handbook therefore not only depicts the current status of the movement but also looks back to its roots and past and looks ahead to its future. A very mixed picture emerges from this perspective on the movement's development over time. The initial founding boom and wealth of activities after 2007 was followed by the disappointing Climate Change Conference in Copenhagen in late 2009 and a related crisis and reorientation in the movement. An important current issue is the cooperation with other movements, such as Occupy.

It is important to adopt a variety of perspectives if the climate movement is to be captured in all its diverse facets. Activists as well as scholars have been asked to contribute to the handbook. The insights provided by activists enable the reader to grasp the ways in which movement actors think and offer valuable detailed and insider knowledge about the climate movement's activities, debates, and plans. At the same time, the perceptions offered by activists necessarily retain a degree of subjectivity with regard to questions such as the influence exercised by the climate movement. In their turn the scholars who have contributed to the handbook come from a broad range of disciplines and bring with them experience in movement and NGO research. Views provided from the perspectives of political science and sociology are complemented by authors working in the fields of anthropology, ethnology, or human geography. Theoretically and philosophically oriented authors have also participated. The aim of bringing together such a diversity of authors was to enable the handbook to provide an all-in-all authentic and critical picture of the climate movement. Nonetheless, it is important to remember that scholars' approaches to social movements are just as likely to incorporate certain normative elements as are those of activists (see Teune 2008: 541; Rucht 2011: 34).

The following contents are dealt with in the handbook:

In the second chapter, Donatella della Porta and Louisa Parks identify categories and mechanisms of framing processes within the climate movement in order to explain changes in the focus of many groups as well as activists.

Melanie Müller and Heike Walk investigate in Chapter 3 the democratic structures and the options for participation in the field of international climate politics. Accordingly, they find numerous barriers which make it hard for climate groups and activists to take part. From these findings, they show how the current climate regime could be complemented by mechanisms that push participation.

In Chapter 4, Phillip Bedall and Christoph Görg use the approach of societal relationships with nature, to take a closer look at the repoliticization of climate change that can be observed in recent times. This politicization is based on the criticism of society and the economy of the radical wing of the climate movement. They show how it is largely a reaction to the alignment of positions in the context of the Kyoto Protocol and to the resulting affirmative way of doing politics by actors of civil society that engage themselves in climate politics.

Vito De Lucia argues in Chapter 5 that even positions that were meant critically often unintentionally produce hegemony effects that support the prevailing political and economic system. This mainly occurs by means of supporting technology transfer and the orientation towards scientific figures.

In Chapter 6, Andreas Niederberger uses ethical and philosophical criteria to ask what a just reaction to climate change could look like along with how just the concept of climate justice is from this theoretical perspective.

Based upon these theoretical chapters, eight chapters follow that focus on different national climate movements. These chapters are set up in a similar way and defer to the structure, the origin, the development as well as the influence of the respective national movement. Elaine Graham-Leigh investigates the climate movement in Great Britain as Georg Kössler investigates the one in Germany. Subsequently, Brian Tokar deals with the climate justice movement in the US and beyond as Hans Baer presents the climate movement in Australia. Besides, Dörte Segebart and Claudia König look at the climate movement in Brazil while Arne Harms and Oliver Powalla look at the climate movement in India. Patrick Schroeder reports on Chinese civil society organizations in the field of climate policy. Finally, Patrick Bond elaborates on the African components of the climate justice movement. On the whole, the handbook covers six continents, all of them witnessing significant active and visible – though different – climate movements.

After the chapters on the national submovements, a series of short portraits of important NGOs and activists of the climate movement follow. Here, organizations were chosen as 350.org, Climate Action Network, Rising Tide and La Via Campesina as well as activists such as Bill McKibben, Naomi Klein, and Tim DeChristopher. Although attention was paid to include different organizational types and different countries, the sample of presented persons and groups is not representative. Anyhow, the portraits allow a concrete, fragmentary glance on the basis of the climate movement. They give a human face to the movement and show the person engaged in it. The chapters with the portraits are written by authors who are also involved in other chapters of this handbook.

Subsequently, several chapters on activities and structures of the climate movement follow along with the arenas they are active in. From a comparative perspective to other movements, Achim Brunnengräber analyzes the modified understanding of the role of present climate NGOs, which in parts turn away from the United Nations' (UN) climate negotiations and orient towards more radical forms of protests on the national and local level.

Further, Johannes Kruse investigates the World Peoples Summit on Climate Change, which was an alternative climate conference in the year 2010, where South American states together with actors of the climate movement mapped out ambitious proposals for the resolution of the climate crisis.

In his contribution, Matthias Dietz analyzes numerous internal debates and conflicts of the climate movements, which allow conclusions to be drawn about their way of thinking, the characteristics, and the development of the movement.

Silja Klepp presents specific debates on climate change and climate justice in the Pacific in which she puts her focus on the island of Kiribati, which is endangered by the rise of the sea level. In so doing, she covers aspects of adaptation to climate change.

Finally, some chapters on actors and activities that are to be found in the periphery of the climate movement follow thereafter. They are not a direct part of the movement or do not understand themselves as a part of the movement, but they feature sector-wise, project-based, institutional, or other connections to the climate movement. Accordingly, this may be of a cooperative or confrontational kind. Jonas Hein and Heiko Garrelts look at the so-called carbon offsets, which are highly controversial within the climate movement. Carbon offsets are projects to reduce climate-damaging emissions that are mostly financed by Western companies in countries of the Global South, in order to reduce their own emissions in an arithmetical way. Johannes Herbeck takes a look at humanitarian organizations (which themselves are in a state of transition) and their approach to climate change.

In the concluding chapter, the findings of this manual are compiled and reflected as future fields of research are outlined.

Notes

1 We would like to cordially thank all the authors for their committed collaboration. We would like to direct special thanks towards Frank Nullmeier who had the idea of a handbook and who actively supported its realization. In addition, we would like to acknowledge Michael Flitner's constructive feedback throughout the project.

2 Moral panic means that "actions that contribute to environmental damage are increasingly regarded as both violation of ethical codes and reflecting a lack of 'civic virtue': people (consumers) must make proper ethical choices or humanity is doomed" (Lipschutz and McKendry 2011: 373).

3 Goodman (2009: 510–511) highlights "the South's current capacity to exacerbate the climate crisis, which translates into Northern vulnerability to carbon-intensive Southern development."

4 Also, more recent definitions of social movements resemble the description of Rucht to a large degree (e.g., see della Porta and Diani 2006a: 20–27; Teune 2008: 529).

5 Brunnengräber refers to the example of the various new social movements in Germany (one-world, peace, environmental, and women's movements) which paved the way for those established NGOs which in their own turn "became increasingly integrated in Germany's political system and, in many cases, moved away from their original objectives and aspirations." This is especially apparent in the case of the political party Bündnis90/Die Grünen. Initially founded as a "movement party," it found itself "the target of protest from newly forming anti-globalization movements" when the party was in power from 1998 to 2005.

References

Altvater, E. (2005) *Das Ende des Kapitalismus wie wir ihn kennen. Eine radikale Kapitalismuskritik*, Münster: Westfälisches Dampfboot.

Altvater, E. (2006) Das Ende des Kapitalismus, *Blätter für deutsche und internationale Politik*, 51(2): 171–182.

Baer, H. (2009) A Field Report on the Critical Anthropology of Global Warming: A View from a Transplanted American Downunder, *Dialect Anthropol*, 33(1): 79–85.

Baer, H. and Burgmann, V. (2012) *Climate Politics and the Climate Movement in Australia*, Melbourne: Melbourne University Publishing.

Baer, H. and Reuter, T. (2011) The Global Movement for a Safe Climate and Environmental Sustainability, *The Australian Journal of Anthropology*, 22(2): 255–256.

Bailey, I., Gouldson, A., and Newell, P. (2011) Ecological Modernisation and the Governance of Carbon: A Critical Analysis, *Antipode*, 43(3): 682–703.

Banse, H. (2012) Vom Ende der Globalisierungsbewegung – und dem, was kommt: ein Rück- und Ausblick, *Forschungsjournal Soziale Bewegungen*, 25(1): 51–60.

Banse, F. and Habermann, F. (2012) Vom Ende der Globalisierungsbewegung – und dem, was kommt: ein Ruck- und Ausblick, *Forschungsjournal Soziale Bewegungen*, 25(1): 51–60.

Batliwala, S. (2002) Grassroots Movements as Transnational Actors: Implications for Global Civil Society, *Voluntas*, 13(4): 393–409.

Beck, S. (2009) *Das Klimaexperiment und der IPCC. Schnittstellen zwischen Wissenschaft und Politik in den internationalen Beziehungen*, Marburg: Metropolis.

Bedall, P. (2011) NGOs, soziale Bewegungen und Auseinandersetzungen um Hegemonie. Eine gesellschaftstheoretische Verortung in der Internationalen Politischen Ökonomie, in Brunnengräber, A. (ed.) *Zivilisierung des Klimaregimes. NGOs und soziale Bewegungen als Akteure der nationalen, europäischen und internationalen Politik*, Wiesbaden: VS Verlag.

Beinin, J. (2012) Soziale Bewegungen und der »Arabische Frühling«: Rolle der ägyptischen Arbeiterbewegung, in Bundeszentrale für politische Bildung (ed.) *Arabische Zeitenwende. Aufstand und Revolution in der arabischen Welt 2012*, Schriftenreihe Band 1243, Bonn: Bundeszentrale für politische Bildung.

Boese, D. (2011) *Wir sind jung und brauchen die Welt. Wie die Generation Facebook den Planeten rettet*, München: oekom.

Bond, P. (2012) *Politics of Climate Justice: Paralysis Above, Movement Below*, Scottsville, South Africa: University of Kwazulu-Natal Press.

Brand, U., Bullard, N., Lander, E., and Müller, T. (eds.) (2009) *Contours of Climate Justice: Ideas for Shaping New Climate and Energy Politics*, Critical Currents No. 6, Uppsala: Dag Hammarskjöld Foundation.

Brunnengräber, A. (2005) Gipfelstürmer und Straßenkämpfer. NGOs und globale Protestbewegungen in der Weltpolitik, in Brunnengräber, A., Klein, A., and Walk, H. (eds.) *NGOs im Prozess der Globalisierung. Mächtige Zwerge – umstrittene Riesen*, Bonn: Bundeszentrale für politische Bildung.

Brunnengräber, A. (2008) Klima-Kapitalismus der Europäischen Union, in Altvater, E. and Brunnengräber, A. (eds.) *Ablasshandel gegen Klimawandel. Marktbasierte Instrumente in der globalen Klimapolitik und ihre Alternativen*, Hamburg: VSA-Verlag.

Brunnengräber, A. (ed.) (2011) *Zivilisierung des Klimaregimes*, Wiesbaden: VS Verlag.

Brunnengräber, A. (2012) Ein neuer Bewegungszyklus: von der NGOisierung zur Occupy-Bewegung, *Forschungsjournal Soziale Bewegungen*, 25(1): 42–50.

Bulkeley, H. (2011) Cities and Subnational Governments, in Dryzek, J., Norgaard, R., and Schlosberg, D. (eds.) *Oxford Handbook of Climate Change and Society*, Oxford: Oxford University Press.

Bulkeley, H. and Newell, P. (2010) *Governing Climate Change*, London: Routledge.

Byrd, S. (2010) Transnational Movement Innovation and Collaboration: Analysis of World Social Forum Networks, *Social Movement Studies*, 9(4): 355–373.

Carrington, B. (2010) Climate Justice: The Emerging Movement against Green Capitalism, *The South Atlantic Quarterly*, 109(2): 313–339.

Castaeda, E. (2012) The Indignados of Spain: A Precedent to Occupy Wall Street, *Social Movement Studies*, 11(3): 309–320.

Chatterton, P., Featherstone, D., and Routledge, P. (2012) Articulating Climate Justice in Copenhagen: Antagonism, the Commons, and Solidarity, *Antipode*, 45(3): 602–620.

Cottle, S. (2011) *Transnational Protests and the Media*, New York: Lang.

Dawson, A. (2010) Climate Justice: The Emerging Movement against Green Capitalism, *South Atlantic Quarterly*, 109(2): 313–338.

Dayaneni, G. (2009) Climate Justice in the US, in Brand, U., Bullard, N., Lander, E., and Müller, T. (eds.) *Contours of Climate Justice: Ideas for Shaping New Climate and Energy Politics*, Critical Currents No. 6, Uppsala: Dag Hammarskjöld Foundation.

De Lucia, V. (2009) Hegemony and Climate Justice: A Critical Analysis, in Böhm, S. and Dabhi, S. (eds.) *Upsetting the Offset: The Political Economy of Carbon Markets*, London: Mayfly.

della Porta, D. (2006a) *Globalization from Below: Transnational Activists and Protest Networks*, Minneapolis: University of Minnesota Press.

della Porta, D. (2006b) *The Policing of Transnational Protest*, Aldershot, Hampshire: Ashgate.

della Porta, D. and Diani, M. (2006) *Social Movements: An Introduction*, Malden, MA: Blackwell.

della Porta, D. and Rucht, D. (2012) *Meeting Democracy: Power and Deliberation in Global Justice Movements*, Cambridge: Cambridge University Press.

Diesendorf, M. (2010) The Climate Action Movement in Australia, *Social Alternatives*, 29(3): 56–62.

Dietz, K. and Brunnengräber, A. (2008) Das Klima in den Nord-Süd-Beziehungen, *Peripherie*, 28(112): 400–428.

Dietz, M. (2010) Gibt es eine internationale Klimabewegung? Ein skeptischer Befund, *Forschungsjournal Neue Soziale Bewegungen*, 23(4): 6–18.

Dietz, M. (2012) *Die Klimabewegung. Eine junge und noch wenig bekannte soziale Bewegung*, eNewsletter Wegweiser Bürgergesellschaft, 7, April 13. Accessed 01/15/2013 from www.buergergesellschaft.de/filead-min/pdf/gastbeitrag_dietz_120413.pdf.

Dorsey, M. (2007) Climate Justice and Power: Tales of Skeptic Tanks, Weather Gods, and Sagas for Climate (In)justice, *Capitalism Nature Socialism*, 18(2): 7–21.

Dryzek, J., Norgaard, R., and Schlosberg, D. (eds.) (2011) *Oxford Handbook of Climate Change and Society*, Oxford: Oxford University Press.

Evans, G. (2010) A Rising Tide: Linking Local and Global Climate Justice, *Journal of Australian Political Economy*, 66(11): 199–221.

Glover, L. (2006) *Postmodern Climate Change*, New York: Routledge.

Goodman, J. (2009) From Global Justice to Climate Justice? Justice Ecologism in an Era of Global Warming, *New Political Science*, 31(4): 499–514.

Grittmann, E. (2012) Visuelle Konstruktionen von Klima und Klimawandel in den Medien. Ein Forschungsüberblick, in Neverla, I. and Schäfer, M. (eds.) *Das Medien-Klima*, Wiesbaden: VS Verlag.

Henn, J. (2011) Climate Movement Gives UN Climate Talks a "Reality Check," *Huffington Post*, September 12. Accessed 01/15/2013 from www.huffingtonpost.com/jamie-henn/climate-movement-gives-un_b_1139209.html.

Herkenrath, M. (2011) *Die Globalisierung der sozialen Bewegungen. Transnationale Zivilgesellschaft und die Suche nach einer gerechten Weltordnung*, Wiesbaden: VS Verlag.

Hulme, M. (2009) *Why We Disagree about Climate Change: Understanding Controversy, Inaction and Opportunity*, Cambridge: Cambridge University Press.
Kern, K. and Bulkeley, H. (2009) Cities, Europeanisation and Multi-level Governance: Governing Climate Change through Transnational Municipal Networks, *Journal of Common Market Studies*, 47(2): 309–332.
Kraushaar, W. (2012) *Der Aufruhr der Ausgebildeten: vom Arabischen Frühling zur Occupy-Bewegung*, Hamburg: Hamburger Edition.
Lever-Tracy, C. (ed.) (2010) *Routledge Handbook of Climate Change and Society*, London: Routledge.
Lipschutz, R. and McKendry, C. (2011) Social Movements and Global Civil Society, in Dryzek, J., Norgaard, R., and Schlosberg, D. (eds.) *Oxford Handbook of Climate Change and Society*, Oxford: Oxford University Press.
Manzo, K. (2009) Imaging Vulnerability: The Iconography of Climate Change, *Area*, 42(1): 96–107.
Mattoni, A. (2009) Organisation, Mobilisation and Identity: National and Transnational Grassroots Campaigns between Face-to-face and Computer-mediated-communication, in Baringhorst, S., Kneip, V., and Niesyto, J. (eds.) *Political Campaigning on the Web*, Bielefeld: transcript Verlag.
Mercea, D. (2012) Digital Prefigurative Participation: The Entwinement of Online Communication and Offline Participation in Protest Events, *New Media & Society*, 14(1): 153–170.
Moser, S. (2007) In the Long Shadows of Inaction: The Quiet Building of a Climate Protection Movement in the United States, *Global Environmental Politics*, 7(2): 124–144.
Müller, T. (2008) The Movement Is Dead, Long Live the Movement, Turbulence: Ideas for Movement 2008: 48–55. Accessed 06/25/2012 from http://turbulence.org.uk/turbulence-4/the-movement-is-dead-long-live-the-movement/.
Neuner, M. (2011) Klimabewegungen in Russland. Chancen und Probleme ihrer transnationalen Vernetzung, in Brunnengräber, A. (ed.) *Zivilisierung des Klimaregimes*, Wiesbaden: VS Verlag.
Newell, P. (2000) *Climate for Change: Non-state Actors and the Global Politics of the Greenhouse*, Cambridge: Cambridge University Press.
Newell, P. and Paterson, M. (2010) The Politics of the Carbon Economy, in Boykoff, M. (ed.) *The Politics of Climate Change: A Survey*, London, New York: Routledge.
North, P. (2011) The Politics of Climate Activism in the UK: A Social Movement Analysis, *Environment and Planning*, 43(7): 1581–1598.
Paterson, M. (2011) Selling Carbon: From International Climate Regime to Global Carbon Market, in Dryzek, J., Norgaard, R., and Schlossberg, D. (eds.) *Climate Change and Society*, Oxford: Oxford University Press.
Pusey, A. and Russell, B. (2010) The Climate Crisis or the Crisis of Climate Politics? Perspectives 2010. Accessed 10/22/2012 from www.anarchiststudies.org/node/423.
Reimann, K. (2002) Building Networks from the Outside: International Movements, Japanese NGOs and the Kyoto Climate Change Conference, Political Science Faculty Publications, Paper 6.
Rigg, K. (2012) The Arab Youth Climate Movement should give us hope. If the winds of change sweeping the Arab region collided with the slow-moving front of climate negotiations, anything is possible, *Guardian*, 9 November. Accessed 01/15/2013 from www.guardian.co.uk/environment/blog/2012/nov/09/arab-youth-climate-movement.
Roberts, J. (2009) *How the Internet Is Changing the Practice of Politics in the Middle East: Political Protest, New Social Movements, and Electronic Samizdat*, Lewiston, NY: The Edwin Mellen Press.
Röhr, U., Spitzner, M., Stiefel, E., and von Winterfeld, U. (2008) *Gender Justice as the Basis for Sustainable Climate Policies: A Feminist Background Paper*, Bonn: Genanet.
Rosenbaum, W. and Mautz, R. (2011) Energie und Gesellschaft. Die soziale Dynamik der fossilen und der erneuerbaren Energien, in Gross, M. (ed.) *Handbuch Umweltsoziologie*, Wiesbaden: VS Verlag.
Routledge, P. (2011) Translocal Climate Justice Solidarities, in Dryzek, J., Norgaard, R., and Schlossberg, D. (eds.) *Climate Change and Society*, Oxford: Oxford University Press.
Rucht, D. (1994) *Modernisierung und neue soziale Bewegungen: Deutschland, Frankreich und USA im Vergleich*, Frankfurt/Main: Campus.
Rucht, D. (2008) Anti-Atomkraftbewegung, in Rucht, D. and Roth, R. (eds.) *Die sozialen Bewegungen in Deutschland seit 1945. Ein Handbuch*, Frankfurt/Main, New York: Campus.
Rucht, D. (2011) Zum Stand der Forschungen zu sozialen Bewegungen, *Forschungsjournal Soziale Bewegungen*, 24(3): 20–47.
Rucht, D. (2012) Akzeptanzbeschaffung als Legitimationsersatz: Der Fall Stuttgart 21, in Geis, A., Nullmeier, F., and Daase, Christoph (eds.) *Der Aufstieg der Legitimitätspolitik. Rechtfertigung und Kritik politisch-ökonomischer Ordnungen*, Sonderband 27 des Leviathan, Baden-Baden: Nomos.

Rucht, D. and Roth, R. (2008) Einleitung, in Rucht, D. and Roth, R. (eds.) *Die sozialen Bewegungen in Deutschland seit 1945. Ein Handbuch*, Frankfurt/Main, New York: Campus.

Russell, B., Pusey, A., and Sealey-Huggins, L. (2012) Movements and Movements for Climate Justice: From Copenhagen to Cancun via Cochabamba, *ACME*, 11(1): 15–32.

Schalatek, L. (2011) Zwischen Geschlechterblindheit und Gender Justice, in Brunnengräber, A. (ed.) *Zivilisierung des Klimaregimes. NGOs und soziale Bewegungen in der nationalen, europäischen und internationalen Klimapolitik*, Wiesbaden: VS Verlag.

Schultze, K. (2008) Studentische Bewegungen und Protestkampagnen, in Rucht, D. and Roth, R. (eds.) *Die sozialen Bewegungen in Deutschland seit 1945. Ein Handbuch*, Frankfurt/Main, New York: Campus.

Skinner, E. (2011) *Gender and Climate Change: Overview Report*, Sussex, UK: Institute of Development Studies.

Skrimshire, S. (2007) Approaching the Tipping Point: Climate Risks, Faith and Political, *European Journal of Science and Theology*, 4(2): 9–22.

Teune, S. (2008) "Gibt es so etwas überhaupt noch?" Forschung zu Protest und sozialen Bewegungen, *Politische Vierteljahreszeitschrift*, 49(3): 528–547.

Tokar, B. (2010) Toward a Movement for Climate Change. Accessed 02/08/2010 from www.social-ecology.org/wp/wpcontent/uploads/2009/12/brian1.Toward_a_Movement.pdf.

Unmüßig, B. (2011) NGOs in der Klimakrise. Fragmentierungsprozesse, Konfliktlinien und strategische Ansätze, in Brunnengräber, A. (ed.) *Zivilisierung des Klimaregimes. NGOs und soziale Bewegungen in der nationalen, europäischen und internationalen Klimapolitik*, Wiesbaden: VS Verlag.

Vidal, J. (2012) Doha climate conference diary: Qatar's first environmental march, *Guardian*, December 3. Accessed 01/15/2013 from www.guardian.co.uk/environment/blog/2012/dec/03/doha-climate-conference-diary.

Viehöver, W. (2011) Die Politisierung des globalen Klimawandels und die Konstitution des transnationalen Klimaregimes, in Gross, M. (ed.) *Handbuch Umweltsoziologie*, Wiesbaden: VS Verlag.

Weingart, P., Engels, A., and Pansegrau, P. (2008) *Von der Hypothese zur Katastrophe: Der anthropogene Klimawandel im Diskurs zwischen Wissenschaft, Politik und Massenmedien*, second edition, Opladen: Budrich.

Wissenschaftlicher Beirat der Bundesregierung Globale Umweltveränderungen (WBGU) (2011) *Hauptgutachten. Welt im Wandel. Gesellschaftsvertrag für eine Große Transformation*, Berlin: WBGU.

PART I

Theoretical prospects of the climate movement

2

FRAMING PROCESSES IN THE CLIMATE MOVEMENT

From climate change to climate justice[1]

Donatella della Porta and Louisa Parks

1 Introduction

The climate movement has changed greatly in recent years. Since the wide-ranging umbrella group the Climate Action Network was created in 1989 "to coordinate the NGO response" (Busby 2010: 107) to the work of the Intergovernmental Panel on Climate Change (the IPCC) there may be said to have been a definite current of radicalization in the movement. With what has generally judged to be the failure of the body's 15th conference of the parties, held in Copenhagen in 2009, the movement entered a period of earnest internal debate. Rising Tide North America published a reaction entitled 'the climate movement is dead. Long live the climate movement' (Rising Tide North America 2010); the climate camps in the UK began to question whether they had become 'media savvy' at the cost of diluting their message; and an alternative summit was convened in the city of Cochabamba by Evo Morales in Bolivia. Debates raged over the actions that the movement should use (direct rather than symbolic), the rejection of the 'green economy' solution (where the current global economic system is seen as integral to the destruction of the earth's climate), and the perceived tendency of the movement to isolate itself from the linked and pertinent issues of other movements (namely human rights and social justice, thus the move towards ideas of climate justice).

To make sense of change in the climate movement, theories developed within social movement studies can provide useful handles. The discipline of social movement studies developed in the context of a focus on some specific movements in a specific geographical area and historical era: new social movements, so considered because they addressed issues of reproduction rather than redistribution in societies which were considered as post-industrial. The field also grew under specific contextual conditions: in so-called advanced democracies, characterized by party governments, nation states, and mature welfare systems. Influential works, such as Charles Tilly's, located movements in the development of capitalism and the construction of the state (Tilly 1978, 1984). The most analyzed were in fact the environmental and women's movements, while ethnic, right-wing, or religious movements – and even labor movements – were rarely addressed within this frame.

The very definition of social movements tended to fit the same specific movements that had inspired the development of the field, as they took networked forms, had well-defined collective identities, and used contentious repertoires of action (della Porta and Diani 2006: chapter 1).

These elements allowed scholars to distinguish (more) permanent movements from (more) temporary coalitions.

Yet the very definition of social movements requires adaptation in the face of recent developments as well as the application of social movement studies in other geo-political and historical areas. In particular, reflection on the very tool-kit of social movement studies is needed when we broaden our analysis to include much transnational collective action. The growing importance of political globalization, with the increasing power of some international organizations, has in fact increased attempts to put pressure on multilevel governance beyond national borders. Transnational social movements, when compared with national ones, tend to work more through temporary coalitions and through campaigns that combine various forms of contention and noncontentious actions, bringing about the development of multiple, tolerant identities (della Porta 2005). New technologies facilitate cross-fertilization in action between social movement organizations which retain specific core concerns, but bridge them with others through the adoption of broker issues (such as global justice or environmental justice) (della Porta 2007). In definitional terms, this leaves the field open as to how strong a common identity, how dense a network, and how contentious a form of action an actor needs to use in order to be recognized as a social movement rather than a coalition.

For sure, transnational dynamics become more relevant in transnational protests, such as those on climate change. The multiplication of transnational coalitions active on issues of peace, human rights, and environment call in fact for more attention than thus far granted to be paid to the characteristics of such forms of collective action, that put together what we have been accustomed to consider as separate movements. We need, that is, more reflection on specific tools that might allow us to investigate the different characteristics of these coalitions, as well as their chances of success.

Given the special timing of transnational campaigns, influenced as they are by the complex evolution of multilevel decision making, we also need to reflect more on the dynamics of their evolution. Not only in fact do transnational campaigns alternate (short) moments of visibility through protest action with long moments of latency (or less visible lobbying), they also need to follow policies and the like through the original international decisions to national ratification and often up again. Continuous upward and downward shifting (Tarrow and McAdam 2005) are extremely important at the onset as well as during the development of those campaigns.

Global in aspiration, transnational collective actors are however also often divided along geographical lines. The power and influence of Northern actors is often criticized by Southern ones. More and more – as in the climate change campaigns – generational divides emerge as well. How much diversity is positively enriching and just when diversity becomes divisive instead is a related, central issue of practical and theoretical concern.

In what follows, we shall build on this general discourse by focusing on framing as a central concept in the adaptation of social movement studies to transnational campaigns, in this case that on climate change. Frames will also be used as an illustration of the contribution social movement studies can give to understand protests on climate change, but also vice versa in terms of the contribution that research on climate change can make to update social movement studies.

Understanding and mapping the soul-searching exercises gripping the climate movement now can be usefully guided by drawing on the literature on framing. It is of course true that in recent years social movement theory too has come a long way in terms of developing explanations and analytical frameworks considering movements in their entirety and in a wider context encompassing revolution (for an overview see della Porta and Diani 2006). In light of the recent seismic shift in the climate movement, however, the literature on framing can provide a useful prism through which to make some sense of the developments. Frames are 'cognitive schemata' used to make sense of situations, to attribute blame, identify solutions, and motivate participation (REF). Framing work includes a number of processes via which movement actors change and adapt collective action frames to new social and political realities. Although this short contribution cannot hope to provide anything like an exhaustive analysis of such work, it will point out how the broader trends and trajectories of the climate movement in recent years can be fruitfully viewed through the language of framing processes. It will also show some future possible developments in the framing work literature that would appear to be a necessary in light of how we take part in dialogue and knowledge creation in today's world of the web 2.0.

The chapter begins by describing the literature on frames as found in the scholarly literature on social movements, paying particular attention to those mechanisms of framing work that are most useful for viewing the recent history of the movement. Then, an interpretation of the general discursive shifts in the movement is offered. By way of conclusion, we reflect on how the framing literature can help us to make sense of the ongoing debates in the movement, as well as on how those same debates can contribute to an updating of the literature on framing work as mentioned above.

2 A brief history of frames

One thread of social movement theory that is particularly interesting in terms of interpreting the recent discursive shift away from a climate change and towards a climate justice movement is found in the literature on framing. The concept of the frame was originally developed by Erving Goffman, who saw frames as keys used to bring into focus different tasks and issues in interpersonal encounters, that is a frame or "a particular definition *is in charge* of a situation" (Gamson 1985: 616, emphasis in original). The concept was first theorized in relation to social movements by Snow and Benford (see Hunt et al. 1994). In movements, frames attach characteristics and definitions to people and issues in space and time – they attribute blame, and outline alternative paths and means of achieving goals. They perform the role, that is, of interpreting the significance of a person, an event, or a symbol. Each frame spells an attempt to align individual and collective identities, thus highlighting the role of social movements as constant builders and interpreters of situations rather than as carriers of fixed identities (Snow 2004). These framing *processes* (since the image is of actors constantly shifting and building on frames) are explicitly rooted in a constructionist perspective: "The framing perspective is rooted in the symbolic interactionist and constructionist principle that meanings do not automatically or naturally attach themselves to the objects, events, or experiences we encounter, but often arise, instead, through interactively based interpretive processes" (Snow 2004: 384).

Framing, in other words, requires work, and the literature on the subject also develops various categories of frames as well as framing work that aid analysts in making sense of movement discourses. The information contained in a frame may thus be: (i) diagnostic – that is information identifying a problem; (ii) prognostic – information on how this problem should best be solved; or (iii) motivational – encouraging action to draw attention to or solve this problem. According to Benford and Snow,

> Collective action frames are constructed in part as movement adherents negotiate a shared understanding of some problematic condition or situation they define as in need of change, make attributions regarding who or what is to blame, articulate an alternative set of arrangements, and urge others to act in concert to affect change.
>
> *(Benford and Snow 2000: 615)*

Benford and Snow are also explicit about the many different techniques that social movements may employ in framing: "Frames are developed, generated, and elaborated on not only via attending to the three core framing tasks discussed above, but also by way of three sets of overlapping processes that can be conceptualized as discursive, strategic, and contested" (Benford and Snow 2000: 623). Discursive processes are described as utterances, that is, speech acts and written communications. In framing processes, such utterances contribute by articulation – "the connection and alignment of events and experiences so that they hang together in a relatively unified and compelling fashion" (Benford and Snow 2000: 623), and amplification – stressing the importance of certain issues, events, or beliefs in order to make them more salient. Logically, a coherent argument will produce a frame that is more acceptable and therefore more likely to be acted upon (Gerhards and Rucht 1992). Strategic processes are therefore aimed at building frames to achieve a specific purpose. Here the authors identify four tactics: frame alignment of bridging, frame amplification, frame extension, and frame transformation.

Frame bridging or alignment is especially interesting when looking at movements and issues of a transnational character (see for example Ruzza 2004; Parks 2009). Indeed, it has been described as the most widely employed strategic framing action within movements, defined as "the linking of two or more ideologically congruent but structurally unconnected frames regarding a particular issue or problem" (Benford and Snow 2000: 624). Another strategic framing process that is very important in transnational movements is diffusion. While frame bridging deals with the fusion of two or more frames, diffusion looks at how a frame travels among different groups or the same groups at different levels. Indeed, diffusion is most often employed as a concept to look at frames moving across national borders. Once again, Snow and Benford (1999) are at the forefront in the field, detailing how frames elaborated at the national level can be transferred over national boundaries via diffusion, meaning the "flow of social practices among actors within some larger system" (Strang and Meyer 1993, cited in Soule 2004). Frame diffusion, which may occur both actively through the deliberate efforts of movement actors, or more passively – even undesired – through external channels like the media,[2] which may diffuse frames of their own accord (della Porta and Kriesi 1999), takes place when a frame (the 'innovation' in the language of diffusion) is useful to both parties involved ('transmitter' and 'adopter' respectively), when both share some basic cultural or structural characteristics and are in some way linked (Snow and Benford 1999: 24). Two main models are envisaged for diffusion: the hierarchical model and the proximal model, within which diffusion may take place either through direct or indirect links (Soule 2004). In the former, diffusion occurs in a trickle-down form, with a leading individual or organization diffusing frames downwards to lower organizations or actors. In the latter, organizations or individuals "mimic others who are spatially or culturally relevant to them" (Soule 2004: 295).

Concerning the conscious and deliberate use of diffusion, Snow and Benford (1999) see the important factor in the process not in the actual mechanical act of diffusion but in the manipulation and interpretation of a frame in order to fit a new societal context. They develop a typology of diffusion accordingly. Reciprocation occurs when both the transmitter and the adopter actively take an interest in the process. Where only the adopter takes an active interest, adaptation takes place, while accommodation describes the opposite situation. Finally, 'contagion' describes diffusion between two passive actors, although there is little empirical evidence of such processes.[3]

The framing processes under way in the climate movement mentioned in the introduction form part of this natural and ongoing process of definition that takes place in all movements. The crisis marked by the perceived failure of the Copenhagen talks appears as the crucial trigger for this particularly intensive period of framing work. It shifted the view of the movement from climate change to climate justice and broadens the movement's view in two main ways. First, it broadens the view of the problem it wishes to solve. Climate justice as a frame allows for actions that can be pursued through much more local and active channels, whereas it may be argued that climate change could only effectively be pursued through pressuring international organizations and national governments. Thus, climate justice as a frame brings the movement back to its radical roots – this target can be pursued by concrete action at local levels. Second, the climate justice frame is broader in its appeal, allowing the movement to (re)connect to other movements that its framing work identifies as linked and pertinent to its own goals. As will be demonstrated below, frame bridging work is taking place to begin connecting climate issues with the social justice and anti-war movements (thus perhaps moving back towards the original 'Greenpeace' ethos). In a nutshell, the crisis of the climate change movement in the failure of international negotiations has sparked a necessary reframing period to revitalize the movement around the frame of climate justice. While framing processes are by their very nature ongoing, it seems that the master frame of climate change has shifted to climate justice for a considerable period to come. The following takes a closer look at the mechanisms used to make this shift.

3 Interpreting recent developments in the climate movement

Much of the recent discussion within the climate movement has focused on the two perceived 'camps' within the same; that is, the more radical, direct-action oriented elements of the movement (Rising Tide, the Climate Camps), which we shall label (for the sake of simplicity) the climate justice stream,[4] as opposed to coalitions such as Stop Climate Chaos! steered by the more established NGOs such as Friends of the Earth, Greenpeace, and others, which we shall call the climate change stream. This cleavage between more radical and more moderate wings of the movement is of course a fairly common one (seen for example to different extents in different branches of many movements, such as the workers' movement or the global justice movement's infamous 'black bloc' to name but two). Nevertheless, it is a historically pronounced one in the environmental movement generally, dating back to the distinctions between 'environment' and 'conservation' (Dalton 1994). Using the categories of diagnostic, prognostic, and motivational framing allows us to sketch a comprehensive yet concise portrait of their different viewpoints.[5]

The diagnostic frames of the two sectors of the climate movement share the basic understanding that humankind is responsible for changes in the climate of the planet that will have catastrophic effects should they go unchecked. They both therefore attribute blame to the same area. Where the story gets interesting is where diagnostic and prognostic frames are stated together, broaching the joined issues of the cause and solution to this disastrous scenario. The more climate justice part of the movement, which stated its understanding of the solution to the crisis more vocally following the failure of COP15 in Copenhagen,[6] sees the solution in deep-seated lifestyle changes and the end of the capitalist economic system. Rising Tide, in its aforementioned discussion paper reacting to the Copenhagen debacle in late 2009, notes that "politicians and corporations will not solve the climate crisis" (Rising Tide North America 2010). The Climate Camp Reader 'Criticism Without Critique' provides more detail:

> [I]t is apparent that there is a need for two things. Firstly, a greater visibility for the anarchist roots within the day to day life of the CCA process and proposals. Secondly, and just as important, a more open and explicit critique of capitalism and how it is the root cause of climate change ... the power of the Camp has always been the promise of a genuine alternative action.
>
> *(Shift Magazine and dysophia 2010: 4)*

This highlights the divergence between the different types of groups active in the movement, or indeed within the same actions as the case of the Camp for Climate Action shows, in terms of diagnosis and prognosis. While the climate justice section of the movement sees the root cause of environmental meltdown not merely in humankind, but in *capitalist* humankind, and therefore the solution in an end to this system, the more established NGO-based climate change part of the movement, albeit critical of the global economic system, accepts its existence and seeks to encourage changes in it to mitigate climate change (for a similar reading of this split see Tokar 2010; also Dietz 2010). The Stop Climate Chaos! UK coalition, for example, state they are "The UK's largest group of people dedicated to action on climate change and *limiting its impact* on the world's poorest communities"[7] and that they "demand practical action by the UK to keep global warming as far as possible below the 2 degrees C danger threshold."[8] The international Climate Action Network, similarly, is "working to promote government and individual action to limit human-induced climate change to ecologically sustainable levels."[9] This more moderate vein in the movement is thus dedicated to solutions, or better limitation, achieved within the existing institutional sphere by pushing the 'green economy' agenda. In the same vein, the difference between the two strands within the movement is highlighted in the more moderate groups' focus on achieving technical goals (such as certain percentages of carbon emissions cuts or limitations to a certain number of degrees warming of the global climate) and the more radical climate justice discourse on fundamental change of our global economic system and lifestyles.

The difference between the two streams is also, as may well be imagined, played out in terms of motivational frames and to some extent the actions to be used to secure changes. In a nutshell, while both streams use direct action techniques, they do so in more and less radical ways with a view to achieving the radical and institutional changes touched on above. What is labeled here as the climate justice part of the movement is explicit in the reasoning behind the need for radical and sometimes illegal direct action.

> The climate movement needs to shift gears from what has been a largely symbolic movement to one that is directly disrupting destructive industries.... Our movement needs to judge its progress not by how many media hits our action got nor how many people read our blogs, but by how many power plants we've put out of business.
>
> *(Rising Tide North America 2010: 10)*

This need to 'shift gears' indicates that direct action of this kind is indeed new to a movement best known for a strategy relying on media stunts carried out by small numbers, what has been called a "movement without protest" (della Porta and Diani 2004). The need to engage in new kinds of direct protest with concrete effects for solving the climate disaster is expressed in an explicit discourse directed against the established NGOs working within rather than against the current institutions and economic system.

> These groups, ostensibly fighting on our behalf, have chosen to ally themselves with the very system that we must dismantle in order to avert climate chaos.... The fundamental message coming from mainstream groups is: All we have to do is switch our current society from carbon intensive fuels to low carbon fuel sources and implement energy efficiency measures. We don't have to fundamentally change our lifestyles, or this economic system.
>
> *(Rising Tide North America 2010: 6)*

Similar frames are also found within the Camp for Climate Action in the UK, where founding members of the camps felt the original reasons of their action had been slowly changed and taken over by others more committed to the sorts of actions and 'green economy' solutions advocated by the larger and more moderate NGOs.

> [W]e had opened up an exciting political space from where to challenge the status quo, but it was being filled with a message that was no longer our own. We were becoming a hip, media-savvy campaign of flash mobs and publicity stunts, lobbying for tighter government control of our lifestyles. "Friends of the Earth with D-locks" as one of our contributors wrote in the first issue of Shift.
>
> *(Shift Magazine and dysophia 2010: 7)*

Even where the repertoire of contention of the climate change stream of the movement may then certainly be termed as unconventional and even contentious, we can therefore see a clear juncture where the climate justice stream of the movement clearly calls for more radical, hands-on, and immediately effective action rather than symbolic events designed to raise awareness, change habits, and increase pressure on institutional actors. In this sense the climate justice section of the movement actually seems to be taking a step back in time and coming closer to more traditional direct action tactics and civil disobedience as seen in the antinuclear movement and US civil rights movements, for example.

Indeed, this step back in time to the traditions of other movements in terms of contentious strategies is also mirrored in the climate justice stream's particular efforts at frame bridging the climate problem with other movements' grievances. Here, the same may also be said of the climate change groups, and the frame bridges reflect similar (but less obvious perhaps) divisions in other movements between more radical and more moderate strands. For example, Chawla (2009) notes a significant paragraph of the declaration issued at the end of the first Climate Justice Summit in November 2000 in The Hague:

> We affirm that climate change is a rights issue. It affects our livelihoods, our health, our children and our natural resources. We will build alliances across states and borders to oppose climate change inducing patterns and advocate for and practice sustainable development.

We see therefore that the frames of climate change are bridged with frames concerning human rights in order to produce a 'climate justice' frame, in this case linked with a moderate plan of action towards sustainable development.[10] The framing work undertaken in the more radical climate justice discourse, is however more explicit and moves further than rights issues in the bridges it builds.

Tokar (2010) provides us a pithy overview of the ways in which climate change is linked with problems tackled by other movements, namely the antiwar and global justice movements, highlighting that

> This [climate justice] movement is sharply focused on the social justice implications of the global climate crisis, highlighting the voices of those already massively affected by the heating of the earth. It is linked to antiwar efforts, demonstrating how continuing US military adventures, including the wars in Iraq and Afghanistan, are without question the most grotesquely energy-wasting activities on the planet.
>
> *(Tokar 2010: 10)*

Returning to the example of Rising Tide North America, serious discussion of these frame bridges and how they should be rolled out is offered.

> A notable missing link within our climate movement, even in circles that are more concerned with social justice issues, is a connection with the antiwar movement. Not only is the US military the biggest consumer of fossil fuels in the world, it is often the preferred tool to secure energy supplies for the US.... The climate movement must stand in solidarity with peoples displaced by the impacts of rich nations' emissions and false solutions. This means active opposition to racist anti-immigrant policies while joining in struggle and building solidarity with communities displaced by pollution and economic policies.... As one movement among many fighting for a just future, we must always remember that our struggles are connected.... Wherever possible we need to find common cause with other movements and extend our solidarity to them.
>
> *(Rising Tide North America 2010: 12–14)*

Frank framing discussions such as these demonstrate that a process of active frame diffusion is taking place within this more radical part of the movement.

Finally, attempts towards the diffusion of this new frame of climate justice, and through this the joining of struggles, is also evident in the numbers of coalitions, networks, and summits of groups active on climate issues. Chawla notes, in relation to the more moderate stream, that "climate justice movements collaborate closely with NGOs that in turn incorporate the movement's principles into proposals they submit to the UNFCCC Secretariat" (Chawla 2009). Thus the broadening value of the climate justice frame, as mentioned earlier, may help to rejuvenate both more radical *and* more moderate parts of the movement. Just as more radical and more moderate wings within movements are common, so this frame which allows more linkages to be built will apply to those parts in other movements, as illustrated in the previous quote. Whether the two parts of the movement can overcome their considerable differences as to what they believe the best course of action to stop climate change is of course another question. On the basis of the discourses in the documents drawn on for this analysis, the gap between the two seems wide, although there are signs that some are attempting to bridge it.[11]

Tokar, on the more radical stream, underlines that

> In the fall of 2008, US organizations actively working for climate justice both nationally and internationally, including Indigenous Environmental Network, Global Justice Ecology Project and Rising Tide North America, launched the Mobilization for Climate Justice. The Mobilization was founded to link the climate struggle in the US to the growing international climate justice movement, with an eye toward building for actions around the 2009 UN climate summit and beyond. Its objective was to provide a justice-based framework for organizing around climate change that opened space for leadership by representatives of communities in the US that are most impacted by climate change and the fossil fuel industry.
>
> *(Tokar 2010: 9)*

And again, speaking of the importance of the climate issue at the 2009 World Social Forum, Smythe and Byrd (2010) note that

> The issue of climate change intersected with many other themes and activities during the Belem WSF – from trade to tourism, women's and indigenous rights, food sovereignty

> and housing. Climate Justice Now!, a network of over 200 organizations from both the global south and north, held extensive meetings focusing on climate and ecological debt, the Amazon, and false solutions to the climate crisis. They also used the Forum as an opportunity to empower participation by more southern organizations, especially indigenous groups from the Amazon and the rest of South America.
>
> *(Smythe and Byrd 2010)*

Clearly the challenge for diffusion here is also found in the fact that those most affected by climate change are in the Global South, while the strongest and most established climate movements originate in the Global North. This aspect contributes to the difficulties of reconciling the two streams within the movement. To put it into the theoretical language, the two streams share a basic diagnosis but diverge considerably in terms of the prognoses and motivations. Frame amplification of the climate justice frame appears necessary to a larger extent than detected in order to bring the moderate stream of the movement into the fold. Some efforts at reciprocal diffusion of the climate justice frame can be seen within these networks, and stronger indications still in events such as the summit convened in Cochabamba in Bolivia in 2010 (see Klein 2010). A particularly interesting piece of anecdotal evidence in this direction can also be found in the importance of former Climate Camp members in the establishment of the Occupy LSX camp in London,[12] and the distillation of the camp's central preoccupations in climate justice and internet freedoms (Kaldor and Selchow in press). Similar connections are also being observed in the US.[13] If this move away from a specific climate justice movement and towards a wider multi-issue movement seeking justice for inequalities throughout the globe is anything to go by, the frames of solidarity with other movements will indeed take hold and the climate movement's nascent bridging work should take off.

4 Conclusions

The climate movement has gone through a significant period of reframing in recent years. This seems to have taken hold more within the more radical stream of the movement, which has embraced the climate justice frame and is seeking to broaden its action repertoire and link with other movements. Whether this frame can take hold in the more moderate part of the movement remains to be seen, and will depend on the will of movement actors to tolerate different methods for acting in the same area, and indeed on whether the movement becomes part of a wider and more generalized social movement for global justice as seen in the recent Occupy movement.

Our brief and somewhat anecdotal glance at recent frames and framing work taking place in the climate movement has shed light on how this particular literature from the growing library of social movement theory can help us make sense of the changes within the movement. Focusing on the sometimes competing and sometimes complementary diagnostic, prognostic, and motivational frames found within the two main streams of the climate movement, labeled here as the 'climate change' and 'climate justice' strands, we have attempted to show how each develops its understandings, suggested solutions, and actions in line with two different readings of how climate change should best be tackled. For the more moderate strand this should essentially be achieved via progressive improvements within the current institutional frameworks, while for the more radical strand the unique solution lies in the dismantling of the global capitalist system. Looking at how these frames unfold also highlights the important point that frames are indeed developed "through interactively based processes" rather than somehow fixed and transmitted down the line (Snow 2004: 384). Movements are dynamic, fluid, and changing things, as the divergent paths emerging within the climate movement amply demonstrate, and framing can indeed be a hotly contested issue in itself. Preoccupations with openness, democracy, and above all representation within the more radical climate justice sector of the movement also pay testament to this, with various groups discussing the importance of having no 'spokesperson' who claims to represent the movement throwing up parallels with the debates raging over the frames the movement should adopt, and the solidarity it should display.

The short analysis also pointed to the work of frame bridging as especially pertinent in recent years with the move away from 'climate change' and towards the 'climate justice' frame among some in the movement, with a number of sources indicating the explicit goal of linking the climate movement with the struggles of other movements, most prominently the antiwar and global justice movements. While the literature on framing can certainly help us understand the discursive shifts within the movement, other theoretical approaches are of course needed to interpret recent changes – in particular, political process theory would seem to explain the need for the discursive shift in light of the failure of the climate negotiations in Copenhagen. Network analyses would also certainly be useful to provide evidence of the movement's linking to others through frame bridging: Is this pure talk or are concrete links being forged?

A final question is that of what these very public debates over framing work can teach scholars of social movements and framing in particular? It seems clear that the current scholarly work on frame bridging does not quite do justice to how these frames were developed in contentious dialogue across national and movement borders. General excitement in the field of social movement studies about the possibilities for social mobilization provided by modern communication technologies and the Internet especially have often been limited to comments about how it may be used as an organizational tool by resource-poor actors (Mattoni 2009). But a more recent and cutting edge body of work also considers the potential of interactive technology (the web 2.0) in terms not only of how it facilitates exchanges across the globe but also in terms of how its users create commons and collective knowledge (Fuster Morell 2011). The development of the climate justice frame in public and open debate, and in a manner that has meant it is not associated with any one particular group or charismatic leader seems to reflect this commons culture of shared ownership that has been born with the coming-of-age of the online generation (della Porta and Mattoni 2012). The case of the climate movement therefore underlines a need to bring the framing literature into a similar dialogue with those strands of the media and communications studies literature that deal in particular with social movements if we are to prove the importance of new technologies for framing processes in today's world.

Notes

1 Our particular thanks for insights and guidance go to the editors of the volume.

2 The label of 'indignados' given the Spanish 15M-Movement by the mass media, never used by those involved in the movement before, springs to mind here.

3 There are clear parallels here with the similar ideas elaborated by Tarrow and McAdam (2005). Their concepts of 'scale shift' and 'brokerage' may also be referred to here, but for simplicity's sake we stick to the language employed in the specific literature on framing.

4 According to Chawla

> [t]he first-ever Climate Justice Summit took place in The Hague, Netherlands, in November 2000 at the same time as the Sixth Conference of the Parties to the U.N. Framework Convention on Climate Change UNFCCC. More than 500 grassroots leaders from Asia, Africa, Latin America, and North America gathered to build bridges across borders and thematic issues.
>
> *(Chawla 2009)*

The climate justice discourse is thus older than may be immediately understood from these labels, but since it can generally be identified with a stream within the climate movement that has become steadily more radical and connected to other movements, we feel the label is a fair one.

5 The general analysis presented here is not intended to be exhaustive. It relies mainly on material from Rising Tide and the Climate Camp for frames in the more 'radical' sector of the movement, and on the Climate Action Network and Stop Climate Chaos! coalitions for that sector of the movement based on more established NGOs.

6 Although as Tokar notes,

> For over a year prior to the Copenhagen meetings, activists concluded that this summit would likely fall far short of what the world needs to prevent unprecedented climate disruptions, and pledged a commitment to direct action against the root causes of climate change. In the summer of 2009, activists from more than 20 countries, including several from the global South, gathered as part of an emerging Climate Justice Action network, and agreed on an ambitious alternative agenda to the increasingly business-dominated deal-making at the UN level.
>
> *(2010: 7–8)*

7 See www.stopclimatechaos.org/we-are, accessed 06/08/2012, emphasis added.

8 See note 5.

9 See www.climatenetwork.org/about/about-can, accessed 06/08/2012.

10 As is obvious here, the frame of 'climate justice' is by no means exclusive to the more radical section of the movement we identify it with here. These labels are merely intended to facilitate the differentiation of what are in reality very intermingled streams within the movement.

11 See for example the open letter by 1 Sky, available here: www.1sky.org/openletter and a response by a large number of grassroots groups available here: http://understory.ran.org/2010/10/27/grassroots-organizing-cools-the-planet-a-letter-from-the-grassroots-to-1-sky/, accessed 10/17/2012.

12 Conversation with activist involved in Occupy Times of London, June 2012.

13 See for example articles on Occupy and the climate movement at 350.org, www.350.org/en/node/27881, and the solidarity statement of Rising Tide North America with Occupy Oakland, http://occupyoakland.org/2012/02/rising-tide-north-america-states-solidarity-with-occupy-oakland/, both accessed 10/17/2012. Our particular thanks to Matthias Dietz for directing us to the American evidence.

References

Benford, R. and Snow, A. (2000) Framing Processes and Social Movements: An Overview and Assessment, *Annual Review of Sociology*, 26: 611–639.

Busby, J. (2010) *Moral Movements and Foreign Policy*, Cambridge: Cambridge University Press.

Chawla, A. (2009) Climate Justice Movements Gather Strength, in World Watch Institute (ed.) *State of the World 2009*. Accessed 10/15/2012 from www.worldwatch.org/stateoftheworld.

Dalton, R. (1994) *The Green Rainbow: Environmental Groups in Western Europe*, Yale: Yale University Press.

della Porta, D. (2005) Multiple Belongings, Tolerant Identities and the Construction of 'Another Politics,' in della Porta, D. and Tarrow, S. (eds.) *Transnational Protest and Global Activism*, New York: Rowman and Littlefield.

della Porta, D. (ed.) (2007) *The Global Justice Movement: Cross National and Transnational Perspectives*, Boulder: Paradigm.

della Porta, D. and Diani, M. (2004) *Movimenti senza protesta? L'ambientalismo in Italia*, Bologna: Il Mulino.

della Porta, D. and Kriesi, H. (1999) Social Movements in a Globalizing World: An Introduction, in della Porta, D. and Kriesi, H. (eds.) *Social Movements in a Globalizing World*, New York: St. Martin's Press.

della Porta, D. and Mattoni, A. (2012) Cultures of Participation and Social Movements, in Delwiche A. and Jacobs Henderson, J. (eds.) *The Participatory Cultures Handbook*, Routledge: London.

Dietz, M. (2010) Gibt es eine internationale Klimabewegung? Ein skeptischer Befund, *Forschungssjournal Neue Soziale Bewegungen*, 23(4): 6–18.

Fuster Morell, M. (2011) An Introductory Historical Contextualization of Online Creation Communities for the Building of Digital Commons: The Emergence of a Free Culture Movement, Proceedings 7th International Conference on Internet, Law & Politics (IDP 2011): Net Neutrality and other challenges for the future of the Internet. Accessed 10/15/2012 from http://edcp.uoc.edu/symposia/idp2011/proceedings/.

Gamson, W. (1985) Goffman's Legacy to Political Sociology, *Theory and Society*, 14(5): 605–622.

Gerhards, J. and Rucht, D. (1992) Mesomobilization: Organization and Framing in Two Protest Campaigns in West Germany, *The American Journal of Sociology*, 98(3): 555–596.

Hunt, S., Robert B. and Snow, D. (1994) Identity Fields: Framing Processes and the Social Construction of Movement Identities, in Laraña, E., Hank, J., and Gusfield, J. (eds.) *New Social Movements: From Ideology to Identity*, Philadelphia: Temple University Press.

Kaldor, M. and Selchow, S. (eds.) (in press) *Subterranean Politics in Europe*, Basingstoke: Palgrave Macmillan.

Klein, N. (2010) A New Climate Movement in Bolivia, *The Nation*, May 10.

Mattoni, A. (2009) Organisation, Mobilisation and Identity: National and Transnational Grassroots Campaigns Between Face-to-face and Computer-mediated-communication, in Baringhorst, S., Kneip, V., and Niesyto, J. (eds.) *Political Campaigning on the Web*, Bielefeld: transcript Verlag.

Parks, L. (2009) In the Corridors and in the Streets: A Comparative Study of the Impacts of Social Movement Campaigns in the EU, Doctoral thesis, Florence: European University Institute.

Rising Tide North America (2010) The Climate Movement is Dead: Long Live the Climate Movement. Accessed 06/09/2012 from www.risingtidenorthamerica.org.

Ruzza, C. (2004) *Europe and Civil Society: Movement Coalitions and European Governance*, Manchester: Manchester University Press.

Shift Magazine and dysophia (2010) Criticism without Critique: A Climate Camp Reader. Accessed 06/08/2012 from http://dysophia.files.wordpress.com/2010/01/cca_reader.pdf.

Smythe, E. and Byrd, S. (2010) World Social Forum Activism in Belém and Beyond, *Journal of World Systems Research*, XVI(1): 94–105.

Snow, D. (2004) Framing Processes, Ideology, and Discursive Fields, in Snow, D., Soule, S., and Kriesi, H. (eds.) *The Blackwell Companion to Social Movements*, Oxford: Blackwell.

Snow, D. and Benford, R. (1999) Alternative Types of Cross-National Diffusion in the Social Movement Arena, in della Porta, D. and Kriesi, H. (eds.) *Social Movements in a Globalizing World*, New York: St. Martin's Press.

Soule, S. (2004) Diffusion Processes Within and Across Movements, in Snow, D., Sarah, S., and Kriesi, H. (eds.) *The Blackwell Companion to Social Movements*, Oxford: Blackwell.

Tarrow, S. and McAdam, D. (2005) Scale Shift in Transnational Contention, in della Porta, D. and Tarrow, S. (eds.) *Transnational Protest and Global Activism*, Lanham: Rowman and Littlefield.

Tilly, C. (1978) *From Mobilization to Revolution*, Reading, MA: Addison-Wesley.

Tilly, C. (1984) Social Movements and National Politics, in Bright, C. and Harding, S. (eds.) *State-Making and Social Movements: Essays in History and Theory*, Ann Arbor: University of Michigan Press.

Tokar, B. (2010) *Toward a Movement for Climate Justice*. Accessed 09/15/2012 from www.social-ecology.org/2008/06/toward-a-movement-for-peace-and-climate-justice-2/.

3

DEMOCRATIZING THE CLIMATE NEGOTIATIONS SYSTEM THROUGH IMPROVED OPPORTUNITIES FOR PARTICIPATION

Melanie Müller and Heike Walk

1 Introduction

The conversion of the carbon-based economic model into one of sustainable economic growth requires not only the development of economic and technological potential; it is first and foremost a political and societal project. In addition to the mitigation of climate change through significant reductions in CO_2 emissions, adaptation to its consequences through protective measures is increasingly the focus of public debate. However, the implementation of such measures is difficult and requires the agreement of diverse actors and interests. Thus, the democratization of the climate negotiations system through the expansion and institutionalization of opportunities for citizens and civil society groups to participate is of great importance. Such participation should help to prevent one-sided economic interests from dominating, as well as help those states and populations that are weaker and more affected by climate change to be included. Calls for more participation are gaining support from ever more prominent actors, such as the recent example of the latest release from the German Advisory Council on Global Change (WGBU) "World in Transition – A Social Contract for Sustainability." At first glance, the call for increased participation appears to be extremely feasible and straightforward. On closer examination, however, things get somewhat more complicated, as participation puts considerable demands on the ways in which public opinion is informed and how cooperation takes place locally, nationally, and globally, requiring intervention in existing systems. The consistent evaluation and institutionalization of opportunities for participation puts the logic of state institutions into question, both in terms of hierarchical decision-making processes and in terms of representative political models.

These complicated processes and requirements, which are necessary for participation in democratic systems, can be particularly clearly observed in the international climate negotiations. Accordingly, there has been an increase in the number of publications that address the question of democracy in connection with solving the climate problem (Shearman and Wayne Smith 2009; Welzer et al. 2010; Saretzki 2011). Some sketch out the domain of the problem and point out existing difficulties, while others attempt to address the generalized attribution of guilt in democratic systems as well as the associated lobbying activity and selective representation of interests. Still others enthusiastically praise authoritarian climate regimes. The latter generally do not lead anywhere, as the analyses are generally superficial and draw premature conclusions. Democratic systems cannot be blamed for climate change per se (and if so, only to the extent that they have been slow to respond), but nor can authoritarian systems solve the climate issue. Rather, it makes more sense to concern ourselves with practical prospects for participation and obstructions in democratic systems and the identification of alternatives (Leggewie 2011).

In accordance with this, this article seeks to pay particular attention to precisely these questions and problems concerning democracy and participation in the climate negotiations since, after almost 20 years of discussion about the opportunities and limits associated with opening up international politics, actual power structures have hardly changed at all. Our aim is to highlight these obstructions to participation and the resulting drawbacks and missed opportunities. Although many political programs advocate the participation of citizens and cooperative politics, it can still be seen that ministries, administrative departments, and politicians continue to harbor concerns about the effectiveness of political decision-making processes based on the broadening of participation, as well as diffuse fears of loss of power. Specialist administrations frequently question the specialist knowledge and long-term reliability of the citizenry. Conversely, many citizens speak of the inadequate implementation of participatory outcomes and the fact that such processes of participation are merely fig leaves.

Although scientific analyses do address the need for a capacity for societal change and an increased willingness to commit in order for participatory processes to be carried out, these are generally not discussed in connection with capacities for democratization (see Klein et al. 2010). Here we find a considerable deficiency in the discussion of the dilemmas of democracy. We seek to examine this by means of a theoretical discussion of the various forms and functions of participation as well as of the empirical underpinning of the current activities of civil society in the context of the climate negotiations. In addition to affecting access, resource endowments, the setting of agendas, and most importantly the legitimacy of decision-making, the dilemmas of democracy also affect the democratization of international climate governance through rules governing appropriate procedures as well as the transnational integration of processes for decision-making and informing public opinion. Indeed, until now there have only been extremely selective opportunities for participation for individual well-organized groups, while the citizenry acts more as a democratic paper tiger, occasionally registering their presence through protest.

2 What does participation actually mean and why do we need it?[1]

Civil society groups justify calls for increased participation by reference to improved democracy and more transparent democratic structures. Political decision-makers, on the other hand, justify measures for increasing participation on the basis that they are a necessary requirement for a cohesive society and can counter political apathy (see Gabriel 1999; Abromeit 2002). Of course, there are still critics who see (broader) participation as an added complication (that is to say an increase in time and cost) to the already difficult process of decision-making. In the analysis, the notion of participation attributes particular significance to the involvement of various civil society and private groups. The thesis behind this is that a functioning democracy requires the dedicated participation of the citizenry. Proponents of this thesis include Carol Pateman (1970), Benjamin Barber (1994), Archon Fung and Erik Olin Wright (2003), Ansgar Klein (2001), and Heike Walk (2008).

Participation does not necessarily have to be a process of cooperation, however. On the contrary, exchanges involving conflict between different opinions can be a fruitful part of the political process. We will examine this in greater detail later. Instead, it is important to recognize the great significance of political education and other efforts at educating the public in connection with processes of democratization. In doing so, democratic rights are closely tied to demands for individual self-determination and collective empowerment. According to the views of Pateman, this requires a targeted broadening of participative structures and educational elements at the local level in order to develop the necessary individual and collective democratic qualities:

> Society can be seen as being composed of various political systems, the structure of authority of which has an important effect on the psychological qualities and attitudes of the individuals who interact within them; thus, for the operation for a democratic polity at national level, the necessary qualities in individuals can only be developed through the democratisation of authority structures in all political systems.
>
> *(Pateman 1970: 35)*

Public discussion and communication thus becomes a significant element in contributing to common welfare and promoting the democratic process. In order to achieve this, sophisticated procedures for the regulation of communication are considered necessary.

According to this interpretation, the development of learning processes and capacity-building is associated with the concept of empowerment (Stark 1996). Knowledge is, and generates, power. At the level of the individual, empowerment means that people move out of a situation of powerlessness and helplessness and (re)discover their strengths. At the level of the group, empowerment is paraphrased as the process by which individuals in a group, through group decisions and activities, become aware of their increased strengths and capacity to act. They serve to promote the development of envisioned solutions to problem areas and the development of possibilities for implementation. At the structural level, empowerment can result in changes in ossified structures and conditions, thus providing impetus for organizational development and political reform.

The problems of democratic systems result primarily from a lack of willingness to provide decision-making freedoms for endeavors at empowerment, as well as an unwillingness to acknowledge empowerment as an asset to political decision-making. The relationship between the state – that is to say government, ministries, parliament, administrative departments, public authorities – and civil society is changing. With the new governance systems and regimes, which rest on private organizations and include them in consultation processes, the political process is extending far out into the social sphere. This necessarily results in a change in, and reconstitution of, interests and the forms of organization which these interests take, in the actors and their actions, as well as in the kinds of negotiations taking place between these actors. In a number of cases, civil society organizations are more or less built using public funds, in order for European and national authorities to be able to find communication and negotiation partners, draw upon expertise and acquire policy advice, or to legitimately arrive at the decisions which they expect expanded participation to lead to. Time and time again, state-level actions and the implementation of programs are delegated to private or nongovernmental organizations.

Civil society organizations, professionalized methods of campaigning, media relations, fund-raising, lobbying, the acquisition of expertise, and political participation are thus not merely activities of the citizenry, by which citizens temporarily take their interests into their own hands, but are actually new professional fields and thus additional forms of organization, expertise, and knowledge, which surround formal politics. Accompanying this development is the formation of multiple new channels for the developing informed political opinion. This presents representatives of civil society organizations with completely novel challenges. They must have the ability to take part in such governance mechanisms, acquire and maintain access to the various official decision-makers, maintain an overview of the area and all participating interest groups in order to notice shifts, apply for funds, make sure they receive the funds, and ensure that they do not become sidelined in the process of competing with other civil society organizations. Participation in international systems thus has far-reaching consequences for society itself, in that it creates new professions, differentially regulates the distribution of public and private funds, reshapes the formation and expression of interests, and reorganizes processes of negotiation.

At the same time, after 15 years of climate negotiations, it is clear that only limited progress has been made with respect to the transparency of participating actors, empowerment, and missing regulations concerning participation. As before, there are unclear power structures and insufficient possibilities for monitoring. There are now a number of studies that explore the problem of selective participation (Papadopoulos 2004; Blatter 2007; Walk and Demirovic 2011). At a first glance, the climate change regimes favor consultative participation and orientate themselves for the most part towards deliberative models of democracy, that is to say a discursive formulation of a consensus opinion, whereby organized civil society is targeted. The encouragement of democratic competence and democratic communication behaviors does not, however, play a role in the climate negotiations, and opposition movements and protest groups are likewise uninvolved in the negotiations. The initial problem of unequal participation has not been solved by the opening up of the structures, but has rather been shifted. Power relations have thus become more complex and simultaneously less transparent. While the dividing line previously tended to run between governments and civil society, civil society groups now compete with each other more than ever for influence on the political landscape. Well-organized organizations, as well as groups, who already have a strong connection to the government's position, are usually at an advantage. Important possibilities for political innovation, which lie in the resolution of conflicts between opposing groups, remain hidden from view. At the same time, decision-making structures remain untouched, that is to say, the emerging requirements for participation fit into the existing political system very well.

3 Challenges for the theory of democracy

One challenge to the removal of barriers to participation hence lies in identifying possibilities for institutionalizing the participation of special interest groups with occasionally divergent interests and the development of sophisticated procedures for the involvement of these differing groups as well as possibilities for integrating new forms of participation.

Roth (2005) has proposed a number of ways to improve structures which could be useful in this context: The diverse political activities and activists can be roughly divided into three groups which should be taken into account in the climate negotiations: *First*, there are political activities that can be described as lobbying. These are described as legal, partly institutionalized, and mostly temporary forms of participation with a high degree of legitimacy. *Second*, there are political activities that are often characterized as counter-public. This includes legal forms of protest such as approved demonstrations or social fora which present a space for discourse or promote political education. Though these forms of participation are not institutionalized, they have a high degree of legitimacy. *Third*, there are activities that are better described as self-help. In doing this the broad spectrum of people affected is restricted to organized self-help groups. Naturally, the more abstract the problem in question is, the more difficult the identification of these three groups becomes. Nevertheless, this rough structuring does make analysis easier.

In addition to structuring public opinion, it makes sense to engage with the various different functions of participation and to explore the question of which ideas are being pursued with the participation of civil society groups. The more precisely the interests connected to the participation are disclosed the better the participation procedures can be conceptualized. In doing so, the analysis can take the various following starting points into account:

- *Broad participation of all relevant groups if possible:* Identification of all relevant stakeholders and establishment of the criteria used to select participants.
- *Recognition of power of decision*: Identification of decision-making structures.
- *Disclosure of conflicting aims among groups of actors*: Assessment of the transparency of the various interests of the actors.
- *Recognition of power hierarchies*: Identification of interest groups and structures of dominance.
- *Confrontation vs. consensus*: Identification of the various political modes of action, communication, and cooperation.
- *Encouragement of democratic processes*: Assessment of contribution to the encouragement of society's ability to shape politics.
- *Societal learning*: Identification of learning and emancipatory processes.

Taking all of these approaches into account is, of course, very demanding and calls for a comprehensive empirical analysis. With the help of research that has been carried out in the area of climate politics in recent years, it is possible to carry out an analysis from the point of view of the actors. Though this interpretation does not provide a complete picture, it does, however, offer an insight into the possibilities for deepening knowledge which arise when taking a targeted approach towards locating barriers to participation.

4 Structures of participation in international climate policy

Various forms of participation are used in the area of climate. These have developed in different ways at different levels and have led to some groups being subjected to strongly selective involvement at the very least. At the international level, the participation of civil society groups takes place primarily through the participation of NGOs in international conferences. They are able to register themselves for participation in the negotiations and to accredit their staff. However, the term 'civil society' applies not only to 'NGOs,' but also to other interest groups, research institutes, and trade associations. With the rising importance of climate policy, the number of registered organizations has also risen. At the first COP in 1995, there were just less than 200 organizations which registered for the negotiations. In the years since then, the number has continually risen, and by the 14th conference of the parties in Poznan, the number of organizations exceeded 1,000 for the first time. The highest number of participating organizations occurred at the Copenhagen conference, for which over 1,400 organizations registered. A total of 25,000 people took part in the conference, which was expected to be one of the most important conferences in the history of the international climate negotiations. After the disappointment of the proceedings in Copenhagen, approximately 12,500 participants still traveled to Mexico, and in 2011 more than 13,000 found their way to Durban, South Africa (UNFCCC 2012). While the first years of the international negotiations on climate policy were dedicated primarily to the gathering of evidence and the struggle to get anthropogenic climate change recognized, from the end of the 1990s, the negotiations became increasingly restricted to marketable solutions and detailed technocratic questions (Brunnengräber et al. 2008). Many NGOs reacted to this development and shifted their focus in accordance with the official agenda.

At present, the inclusion of NGOs takes place in the context of their function as consultants. Their main task at the international level is to lobby. There are, however, a few powerful, internationally active NGOs that have, to some extent, been granted the right to speak at the negotiations, and delegates of these nongovernmental organizations are given the floor during the negotiations. These procedures are often emphasized by political decision-makers whenever they find that they are being criticized for being insufficiently inclusive. The fact that these organizations only represent a small fraction of civil society thus remains ignored.

All that remains for the other NGOs that have registered for the negotiations is 'observer status.' They can also run side events during the negotiations and use these to make their positions known and raise controversial issues. Whether these positions actually have any influence on the government positions, however, is generally more dependent on the lobbying activities of the groups concerned (Müller 2010a, 2010b). By the time of the negotiations in Copenhagen, a crack had appeared in cooperation with civil society. As the two weeks of negotiations had been organized very badly, and the pressure on the delegations in the conference center was continuously increasing, the majority of the civil society representatives were excluded from participation in the conference towards the end. On the last Friday of the conference, a mere 90 or so delegates from civil society organizations were allowed to take part (Fisher 2010: 14). In the run-up to the negotiations, civil society organizations had already announced that there would be massive protests. However, the exclusionary behavior of the authorities in Copenhagen may well have further contributed to the protests (Bergfeld 2011). Critical organizations in particular, such as staff of the Bund für Umwelt- und Naturschutz Deutschland, which as a part of Friends of the Earth International had participated in the call for the protests, were refused reentry into the conference center (Klimaretter 2009).

Besides these political conflicts, a structural problem is apparent with regard to inclusion at the international level which above all affects those NGOs with limited finances, who can therefore either not participate in the negotiations or only do so to a very limited extent. This applies to groups from what is termed the Global South which, in contrast to their counterparts in the North, often have too few financial resources and personnel in order to be able to continuously follow the negotiations and to put their positions forward. The imbalance of power between the 'developed' countries as opposed to the 'developing' countries is thus also reflected at the level of civil society. About 25 organizations from Brazil registered for the negotiations, for example, and 17 organizations came from Kenya. However, only one NGO registered from each of the smaller developing countries such as Zambia and Senegal. In contrast, European states such as Germany mustered over 100 organizations (UNFCCC 2011). Especially in view of the different positions adopted by civil society organizations in the North and in the South, equal participation would be all the more urgent (Unmüßig 2011). In such a situation as this, it is indeed honorable when development-oriented NGOs attempt to strengthen the position of the Global South. Many activists from those countries which 'should be helped,' however, often take a skeptical view of this proxy position. Some South African NGOs still remembered the behavior of international partners who attempted to act in the role of advocates for the organizations at the 2002 WSSD in Johannesburg. For some South African NGOs, this looked like an attempt to dominate the local agenda and to push their own positions while disguising them as an expression of collective solidarity. South African organizations in particular have therefore become a lot more careful when it comes to cooperation with international partners, and for them the right to self-representation at the national as well as the international level is an important concern (Interview, see Adam 2011).

5 Alternatives: national ministries and the organization of counter-summits

It is often the case that, for some organizations, processes of inclusion work somewhat better at the national level than direct access to the climate conferences does. During the negotiations, as well as in advance, individual NGOs are invited by their own governments or their own delegations to meet in order to discuss the progress of the negotiations and the positions taken by the ministries concerned. In Germany, a process of continuous exchange between individual civil society groups and the ministry has developed. Within the Federal Ministry for the Environment, Nature Conservation and Nuclear Safety the department responsible regularly meets with a number of NGOs before the big rounds of negotiations in order to discuss the progress of the negotiations as well as the position to be taken by Germany at the European level. A true setting of agendas is, however, not possible for the organizations (Müller 2010a, 2012). This co-opting by civil society actors can be even more clearly seen in countries with only weakly developed ministerial structures for environmental matters. In such cases, it is often the civil society activists who have extensive, often professional, experience in environmental matters and who are thus brought into political decision-making processes as advisors. Chile, for example, very specifically involves its environmental policy activists in the political processes of the Environment Ministry (Müller 2010b). The same goes for the Philippines, which has included several representatives from NGOs in their delegation, who take part in the working groups on political decisions.

Many of the civil society organizations that are closely involved in the advisory process in this way receive criticism from their fellow campaigners for their proximity to the technocracy and the realpolitik of the discussion process. They are accused of having lost their distance from governmental positions, a distance which is necessary in order to remain critical. At the international level, these criticisms have even resulted in splits. The Climate Action Network (CAN), an amalgamation of various climate-related NGOs has been heavily criticized since 1997 for operating more of a lobbying-orientated strategy. Out of this criticism arose Climate Justice Now!. This network takes a very skeptical view of collaboration with governments and ministries as well as of policies at the level of the UN, and pursues a strategy of confrontation (Bond 2011). Even some of the activists who work together with the ministries hold partly critical opinions concerning their lobbying work. The slow progress of the processes of negotiation has left many with doubts as to whether there is any sense in lobbying when no firm agreements are produced anyway, and when it is hardly possible to challenge free market policy positions anymore. In the run-up to the 2011 McPlanet congress in Berlin, the debate concerned the weak and largely uncritical bias of the German environmental movement (Weber-Steinhaus 2012).

More radicalism is often seen at the counter-summits, which have been held regularly since 1995 by environmental policy groups and initiatives at the venue of the respective climate conference. At these events, there is rather more of a focus on controversial and critical topics such as questions concerning climate justice between North and South or vulnerable groups (see Tokar in this book). The most recent example is that of the South African negotiations in Durban, where local civil society set up a "People's Space" for the civil society organizations where the detailed technical questions of the international conference were discussed, as were alternatives to existing systems. The meetings took place at the same time as the 14-day conference of the UNFCCC and were held at the University of Kwa-Zulu Natal. Besides the South

African nongovernmental organizations, academics from the university also supported the People's Space, and promoted an ambitious treaty in South African and international newspapers and magazines (Bond 2011).

Unfortunately, in just under 20 years of the negotiations marathon, there has been hardly any effort to unify this counter-summit with the official agenda and thus to enrich the official negotiations system with an alternative program. There is, however, an important exception: The "People's Conference on Climate Change and Mother Earth" which took place in 2010 in Cochabamba, Bolivia, was significant. A total of 35,000 participants from 142 countries took part in the alternative summit, which was supported by the Bolivian government and which had an anticapitalist focus. The critical points of the conference were expressed in a "People's agreement" and were ultimately brought into the official climate negotiations by Venezuela (Russell 2012). Without the support of the Bolivian and Venezuelan governments, the integration of these demands would barely have been thinkable. The states that dominate the international climate process, most of all the European Union, have made minimal efforts to accommodate suggestions from civil society in the negotiations.

6 Few opportunities at the European level

A similar situation to that occurring at the international level can be found at the level of the European Union, where lobbying often presents even greater challenges to civil society actors. Involvement at the EU level occurs primarily via the mediation of interests by associations and lobby groups within EU institutions, principally through the Commission and the many working groups and committees which have formed around the Commission. Depending on the policy area and the significance of the policy decision, there are different processes, each having a distinct political organ at the European level with important rights of co-determination and decision-making at its disposal (Schmidt 2006). Thus, the European procedures are often only comprehensible for those organizations who are already familiar with the area in question and who have sufficient finances and personnel at their disposal. Despite the high hurdles, a number of civil society organizations with an environmental focus have active lobbying operations in Brussels.

It is, however, primarily the influential and cooperative NGOs that are able to gain access to the administration of the EU. Here, again, the majority of the cooperation taking place at the European level is best understood in terms of the context of a strategy for increasing the efficiency of political institutions. The selective participation of societal groups is primarily intended to improve political results. Thus, the focus is neither possibilities for democratization nor the broadening of the scope of the negotiations and decision-making powers of civil society actors. By bringing in civil society interests, the basis for the legitimacy of the Commission as well as the expertise and information available for various political committees and institutions are increased. A series of studies has demonstrated the selective participation of environmental groups in EU committees (Hey and Brendle 1994; Swyngedouw et al. 2002). To summarize: The European Union is a long way from possessing transparent and reliable political decision-making structures and is often difficult to understand even for skilled connoisseurs of Brussels. A recent study undertaken by the Otto-Brenner-Stiftung shows that nontransparent decision-making systems endanger both the political system and democracy itself (Kolbe et al. 2011).

The rights to participation of civil society organizations have, however, been strengthened by the coming into force of the Treaty of Lisbon in 2009. The extent to which the new possibilities for inclusion are going to change the dynamics of the involvement of civil society, or have already done so, has been little explored as yet. To date, it appears that protest movements have not regarded the European Union to be a target worth their while. Ultimately, there has been little in the way of political protest concerning environmental and climate policy in Brussels in comparison to the protests of European farmers, for example. On the one hand, this can to an extent be explained by the general difficulties for organizations at the European level, an example of this being the poor institutional infrastructure of most civil society organizations in Brussels, and on the other hand the problematic conditions for the organization of protests in all European countries, the organizations' lack of networking at the European level and a lack of knowledge of the structures of the European Union (Roose 2003). In contrast to lobby groups, the European level is clearly not yet of great relevance to climate and environmental political protest groups and is rarely used by social movements in general. Organized protest occurs much more at the national level, thus having a 'bottom-up' influence over decisions taken at the European level. However, this also changes the framing of the social movements and increasingly addresses European politics, even if only at the national level (della Porta and Caiani 2009).

As far as the problems of democracy in the context of a lack of opportunities for participation is concerned, the international and European levels present us with a disappointing picture: There is certainly no sign of broad participation by all relevant groups. Neither are many stakeholders included, nor are there broadly based and defined selection criteria. In most cases, the NGOs that take part are those who see eye-to-eye with the government, that is to say, those which have accepted the discourse of the negotiations in question. In terms of climate policy, this means that the only topics that get discussed are the ones that are on the official agenda of the climate negotiations. New topics, which are disputed and subject to controversy, generally find no place in the discussions. On the contrary, the climate negotiations processes are characterized by language which is strongly orientated towards science and technology. Political issues are hardly discussed during the negotiations any more. The technocratic decision-making mechanisms are specialized to such an extent that not only non-specialists, but also experts who work in similar areas, find them barely comprehensible. One of the most important functions of civil society, namely that of exercising influence over the form of the agenda, is thereby curtailed. As the above examples show, the NGOs have few decision-making or co-determination rights, meaning that the decision-making structures of the negotiations system are hardly, if at all, influenced by the participation of NGOs.

7 The democratic potential of the counter-public, protest, and local involvement

The selective participation of NGOs has contributed at neither the international nor the European level to processes of democratization in terms of the development of the democratic skills and functions of the citizenry through the encouragement of mutual learning and communication processes. A broadening of the rights of citizens to self-fulfilment and self-determination has not been observed at the international level. This could change through the establishment of climate camps which have been organized by radical environmental activist groups across Europe since 2006. In Great Britain and the US, as well as in Germany, climate camps have been organized by leftist activists who have a dismissive attitude regarding the hitherto market-orientated handling of climate change and who instead opt for a combination of knowledge exchange (through workshops), networking, and publicity-generating direct action. The organizers of these camps also took part in the massive protests in Copenhagen. Following the disappointing negotiations, the movements put more of a focus on a local approach, which should above all take local energy problems into account. During 2011, three climate camps with an overriding focus on energy-related issues took place in Germany alone. It remains to be seen to what extent the climate camps will be able to influence the public and policy discussions and what form the communication between the civil society actors will take. So far, no clear overall cross-organizational identity has developed within the movement, although the groups taking part have converged strongly and are on their way towards developing shared aims and a shared identity through participatory and grassroots approaches (Indymedia 2011). The development of democratic skills and the democratizing functions of the citizenry through the encouragement of mutual processes of learning and communication can be most effective at the local level. This is because such processes of cooperation contribute in a very unique way to the integration of everyday problems into the political system, to the harnessing of additional information and knowledge resources, and to the setting into motion of emancipatory processes. Energy cooperatives are a good example of this. The first initiative in Germany, 'die Stromrebellen von Schönau' (the electricity rebels of Schönau), who took over their local power network in a spectacular fashion, first convinced the locals of their district and residents of the broader region, and paved the way for a broad public discussion. Through strong local and regional networking and the initiation of an intense debate concerning possibilities for a green, decentralized energy supply owned by the citizens themselves, the initiative quickly became an example for other networks and communities and motivated a great number of citizens to get involved in the energy sector (Graichen 2003).

Participatory projects which satisfy the standard demands of open participation in the sense of a maximization of citizens' opportunities for self-fulfilment and self-determination are most easily identified at the local level. Although the increased consolidation of powerful interest groups and an increase in the number of market-orientated actors at this level of the climate sector can be seen (accompanied by the simultaneous dwindling in the presence of actors representing social solidarity and societal aims), efforts concerned with the support of democratic skills and democratic communication practices are still most clearly pronounced at the local level.

There are also a few examples in which strong public opinion has stopped political projects for which plans had already been made. The most recent example of this is the Tar Sands Movement in the US, which argued against the construction of a pipeline for the extraction of tar sands. The extraction of tar sands is not only energy-intensive but also has immense environmental impacts. Following massive protests and the encirclement of the White House, US President Barack Obama announced a review of the construction of the oil pipeline, which was to take 12–18 months. Some experts interpret this review as signifying a moratorium on the project (Tar Sands Action 2011).

Particularly with regard to the development of strategies for adaptation to climate change, increased incorporation of indigenous knowledge into political programs would be an important step. While indigenous groups have contributed the least to climate change, they are, nevertheless, affected to a large degree by climate change. This is largely due to the fact that these groups are more dependent on the use of ecosystems, biodiversity, and local farming. When climate change alters local conditions, it is these communities which are often the first to feel the effects. Conversely, indigenous communities often have a large amount of knowledge at their disposal. They know their territory well, have already developed their own adaptation strategies and pass on their experiences from one generation to the next (Nyong et al. 2007). In recent years many tailor-made political programs, such as the Clean Development Mechanism (CDM) and programs taking place within the context of REDD (Reducing Emissions from Deforestation and Degradation), have missed opportunities to actively integrate this knowledge beforehand (Raygorodetsky 2011). The cultural characteristics of local communities have also been neglected. Truly sustained inclusion of local cultures and groups, not only in order to legitimize political programs, but as an end in itself, would have prevented the failure of a number of development projects in recent decades (Escobar 1994).

It is precisely the systems of climate negotiation that offer a great deal of potential to think about opportunities with respect to participatory approaches. It is not only in Germany that the numbers of those supporting increased participation and shaping of the system by societal actors are constantly increasing. Other countries have also seen an increasing understanding that a reconfiguration of the energy system and a climate-friendly infrastructure is not possible without societal involvement (Adger 2003). There is, therefore, an urgent need for a continuing scholarly discussion about practical possibilities for NGOs to be actively involved in the shaping of (inter)national systems of climate negotiation and to have a voice in them. An example of one possibility would be additional chambers, for a, and panels with differentiated voting rights and clearly established participant rights (there is a particular need for juridical input here). At the same time, the international negotiations could make use of new participatory models in which international decision-makers not only cast their votes, but also work through the ideas and projects of societal interest groups in discursive workshops. Shared knowledge and the inclusion of different perspectives generally lead to long-term solutions, but also result in innovative and creative ideas and can assist in speeding up the search for solutions in the otherwise very structured and drawn-out parliamentarian or ministerial process (Roth 2010). By these means, unforeseeable side-effects could also be better taken into account. Integrating local knowledge into political programs can improve them even in advance. Besides these functional aspects, however, genuinely broad participation, in the sense of the shaping of agendas and participation in decision-making, remains an intangible commodity and should be viewed as a democratic ideal in political decision-making procedures.

Note

1 Some passages are based on an older contribution by Walk: "Partizipative Governance. Beteiligungsformen in der Klimapolitik" in Walk and Demirovic (2011).

References

Abromeit, H. (2002) *Wozu braucht man Demokratie? Die postnationale Herausforderung der Demokratietheorie*, Opladen, Leske + Budrich.

Adam, F., Greenpeace South Africa (2011) Interview during the 17th Conference of the Parties in Durban, December 2011.

Adger, W. N. (2003) Social Capital, Collective Action and Adaptation to Climate Change, *Economic Geography*, 79(4): 387–404.

Barber, B. (1994) *Starke Demokratie*, Hamburg: Rotbuch.

Bergfeld, M. (2011) The Birth of the Climate Justice Movement, *Essex Graduate Journal of Sociology*, 10.

Blatter, J. (2007) Demokratie und Legitimation, in Benz, A., Lütz, S., Schimank, U., and Simonis, G. (eds.) *Handbuch Governance. Theoretische Grundlagen und empirische Anwendungsfelder*, Wiesbaden: VS Verlag.

Bond, P. (2011) *Politics of Climate Justice: Paralysis Above, Movement Below*, University of KwaZulu-Natal: UKNZ Press.

Brunnengräber, A., Dietz, K., Hirschl, B., Walk, H., and Weber, M. (2008) *Das Klima neu denken. Eine sozial-ökologische Perspektive auf die lokale, nationale und internationale Klimapolitik*, Münster: Verlag Westfälisches Dampfboot.

della Porta, D. and Caiani, M. (2009) *Social Movements and Europeanization*, Oxford: Oxford University Press.

Escobar, A. (1994) *Encountering Development: The Making and Unmaking of the Third World*, Princeton: Princeton University Press.

Fisher, Dana R. (2010) COP15 in Copenhagen: How the Merging of Movements Left Civil Society out in the Cold, *Global Environmental Politics*, 10(2): 11–18.

Fung, A. and Elin Wright, O. (2003) *Deepening Democracy: Institutional Innovations in Empowered Participatory Governance*, London, New York: New Left Books.

Gabriel, Oskar W. (1999) Demokratie in der Vereinigungskrise? Struktur, Entwicklung und Bestimmungsformen der Einstellungen zur Demokratie im vereinigten Deutschland, *Zeitschrift für Politikwissenschaft*, 9(3): 827–861.

Graichen, P. (2003) *Kommunale Energiepolitik und die Umweltbewegung: Eine Public-Choice-Analyse der "Stromrebellen" von Schönau*, Frankfurt: Campus-Verlag.

Hey, C. and Brendle, U. (1994) *Umweltverbände und EG. Strategien, politische Kulturen und Organisationsformen*, Opladen: Westdeutscher Verlag.

Indymedia (2011) Klimacamps und Energiekampfe 2011. Accessed 12/19/2011 from http://de.indymedia.org/2011/07/311642.shtml.

Klein, A. (2001) *Der Diskurs der Zivilgesellschaft. Politische Hintergründe und demokratietheoretische Folgerungen*, Opladen: Leske + Budrich.

Klein, A., Olk, T. und Hartnuß, B. (2010) Engagementpolitik als Politikfeld: Entwicklungserfordernisse und Perspektiven, in Olk, T., Klein, A., and Hartnuß, B. (eds.) *Engagementpolitik: Die Entwicklung der Zivilgesellschaft als politische Aufgabe*, Wiesbaden: VS Verlag.

Klimaretter (2009) Umweltschützer gelten von nun an als Risiko. *Klimaretter.info*. Accessed 07/02/2013 from www.klimaretter.info/kopenhagen/nachricht/4652-umweltschr-gelten-von-nun-an-als-risiko.

Kolbe, A., Königsberger, H., and Osterberg, S. (2011) Marktordnung für Lobbyisten. Wie Politik den Lobbyeinfluss regulieren kann. Studie für die Otto-Brenner-Stiftung, *OBS-Arbeitsheft* 70, Frankfurt/Main.

Leggewie, C. (2011) 2050: Die demokratische Frage heute, in Schüttemeyer, S. (ed.) *Politik im Klimawandel*, Baden-Baden: Nomos.

Müller, M. (2010a) Die Beteiligung von Nichtregierungsorganisationen (NRO) in der internationalen Klimapolitik. Eine empirische Analyse des Politikformulierungsprozesses am Beispiel Deutschlands, Diplomarbeit, Universität Potsdam.

Müller, M. (2010b) Zwischen Kooperation und Instrumentalisierung. Die Zivilgesellschaft in der internationalen Klimapolitik, *Welttrends – Zeitschrift für internationale Politik*, 73: 65–73.

Müller, M. (2012) Lobbyarbeit, Protest und ungleiche Teilhabe. NGOs und soziale Bewegungen in der internationalen Klimapolitik, in Christ, T. (ed.) *Zugänge zur Klimadebatte in Politikwissenschaft, Psychologie und Soziologie. Beiträge zur sozialwissenschaftlichen Nachhaltigkeitsforschung*. Band 4. Marburg: Metropolis-erlag.

Nyong, A., Adesina, F., and Osman Elasha, B. (2007) The Value of Indigenous Knowledge in Climate Change Mitigation and Adaptation Strategies in the Sahel Zone, *Mitigation and Adaptation Strategies for Global Change*, 12(5): 787–797.

Papadopoulos, Y. (2004) Governance und Demokratie, in Benz, Artur (ed.) *Governance – Regieren in komplexen Regelsystemen. Eine Einführung*, Wiesbaden: VS Verlag.

Pateman, C. (1970) *Participation and Democratic Theory*, Cambridge: Cambridge University Press.

Raygorodetsky, G. (2011) Why Traditional Knowledge Holds the Key to Climate Change, United Nations University. Accessed 12/19/2011 from http://unu.edu/articles/global-change-sustainable-development/why-traditional-knowledge-holds-the-key-to-climate-change.

Roose, J. (2003) *Die Europäisierung von Umweltorganisationen. Die Umweltbewegung auf dem Weg nach Brüssel. Studien zur Sozialwissenschaft*, Wiesbaden: Westdeutscher Verlag.

Roth, R. (2005) Transnationale Demokratie. Beiträge, Möglichkeiten und Grenzen von NGOs, in Brunnengräber A., Klein, A., and Walk, H. (eds.) *NGOs im Prozess der Globalisierung Mächtige Zwerge – umstrittene Riesen*, Wiesbaden: VS Verlag.

Roth, R. (2010) *Bürgermacht. Eine Streitschrift für mehr Partizipation*, Hamburg: Edition Körber-Stiftung.

Russell, B. (2012) Movements and Moments for Climate Justice: From Copenhagen to Cancun via Cochabamba, *ACME: An International E-Journal for Critical Geographies*, 11(1): 15–32.

Saretzki, T. (2011) Der Klimawandel und die Problemlösungsfähigkeit der Demokratie, in Schüttemeyer, S. (ed.) *Politik im Klimawandel*, Baden-Baden: Nomos.

Schmidt, S. (2006) *Die Europäische Union in der Vergleichenden Politikwissenschaft*, Wiesbaden: VS Verlag.

Shearman, D. and Wayne Smith, J. (2009) *The Climate Change Challenge and the Failure of Democracy*, Westport, Conn.: Praeger.

Stark, W. (1996) *Empowerment: neue Handlungskompetenzen in der psychosozialen Praxis*, Freiburg i. Br.: Lambertus.

Swyngedouw, E., Page, B. and Kaika, M. (2002) Sustainability and Policy Innovation in a Multi-Level Context: Crosscutting Issues in the Water Sector, in Heinelt, H., Getimis, P., Kafkalas, G., Smith, R., and Swyngedouw, E. (eds.) *Participatory Governance in Multi-Level Context: Concepts and Experience*, Opladen: Leske + Budrich.

Tar Sands Action (2011) [Blog of the Tar Sands Movements]. Accessed 12/12/2011 from www.tarsands-action.org/.

UNFCCC (United Nations Framework Convention on Climate Change) (2012) Participation Breakdown. Accessed 03/29/2012 from http://unfccc.int/files/parties_and_observers/ngo/application/pdf/participation_breakdown_cop1–17.pdf.

Unmüßig, B. (2011) NGOs in der Klimakrise: Fragmentierungsprozesse, Konfliktlinien und strategische Ansätze, in Brunnengräber, A. (ed.) *Zivilisierung des Klimaregimes – NGOs und soziale Bewegungen in der nationalen, europäischen und internationalen Klimapolitik*, Wiesbaden: VS Verlag.

Walk, H. (2008) *Partizipative Governance. Beteiligungsrechte und Beteiligungsformen im Mehrebenensystem der Klimapolitik*, Wiesbaden: VS Verlag.

Walk, H. and Demirovic, A. (2011) *Demokratie und Governance. Kritische Perspektiven auf neue Formen politischer Herrschaft*, Münster: Verlag Westfälisches Dampfboot.

Weber-Steinhaus, F. (2012) *Kampf der Langweiler. Ökobewegung in der Krise, taz*, 19 April. Accessed 08/31/2012 from www.taz.de/!91824/.

Welzer, H., Soeffner, H.-G., and Giesecke, D. (2010) *Klimakulturen. Soziale Wirklichkeiten im Klimawandel*, Frankfurt/Main: Campus.

4

ANTAGONISTIC STANDPOINTS

The climate justice coalition viewed in light of a theory of societal relationships with nature

Philip Bedall and Christoph Görg

1 Introduction: climate change as a societal problem

Over the last few years climate change has seen a remarkable process of (re-)politicization that has come to envelop even the political institutions created to address it. Not least since the heated debates that took place at the UN climate conference at Copenhagen in 2009 and the subsequent laborious negotiations within the UNFCCC, it has become apparent that the 'Rio institutions' are in crisis (Park et al. 2008). It has also become clear that climate change is a complex set of conflicts characterized by numerous interlocking lines of division and dissent. Some of these have to do with how we arrive at an adequate understanding of the climate crisis, how we interpret it and frame it given its scientific, economic, and political aspects – ranging from doubts about the very existence of anthropogenic climate change to disagreement over the use of market-based instruments. These debates are closely connected in turn to substantial issues of societal development, the eruption of conflicts of interest along with their associated power relations, and, not least, the intermeshing of the climate crisis with the economic and financial crisis.

Climate change is embedded in complex societal conflicts, analysis of which poses a number of challenges for the social sciences. Conflicts related to climate change are rooted in *divergent interests in society*. They have to do, for example, with the dependence of certain industrial sectors on the use of fossil fuels and with the greenhouse gas emissions associated with that use. They also have to do with the situation of social groups who are particularly hard hit by the impacts of climate change. Yet neither of these facets – more narrowly defined economic interests on the one hand and social vulnerabilities on the other – exists independently of its interpretation and construction through language. They have to be created *discursively* and become established and accepted as more or less dominant interpretations. Further still, the precise *nature of the problem* itself – its societal causes as well as its consequences for society at global, regional, and local level – is highly disputed. Meanwhile, the interests of various stakeholder groups and the complex power relations associated with them are reproduced in the course of these conflicts. The lack of clarity certainly cannot be reduced to the actions of 'climate skeptics' who cast doubt on the existence of climate change, often with huge financial support from the fossil fuel industry. As climate scientist Mike Hulme (2009) has noted, the sheer complexity of the issues involved – often due to the multiple points of interaction between climate and society – means that many uncertainties cannot be resolved, and these then become focal points for disagreement. The anthropogenic causes of climate change along with their historical roots in the industrial revolution and uneven global capitalist development are one of the sources of this politicization, given that they form the point of departure for every debate about political responsibility and for justice-related issues (as Engels and Weingart noted as far back as 1997).

For this reason a precise analysis of the social *construction* of the problem of climate change is indispensable. Such an analysis must consider the broad range of discursive processes involved, from the actual climate sciences themselves – with the knowledge needs and uncertainties associated with their various disciplines – through to political and economic discourses that determine what is accepted as part of the problem of climate change and what is not. At the same time, however, we need to assume that climate change itself *impacts back* on societies around the world. So it cannot simply be a matter of analyzing how the problem is discursively constructed and framed; we also need to take account of the *material dimensions* of the climate crisis. For example, what are its concrete impacts on the living conditions of various social groups and on the potential profits available to certain economic sectors? A whole host of conflicts need to be taken into account here alongside the overarching dynamic of societal development as it is manifesting in the multiple crises currently in progress. It is only by taking account of this overarching dynamic that it becomes possible to consider whether certain processes are more an expression of these critical trends or whether the actors involved are in a position to transform and shape the logic of such processes, to name just the two extremes of the spectrum of possible answers.

In order to address the full range of these challenges we need a variety of explanatory approaches, from sociology of science (Jasanoff and Martello 2004), the theory of hegemony (Laclau and Mouffe 1985; also see Howarth et al. 2000; Methmann et al. 2013), and the analysis of neoliberal globalization and its multi-scale intermeshing of global, national, and local processes (Harvey 1996; Swyngedouw 2010) through to critical state theory (Brand et al. 2008; Wissen 2011; for an overview, see Bedall 2011) and others besides. What follows, however, is not a comprehensive analysis (see e.g., Brunnengräber et al. 2008). Our aim instead is to analyze changes that have become apparent in the group of actors involved in international climate politics since the mid-2000s. The appearance of new actors, especially networks consisting of social movements, nongovernmental organizations (NGOs), and other civil society groups, has meant that the link between climate change and societal development in the global context – and thus its embeddedness in the crisis-ridden development of globalized capitalism – has come to be an issue of public debate. In contrast to the prevailing – hegemonic – policy approach to climate change it is possible to identify an antagonistic discourse coalition[1] which is advancing the demand for *climate justice* along with a specific programmatic agenda. This coalition is also calling into question the terrain itself on which political debate is conducted, i.e., the state-level or intergovernmental character of climate politics, and is calling for fundamental societal transformation.[2]

The climate justice coalition, we suggest in the following, has played an important role in shifting the discursive framework. By contributing actively towards (re-)politicizing climate change, it has helped ensure that global warming is no longer addressed as an isolated environmental problem but rather as a 'crisis of society' (*Gesellschaftskrise*, Brunnengräber 2002), a far-reaching crisis of societal development and of the relationships with nature that sustain it. Tackling it effectively therefore requires a transformation capable of revealing and combating the societal roots of the problem and its diverse interconnections with other crises.

In order to understand the role of the climate justice coalition, we begin by introducing a theoretical approach that makes it possible to grasp both the discursive elements of the (re-)politicization of the climate problem as well as the physical-material implications of the climate crisis for societal development, namely, the concept of societal relationships with nature (section 2). In a further step, we elucidate the terrain of conflict itself. This involves looking in particular at nonstate actors such as NGOs and social movements (section 3). Special attention will be paid to the way in which, since the adoption of the Kyoto Protocol, an affirmative style of political engagement has emerged within this group of actors which (re-)produces the (hegemonic) consensus of market-based approaches to climate change. This part of the paper forms the point of departure for understanding the emergence of the new discourse coalition aligned in antagonistic opposition to this consensus. The antagonistic actors are analyzed in more detail in the following section (4) and then examined in terms of their role in the crisis of societal relationships with nature.

2 The concept of societal relationships with nature – a political ecology approach

The theoretical construct of societal relationships with nature is located in the critical theory tradition of the Frankfurt School and is conceived as a critical development of Marxist theory.[3] Taking Marx as a starting point, it takes as given the key role of society's material interaction with nature in societal development but strongly emphasizes the "non-ontological character" of historical materialism (Schmidt 1993, our translation): it is not a matter of asserting an ontological priority of (external or internal) nature but of analyzing the dialectical mediation between nature and society. Crucial for doing this are the concepts of labor, technology, and economic *production* on the one hand and the linguistic-discursive (conceptual as well as cultural) *construction* of nature on the other. The former is responsible for an ongoing transformation of preexisting conditions that leaves neither 'relations of nature' nor social relations untouched. These processes of transformation are currently occurring on a global scale. The second trend, by contrast, is considerably less developed in Marx and is discussed especially in relation to the techno-scientific appropriation of nature in critical theory (Marcuse 1979; Horkheimer and Adorno 1987; see below). The important point is that nature is subject to a double mediation by and through societal development: as a scientific (and cultural) construction and as a product of labor (including its economic and technological dimensions).

The analytical point of departure and reference for a theory of societal relationships with nature, then, is not a nature or wilderness untouched by human hands – nor indeed a climate system demanding respect on the basis of having cast-iron laws and regularities. Rather, the starting point is always social construction, be it of a linguistic, economic, or technological kind. Everything we know about the climate system is dependent on technical apparatuses (for measurement) and on scientific models and theories. Without these, there can be no climate change and no understanding of global warming or its causes based on human actions (Jasanoff 2004; Hulme 2009). This is not to say that climate change has no material reality, however – on the contrary. In the critical theory of Adorno in particular the concept of nature resolutely retains a material dimension relating to the biophysical conditions of our existence. This material dimension becomes apparent precisely when 'nature' puts up resistance to our constructions. This may occur, for example, when our scientific or conceptual constructions prove to be plain wrong in practical terms, when the ways we appropriate nature on their basis give rise to socio-ecological problems; or it may happen when economic and technological constructions undermine nature's ability to function – on which they themselves depend.

Thus it is precisely in its linguistic-discursive and economic-technological construction (or through it) that nature reveals its material properties. This happens, for example, in the form of viable ecosystems and in the globally interconnected components of the climate system of which we were previously unaware or had not taken proper account, a climate system whose mode of functioning may be undermined by certain forms of appropriation.

This materiality is itself constitutive of societal development, whether in the form of 'natural' resources used to produce a secure supply of food or as a source of energy. Just how constitutive it is often only becomes apparent when the material resources on which we depend become scarce or when doubts arise regarding the ability of ecosystems to function properly (MA 2005). The concept of societal relationships with nature thus gives expression to a relationship of mutual constitution: even as societies produce and appropriate nature for their own purposes, thereby transforming it, they themselves remain dependent on these material foundations on which their existence depends.[4] The climate problem can be understood neither as a disturbance affecting nature and its laws that are supposedly ontologically pregiven nor as an environmental problem that can be managed in a purely technical manner. Rather, it implies a comprehensive *crisis of societal relationships with nature*, including its scientific and socio-cultural dimensions. The regional climate has long been subject to alteration by human beings, just as humans themselves have been influenced by climatic fluctuations (Behringer 2007). But it is only since the onset of capitalist industrialization that these interactions have mutually reinforced one another to such an extent that we neither understand the complexity of the processes let alone have them under control. Although the current era in Earth's history has been declared the Anthropocene period (Crutzen 2002), i.e., a period shaped indelibly by humans, what this fails to acknowledge is that we don't really have any understanding of the nature – society interactions embedded within it. We can neither control nature's impacts on society nor – even less so – the transformation of nature by society: ecological damage is causing us more and more problems and we are still far from having the societal causes of climate change under control. The roots of the ecological crisis lie in a particular construction of nature, namely, in the *ideology of the domination of nature*, which ignores the fact that humans have no genuine control over nature.

The *critique of the domination of nature*[5] is the key conceptual motif of this version of political ecology. This idea holds that various forms of social domination – whether by class, sex, or race – have become inscribed onto our relationships with nature (Görg 1999, 2003). These relations of domination are by no means total but always relative and subject to struggle; neither are they at all absolute, because they can never achieve their goal of total control over the other. Furthermore, these different forms of domination cannot be reduced to any one model of domination, be it class domination (as asserted in Marxism) or the domination of nature.[6] The forms of domination cannot exist independently of each other either, but rather mutually influence one another. At the same time, they generate contradictions and are articulated in crises and conflicts.

Social movements can be interpreted, in abstract terms, as expressions of these societal contradictions and emerging crises. However, they cannot be reduced to the structural dimensions of society or be understood as having emerged from them. Social movements are made up of actors who cannot be reduced to the structural conditions of their emergence nor to established forms of politics but rather, conversely, give the latter political meaning and thereby seek to generate new forms of political articulation (Görg 1992). The extent to which they are bound up with hegemonic forms of political engagement or stand in critical or even antagonistic opposition to them is a different question that can only be answered empirically; it is to this question that we turn in the following.

As a key element in the crisis of societal relationships with nature, climate change throws a spotlight on the contradictions entailed by industrial societal development – in particular the contradictions of the Fordist phase of capitalist development with its vastly increased consumption of resources and its ecological side effects (Altvater 1992). Since the 1970s social movements have articulated various facets of these contradictions, contributing towards making various dimensions of the crisis themselves the object of societal debate and social struggle. In the following we examine the social construction of climate change, focusing especially on civil society actors and basing our discussion on these two key basic assumptions regarding the character of the climate problem and the role of social movements.

3 NGOs, social movements, and the neoliberal appropriation of climate policy

The central location for conflicts over hegemony is the terrain of civil society, understood here in Antonio Gramsci's sense as the "extended state" (see Gramsci 1971: 271). It is on this terrain that the key struggles and debates over the direction of societal development and thus of how to shape societal relationships with nature take place. Thus civil society can be conceived of as the terrain where struggles and debates over the hegemonic construction of climate change take place. This is where social movements, NGOs, and other actors are engaged in many different, often keenly contested struggles over interpretation that bear on how problems are constructed and what potential solutions arise out of these constructions. In this context it is necessary to distinguish between affirmative and critical, or antagonistic, strands. The civil society groups involved in climate politics present a heterogeneous picture in terms of their demands, the strategies they pursue, and, associated with these, the structures of the networks formed by them. Both strategies and networks change over time. A variety of positions are adopted in relation to the hegemonic discourses and the actors who sustain them.

It is possible to identify *three phases* relating to the way the field of NGO and social movement actors has been structured since negotiations over the draft of a framework convention on climate change within the Intergovernmental Negotiation Committee (INC) began in 1991. The first marker point was the adoption of the Kyoto Protocol in 1997. The growing politicization of climate politics from 2005 onwards also had a pronounced impact on the structuring of the field of actors. It is this politicization of climate change that opens up a terrain on which antagonistic demands are increasingly articulated towards hegemonic climate politics. Linked by the demand for *climate justice*, networks of a new type of actor now appear on the stage whose political style differs markedly from the affirmative type that previously dominated the field of actors. In the following we describe the first two phases of this structuring process before proceeding in the next section to describe the specific emergence of the climate justice coalition.

3.1 Pre-Kyoto – critical stances within an open-ended process

Looking back first to the time prior to the adoption of the Kyoto protocol, the negotiation process was still open-ended at that point. There was no agreement on how climate change should be dealt with politically, whether through legally binding regulations or market-based instruments. Alongside primarily environmental NGOs, a large number of development NGOs and social movements also made an appearance on this politically contested terrain. One of the key motivations for civil society organizations to connect up with one another in the Climate Action Network (CAN) from 1989 onwards was to forge a link between environmental and development issues[7] (Walk and Brunnengräber 2000: 148).

Prior to the adoption of the Kyoto protocol, it was striking that the majority of environmental and development groups had a critical stance towards market-based solutions.[8] At the start of the UN climate negotiations CAN even demanded that negotiators should "avoid any emission trading schemes which only superficially address climate change problems, perpetuate or worsen inequities hidden behind the problem, or have an negative ecological impact" (The NGO Alternative Treaties at Rio de Janeiro, quoted in Brunnengräber 2009: 183). At the start of climate negotiations over the draft of a climate convention many NGOs rejected other market-based mechanisms as well, such as the Clean Development Mechanism (CDM) and the Joint Implementation Mechanism. The Joint Implementation Mechanism was criticized not only by NGOs from the Global South (Rahman and Roncerel 1994: 256, 261) but also by organizations such as Friends of the Earth, Greenpeace, and CAN (Arts 1998: 142). Indeed it was an intervention by NGOs that prompted the G77 nations and China to withdraw their initial support for the Joint Implementation Mechanism in the context of the INC negotiations (Arts 1998: 144; Newell 2000: 142).

3.2 Post-Kyoto – an affirmative stance on hegemonic structures, processes, and issues

The adoption of the Kyoto protocol in 1997 along with its practical outcome (i.e., agreement on specific reduction targets and mechanisms) led to a shift in the terrain of engagement among NGOs and social movements. The specific and *strategically selective* (Brunnengräber 2013) policy approach to the problem of climate change manifested in the protocol was now rarely questioned. On the contrary, civil society actors were integrally involved in (re-)producing the emerging hegemonic consensus.

This adjustment of positions in line with those taken in the official negotiation process was an expression of the emergence of an affirmative style of political engagement on the part of the civil society actors involved in climate politics. Walk and Brunnengräber use the term "*combative cooperation*" to describe this stance (*konfliktive Kooperation*, Walk and Brunnengräber 2000: 276, emphasis in original): it is a political style geared towards 'political and conformist restrictions' that can be described as "*sceptically approving* and *critically observing*" (Walk and Brunnengräber 2000: 276, our translation, emphasis in original). The intention to participate in the official negotiation process determined the limits of these actors' political demands and protests. From this point on, fundamental criticism of the kinds of issues being discussed as part of the process and procedures was rarely expressed. This affirmative style of political engagement confirms Lohmann's observation that the creation of CO_2 markets was "no corporate conspiracy, but rather a joint invention of civil society, business and the state. Non-governmental organizations (NGOs) have been nearly as prominent in its development as private corporations" (Lohmann 2006: 58). Civil society organizations thus helped to (re-)produce the hegemonic consensus. They acquired a legitimating function by making reference to the international process of negotiations in their demands and strategies (see Walk and Brunnengräber 2000: 182; Brunnengräber et al. 2001; Paterson 2010).

The hegemonic consensus consisted in the universalization of a specific approach to climate change and the exclusion or marginalization of perspectives critical of this approach. In the following we discuss two aspects in particular that serve to illustrate the *(re-)production of this consensus by civil society actors*: first, the emerging *environmental* emphasis of the organizations engaging with the negotiation process and, linked to this, the tendency to adopt a naturalizing understanding of the problem and exclude a North–South perspective,[9] and, second, the focus on *market-based* mechanisms.

3.3 Environment vs. global justice

While the UN Conference on Environment and Development held in Rio de Janeiro in 1992 marking the birth of the framework convention on climate change brought together an impressive number of both environmental and development NGOs, over the subsequent years the composition of NGOs engaging with the UN climate negotiations saw a marked change. Particularly after the Kyoto negotiations of 1997 a majority of NGOs turned away from the UN negotiations, although it was above all the more development-oriented organizations that withdrew from the climate process, placing their focus more squarely on issues of poverty and trade, and in particular on the WTO process (Unmüßig 2011: 3). As a consequence, "a clear environmental emphasis" (Walk and Brunnengräber 2000: 149, our translation) became apparent among the NGOs still engaging with the international climate negotiations. The social dimension of climate change – that is, discussion of socio-economic inequalities regarding the causes of climate change, differing degrees of impact and options for action, as well as issues to do with distribution and justice – shifted into the background. At the UN Rio conference the link between environment and development had been embedded in the very name itself, whereas subsequently much less common debate took place initially about social and ecological issues.[10] Once the Kyoto protocol had been adopted, the environmental organizations taking part in the negotiations were virtually on their own (Unmüßig 2011: 3).

The environmental focus on climate change entails a naturalizing perspective on the issue. The main emphasis is on greenhouse gas emissions and the need to regulate them. This interpretation of the problem focuses on the output of the energy system[11] while the input side – energy production – is bracketed out of the equation (Brunnengräber 2013: 73). Accordingly climate change is understood by the actors not as an expression of a crisis of the prevailing relationships with nature as mediated by society but rather as a crisis to do with a climate system that exists outside society. What this understanding of climate change fails to acknowledge, however, is that the crisis has a social dimension – one that can be identified in specific phenomena occurring locally, regionally, and nationally as well as at a specific time in history (BUKO 2009: 3). After Kyoto this understanding of climate change as a socio-ecological crisis and thus of specific societal relationships with nature had hardly any advocates among civil society actors.

This environmental emphasis led to the marginalization of a perspective on climate change that highlights issues of global justice in North–South relationships. One reason for this was the composition of the field of actors: in contrast to organizations from the Global North, many of which are focused principally on the environment, it is groups from the Global South especially that emphasize "much more strongly what the Kyoto instruments mean in terms of development policy along with the issue of social *justice*"[12] (Brunnengräber et al. 2008: 105, our translation, emphasis in original; also see Pettit 2004: 102; Rahman and Roncerel 1994: 244, 246–247). However, of the NGOs engaging with the negotiations, those from the Global South are in a clear minority compared with NGOs from the Global North (Missbach 1999: 274). Those from the Global South affected by development policy rarely have a voice in the negotiations themselves (Sagar and Banuri 1999). A lack of resources and of knowledge about how the UN bureaucracy works also contributes to NGOs from the South being excluded from the negotiation process (Newell 2000: 138). As a result, the organizations represented at the negotiations do not constitute a representative selection of active climate groups (for more on this, see Conca 1995: 444–445). With regard to climate politics after Kyoto, Unmüßig even suggests that "apart from transnational networks such as the Climate Action Network or Friends of the Earth International, civil society actors from the countries of the global South [have been] virtually absent from the negotiations" (Unmüßig 2011: 4).

In addition to unequal access to the negotiations, the marginalization of issues to do with global justice also manifests in the structure and positioning of civil society networks. A balanced North–South ratio cannot be assumed to exist even within the networks themselves. In CAN, for example, it is the large, Western European and North American NGOs that predominate over Southern or Eastern European organizations (Brunnengräber 2009: 180). NGOs from the Global North such as Greenpeace, WWF, and the Environmental Defense Fund are those that adopt a key role (Missbach 1999: 274). As Newell puts it: "The amalgamation of NGO positions favours the more predominant Western interpretation of the issue, given the greater presence of Western groups at international meetings" (Newell 2000: 139).

3.4 *Solving the climate problem via the market*

Once the Kyoto protocol was adopted, voices critical of or clearly opposed to market mechanisms were raised only on isolated occasions. The majority of NGOs now turned their attention to debates over the design of such instruments, thereby (re-)producing the hegemonic consensus that the solution to the climate problem lies in the market. This made apparent the hegemony of neoliberalism in global climate governance (see Brunnengräber 2006: 219–220; Newell and Paterson 2010: 23–31; Bedall 2013: 203–206). Where there had initially been widespread rejection of these instruments on principle, they were now accepted by a majority of actors (Walk and Brunnengräber 2000: 150–154; Lohmann 2006: 53, 59; Brunnengräber 2009: 125, 183–184; Bond and Dorsey 2010: 287). For most NGOs the focus was now on issues associated with the technical details (Walk and Brunnengräber 2000: 150; Unmüßig 2011: 3–4). According to Oberthür and Ott, "representatives of civil society find themselves caught up in the complexity trap.... [T]hey [are] trapped in the microcosm of technical details" (Oberthür and Ott 1999: 300). The positions adopted by the majority of NGOs involved in CAN vary according to the different market mechanisms enshrined in the Kyoto protocol, which they regard as being "the only solution"[13] (Brunnengräber et al. 2008: 97, our translation).

The transformation from critique to acceptance also found expression in the demands that came to be formulated by the NGOs. For example, an integral part of the three-pronged approach called for by CAN is the creation of a market for CO_2 emissions[14] (de Lucia 2009: 235). And although Greenpeace voices criticism about the way the market mechanisms are designed it no longer questions the mechanisms themselves (Lohmann 2006: 58; de Lucia 2009: 235). Furthermore, a few NGOs are themselves actively involved in designing and implementing the mechanisms. WWF, for example, one of the most influential members of CAN, is one of the founders of the 'Gold Standard,'[15] a label that attests the compatibility of CDM projects with environmental standards[16] (Lohmann 2006: 182, 296; de Lucia 2009: 234; Pearse 2010: 177). Acceptance of market mechanisms has perhaps become most obvious in the emerging alliances forged between NGOs and companies or corporate federations which engage in joint lobbying for market-based regulatory measures (see Newell 2000: 135–136; Walk and Brunnengräber 2000: 175–178; Dorsey 2007: 10).

4 Climate justice – origins of an antagonistic programmatic agenda

From the mid-2000s onwards the field of international climate politics has seen a process of (re-)politicization. This was prompted first by the negotiations, begun in 2005, over a successor agreement to the Kyoto protocol which was due to expire in 2012.[17] In these negotiations on global climate governance 'post-2012,' the renewal of what had been, up to that point, a hegemonic consensus is subject to contestation. Another reason for this renewed politicization is the increasingly powerful impact of multiple crises: the ecological crisis, the financial crisis, and heightened social inequalities (see, e.g., Brand 2009: 107–108). These crises are contributing towards a temporary destabilization of the neoliberal hegemony beyond the field of climate policy. The hegemonic consensus is coming unstuck.

> Terms such as 'market,' 'growth,' 'prosperity' and 'stability' that had been firmly embedded in the neoliberal consensus are being uprooted from this context and their meaning seriously challenged. It is natural that this should have an impact on the discourse of climate politics, where such terms occupy a similarly prominent position.
>
> *(Methmann 2011: 120, our translation)*

The multiple crises mentioned above can be interpreted in terms of the entry of the *excluded Other* into the hegemonic discourse. The market-based approach to climate change, which had initially appeared to be universal and yet is based on exclusion and marginalization, is thus revealing its particularity.

Once politicized, the hegemonic (neoliberal) consensus started to break down in civil society. An antagonistic programmatic agenda linked to the demand for climate justice came to be developed within the field of actors consisting of NGOs and social movements. Two processes linked to the heightened political profile of the debate constitute the background to this development. We discuss these in greater detail in the following. *On the one hand* perspectives that have been excluded or marginalized by the affirmative style of politics – such as issues of global justice or the critique of or alternatives to a market-based approach to climate politics (see above) – have come to be capable of articulation and are increasingly being articulated. *On the other hand* actors keen to emphasize the social aspects of climate change began to turn their attention to international climate politics. This includes actors from the development sphere (such as international NGO Oxfam) which had diverted much of their attention away from the climate change issue once the neoliberal approach had become established (Ott et al. 2008: 94; Unmüßig 2011: 4). In addition, though, the antiglobalization movement also began to address the climate change issue – one to which they had previously paid barely any attention – in the context of this (re-)politicization (Müller 2008: 48).

The demand for climate justice has been put forward by individual actors since the end of the 1990s. In a sense, the formulation contained in the UNFCCC – that the parties to the convention should protect the climate system "on the basis of equity and in accordance with their common but differentiated responsibilities" (UNFCCC 1992: Art. 3) – constitutes such a reference to the demand for justice in matters of climate change mitigation and adaptation, albeit a highly imprecise and uncritical one. However, it was only when the issue acquired a more obvious political profile from the mid-2000s onwards that the process of developing an antagonistic programmatic agenda around climate justice occurred with greater intensity. This agenda is no longer focused on developing countries' various obligations to reduce greenhouse gas emissions but rather addresses the whole range of societal origins and social consequences of climate change.

One key point of reference relating to the demand for climate justice historically speaking are the conflicts over environmental justice[18] which erupted from the early 1980s at local and national level in the US (see Bullard 1990, 1994; Cole and Foster 2001; Checker 2005: 20–24). From the late 1990s onwards the concept of environmental justice gained currency beyond the US (see Roberts 2007; Walker 2009); this is manifested especially in the greater attention paid to the issue of climate change. The demand for climate justice was first formulated by indigenous groups in the US[19] (Kaufmann and Müller 2009: 195; Tokar 2010: 46), who subsequently introduced it into international politics (Tokar 2010: 30). The point being made in the demand for climate justice is that the people who are and will be affected most by climate change are those who have contributed least to its causes.

A report by US NGO CorpWatch (Bruno et al. 1999) played a significant role in the emergence of the demand for climate justice in that it prompted a redefinition of climate change by taking on board issues of justice (Roberts and Parks 2009: 395; Dawson 2010: 327). Climate justice is conceived of as a point of reference for a broad-based movement. A series of alternative summits staged in parallel to the UN climate negotiations[20] contributed towards spreading word of this demand and enhancing networking among the actors concerned (see Pettit 2004; Dorsey 2007: 13–20; Goodman 2009: 502–504). In particular, the Bali Principles of Climate Justice from 2002 and the Durban Declaration on Carbon Trading from 2004[21] serve to embed climate justice within an antagonistic programmatic agenda. The declarations that followed constitute key points of reference for numerous climate-related movements and NGOs.

Demands for climate justice came strongly to the fore among the NGOs accredited to participate in the negotiations not least after oppositional groupings broke away from CAN during the climate negotiations in Bali in 2007, prompting the creation of the 'Climate Justice Now!' network (CJN! 2007; see also Gibson 2011: 103–143; Byrd 2012: 75–79, 107–110). The role of the new network was boosted considerably in the run-up to the UN negotiations held in Copenhagen in 2009 when the UNFCCC recognized it as an additional 'focal point' for environmental NGOs (CJN! 2010) and thereupon allocated it around 40 percent of the accreditations granted in this camp (Reitan 2011: 60). It was the mobilization of NGOs and social movements in the context of the 15th Conference of the Parties (COP15) in Copenhagen in particular that further cemented the demand for climate justice within the field of actors and made it a "key framing and mobilising discourse" (Chatterton et al. 2013).

Following the G8 summits in Gleneagles (2005) and Heiligendamm (2007) there was a noticeable increase in potential to mobilize around the climate issue among antiglobalization activists from the industrialized countries of the Global North. This movement (see Notes from Nowhere 2003; Juris 2008) turned its attention to the climate change issue when G8 governments used it to try to wrest back a legitimacy damaged by negative media reporting.[22] By setting up protest camps with a climate change theme – so-called climate camps – activists attempted to set the agenda and location of the debate themselves, rather than merely reacting to the summits (Schlembach 2011: 197; see also Frenzel 2009). Prior to Copenhagen this internal dynamic in the movement led, besides CJN!, to the creation of other transnational networks committed to the demand for climate justice (Routledge 2011: 385), namely, Climate Justice Action (CJA 2009a) and Never Trust A COP (NTAC 2009). The appearance of these networks marks an observable shift in climate politics from institutionalized to less institutionalized actors (see Gibson 2011). Their formation additionally signaled a new "cycle of movement" (Brunnengräber 2012, our translation) within civil society that extends beyond climate politics.

The alternative climate summit convened in Cochabamba by the Bolivian government under Evo Morales in April 2010 constituted a further significant step in the process of agenda setting and networking within the climate justice coalition that was evolving in this context (Building Bridges Collective 2010; Killoran-McKibbin 2010; Müller 2010). This coalition of actors organized (as in Copenhagen) alternative civil society forums as places of mutual exchange in parallel to the negotiations held in Cancùn and Durban (COP 16 and 17). Within the official negotiation process the CJN! network continued to put forward demands articulated in antagonistic opposition to the hegemonic (neoliberal) approach to climate change. The level of civil society mobilization around the post-Copenhagen climate summits was much lower, however. The reason for this – apart from the way the negotiations themselves were developing – lies in the locations chosen to hold them, which made it difficult for activists from Europe to participate in protests. The phenomenon of climate camps has so far (to 2013) remained a regular fixture of locally based climate debates.

4.1 "Social change not climate change!" – climate change as a crisis of societal relationships with nature

The demands and strategies advanced within the field of NGO and social movement actors are fundamentally subject to contestation. They need to be understood as the result of "debates, discussions and conflicts" (Schlembach 2011: 211). At the same time, though, hegemonic interpretations are emerging within civil society as well, even if they remain marginal or constitute a critical minority in relation to the official negotiation process. It is also possible to identify, in relation to the demand for climate justice and its associated programmatic agenda,[23] a hegemonic understanding that is continually being (re-)articulated. Yet climate justice is not a static concept (Moore and Russell 2011: 18) but rather consists of particular perspectives that facilitate the creation of links to other issues and concerns: "Climate Justice is [. . .] still evolving as we write this. Climate Justice is a fluid framework developed by social movements around the world, with identifiable roots from the movements whose shoulders we stand on" (Moore and Russell 2011: 18).

We refer to an interpretation as hegemonic when it displays a high "degree of discursive distribution" (Nonhoff 2006: 141, our translation), i.e., when a specific collection of demands is repeatedly articulated by a large number of actors. Climate justice is characterized by just such a specific linking of a set of demands. These include the following: "leave fossil resources in the ground; acknowledge the North's ecological debt to the South and make reparations; engage in the struggle for energy, resource and food sovereignty; and reduce overconsumption and overproduction, especially in the global North" (Kaufmann and Müller 2009: 194). These demands (see also Angus 2009) reflect an understanding of climate change as a crisis of the way society deals with nature – a crisis of societal relationships with nature. Just as in the theoretical approach discussed above, the understanding of crisis that underpins the climate justice agenda sees society and nature as existing in a constitutive relation to one another. The causes of climate change are understood to be an unjust economic system and the nonsustainable patterns of consumption engaged in by an elite global minority (Pettit 2004: 104). With its specific understanding of climate change as being (re-)produced in society through patterns of production and consumption, the climate justice agenda sets itself apart from naturalizing perspectives in the broader field of actors. At the same time, a naive realistic understanding of the climate problem is rejected by acknowledging society's dependence on the materiality of climate change (and nature), one from which it cannot escape by technological means. As such the climate justice agenda articulates a critique of the ideology of domination of nature: the climate crisis, it says, cannot be overcome with technological "fixes" (see International Climate Justice Network 2002; CJA 2009b).

4.2 Interlinking social and ecological issues

Just as activists speak of environmental justice for strategic reasons, in order to overcome the division between environmental and social justice activists (Roberts 2007: 298–299), the aim of climate justice is also to interconnect both sets of issues. The principal focus is not on the output of the energy system (emissions) but on "our inherited systems of materials extraction, transport and distribution, energy-generation, production of goods and services, consumption, disposal and financing" (Bond 2012: 194–195). It is this point of interconnection in particular that distinguishes those actors advancing the environmental and climate justice agenda from those who are involved in an environmental movement focused primarily on ecological issues (Sandler and Pezzullo 2007).

In articulating the social dimension of climate change, the climate justice agenda is simultaneously articulating a North–South perspective and thus one of global justice. In this sense, the climate justice agenda serves to integrate previously excluded or marginalized issues and actors into the climate debate. In doing so it is counteracting a form of exclusion that has its roots not only in relations of domination between societies but also in such relations *within* societies – in "vast inequalities ... along the lines of class, gender and race for example in terms of contribution to the problem on the one hand, and vulnerability to its effects on the other" (Newell and Paterson 2010: 157). Resolving the climate crisis in a way that is just and rejects all forms of domination entails including people and practices that had previously been excluded:

> Indigenous Peoples, peasant communities, fisherfolk, and especially women in these communities, have been living harmoniously and sustainably with the Earth for millennia. They are not only the most affected by climate change, but also its false solutions, such as agrofuels, megadams, genetic modification, tree plantations and carbon offset schemes. Instead of market led schemes, their sustainable practices should be seen as offering the real solutions to climate change.
>
> *(CJN! 2008)*

4.3 Critique of market orientation

Inherent in the climate justice agenda is a critique of the prevailing market-based orientation of international climate politics. This critique is expressed through a fundamental rejection of market mechanisms, which are seen as part of the problem and not as part of the solution (Durban Group for Climate Justice 2004; on critique see also Carbon Trade Watch 2003; Lohmann 2006). The Delhi Climate Justice Declaration states, for example: "We reject the market based principles that guide the current negotiations to solve the climate crisis: Our World is Not for Sale!" (India Climate Justice Forum 2002). The UNFCCC Kyoto protocol is seen as a project devised by the Global North to buy itself out of any obligation to make fundamental changes to its mode of production and consumption (Pettit 2004: 103). The climate justice coalition actors are united in the conviction that "present doxas are insufficient to resolve the crisis of climate change, and other paths are both necessary and possible" (Dorsey 2007: 20). They are demanding change that occurs without reproducing relations of domination such as those inscribed onto the state and the market (Dorsey 2007: 20). In this sense, the aim of climate justice is to generate "a new postcapitalist ideology" (Foti 2009).

At heart, the climate justice agenda comprises both a reactive and a proactive agenda (Pearse 2010: 185). On the one hand it is directed against a set of technological and market-based approaches to climate change. These 'false solutions' include such measures as "carbon offsetting, carbon trading for forests, agrofuels, trade liberalization and privatization pushed by governments, financial institutions and multinational corporations" (CJN! 2007). On the other hand there is a push for a range of alternative approaches – 'real solutions' – that are directed against the hegemony of the market and are simultaneously designed to facilitate social change: "In contrast to such corporate, market and state-driven solutions, the movement advocates community-led responses to the climate crisis and the promotion of localized, community sovereignty over food, energy and other resources" (Long et al. 2010: 225).

4.4 The political strategy of the climate justice coalition

The climate justice coalition's understanding of crisis is made manifest in its organizational structure. Previously, the type of network that had predominated was the "transnational advocacy network" (see Keck and Sikkink 1998: 8–10); it was characterized by a hierarchical network structure and a certain paternalistic approach.[24] The networks built around climate justice display clear changes by contrast. We highlight three aspects in this regard[25] (see also Reitan 2011: 58). *First*, the new networks are of a hybrid character regarding the actors that constitute them. The actors in these networks are highly heterogeneous (Reitan 2011: 52), whether with regard to the issues on which they focus (such as indigenous or gender rights and climate change) or to the degree of their institutionalization or their strategic orientation towards the official negotiation process in the context of UNFCCC (inside vs. outside, see Dryzek and Stevenson 2010: 20–22; Long et al. 2010: 226). Even some governmental actors have become part of these hybrid network structures.[26] Key to these networks is an "emphasis on and affirmation of difference" (Hierlmeier 2006: 160, our translation) in the formation of broad-based alliances.[27] Based on their social-ecological understanding of the crisis, the networks are forcefully advocating concrete solidarity between primarily ecological (green) and primarily social (red) groupings. The inside-outside difference is not understood in terms of a mutually exclusive opposition but rather as something that can potentially be transcended by working together in alliances.[28] Accordingly, one of the founding documents of the Climate Justice Now! network states: "We will take our struggle forward not just in the talks, but on the ground and in the streets – Climate Justice Now!" (CJN! 2007). This solidarity was expressed clearly in a campaign of mass civil disobedience during COP 15 (CJA 2009b; CJN! 2009). Accredited groups from the CJN! network in particular, along with activists from outside the negotiation hall, attempted to hold a "Climate Justice Assembly" on conference premises that had been secured by the police. The aim was to shift the agenda of the negotiations towards the demands made by social movements, especially those from the Global South, which have been marginalized in climate politics to date.

Second, the networks are largely polycentric: social debates are conducted in many different locations where it is apparent that the crisis is being (re-)produced, i.e., where it is clear that relationships with nature are in crisis. The networks act on different spatial scales to address these crises: Networks working in conjunction with the international negotiations can be distinguished from those focusing on local conflicts. Diverse conflicts are united beneath the banner of the demand for climate justice (Bedall and Austen 2010) – climate camps focusing their attention on key greenhouse gas emitters such as coal-fired power stations or on the business policy of certain companies (see Frenzel 2009), the struggles of peasant farmers in Latin America against being driven from the land on which they subsist and for their right to food sovereignty, and the many other grassroots-oriented groups engaged in local struggles (see, e.g., Rising Tide 2002; Evans 2010; K!BN 2010).

Third, the networks are characterized by an oppositional orientation towards multiple relations of domination, including the state in its hegemonic (neoliberal) form as well as the way societal relationships with nature are conducted – expressed not least in climate politics – and the domination over nature manifested in this approach. At the same time it is possible to detect a change in emphasis compared to the antiglobalization movements: the latter have been characterized by their *resistance* to international institutions such as the WTO, the World Bank, and the IMF. The networks, by contrast, are prepared to *engage* with these institutions – in the case of international climate politics this means with the official process of the UNFCCC, where parts of the networks act as accredited organizations (see also Reitan 2011: 66). Rather than rejecting these institutions, the focus is instead on transforming them:

> [W]hen the crisis of neoliberalism irrupted, it became apparent that ... antagonism against institutions as an end in itself is a dead end.... Moments of antagonism are

> either part of ongoing processes of building autonomy and constituting new forms of power, or they risk dissipation, or even worse, backlashes.
>
> *(Turbulence 2009: 6–7, quoted in Reitan 2011: 66)*

The strategy of critical engagement with political institutions is expressed, for example, in individual Latin American countries (especially the ALBA countries) where indigenous movements have been centrally involved in transforming the constitutions of some states.[29] In the case of the climate justice networks it is these movements that are energizing such a strategy.

Critical engagement with institutions is a double-edged approach in terms of political practice – a balancing act between cooperation and resistance. In international climate politics, however, it appears to many activists to be necessary in order to avert the negative consequences of an international climate agreement. This is not a form of "abstract idealism" but is much more a matter of averting situations in which "people lose legal rights and ownership over land and resources that are vital to their self-determination and survival" (Long et al. 2010: 239).

5 Concluding remarks

Climate change is a complex socio-ecological problem that is closely connected with many other crises. Politicization of the issue over the last few years has also highlighted the heterogeneous, multiple, and multiscale lines of conflict that permeate it. The climate justice coalition is not merely an expression of these problems. Instead, by resisting hegemonic patterns of interpretation, it has contributed towards naming these complex conflicts in the first place by calling into question the hegemonic framings of the problem and articulating issues so far excluded from the debate. The present paper has highlighted the role civil society actors – NGOs and social movements – have played in shifting the discursive framework in international climate politics, a shift that has been observable over the last few years. We have shown how the actors involved in various historical phases either (re-)produce or challenge the hegemonic discourse of climate politics. Whereas at the start of the UNFCCC climate negotiations a host of critical standpoints were represented within the field of actors examined here, the adoption of the Kyoto protocol was initially followed by the emergence of an affirmative style of political engagement that (re-)produced the hegemonic consensus on dealing with climate change via the market. From the mid-2000s onwards, a further (re-)politicization of climate change has taken place in conjunction with the negotiations over climate policy 'post-2012' and the articulation of multiple crises. It is on this politicized terrain that perspectives are (increasingly) being articulated which had previously been excluded or marginalized: an understanding of climate change as a social crisis and, linked to this, the debate around issues of global justice and the critique of or alternatives to the market-oriented thrust of climate politics. It is possible to identify the formation of an agenda oriented in opposition to the hegemonic consensus, which is being driven forward by a specific coalition of actors. Within this, it is the demand for climate justice that gathers together and coherently presents a range of oppositional demands.

However much these demands may have contributed towards breaking down the hegemonic (neoliberal) consensus, it is not possible at this point to say to what extent this consensus is already transformed and the demands made have already had an impact on official governmental and intergovernmental politics. Finding an answer would require closer analysis of the international negotiations themselves and the role of the climate justice coalition in them. We are inclined to assume, though – much as in other processes – that struggles and debates over *post-neoliberalism* (Brand and Sekler 2009) have begun without any clear shape or structure having emerged thus far. Similarly, we cannot say either whether it is possible to refer to the climate justice coalition – in the terms of social movements research (e.g., Raschke 1985: 76–78; Tarrow 2003: 4–5) – as *a* social movement in the singular ('the' climate justice movement or even 'the' climate movement). It is difficult to judge the specificity of the issues and organizational forms and the extent of their temporal continuity, as many activities are geared towards one-off events and occur within a specific time horizon.[30] This is a contested issue among both activists and academics[31] and cannot be discussed in detail in the present paper.

It is worth highlighting, nonetheless, that the long unchallenged hegemonic consensus has been broken by the climate justice agenda. The set of demands around climate justice has successfully stood up to the neoliberalization of climate change: It has rendered visible (again) issues that have thus far been excluded and has developed an oppositional stance to the predominant way in which societal relationships with nature are conducted. The notion of dominion over nature has thus at least been subject to challenge.

Notes

1 Drawing on Hajer (1995), 'discourse coalition' is understood here as a group of subjects who jointly advance a particular set of demands and thus make them their own (see also Nonhoff 2006: 188–202). With Ernesto Laclau and Chantal Mouffe every discourse includes an antagonism. There is always an excluded other vis-à-vis the common of the discourse. This antagonism – or rather the excluded other – bears witness to the possibility of an alternative (Laclau and Mouffe 1985: 125).

2 The term 'climate justice coalition' is used in this paper as an analytical term that draws together a heterogeneous range of actors who have in common the fact that they are putting forward the demand for climate justice and a specific programmatic agenda associated with it. To what extent this is a unitary, clearly identifiable, and relatively stable phenomenon is deliberately left unanswered here – and with it also the question as to the identity of a climate movement and the role of the climate justice coalition within it. We return to this issue at the end of the paper.

3 The Frankfurt School comprises above all Theodor W. Adorno, Max Horkheimer, Herbert Marcuse, and Walter Benjamin. See Görg (2003) and Becker and Jahn (2006) as contributions to locating the approach in a broader context. For a history of the Institute for Social Research, see Demirovic (1999).

4 The dependencies should not be understood as monocausal. Societies can decide to overuse their natural resources, thereby exposing themselves to the risk of collapse. It is probably extremely difficult to put a figure on the amount of leeway beyond which these dependencies make their presence felt (as is currently being attempted with planetary limits; see Rockström et al. 2009). But this doesn't mean that there are no dependencies, as peak oil and peak soil (Fritz 2010) are making us painfully aware.

5 This motif is based on two elements that cannot be elaborated further in the present context: one is the *critique of science and the idea of scientific progress* (Marcuse 1979) and the other is the *critique of the idea of progress itself* (Benjamin 1980; Horkheimer and Adorno 1987).

6 The assumption that Horkheimer and Adorno consider the domination of nature to be a primary form of domination led to their work being misapprehended for a long time; on this point, see Honneth (1989), Smith (1990); for a critique of this view, see Görg (1999, 2011).

7 This aim was difficult to achieve from the very beginning: the development NGOs openly articulate a "more radical political stance and one critical of broader societal conditions," thus differing markedly from NGOs such as WWF and Greenpeace with their stronger environmental orientation which "represent primarily moderate positions" (Walk and Brunnengräber 2000: 148, our translation). The latter have adopted a leading role in the networks of climate groups.

8 Within CAN it is NGOs from the US that are pushing for market-based approaches, whereas European NGOs often speak out in favor of governance-based measures at national level (Oberthür and Ott 1999: 76). In particular, the US NGOs Environmental Defense Fund and Nature Conservancy contributed towards enhancing the legitimacy of emissions trading when it was established in the US and in Europe (Pearse 2010: 176).

9 Our aim here is to highlight a *trend* among the *civil society* actors engaging with international climate politics. North–South conflicts and related instruments are discussed at various points during the *official international negotiations* over a post-Kyoto agreement. The most obvious examples of this are calls for

technology transfer and the creation of a fund to support the countries of the Global South. The extent to which such demands are capable of breaking the hegemonic political consensus on how to tackle climate change cannot be addressed within the confines of this article.

10 On the ecological bias of the debate on sustainability, see Görg and Brand (2002). No connection was made either between the issues of environment and development in the antiglobalization movement which developed from the late 1990s onwards. During the first decade of its existence this movement only rarely paid attention to the issue of ecology, focusing much more on social aspects in relation to issues of distribution or justice (Unmüßig 2011: 3).

11 Symptomatic of this kind of interpretation is the demand expressed by the organization 350.org to limit the volume of atmospheric CO_2 to a specific amount: a maximum of 350 parts per million.

12 The focus of NGOs from the Global South on development-related issues can be understood as a consequence of their specific concern with the national context. The climate issue as an ecological problem is not politically relevant to them unless there is an equal emphasis on issues of justice and thus of economic and social issues (Walk and Brunnengräber 2000: 146).

13 CAN continues to hold onto the market mechanisms even though George Bush authorized the US pullout from the Kyoto protocol in 2001. In this way it continues to contribute towards the "global consensus that sees market-based mechanisms as the 'only show in town'" (Lohmann 2006: 50).

14 The three elements of this approach are, first, mitigation through the Kyoto mechanisms, second, mitigation through decarbonization, and, third, adaptation.

15 The Gold Standard is currently supported by more than 80 NGOs, including members of CAN such as Care International, Germanwatch, and Greenpeace International (see GS 2012).

16 According to criticism expressed by Larry Lohmann, projects awarded the Gold Standard are not necessarily sustainable, as they lose sight of social aspects (Lohmann 2006: 296).

17 The conference of the parties to the Kyoto protocol held in 2005 in Montreal (COP 11/MOP1) adopted a resolution to begin negotiations over a second commitment period in accordance with their commitment as stated in Article 3.9 of the Kyoto protocol (see UNFCCC 1997: Art. 3.9).

18 The Climate Justice agenda is based, as Dawson puts it, "on the deep and powerful roots of the environmental justice movement" (Dawson 2010: 316).

19 Tokar attributes the first articulation of the demand for climate justice in the mid-1990s to the founder of the Indigenous Environmental Network, Tom Goldtooth (Tokar 2010: 45–46).

20 For an overview of significant events, see, e.g., Goodman (2009: 502–503) and Bond (2012: 187).

21 The Bali Principles were formulated by a group of 14 NGOs from the Global North and South (International Climate Justice Network 2002) in the context of negotiations aimed at preparing the world summit in Bali (United Nations World Summit on Sustainable Development). The Durban Group for Climate Justice 2004 is an outcome of the Durban Climate Justice Summit organized by Carbon Trade Watch. The declaration was adopted by 20 organizations from Europe, the US, Latin America, India, and Africa (Goodman 2009: 502; Tokar 2010: 46), including Carbon Trade Watch, the Indigenous Environmental Network, and the Global Justice Ecology Project (US).

22 The outcomes from the summit itself remained vague when the G8 eventually agreed to "seriously consider" reducing global CO_2 emissions by 50 percent by the year 2050, recognizing the negotiation process within the UN and accepting the reports of the scientific advisory body IPCC. Nonetheless the G8 managed through the media to convey the impression that they were seriously tackling the challenge of climate change.

23 This includes acknowledgment of the fact that different versions of the demand for climate justice exist alongside the hegemonic understanding. See, e.g., de Lucia (2009: 231–232) and Bond (2012: 196–197).

24 Actors from the Global North endowed with plentiful resources dominate in these networks and raise their voice in international politics on behalf of others, while the people on whose behalf they speak and act have barely any influence at all.

25 Reitan (2011) looks at instances of this new type of network beyond the climate policy field as well.

26 This includes especially the governments of the ALBA states of Latin America – especially Bolivia – who are advancing strategic alliances with civil society actors using post-neoliberal discourse (Dryzek and Stevenson 2010: 21). In addition to this, individual actors such as the Third World Network are also working with the Group of 77 state alliance and China in order to strengthen perspectives from the Global South (Long et al. 2010: 239).

27 The tension between the different actors is not resolved in this process. In contrast to the history of social conflicts on the Left, however, there is no prominent division into different 'sects.' Instead, the

aim of the actors in these hybrid networks is to use their practical resistance to link their diversity in the service of creating a different world (for more on this, see also Notes from Nowhere 2003; Holloway 2006).

28 *Inside* the UNFCCC there are networks consisting of groupings operating as accredited observer organizations, providing critical support for the negotiation process and trying to exert influence on its outcomes. *Outside* the official process, networks are trying to create an "alternative political space" (Long et al. 2010: 226) in which activists seek to promote their own – often marginalized – issues.

29 In certain Latin American countries movements have been involved in developing concepts of statehood that are aimed at structural transformation and that differ clearly from the previously hegemonic neoliberal state. The most prominent are plurinationality in Bolivia (on the concept of plurinationality, see Radhuber 2012: 76–79) and the constitution of Ecuador oriented towards the good life (*sumak kausay*) (see Acosta 2009).

30 For example, the activities of the CJA und NTAC networks associated with the COP 15 UN Climate Conference concentrated essentially on mobilizing for COP 15. NTAC ceased to continue its work and the number of active members of CJA progressively declined after the conference. As a result, there have been no further international meetings or network activities since 2010.

31 On this debate see Agyeman et al. 2007; Bedall et al. 2011; Bricke and Müller 2011; Dietz 2011; Chatterton et al. 2013. What is striking in connection with the post-Kyoto negotiations is surely the previously unknown quality of mobilization of social movements in the field of international climate politics.

References

Acosta, A. (2009) The Rights of Nature, New Forms of Citizenship and the Good Life – Echoes of the Constitución de Montecristi in Ecuador, in Brand, U., Bullard, N., Lander, E., and Müller, T. (eds.) *Contours of Climate Justice: Ideas for Shaping New Climate and Energy Politics*, Uppsala: Dag Hammarskjöld Foundation.

Agyeman, J., Bulkeley, H., and Nochur, A. (2007) Climate Justice, in Isham, J. and Waage, S. (eds.) *Ignition: What You Can Do to Fight Global Warming and Spark a Movement*, Washington DC: Island Press.

Altvater, E. (1992) *Der Preis des Wohlstands – oder Umweltplünderung und neue Welt(un)ordnung*, Münster: Westfälisches Dampfboot.

Angus, I. (2009) *The Global Fight for Climate Justice: Anticapitalist Responses to Global Warming and Environmental Destruction*, London: Resistance Books.

Arts, B. (1998) *The Political Influence of Global NGOs: Case Studies on the Climate and Biodiversity Conventions*, Utrecht: International Books.

Becker, E. and Jahn, T. H. (2006) *Soziale Ökologie. Grundzüge einer Wissenschaft von den gesellschaftlichen Naturverhältnissen*, Frankfurt/Main and New York: Campus.

Bedall, P. (2011) NGOs, soziale Bewegungen und Auseinandersetzungen um Hegemonie. Eine gesellschaftstheoretische Verortung in der Internationalen Politischen Ökonomie, in Brunnengräber, A. (ed.) *Zivilisierung des Klimaregimes. NGOs und soziale Bewegungen in der nationalen, europäischen und internationalen Klimapolitik*, Wiesbaden: VS.

Bedall, P. (2013) 'Climate Justice,' 'Green Economy' or a 'One Planet Lifestyle' – Hegemonic Narratives in Transnational NGOs and Social Movements, in Methmann, C., Rothe, D., and Stephan, B. (eds.) *(De-)Constructing the Greenhouse: Interpretive Approaches to Global Climate Governance*, Abingdon and New York: Routledge.

Bedall, P. and Austen, M. (2010) *Climate Justice – Point of Reference for a Counter Hegemony or Nebulous Empty Phrase?* Accessed 05/14/2013 from http://notesfrombelow.wordpress.com/2010/07/29/climate-justice-point-of-reference-for-a-counter-hegemony-or-nebulous-empty-phrase/.

Bedall, P., Bricke, M., and Tzschiesche, S. (2011) Klima- und Energiekämpfe, in Bois, M. and Hüttner, B. (eds.) *Beiträge zur Geschichte einer pluralen Linken, Heft 3 – Bewegungen, Parteien, Ideen*, Berlin: RLS Papers.

Behringer, W. (2007) *Kulturgeschichte des Klimas. Von der Eiszeit bis zur globalen Erwärmung*, München: C.H. Beck.

Benjamin, W. (1980) Über den Begriff der Geschichte, in Benjamin, W. (ed.) *Illuminationen*, Frankfurt/Main: Suhrkamp.

Bond, P. (2012) *Politics of Climate Justice: Paralysis Above, Movement Below*, Pietermaritzburg: University of KwaZulu-Natal Press.

Bond, P. and Dorsey, M. K. (2010) Anatomies of Environmental Knowledge and Resistance: Diverse Climate Justice Movements and Waning Eco-neoliberalism, *Journal of Australian Political Economy*, 36(66): 286–316.

Brand, U. (2009) Environmental Crisis and the Ambiguous Postneoliberalising of Nature, in Brand, U. and Sekler, N. (eds.) *Postneoliberalism – A Beginning Debate*, Uppsala: Dag Hammarskjöld Foundation.

Brand, U. and Sekler, N. (2009) *Postneoliberalism – A Beginning Debate*, Uppsala: Dag Hammarskjöld Foundation.

Brand, U., Görg, C., Hirsch, J., and Wissen, M. (2008) *Conflicts in Environmental Regulation and the Internationalisation of the State: Contested Terrains*, London and New York: Routledge.

Bricke, M. and Müller, T. (2011) Kurze Geschichte einer kurzen Geschichte. Die Klimagerechtigkeitsbewegung in der BRD, *arranca* #44: Don't look back in anger. Accessed 05/14/2013 from http://arranca.org/ausgabe/44/kurze-geschichte-einer-kurzen-geschichte.

Brunnengräber, A. (2002) Umwelt- oder Gesellschaftskrise? Zur politischen Ökonomie des Klimas, in Görg, C. and Brand, U. (eds.) *Mythen globalen Umweltmanagements: Rio+10 und die Sackgassen nachhaltiger Entwicklung*, Münster: Westfälisches Dampfboot.

Brunnengräber, A. (2006) The Political Economy of the Kyoto Protocol, in Panitch, L., Leys, C., Harris-White, B., Altvater, E., and Gregory, A. (eds.) *Coming to Terms with Nature*, London: Merlin Press.

Brunnengräber, A. (2009) *Die politische Ökonomie des Klimawandels*, München: Oekom.

Brunnengräber, A. (2012) Ein neuer Bewegungszyklus. Von der NGOisierung zur Occupy-Bewegung, *Forschungsjournal Soziale Bewegungen*, 25(1): 42–50.

Brunnengräber, A. (2013) Multi-Level Climate Governance: Strategic Selectivities in International Politics, in Knieling, J. and Leal Filho, W. (eds.) *Climate Change Governance*, Frankfurt/Main: Springer, 67–84.

Brunnengräber, A., Klein, A., and Walk, H. (2001) *NGOs als Legitimationsressource. Zivilgesellschaftliche Partizipationsformen im Globalisierungsprozess*, Opladen: Leske + Budrich.

Brunnengräber, A., Dietz, K., Hirschl, B., Walk, H., and Weber, M. (2008) *Das Klima neu denken. Eine sozial-ökologische Perspektive auf die lokale, nationale und internationale Klimapolitik*, Münster: Westfälisches Dampfboot.

Bruno, K., Karliner, J. and Brotsky, C. (1999) *Greenhouse Gangsters vs. Climate Justice*, Transnational Resource and Action Center. Accessed 05/14/2013 from www.corpwatch.org/downloads/greenhousegangsters.pdf.

Building Bridges Collective (2010) *Space for Movement? Reflections from Bolivia on Climate Justice, Social Movements and the State*, Leeds: Footprint Workers Co-op.

BUKO Working Group on Social Ecology (2009) *Forget About Copenhagen! The Catastrophe is Occurring Already*. Accessed 05/14/2013 from www.buko.info/fileadmin/user_upload/doc/projekte/klima_assoe_engl.pdf.

Bullard, R. D. (1990) *Dumping in Dixie: Race, Class, and Environmental Quality*, Boulder: Westview Press.

Bullard, R. D. (1994) *Unequal Protection: Environmental Justice and Communities of Color*, San Francisco: Sierra Club Books.

Byrd, S. C. (2012) *The UNFCCC and Global Climate Justice Activism: Rival Networks and Organizing Cascades*. Dissertation, Sociology. Irvine: The University of California, Irvine.

Carbon Trade Watch (2003) *The Sky is Not the Limit. The Emerging Market in Greenhouse Gases*. Accessed 05/14/2013 from www.carbontradewatch.org/downloads/publications/skyeng.pdf.

Chatterton, P., Featherstone, D., and Routledge, P. (2013) Articulating Climate Justice in Copenhagen: Antagonism, the Commons and Solidarity, *Antipode*, 45(3): 602–620.

Checker, M. (2005) *Polluted Promises: Environmental Racism and the Search for Justice in a Southern Town*, New York: New York University Press.

CJA (Climate Justice Action) (2009a) *About CJA*. Accessed 05/14/2013 from www.climate-justice-action.org/about/about-cja.

CJA (Climate Justice Action) (2009b) *Reclaim Power! Pushing for Climate Justice*. Accessed 05/14/2013 from www.climate-justice-action.org/mobilization/reclaim-power-pushing-for-climate-justice.

CJN! (Climate Justice Now!) (2007) *What's missing from the climate talks? Justice!* Press release. Accessed 05/14/2013 from www.climate-justice-now.org/cjn-founding-press-release.

CJN! (Climate Justice Now!) (2008) *CJN! Poznan Statement. Radical New Agenda Needed to Achieve Climate Justice*. Accessed 05/14/2013 from www.climate-justice-now.org/cjn-poznan-statement-signatures.

CJN! (Climate Justice Now!) (2009) *Call to Action! Reclaim Power – Join the Yellow (Badge) Bloc. No Decisions*

About Us Without Us – Climate Justice Now! Press release. Accessed 05/14/2013 from www.climate-justice-now.org/call-to-action-reclaim-power-–-join-the-yellow-badge-bloc.
CJN! (Climate Justice Now!) (2010) *A Short History of Climate Justice Now!* Accessed 12/14/2012 from www.climate-justice-now.org/about-cjn/history.
Cole, L. W. and Foster, S. R. (2001) *From the Ground Up. Environmental Racism and the Rise of the Environmental Justice Movement*, New York: New York University Press.
Conca, K. (1995) Greening the United Nations: Environmental Organisations and the UN System, *Third World Quarterly*, 16(3): 441–458.
Crutzen, P. J. (2002) Geology of Mankind: The Anthropocene, *Nature*, 415(6867): 23.
Dawson, A. (2010) Climate Justice: The Emerging Movement against Green Capitalism, *South Atlantic Quarterly*, 109(2): 313–338.
de Lucia, V. (2009) Hegemony and Climate Justice: A Critical Analysis, in Böhm, S. and Dabhi, S. (eds.) *Upsetting the Offset: The Political Economy of Carbon Markets*, London: MayFly.
Demirovic, A. (1999) *Der nonkonformistische Intellektuelle*, Frankfurt/Main: Suhrkamp.
Dietz, M. (2011) Gibt es eine internationale Klimabewegung? Ein skeptischer Befund, *Forschungsjournal Neue Soziale Bewegungen*, 24(4): 6–18.
Dorsey, M. K. (2007) Climate Knowledge and Power: Tales of Skeptic Tanks, Weather Gods, and Sagas for Climate (In)justice, *Capitalism Nature Socialism*, 18(2): 7–21.
Dryzek, J. S. and Stevenson, H. (2010) *The Deliberative Global Governance of Climate Change*. Paper presented at the 'Democratizing Climate Governance' conference Australian National University, Canberra, July 15–16, 2010. Accessed 01/02/2011 from http://deliberativedemocracy.anu.edu.au/Content/pdf%20files/conference%20papers/Dryzek_and_Stevenson_DCG_paper.pdf.
Durban Group for Climate Justice (2004) Climate Justice Now! The Durban Declaration on Carbon Trading. Accessed 05/14/2013 from www.durbanclimatejustice.org/durban-declaration/english.html.
Engels, A. and Weingart, P. (1997) Die Politisierung des Klimas. Zur Entstehung von anthropogenem Klimawandel als politischem Handlungsfeld, in Hiller, P. and Krücken, G. (eds.) *Risiko und Regulierung*, Frankfurt/Main: Suhrkamp.
Evans, G. (2010) A Rising Tide: Linking Local and Global Climate Justice, *Journal of Australian Political Economy*, 36(66): 199–221.
Foti, A. (2009) *Climate Anarchists vs. Green Capitalists*. Accessed 05/14/2013 from www.zmag.org/znet/viewArticle/22367.
Frenzel, F. (2009) *Politics in Motion: The Mobilities of Political Tourists*. Dissertation. Leeds Metropolitan University.
Fritz, T. (2010) *Peak Soil – Die globale Jagd nach Land*, 2nd edn., Berlin: FDGL-Verlag.
Gibson, S. (2011) *Dynamics of Radicalization: The Rise of Radical Activism against Climate Change*. Dissertation, International Studies. University of Miami, Coral Gables.
Goodman, J. (2009) From Global Justice to Climate Justice: Justice Ecologism in an Era of Global Warming, *New Political Science*, 31(4): 499–514.
Görg, C. (1992) *Neue soziale Bewegungen und Kritische Theorie*, Wiesbaden: Deutscher Universitätsverlag.
Görg, C. (1999) Kritik der Naturbeherrschung, *Zeitschrift für kritische Theorie*, 5(9): 73–88.
Görg, C. (2003) *Regulation der Naturverhältnisse*, Münster: Westfälisches Dampfboot.
Görg, C. (2011) Societal Relationships with Nature: A Dialectical Approach to Environmental Politics, in Biro, A. (ed.) *Critical Ecologies: The Frankfurt School and Contemporary Environmental Crises*, Toronto: University of Toronto Press.
Görg, C. and Brand, U. (2002) *Mythen globalen Umweltmanagements: Rio+10 und die Sackgassen nachhaltiger Entwicklung*, Münster: Westfälisches Dampfboot.
Gramsci, A. (1971) *Selections from the Prison Notebooks*, ed. Hoare, Q. London: Lawrence & Wishart.
GS (The Gold Standard) (2012) The only standard endorsed and supported by more than 80 NGOs. Accessed 05/14/2013 from www.cdmgoldstandard.org/our-supporters/ngos.
Hajer, M. A. (1995) *The Politics of Environmental Discourse: Ecological Modernization and the Policy Process*, Oxford: Clarendon Press.
Harvey, D. (1996) *Justice, Nature & the Geography of Difference*, Malden: Blackwell Publishing.
Hierlmeier, J. (2006) *Internationalismus. Eine Einführung in die Ideengeschichte – von den Anfängen bis zur Gegenwart*, 2nd edn., Stuttgart: Schmetterling-Verlag.
Holloway, J. (2006) *Die zwei Zeiten der Revolution. Würde, Macht und die Politik der Zapatistas*, Wien: Turia + Kant.
Honneth, A. (1989) *Kritik der Macht*, Frankfurt/Main: Suhrkamp.

Horkheimer, M. and Adorno, T. W. (1987) *Dialektik der Aufklärung. Philosophische Fragmente*, quoted in Horkheimer, M., *Gesammelte Schriften*, Vol. 5, Frankfurt/Main: Suhrkamp.

Howarth, D., Norval, A., and Stavrakakis, Y. (2000) *Discourse Theory and Political Analysis: Identities, Hegemonies and Social Change*, Manchester: Manchester University Press.

Hulme, M. (2009) *Why we Disagree about Climate Change: Understanding Controversy, Inaction and Opportunity*, Cambridge: Cambridge University Press.

India Climate Justice Forum (2002) *Delhi Climate Justice Declaration*. 1 November 2001. Accessed 05/14/2013 from www.indiaresource.org/issues/energycc/2003/delhicjdeclare.html.

International Climate Justice Network (2002) *Bali Principles of Climate Justice*. Accessed 05/14/2013 from www.corpwatch.org/article.php?id=3748.

Jasanoff, S. (2004) *States of Knowledge: The Co-Production of Science and the Social Order*, London: Routledge.

Jasanoff, S. and Martello, M. L. (2004) *Earthly Politics: Local and Global in Environmental Governance*, Cambridge: MIT Press.

Juris, J. S. (2008) *Networking Futures: The Movements Against Corporate Globalization*, Durham: Duke University Press.

K!BN (Klima!Bewegungsnetzwerk) (2010) Das Scheitern als Chance begreifen. Wie weiter nach Jokenhagen? Accessed 05/14/2013 from http://klima.blogsport.de/aktuelles/das-scheitern-als-chance-begreifen-wie-weiter-nach-jokenhagen.

Kaufmann, S. and Müller, T. (2009) *Grüner Kapitalismus. Krise, Klimawandel und kein Ende des Wachstums*, Berlin: Dietz.

Keck, M. E. and Sikkink, K. (1998) *Activists Beyond Borders: Advocacy Networks in International Politics*, Ithaca: Cornell University Press.

Killoran-McKibbin, S. (2010) The Path from Cochabamba, in Sandberg, L. A. and Sandberg, T. (eds.) *Climate Change – Who's Carrying the Burden? The Chilly Climates of the Global Environmental Dilemma*, Ottawa: The Canadian Centre for Policy Alternatives.

Laclau, E. and Mouffe, C. (1985) *Hegemony and Socialist Strategy: Towards a Radical Democratic Politics*, London: Verso.

Lohmann, L. (2006) *Carbon Trading: A Critical Conversation on Climate Change, Privatisation and Power*, Uppsala: Dag Hammerskjöld Foundation.

Long, S., Roberts, E., and Dehm, J. (2010) Climate Justice Inside and Outside the UNFCCC: The Example of REDD, *Journal of Australian Political Economy*, 36(66): 222–246.

MA (Millennium Ecosystem Assessment) (2005) *Millennium Ecosystem Assessment Synthesis Report*, Washington DC: Island Press.

Marcuse, H. (1979) Industrialismus und Kapitalismus im Werk Max Webers, in Marcuse, H. (ed.) *Kultur und Gesellschaft 2*, Frankfurt/Main: Surhkamp.

Methmann, C. (2011) NGOs in der staatlichen Klimapolitik. Zwischen diskursiver Intervention und hegemonialer Verstrickung, in Brunnengräber, A. (ed.) *Zivilisierung des Klimaregimes. NGOs und soziale Bewegungen in der nationalen, europäischen und internationalen Klimapolitik*, Wiesbaden: VS.

Methmann, C., Rothe, D., and Stephan, B. (2013) *(De-)Constructing the Greenhouse: Interpretive Approaches to Global Climate Governance*, Abingdon/New York: Routledge.

Missbach, A. (1999) *Das Klima zwischen Nord und Süd. Eine regulationstheoretische Untersuchung des Nord-Süd-Konflikts in der Klimapolitik der Vereinten Nationen*, Münster: Westfälisches Dampfboot.

Moore, H. and Russell, J. K. (2011) *Organizing Cools the Planet: Tools and Reflections to Navigate the Climate Crisis*, Oakland: PM Press.

Müller, T. (2008) The Movement Is Dead, Long Live the Movement, *Turbulence: Ideas for Movement* (July 2008): 48–55. Accessed 05/14/2013 from http://turbulence.org.uk/turbulence-4/the-movement-is-dead-long-live-the-movement.

Müller, T. (2010) *From Copenhagen to Cochabamba: Caminamos preguntando 2.0?* Accessed 05/14/2013 from http://notesfrombelow.wordpress.com/2010/05/17/from-copenhagen-to-cochabamba-caminamos-preguntando-2–0.

Newell, P. J. (2000) *Climate for Change: Non-state Actors and the Global Politics of the Greenhouse*, Cambridge: Cambridge University Press.

Newell, P. J. and Paterson, M. (2010) *Climate Capitalism: Global Warming and the Transformation of the Global Economy*, Cambridge: Cambridge University Press.

Nonhoff, M. (2006) *Politischer Diskurs und Hegemonie. Das Projekt "Soziale Marktwirtschaft,"* Bielefeld: Transcript.

Notes From Nowhere (2003) *We Are Everywhere: The Irresistible Rise of Global Anti-Capitalism*, London: Verso.

NTAC (Never Trust A COP) (2009) Never trust a COP! Gegen den COP 15 Gipfel im Dezember 2009 in Copenhagen. Accessed 09/07/2009 from http://nevertrustacop.org/Deutsch/Aufruf.

Oberthür, S. and Ott, H. E. (1999) *The Kyoto Protocol: International Climate Policy for the 21st Century*, Berlin, Heidelberg: Springer.

Ott, H. E., Sterk, W., and Watanabe, R. (2008) The Bali Roadmap: New Horizons for Global Climate Policy, *Climate Policy*, 8(1): 91–95.

Park, J., Conca, K., and Finger, M. P. (2008) The Death of Rio Environmentalism, in Park, J., Conca, K., and Finger, M. P. (eds.) *Crisis of Global Environmental Governance: Towards a New Political Economy of Sustainability*, London/New York: Routledge, 1–12.

Paterson, M. (2010) Legitimation and Accumulation in Climate Change Governance, *New Political Economy*, 15(3): 345–368.

Pearse, R. (2010) Making a Market? Contestation and Climate Change, *Journal of Australian Political Economy*, 36(66): 166–198.

Pettit, J. (2004) Climate Justice: A New Social Movement for Atmospheric Rights, *IDS Bulletin*, 35(3): 102–106.

Radhuber, I. M. (2012) *Der plurinationale Staat. Politik, Ökonomie und die Rolle der natürlichen Ressourcen in Bolivien*, Dissertation, Vienna University, Institute of Political Science.

Rahman, A. and Roncerel, A. (1994) A View from the Ground Up, in Mintzer, I. M. and Leonard, J. A. (eds.) *Negotiating Climate Change: The Inside Story of the Rio Convention*, Cambridge: Cambridge University Press.

Raschke, J. (1985) *Soziale Bewegungen. Ein historisch-systematischer Grundriß*, Frankfurt/Main and New York: Campus.

Reitan, R. (2011) Coordinated Power in Contemporary Leftist Activism, in Olesen, T. (ed.) *Power and Transnational Activism*, London/New York: Routledge.

Rising Tide (2002) The Rising Tide Coalition for Climate Justice Political Statement. Barcelona, February 2002. Accessed 05/14/2013 from http://risingtide.org.uk/about/political.

Roberts, J. T. (2007) Globalizing Environmental Justice, in Sandler, R. D. and Pezzullo, P. C. (eds.) *Environmental Justice and Environmentalism: The Social Justice Challenge to the Environmental Movement*, Cambridge: MIT Press.

Roberts, J. T. and Parks, B. C. (2009) Ecologically Unequal Exchange, Ecological Debt, and Climate Justice: The History and Implications of Three Related Ideas for a New Social Movement, *International Journal of Comparative Sociology*, 50(3–4): 385–409.

Rockström, J., Steffen, W., Noone, K. et al. (2009) A Safe Operating Space for Humanity, *Nature*, 461(282): 472–475.

Routledge, P. (2011) Translocal Climate Justice Solidarities, in Dryzek, J., Norgaard, R., and Schlosberg, D. (eds.) *The Oxford Handbook of Climate Change and Society*, Oxford: Oxford University Press.

Sagar, A. D. and Banuri, T. (1999) In Fairness to Current Generations: Lost Voices in the Climate Debate, *Energy Policy*, 27(9): 509–514.

Sandler, R. D. and Pezzullo, P. C. (eds.) (2007) *Environmental Justice and Environmentalism: The Social Justice Challenge to the Environmental Movement*, Cambridge: MIT Press.

Schlembach, R. (2011) How Do Radical Climate Movements Negotiate their Environmental and their Social Agendas? A Study of Debates within the Camp for Climate Action (UK), *Critical Social Policy*, 31(2): 194–215.

Schmidt, A. (1993) *Der Begriff der Natur in der Lehre von Marx*, Hamburg: VSA.

Smith, N. (1990) *The Production of Nature: Uneven Development: Nature, Capital, and the Production of Space*, Cambridge: University of Georgia Press.

Swyngedouw, E. (2010) Place, Nature and the Question of Scale: Interrogating the Production of Nature. *Materialen der interdiszipliären Arbeitsgruppen IAG Globaler Wandel – Regionale Entwicklung*, Berlin: Berlin-Brandenburgische Akademie der Wissenschaften.

Tarrow, S. (2003) *Power in Movement: Social Movements and Contentious Politics*, 2nd edn., Cambridge: Cambridge University Press.

Tokar, B. (2010) *Toward Climate Justice: Perspectives on the Climate Crisis and Social Change*, Norwegen: Communalism Press.

UNFCCC (1992) *United Nations Framework Convention on Climate Change*. Accessed 05/14/2013 from http://unfccc.int/resource/docs/convkp/conveng.pdf.

UNFCCC (1997) *Kyoto Protocol to the United Nations Framework Convention on Climate Change*. Accessed 05/14/2013 from http://unfccc.int/resource/docs/convkp/kpeng.pdf.

Unmüßig, B. (2011) *NGOs and Climate Crisis: Fragmentation, Lines of Conflict and Strategic Approaches*. Accessed 05/14/2013 from www.kh.boell.org/downloads/110609_-ngos_climate_crisis_–_barbara_unmuessig.pdf.

Walk, H. and Brunnengräber, A. (2000) *Die Globalisierungswächter. NGOs und ihre transnationalen Netze im Konfliktfeld Klima*, Münster: Westfälisches Dampfboot.

Walker, G. (2009) Globalizing Environmental Justice: The Geography and Politics of Frame Contextualization and Evolution, *Global Social Policy*, 9(3): 355–382.

Wissen, M. (2011) *Gesellschaftliche Naturverhältnisse in der Internationalisierung des Staates*, Münster: Westfälisches Dampfboot.

5

THE CLIMATE JUSTICE MOVEMENT AND THE HEGEMONIC DISCOURSE OF TECHNOLOGY

Vito De Lucia

1 Introduction

Climate Justice has grown into a mainstream idea. It has become a banner raised higher for each Conference of the Parties that fails to agree on a stringent and fair climate deal.

The idea of climate justice began taking shape in 1999 (Bruno et al. 1999). Various forums have contributed to the development of the concept, with various degrees of departure from mainstream discourses and approaches.[1] During COP 13 in Bali in 2007, a number of NGOs exited the Climate Action Network to form a new movement called Climate Justice Now!, with a significantly more system-critical orientation. Today however climate justice encompasses a very large ensemble of positions. Indeed, it has become a central discursive element for a wide variety of actors, whose social and political visions are different, sometimes distant, and even conflicting. Mary Robinson's Climate Justice Foundation, 350.org, TckTckTck, Caritas, Pan African Alliance for Climate Justice, The Durban Climate Justice group, Climate Justice Now!, Time for Climate Justice movement: all these actors and organizations operate in the discursive field of climate justice. Over the years, the climate justice label, one may suspect, has been stretched very thin. To be sure, the core of the climate justice movement remains rhetorically system-critical. However, as discussions within the movement show, questions have emerged as to the role of climate justice as a label. Is climate justice a "point of reference for a counter hegemony or [is it a] nebulous empty phrase?" ask Martina Austen and Philip Bedall, in a summary of a discussion held within the context of the BUKO Working Group on Social Ecology in 2010 (Austen and Bedall 2010, 2012). Indeed the question is extremely pertinent, more so today than ever. In Copenhagen, the climate justice idea has "exploded." From a largely critical concept, it has gone mainstream, and has become a central element of the Copenhagen spectacle, a global marketing campaign whose banners, posters, videos, and light plays have been splattered all over the city. Indeed, all over the world.[2] Yet Copenhagen has also seen the sharpest system-critical actions and demonstrations for climate justice to date.[3]

After Copenhagen, hailed as the "last, best chance" for the climate regime to move forward (Walker 2010), the climate negotiations have stalled, and then morphed into the beginning of a new regime premised on voluntary actions rather than on binding targets.[4] The climate justice movement has taken residence in the formal venues of the UN climate regime, quickly expanding its role from external critique to internal critical constituency. Gaining a formal voice however, may have transformed something in the core of the movement. Participation may increase influence, and facilitates pushing the boundaries of the debate towards more radical positions than those embodied by, for example, the Climate Action Network. Yet participation may also entail a legitimation of a climate regime which no longer appears as a legitimate forum, as already forcefully pointed out in 2009,[5] and as further proved by the "Durban Disaster."[6]

Is 2 degrees enough? Is carbon trading a legitimate mitigation instrument? Is carbon capture and storage a real solution? Does clean coal exist? The answers to these questions – and many others related to them – form the boundary line separating system-critical and system-friendly positions. Yet this line may hide ambiguities, and may contain system-legitimating seeds on both sides of it. This is indeed a risk inherent in the fact that climate justice – as an "empty signifier" (Austen and Bedall 2012) – is a contested and ambiguously under-determined concept, as such open to hegemonic appropriation and vulnerable to the mechanics of *trasformismo*. To be sure, the climate justice movement is clear about its anticapitalist rhetoric, about its critique of neoliberalism, and of its project of commodification of nature. Yet, within the field of power of the climate regime, pulls towards the hegemonic center are strong. It is these pulls, and the mechanics of hegemony, which will form the central theme of this article.

2 Hegemony and *trasformismo*

Gramsci (Gramsci 2007a, 2007b) and neo-Gramscian international relations theory (Cox 1981, 1983; Gill 1993) offer a useful theoretical lens to interpret the struggle ongoing within the climate regime's field of power. Within this struggle different conceptions of climate justice are articulated, disarticulated, and negotiated. An outline of key concepts of Gramscian theory is thus in order.

2.1 Hegemony

Hegemony is for Gramsci the mechanism through which a social order endures. While a dominant social group may rule through mechanics of coercion, a cohesive society is constructed on the intellectual and moral leadership of a fundamental ruling class (Benney 1983). Such leadership is necessary to unite a heterogeneous ensemble of social groups – the people of a State, or transnational social formations – "within a common system of values, goals and beliefs" (Hall, quoted in Benney 1983: 193). Gramsci operates a methodological distinction between the social sphere of civil society and the sphere of political society (Morton 2007: 89). While coercion is the prerogative of the latter, it is in the social space of civil society that hegemony is secured, through the relationship established by the dominant social forces with antagonistic ones. This relationship – which Gramsci calls "historical bloc" (Gramsci 2007a: 137)[7] – implies the integration of different class interests, engendering a convergence of economic and political objectives. Hegemony is, ultimately, premised and founded upon the creation and maintenance of a *Weltanschauung*.[8]

It is the element of consent which is central in Gramsci's analysis. However, processes and practices of hegemony take place in a social space which is open to contestations, and where hegemony must be maintained and reinforced assiduously, by "subduing and co-opting dissenting voices through subtle dissemination of the dominant group's perspective as universal and natural, to the point where the dominant beliefs and practices become an intractable component of common sense" (Litowitz 2000: 519).

Ideology becomes in this context a crucial instrument of hegemony, to be "endlessly reinforced" (Litowitz 2000: 519)[9] in the entire field of action of civil society,[10] in order to ensure the spontaneous consent of the great masses. Hegemony *is* this spontaneous consent to the "general direction imposed on social life by the dominant fundamental group" (Gramsci 2007a: 12).[11] As Litowitz observes, class rule is "a matter of colonizing the internal world of the dominated classes, a feat that cannot be accomplished by force but only through messages, codes, and the dissemination of images and information" (Litowitz 2000: 524). Hegemony then requires "the creation of a . . . world view which will serve as a unifying principle for a . . . collective will" (Mouffe 1979: 191). This unifying principle serves to organize and direct the "ensemble of values and ideologies" (Benney 1983: 194) operating within the social space of civil society. In this context, civil society is both "the object and the medium" (Haug, quoted in Brand 2007: 10) of the hegemonic struggle. Colonizing its imaginary allows the containment of antagonism within specific boundaries which will have been *internalized*. Thus hegemony, historically caused, endures.

2.2 *Trasformismo*

Robert Cox distinguishes between two modes of theoretical endeavor. One is oriented towards problem-solving, and "takes the world . . . [with its] prevailing social and political relations and . . . institutions . . . as the given framework for action." Importantly, "[t]he aim of problem solving is to make these relationships and institutions work smoothly by dealing effectively with particular sources of trouble" (Cox 1981: 128–129). A second mode is critical, counterhegemonic, and calls into question these institutions and social and power relations, aiming at decentering that very framework of action (Cox 1981).[12] This distinction is important as regards the operations of the hegemonic forces within the field of power of the climate regime.

Climate governance practices can be understood as attempting to reach coherence between legitimacy and accumulation within the context of a process of ecological restructuring of capitalism (Paterson 2010). Such process is the driver of the now mainstream project of a "green economy." By overcoming the contradiction between natural limits and capital accumulation, climate governance "aims to forestall more radical critiques that argue that capitalism and sustainability are inimical" (Paterson 2010: 345). Antagonistic forces are thus either embraced as potential partners or ejected from the realm of the reasonable. This is indeed the *modus operandi* of what Gramsci calls *trasformismo*.

Trasformismo describes a strategy for assimilating or domesticating radical ideas and the groups and organizations which promote them.[13] It "arises where entire layers of formerly radical political [groups] are absorbed en bloc by a hegemonic ruling political formation" (Law 2009, para 2.3). It aims at avoiding fundamental philosophical confrontations, diverting antagonism by either absorbing it within the prevailing ideological horizon, or by excluding it from the realm of common sense. Dubbed also "minimal hegemony,"[14] *trasformismo* is a strategy typical of historical periods where the legitimacy of a specific social and political project – of a worldview – is waning, the historic bloc falters, and radical contestations arise. It is thus a key mechanism through which the ensemble of values and ideologies dispersed within the social space of civil society are "*transformed* by their articulation to a particular hegemonic principle" (Benney 1983: 194, emphasis in original). This is to be contrasted to what is called "expansive hegemony,"[15] and which "manifests itself 'spontaneously' [when a] given social group is . . . not merely satisfying its own existential requirements" (Gramsci 2007a: 60). The "direct consensus" achieved in a situation of expanded hegemony is premised on the "genuine adoption of the interests of the popular classes by the hegemonic class" (Mouffe 1979: 182).

As a regime of accumulation in an era of ecological crises, the climate regime is in need of legitimacy. Climate justice becomes in this respect a significant element towards the stabilization of the hegemonic project of dominant social forces. However, the distinction between minimal and expansive hegemony, particularly if understood as a "theory of *crisis* in advanced democracies" (Benney 1983: 196), explains why current hegemonic processes cannot take the latter form, and must operate intensely following the strategy of *trasformismo*. At the same time, within this field of power, climate justice represents a potentially crucial instrument of the multiple resistances of social movements and antagonistic forces. It allows resistances to remain *within* the realm of common sense, and yet organize and articulate radical critiques. This double role is both a risk and an opportunity for the climate justice movement, as it is torn between assimilation and counter-hegemonic potential. This view may be further clarified by reference to Laclau and Mouffe's notions of field of discursivity and "empty signifier" (Laclau 1989; Laclau and Mouffe 2001). In their view, as discourse is intrinsically afflicted by indeterminacy,[16] a field of discursivity remains empty at its core, and is characterized by semantic flexibility and multiplicity. In other words, it remains entirely susceptible to semantic colonization, having no essential objectivity, no essential meaning, of its own. At the center of such field stands the empty signifier, which operates as the organizing concept of the entire field. This discursive center is filled with meaning through political antagonism. An empty signifier then is a central *locus* of a hegemonic struggle.[17] Climate justice, understood as an empty signifier, can then be also understood as a central *locus* of the hegemonic struggle occurring within the practices of climate governance. So, then: what climate justice?

3 Articulations of climate justice

3.1 Copenhagen and the mainstreaming of climate justice

The 15th Conference of the Parties to the UNFCCC, held in Copenhagen in 2009, was a crucial turning point for the climate regime and the discourse of climate justice. Dubbed "Hopenhagen" by a UN global marketing campaign, COP 15 was effectively turned into an ad hoc "event,"[18] a spectacle as epic as the challenge posed by climate change.[19] Within this spectacle, climate justice assumed a significant role, and global campaigns such as TckTckTck's "Time for Climate Justice"[20] effectively ferried the notion to a global audience. Yet such mainstreaming ensured that climate justice could become a label brandished by a multiplicity of campaigns and organizations with extremely different ideological and political horizons. To be sure, that part of the climate justice movement aggregated under the Climate Justice Now! Coalition remains system-critical in aspiration and orientation. However, climate justice is now a central discursive element of a wide variety of actors, expression of various, and contrasting, social and political visions, as evident by the enumeration of organizations presented in the introduction.

Climate Justice's core element is grounded on the sharp asymmetry existing between those who have done the least to cause global warming – poor people in poor countries – and the high emitting countries which have benefited from such emissions and are best placed to finance mitigation of, and adaptation to, climatic changes. However, from this core starting point, a multiplicity of meanings arises. In particular, two divergent positions represent what Cox refers to, respectively, as the "problem-solving" and the "critical" endeavors.

3.2 Between problem-solving and systemic critique

Broadly speaking, two main articulations of climate justice can be identified. One is, shall we say, institutional, and to a large degree is captured by article 3 of the UNFCCC, which sets out the principles that the Parties shall follow "in their actions to achieve the objective of the

Convention." Particularly important are the principle of common but differentiated responsibilities and respective capabilities[21] – a centerpiece of the climate regime – and the "right to ... promote sustainable development"[22] A second articulation can be considered the starting point for the climate justice movement. Despite ambiguities and contradictions – inevitable given the large number, and the geographical diversity, of groups and networks "members" of the Climate Justice Now! coalition – this second articulation presents a broadly speaking system-critical orientation, or at the very least is supported by openly system-critical organizations and networks. In many ways the two articulations overlap, conceptually as well as rhetorically. Equity, responsibility for future generations, historical responsibility for climate-forcing emissions, common but differentiated responsibilities, right to development, are shared dimensions and elements of climate justice. Indeed, the very motivation for the Climate Justice Now! coalition to form has been the reluctance, or, better, the failure, of the Parties to the UNFCCC, and in particular of developed country Parties, to give concrete implementation to the principles established in article 3(1) of the UNFCCC. And it is in the "concrete implementation" in the translations of those principles from general textual formulation to specific rules that climate justice becomes a contested notion. However, those principles are also susceptible to radical interpretations, an interpretation which transforms them into a general ethical and political horizon inimical to the overall capitalist framework within which the UNFCCC is inserted. Such interpretations support structural, systemic critiques, which operate in terms of boundary lines.

3.3 The boundary lines of critique: false and real solutions

The founding statement of CJN speaks of "false solutions" as a way to level a radical critique to the extant legal regime. It establishes what can be defined as boundary lines of critique. The most crucial boundary line dividing the two articulations of climate justice is probably represented by carbon trading. However, there are at least two others, both of which remain entangled with hegemonic discourses. Of these, one is expressed through numbers. The other relates to technology.

Carbon trading is both a long-standing policy instrument of the climate regime and the target of equally long-standing critiques on the part of climate justice movements. It is, perhaps, the clearest boundary line of critique, the crucial discriminant when assessing one's position in the constellation of climate politics. Carbon trading cuts in two the landscape of civil society organizations and operates as well as a field of contestation and hegemonic pressures.[23]

The second boundary line of critique has to do with "numbers," a central theme ever present in any debate about climate justice. A bitter struggle is ongoing as regards where the threshold for a safe – or at least tolerable – temperature increase should be placed. As one commentator notes, the "political imagination of those responding to the crisis of climate change has been stifled by a scientific discourse that has fostered an apolitical space and resulted in a carbon consensus." A fundamental compatibility has arisen between autonomous organizations, NGOs, government and businesses, around the shared discourse of "parts per million," facilitating a politics-without-antagonism where "the 'enemy' is a mere thing [CO_2], not socially embodied, named and counted" (Pusey and Russell 2010). Should it be 2C degrees? 1.5C? 1C?[24] At which temperature levels will "Africa burn" (Sheppard 2009)? Indeed, an entire organization has been founded upon one particular concentration of atmospheric CO_2, expressed in parts-per-million: 350.[25] Such numerical value is indeed projected as the symbol of safety.[26] Such "scientific rationality" is a pervasive element of climate justice critiques of the climate regime, formed and developed on the "science and equity" dual platform.[27] As politics is considered ineffectual because subjective, the climate justice critique offers the objectivity of science. Yet, the scientific discourse is, arguably, a central element of capitalism's hegemonic project (De Lucia 2009). The risk then is that of remaining hostage of the truth of science as a mode of power, so that

> the only debate that remains is over what technical ... measures are best.... The politics of these movements have become focused on carbon-cuts and tipping-point timelines, and despite sometimes fiery rhetoric, the methods for affecting change become hardwired to affecting a thoroughly apolitical debate.
>
> *(Pusey and Russell 2010)*

Ultimately, such boundary line operates in one direction: the marginalization of other knowledges and worldviews. At the same time, it reinforces the prevailing worldview, the prevailing configuration of truth, knowledge, and power which supports the hegemonic project.

Relatedly, the last boundary line of critique relates to technology. Technology operates as a demarcation line between what are considered false solutions – geo-engineering, carbon capture and storage, large dams, nuclear, etc. – and what are considered real solutions – wind power, small hydro, solar power, geothermal, etc. In this manner, the central question of technology, as we shall see, remains hidden from view. Technology becomes thus an ideological framework encompassing disparate and opposing groups. This will be the focus of the remainder of the article.

4 Technology as hegemonic discourse

Technology enjoys a hegemonic role in the social imaginary of modernity (Mumford 1974).[28] Within the climate regime, this hegemony is further strengthened by the purported emancipative role of technology, as articulated through the discourse of the right to development, sustainability, and the green economy. Advanced technological innovation is perceived and framed as the *only* path to ecological and climate salvation, all the more so as the urgency of acting in order to maintain global warming within a precautionary threshold increases. Moreover, whether the climate challenge will be met in an equitable way, seems also to rest, to a significant extent, on technology and technological innovation fueling (sustainable) development in developing countries. To the extent that the realization of the right to development responds to the demands of climate justice, technology can then be recast as an instrument of justice (De Lucia 2009), and indirectly, as a human right. Global civil society demands in fact, as Lohmann underlines, that climate change and international development be addressed together in the name of global justice (Lohmann 2008a).

Gramsci's framework for understanding hegemonic struggles, processes of *trasformismo*, and negotiation of consent, provides useful interpretive and critical instruments for analysis. Central to understanding the way the mechanics of hegemony operate, particularly as regards technology as a hegemonic discourse, is the colonization of the imaginary of civil society which allows the containment of antagonism within a specific ideological horizon, and within specific boundaries of the possible.

Technology is invoked and embraced by most actors involved in the shaping of the climate regime (developed States, developing States, IGOs, TNCs, NGOs, ENGOs, Academia, etc.), although the *specific* technological horizon promoted – geo-engineering, carbon capture and storage, wind energy, solar power, nuclear, etc. – varies. The centrality of the discourse of technology is evident in documents and platforms of most orientations. From Kofi Annan's Global Humanitarian Forum to WWF, from Greenpeace to Oxfam, from Climate Action Network to Third World Network, from Friends of the Earth to Climate Justice Now!, the language of technology finds its way into most policy platforms and political statements, through the discourse of energy-efficiency on the one hand, and keywords such as "clean," "low-carbon," "zero-carbon," "climate-friendly," "renewable," "climate smart," implying a reorientation of technology away from fossil fuels and toward wind, solar, geo-thermal, etc. And although in many instances technology is qualified as needing to be appropriate, community-managed, etc., so as to respond to local social needs,[29] the discourse of "salvific" technology remains dominant.[30]

The language of the climate justice movement has increasingly adapted to the institutional context of its utterances and of its "lobbying," monitoring, or criticizing activity (that is, the official forums of the UNFCCC negotiations).[31] The discursive anchors contained within the official legal or negotiating texts have become a dominating feature of the interventions of groups representing the Climate Justice Now! coalition. At the same time however questions such as the commodification of nature and climate debt have entered the official discourse through this very participation to the institutional climate regime.[32] However, it may be useful to keep in mind Slavoj Žižek's discussion of the Occupy movement. Žižek emphasizes how it is crucial *not* to engage in dialogue with the established capitalist structures before having managed to construct a new language.[33] One may consider in this respect this statement of Third World Network:

> [i]f developing countries are to moderate their emissions growth and eventually cut their emissions, while still having the capacity to have economic growth (of the appropriate type, consistent with sustainable development), the key is for them to have access to climate-friendly technology at affordable prices.
>
> *(Third World Network 2008)*

The statement captures in a nutshell the two central, and interrelated, themes of climate justice: the right to development; and technology. The former is to be realized through the latter. The notion of climate justice seems then to flatten around the idea of technology, as both themes are shared across the entire spectrum of the climate justice movement and discourse. The notion becomes, moreover, vulnerable to the binary dichotomization of good/bad, wrong/appropriate,[34] clean/dirty, false/real solutions. Once the dichotomy is in place, the demarcation line can be easily moved. Such binary critique of technology remains too often impermeable to considerations of the wider implications inherent in *any* modern, industrial, globalized technology.

4.1 The row over IPRs, and technology as a human right

The hegemonic role of technology can be further illustrated through the "row" over Intellectual Property Rights (IPRs). The US's position on the matter can be taken as representative of developed countries,[35] interested in the application of the IPRs regime established under the TRIPs.[36] As formulated in the now defunct Waxman – Markey Bill,[37] investments in clean technology in developing countries are seen as an opportunity to open new markets.[38] This approach was already evident in Bali.[39]

BASIC[40] countries, by contrast, reject such IPRs' "fundamentalism," and are adamant about relaxing the IPRs regime with regards to the transfer of climate-friendly technology.[41] This position is echoed by environmental NGOs and think tanks, especially from the Global South, which demand patent protection on climate-friendly technologies be removed. However, a number of States and NGOs go further, linking technology transfer and human rights through the right to development. In a recent submission to the UNFCCC's AGW-LCA, Bolivia – a leading critic of capitalism and supporter of social movements and indigenous peoples – has linked emissions equity, right to development, and technology transfer.[42] The International Council on Human Rights Policy (ICHRP 2008) links directly human rights and technology transfer, and urges human rights advocates to "help to flesh out adaptation agendas in particular countries and identify how transfers of technologies can help alleviate climate-related suffering and head off future global warming" (ICHRP 2008: 89). It also observes that "[w]here poorer countries are not yet locked into carbon intensive economies, technical innovation and transfer should be initiated immediately, to make possible and to promote alternative development paths" (ICHRP 2008: 89). This is the leap-frogging concept (Sauter and Watson 2008):[43] underdeveloped or non-yet developed countries have the opportunities to hop over the high emitting development stage, and instead directly engage in a low-carbon development path. Further, and here there is a link with the IPR row: "the patents and investment on which innovation depends are often controlled privately. Richer states will need to ensure that this area of policy is properly and equitably regulated" (ICHRP 2008: 13). Of the same advice is Friends of the Earth, which in a recent report emphasizes how technology transfer is a crucial aspect of climate mitigation and adaptation: "[s]upporting developing countries in making emissions cuts necessitates large-scale technology transfer of environmental goods and services. Current intellectual property rights stand in the way of this and must be tackled" (Friends of the Earth 2009: 5). The report is a critique of carbon trading. Yet, such critique is significantly built (in part at least) on its shortcomings as regards stimulating technological innovation (Friends of the Earth 2009: 3). The recent campaign PushEurope also demands the removal of intellectual property barriers in the context of "appropriate" technology transfer (PushEurope 2012: 5).

Given this wide embrace of technology – under different agendas and different political goals, to be sure – it seems necessary to stop for a second and reflect. In the remainder of this article I shall try and discuss the dark shadow of technology – even of the "green" kind.

5 The dark shadow of (industrial) technology

The discourse of technology dominates almost the entire political spectrum, to the extent perhaps that the entire space of the imagination is saturated. In the context of climate change in particular, technological innovation and development is framed as the only path to reduce emissions and maintain climatic changes and temperature increases within acceptable levels. The boundary of critique is then measured against whether technologies are environmentally sound, and socially just. As the statements of Third World Network already mentioned illustrates, technology is the central instrument to allow both sustainable development and climate mitigation and adaptation. Thus technology becomes the *Deus ex machina* of the climate regime. As such, and as a matter of human rights, it must be shared, transferred.

This transfer seems *prima facie* unproblematic, as technology is usually understood as a neutral instrument (Winner 1978). Being neutral, the central question hinges on human choice as to the direction, purpose, and effects of technology. Technology transfer is thus usually characterized as implying primarily transfer of know-how, institutional and technical capacity building and financing. In itself,

> technology transfer means a broad set of processes, covering flow of knowledge, experience, equipment amongst stakeholders; technology diffusion and technology cooperation within and across countries, between developed and developing countries

> as well as within/amongst each of the groups; learning to understand, choose, utilize, replicate and adapt technology.
>
> *(IPCC 2000: Foreword)*

Within this context, it's possible to discriminate between bad (brown) and good (green) technology, and orient towards the latter through policy.

Probing further, however, technology transfer shows a distinct ideological footprint. The transfer of technology expands its main modes of extracting, appropriating, organizing, utilizing, and distributing resources – human and natural alike. In this sense, there may be no fundamental difference between "brown" and "green" technologies.

5.1 The "power of the machine"

Alf Hornborg (2001) carries out a compelling analysis of the relation between technology, capitalism, and society. He shows how the idea of a neutral and independent technological realm is an illusion, as technology necessarily rests on specific production and exchange relations. Hornborg identifies an intimate, inevitable link between industrial technology and the unequal exchange relations that facilitate the extraction, appropriation, and accumulation of ecological and social resources in a world system aligned along a core-periphery continuum. It is capitalist accumulation "*which made industrial technology possible to begin with*" (Hornborg 2001: 46).[44] "If specific technologies require and reproduce specific forms of social organization," continues Hornborg – and it is well worth quoting at length –

> it is no less true that industrial technology as a general phenomenon requires and reproduces a specific world order ... technological knowledge is "true" (i.e., "works") only within a restricted social space. The social definition of what is technologically feasible is not external to technology but ... intrinsic to it.
>
> *(Hornborg 2001: 107)*[45]

In other words, machine productivity *in itself*, disjointed from global accumulation practices, cannot exist. The depredation of the natural space, furthermore, is *one* aspect of a larger, systemic socio-ecological appropriation, the instruments of which are world trade and social and power relations. The whole discourse of development, and of "green" development in particular, is founded on a representation of technology as something the "presence or absence [of which] hinges [on the] level of technological know-how" (Hornborg 2001: 120) a culture (a civilization) possesses. Such representation, argues Hornborg, is based on the illusion that an "unevenly distributed industrial technology and its setting of global socioeconomic conditions" (Hornborg 2001: 120) are separable circumstances, and that they can be amended through policy intervention. They are, rather, twin sides of the same coin, with technology being the embodiment of global (and local) socio-economic practices and conditions. Furthermore, (industrial) technologies require the restructuring of the environment in which they operate. Modern, large-scale technology systems, due to their size, require that

> more of the available world – both material and human – is removed from its original context, defined as a "resource," and brought into a functional position.... It is now impossible to maintain many kinds of advanced technologies unless one has a world-wide access to resources.
>
> *(Winner 1978: 208–209)*

Two crucial elements emerge from Hornborg's theoretical construction, and must be highlighted. First, industrial technology, of necessity, *embodies* processes of accumulation, appropriation, and depredation at the root of the (socio-)ecological and climatic crises. Whether environmental efficiency is increased[46] – through green, low-carbon, or clean technologies – is of marginal significance in terms of the overall mechanics of appropriation, as many examples testify.[47] Clean technologies are thus bound to reproduce – even if at different levels of material throughput and emissions – the same project, as they are embodiment of a particular economic, social, and productive organization of society – and of nature. Today's industrial technology, in other words, embodies past and present social, political, and discursive struggles, past and present values as expressed and articulated by the dominant social forces, and "translated into the technical codes we take for granted today" (Feenberg 2005: 57). A technological artifact is then, arguably, a specific embodiment of the dominant model of political economy,[48] it is only possible *because* of it.

Second, the construction of technology as a neutral instrument is a central element of the construction of hegemony. To that purpose, it is crucial that advanced industrial technology be framed as the only horizon of possibility. Organic intellectuals' role in this sense is crucial. However, it is equally crucial that debates occurring within the social space of civil society reinforce such view. In this way, the project of capitalism can be furthered.

6 Conclusions

Understood radically, Hornborg's analysis supports the idea that technology transfer actually implies the expansion of the global capitalist project, as it furthers the transfer of specific power relations, a mode of production, an ideology, and, to the extent that development and the world capitalist system is based on domination, a mode of domination (e.g., Tucker 1999). Machines embody significant aspects of a given social reality – of, in fact, a historical socio-ideological project (Winner 1978; Marcuse 2006).

The language and discourse of technology transfer then is far from neutral. Reducing the critique of technology to a positional analysis measured against a boundary rhetorically constructed on a false/real solution dichotomy does not seem sufficient, not for a critique which aims at being radically system-critical. And even if on such basis it is possible to capture the important dimension of the decentralization of energy production,[49] itself a central counterhegemonic strategy, a certain blindness remains as regards the unequal exchange relations and the concentration of power implicated by so-called real solutions. Solar and wind technologies may be installed in a decentralized manner, giving a sense of "energy independence," but remain hostage of highly centralized extraction and production processes dependent on the same mechanics of power and plunder implicated by false solutions. Furthermore, to the extent that localized practices are constituted and maintained through an enduring mutuality with nature,[50] whose primary feature is that being carried out within objective ecological limits enforced by their context and normatively internalized in those very social practices (Hornborg 2001; Rappaport 2010), industrial technology transfer is a mode of destruction. It destroys the diversity of the local and the particular, replacing them with a universal ethics of technical efficiency (Ellul 1964). It obliterates locally adapted practices embedded in – and embodiment of – local socio-ecological conditions, replacing them with global "megamachines" (Mumford 1974),[51] integrated in, and instrumental to, global capitalist extraction, appropriation, and accumulation flows. For reasons of space, it shall suffice to remand to two exemplary instances provided by Larry Lohmann (2008a, 2008b)[52] and Soumya Dutta (2009).[53] The climate justice movement is certainly aware of these implications, as both the examples just mentioned, and the growing calls for transparent and participatory technology assessment, testify (ETCGroup 2012). However, such calls, which intend to promote a precautionary approach to technology, health, environment, and society, focus primarily on technologies already labeled as false solutions.[54] It is crucial that the even role of "green" technology be actively deconstructed and subject to radical critique, lest climate justice be transformed and articulated to the dominant position. It is crucial to remember that destruction of nature – ecocide (Higgins 2010) – as well as disastrous health effects and uneven development, are equally the result of Tar Sands or "fracking" projects as they are the result of extraction and production processes functional to green technologies (Parry and Douglas 2011). Access to rare earth metals, core components of most "green" industrial technologies, presupposes destructive processes as regards their extraction, chemical processing, contaminant disposal, and transportation (Paul and Campbell 2011). Where there is a bright green light, there is also a long dark shadow.

It is thus that the hegemonic discourse of technology may capture, deflect, and ultimately transform – in the sense we have seen – the counterhegemonic potential system-critical movements gathered around the label of climate justice may possess. System-critical positions calling for alternative models of development and coalescing around critiques of neoliberalism, carbon trading, and the commodification of nature, may be blunted if they remain framed within a technological narrative premised on the false/real, good/bad dichotomy. As we have seen, this dichotomy obscures a crucial dimension of *any* technology developed industrially, and transforms critique, reduces it to a debate which in its entirety exists within the dominant *Weltanschauung*. Notwithstanding the radical work carried out in forums such as the Cochabamba Conference,[55] a shift in the boundaries of critique may be required, so as to reframe critique in a manner cognizant of the dark shadows even green technologies may cast. The role of civil society is that of a funnel, an engine for the construction, negotiation, reconstruction, and maintenance of a system of values which can act as symbolic and narrative glue, and thus capture the consent of the masses towards a specific social project. Considering also how the involvement of social actors in the negotiation and formation of legal regimes is a crucial requisite today for the establishment of a legitimate regime,[56] climate justice, and the hegemonic articulatory practices struggling to fix its meaning, are important elements in the development of the climate regime. Climate justice then is able to both sustain structural critique and to legitimize the hegemonic project currently dominating the climate regime. A significant part of the wider climate justice movement, and most of the climate justice rhetoric emerged from within mainstream social and environmental civil society organizations, remain indeed entangled in dominant, hegemonic discourses of science and technology. Further, the system-critical positions are exposed to hegemonic pressures and "temptations to relapse into pursuit of incremental gains" (Cox 1993), through processes of *trasformismo* that embrace some (watered down) demands posed under the rubric of climate justice, while ejecting those too threatening to hegemonic projects outside of the realm of common sense.

The challenge for the climate justice movement is, then, that of keeping constant critical watch; to avoid lending legitimacy to the hegemonic forces of an ecological form of capitalism. Indeed, the challenge is that of remaining a stereographic movement, alert to the socio-ecological effects of the monographic discourse of carbon emissions, green technologies, and industrial green economy. This may entail finding a new language, rather than trying to reorient the existing one made of terms such as "sustainable development," "green growth," "green economy," "low carbon," and "climate smart."

The UNFCCC no longer elicits trust. Yet the involvement of critical NGOs in the formal processes of the climate regime has increased. The struggle for climate justice – both as regards its meaning and its effective achievement – remains open. And if the critique of carbon trading has spread in a rather successful way, helped by carbon markets' own disarray,[57] the discourse of technology remains treacherously hegemonic.

Notes

1 Though not always pulling in the same directions. See Delhi Climate Justice Declaration (www.indiaresource.org/issues/energycc/2003/delhicjdeclare.html), Bali Principles of Climate Justice (www.indiaresource.org/issues/energycc/2003/baliprinciples.html); Durban Declaration on Carbon Trading (www.durbanclimatejustice.org/durban-declaration/english.html); Climate Justice Now! principles (www.climate-justice-now.org/principles/, all accessed 10/15/2012).

2 An excellent example is the cover of the Midnight Oil song "Beds Are Burning," remade into a climate justice anthem as a key part of the TckTckTck Time for Climate Justice Campaign. See www.youtube.com/watch?v=aBTZOg6l6cA (accessed 12/15/2012). Such campaigns draw inspiration from a number of 1980s campaigns such as the Band Aid project and USA for Africa's "We Are the World" charity single.

3 Particularly the "Reclaim Power" march, organized by Climate Justice Now! and Climate Justice Action; see Watts et al. (2009). As it has been noted (Pusey and Russell 2010) Reclaim Power

> can be viewed as an attempt to create a rupture with the refigurations of capital and governance that are underway. The attempt of demonstrators to enter the UN conference area to host a "peoples summit" was not a call for a "different" set of talks or a "better" agreement. As dissident delegates on the inside disrupted the sessions and participated in an exodus from the proceedings, we witnessed a fundamental challenge to the process of Copenhagen from above and all it entails.

4 This was already prefigured in the so-called "Copenhagen Accord" (see Decision 2/CP.15, in FCCC/CP/2009/11/Add.1; see http://unfccc.int/documentation/documents/advanced_search/items/6911.php?priref=600005735#beg), and was then formalized in the Cancun Agreements in 2010 (see Decisions 1/CP.16, 1/CMP.6, 2/CMP.6, http://cancun.unfccc.int/, all accessed 10/15/2012).

5

> [R]adical social movements and critical NGOs as well as critical intellectuals and some media are increasingly recognising that the UNFCCC in its current form is not an adequate mechanism to deal with this enormous task. Like other international institutions, the UNFCCC is part of a capitalist, Western, white and masculine regime of global resource management. It should no longer be legitimized through the participation of NGOs, social movements and other critical actors. We do not need "sustainable globalization," basically another expression for neo-liberalism and neo-imperialism.
>
> *(Brand et al. 2009: 13)*

6 As 2011's COP 17 in Durban was dubbed by social movements for its – again – failure to deliver any appreciable result in terms of binding commitments or – indeed, in terms of any other measure tracked by the climate justice movement through the mentioned boundaries of critique.

7 The historical bloc is the "unity between spirit and nature" (structure and superstructure). Gramsci further specifies that a historical bloc exists when "the complex, contradictory and discordant *ensemble* of the social relations of the superstructure is the reflection of the *ensemble* of the social relations of production" (Gramsci 2007a: 366).

8 "The foundation of a directive class … is equivalent to the creation of a *Weltanschauung*" (Gramsci 2007a: 384).

9 Poulantzas expressed the same concept thus:

> the State cannot enshrine and reproduce political domination exclusively through repression, force or "naked" violence, but directly calls upon ideology to legitimize violence and contribute to a consensus of those classes and fractions which are dominated from the point of view of political power
>
> *(Poulantzas 1980: 28)*

10 Called "ideological structure" or "hegemonic apparatuses" (Mouffe 1979: 187) and comprised, for example, of "schools, churches, institutions, scholarly exchanges, museums and popular culture" (Litowitz 2000: 519).

11 Direction is the harmonization of the interests of the different groups within the project and under the leadership of the dominant social forces (Gramsci 2007b, Q15 §59: 1822ff.).
12 "Critical theory ... stands apart from the prevailing order of the world and asks how that order came about" (Cox 1981: 128–129).
13 *Trasformismo* is

> the gradual but continuous absorption ... of the active elements produced by allied groups – and even of those which come from antagonistic groups and seemed irreconcilably hostile. In this sense political leadership became merely an aspect of the function of domination – in as much as the absorption of the enemies' *elites* means their decapitation, and annihilation.
>
> *(Gramsci 2007a: 59)*

It is clear that the emphasis in this article is on the effects of *trasformismo* on antagonistic groups.
14 Minimal hegemony "rests on the ideological unity of the economic, political and intellectual elites along with an 'aversion to any intervention of the popular masses in State life'" (Femia 1981: 47).
15 Gramsci talks of "a homogeneous politico-economic historical bloc, without internal contradictions" (Gramsci 2007a: 168).
16 Since its meaning does not derive from any "ultimate fixity," but only by hegemonic articulations of meanings, which remain subject to counterhegemonic articulations.
17 Laclau and Mouffe call it a hegemonic "articulatory practice." Indeed, a discourse is the attempt to dominate a field of discursivity and "to construct a center"; see Laclau and Mouffe (2001: 105, 112).
18 As opposed to a serial multilateral conference, as in fact it was, given it was the annual Conference of the Parties to a Multilateral Treaty, namely the UNFCCC.
19

> "Climate change is one of the epic challenges facing this and future generations. It is time to seal a deal. We need a global movement that mobilizes real change," said UN secretary-general Ban Ki-moon. "[Hopenhagen] is about global action for a global climate treaty and a better future for humankind," he added.
>
> *(Sweney 2009)*

For an example of the Hopenhagen campaign, see www.youtube.com/watch?v=5DNbDbgoCVQ (accessed 10/15/2012).
20 "Time is running out. Join us, as fathers, in the global alliance for climate justice." Thus Kofi Annan urged the world onwards ahead of the Copenhagen climate conference. Annan's Global Humanitarian Forum was one of the key drivers of the TckTckTck campaign "Time for Climate Justice," launched in Copenhagen. For the Forum, which has ceased activity in March 2010, see www.ghf-ge.org/; for the TckTckTck campaign, see www.ghf-ge.org/tcktcktck.php, both accessed 10/15/2012.
21 Responsibility for future generations and equity are also clearly established in article 3(1), albeit in a nonmandatory formulation:

> The Parties *should* protect the climate system for the benefit of present and future generations of humankind, on the basis of equity and in accordance with their common but differentiated responsibilities and respective capabilities. Accordingly, the developed country Parties *should* take the lead in combating climate change and the adverse effects thereof.
>
> *(Article 3(1), my emphasis)*

22 Art. 3(4).
23 For critiques of carbon trading see the now classic Lohmann 2006 as well as Lohmann 2008a. For the hegemonic pressures surrounding carbon trading see De Lucia 2009.
24 The Cochabamba Peoples' Agreement rejected the 2C degree target incorporated in the Copenhagen Accord, and pledged to limit instead: "... the increase in the average world temperature to a maximum of one degree Celsius."
25 See http://350.org/ (accessed 10/15/2012).
26

> 350 means climate safety. To preserve our planet, scientists tell us we must reduce the amount of CO_2 in the atmosphere from its current level of 392 parts per million to below 350ppm. But 350 is more than a number – it's a symbol of where we need to head as a planet.
>
> *(http://350.org/en/mission, accessed 10/15/2012)*

27 For example, "Annex 1 countries need to increase their mitigation ambition as required by science and equity." Oral statement at the 14th Session of the Ad hoc Working Group on Long-term Cooperative Action (AWG-LCA) April 5–8, 2011, Bangkok on behalf of Climate Justice Now!, on file with author.

28 Lewis Mumford for example maintained that a technological compulsiveness affects Western society, which

> has accepted as unquestionable a technological imperative that is quite as arbitrary as the most primitive taboo: not merely the duty to foster invention and constantly to create technological novelties, but equally the duty to surrender to these novelties unconditionally, just because they are offered, without respect to their human consequences. One may without exaggeration now speak of technological compulsiveness: a condition under which society meekly submits to every new technological demand ...
>
> *(Mumford 1974: 186; see also Winner 1978; Feenberg 1999, 2005)*

29 The ETCGroup is a prime example of this type of approach, as well as being a severe critique of "corporate"-led technology development. However, the usual boundary line of critique, distinguishing false and real solutions, applies. See www.etcgroup.org/briefings (accessed 10/15/2012).

30 At least to the extent that technology is one of the central pillars of the UNFCCC (and particularly of the Bali Road Map, as well as the Cancun Agreements), and a key demand of developing countries. Further, "[t]here is little controversy that ... better technologies must form the core of any successful global strategy to reduce greenhouse gas emissions" (Hultman 2011).

31 By 2009 the UNFCCC secretariat allocated to Climate Justice Now! almost half the intervention slots allotted to environmental NGOs (see www.climate-justice-now.org/about-cjn/history/, accessed 10/15/2012).

32 Directly, as evident for example in Nnimmo Bassey's (of Friends of the Earth Nigeria, member of Climate Justice Now!) statements in Durban, see http://unfccc.int/files/meetings/durban_nov_2011/statements/application/pdf/111209_cop17_hls_friends_of_the_earth.pdf (accessed 10/15/2012), or in Khor (2012), where the concept of climate debt is discussed; and indirectly, through the transversal cooperation established between social movements and institutional actors such as the Plurinational State of Bolivia, which filed numerous submissions to the UNFCCC on climate debt. Bolivia also submitted a draft negotiating text based on the outcomes of the Cochabamba Conference, the Peoples Agreement, and the draft proposal for a Universal Declaration of Mother Earth's Rights; see Bolivia's submission to Ad hoc Working Group on Long-Term Cooperative Action under the Convention (AWGLCA), Tenth session, June 1–11, 2010, Bonn, Germany, FCCC/AWGLCA/2010/MISC.2, accessed 10/15/2012 from http://unfccc.int/resource/docs/2010/awglca10/eng/misc02.pdf#page=14.

33 "[I]f we talk now, we have to use some language, but this will be the language of the enemy. We need time to construct our own new language, time to formulate" (Ahmed and Cutrone 2012).

34 Indeed, even economic growth can be distinguished between a wrong and an "appropriate type, consistent with sustainable development" as Third World Network's submission emphasizes (Third World Network 2008). Sustainable development itself is susceptible of near infinite semantic permutations, hence adding to the confusing nature of the statement.

35 All the Western, developed countries are aligned along the same position as the US's. The US is however the main rhetorical target, as it is the most vocal supporter of strict IPR regimes.

36

> Any US funding directed toward assisting developing countries with regard to exporting clean technology should promote the robust compliance with and enforcement of existing international legal requirements for the protection of intellectual property rights as formulated in the Agreement on Trade-Related Aspects of Intellectual Property Rights.
>
> *(Waxman – Markey Bill, Subtitle D, Section 441(a)(10))*

37 See American Clean Energy and Security Act of 2009 (Engrossed as Agreed to or Passed by House) [H.R.2454.EH], accessed 10/15/2012 from http://thomas.loc.gov/cgi-bin/query/D?c111:3:./temp/~c111lUsThj.

38 Hence

> any weakening of intellectual property rights protection poses a substantial competitive risk to US companies and the creation of high-quality US jobs, inhibiting the creation of new

green employment and the transformational shift to the Green Economy of the 21st Century.

(Subtitle D, Section 441(a)(10))

39 See IDEAcarbon press release "Bali Will Set in Motion a Process That Will Define the Structure of the Carbon Markets for Decades to Come," accessed 10/15/2012 from www.prnewswire.co.uk/news-releases/bali-will-set-in-motion-a-process-that-will-define-the-structure-of-the-carbon-markets-for-decades-to-come-153574525.html.

40 BIC is comprised of Brazil, India, and China, BASIC of Brazil, South Africa, India, and China.

41 In a submission to the UNFCCC, G77/China demanded that "[a]ll necessary steps ... be immediately taken in all relevant fora to mandatorily exclude from patenting climate-friendly technologies held by Annex II countries which can be used to adapt to or mitigate climate change" (as quoted in Third World Network 2009a). It is also noteworthy to quote an entire passage highlighting Bolivia's position on the matter:

> Bolivia said that a "business as usual" approach in dealing with intellectual property rights (IPRs) is simply not an option, if developing countries are to meet their enormous mitigation and adaptation challenges. Where IPRs are identified as barriers to the transfer of climate friendly technologies, measures must be available to relax such rights so as to ensure easy and affordable access by developing countries to such technologies. This is a fundamental prerequisite.
>
> *(Third World Network 2009b)*

42

> There is no viable solution to climate change that is effective without being equitable. Deep emission reductions by developed countries are a necessary condition for stabilizing the Earth's climate. So too are profoundly larger transfers of technologies and financial resources than so far considered, if emissions are to be curbed in developing countries and they are also to realize their right to development and achieve their overriding priorities of poverty eradication and economic and social development.
>
> *(Submission by the Republic of Bolivia to UNFCCC's AGW-LCA, accessed 10/22/2012 from www.ecologicaldebt.org/documentos/bolivia250409.pdf)*

43 Which offer the following definition of (environmental) leap-frogging: "[the] possibility that developing countries might be able to skip some of the dirty stages of development experienced by industrialized countries" (Sauter and Watson 2008: 1).

44 Emphasis in the original.

45 Feenberg seems to agree:

> There is no such thing as technology as such. Today we employ this specific technology with limitations that are due not only to the state of our knowledge but also to the power structures that bias this knowledge and its applications. This really existing contemporary technology favors specific ends and obstructs others.
>
> *(Feenberg 2005: 54)*

46 And here we are rather past the "economic" objections related to the Jevons paradox, the rebound effect, or the problems of frontloading.

47 We shall see some in the next section.

48 This insight had already been captured by Herbert Marcuse: "the technological society is a system of domination which operates already in the concept and construction of techniques" (Marcuse 2006: xlvi).

49 False solutions are generally seen to imply also a high level of centralization, economic, political, structural, and technical. This is a central point made by, among others, Abramsky (2009).

50 Without problematizing a problematic word, I imply in this context, and for analytical clarity, that nature is all that is not human, however recognizing that humanity is also one part of nature.

51 The term megamachine describes large socio-technical organization using humans as its components.

52

> On India's Bhilangana river, local farmers run a finely-tuned terraced irrigation system that provides them with rice, wheat, mustard, fruits and vegetables. This ingenious, extremely low-carbon system of agriculture is threatened by a new hydroelectric project designed to

> help power India's heavy industry. Villagers may have to leave the valley, losing not only their livelihoods but also their knowledge of a uniquely sustainable ... technology ...
>
> *(Lohmann 2008b)*

While Lohmann specifically critiques carbon trading, it is crucial to remember that carbon trading is interlinked with technology as both respond to the dynamics of unequal exchange and technological domination, along the lines Hornborg outlines.

53 Soumya Dutta provides us with a macro-example. Starting from the heat-trapping Atmospheric Brown Cloud (ABC) problem, Dutta shows how the official discourse of the Indian government suggests how the ABC is largely a " 'problem' of biomass burning by the large numbers of poorer families in Asia." Dutta's central point is that this is a strategy to "create further space for the 'way of life' of individual cars, big houses, lit up shopping malls, mass production of uniform goods, frequent flying etc – by occupying more atmospheric pollution space to be vacated by Asia's biomass burning poor." This is a strategy, submits Dutta, to "bring this vast number (over 170 million households in India alone) of partly fuel self-dependent families into the commercial fuel and stove market [modern "clean" energy technology], opening more profit avenues for the large corporations" (Dutta 2009: 164, 170).

54 Such as geo-engineering, biotechnology, synthetic biology.

55 Notwithstanding the attempt at charting a new vision, the link between climate justice, climate debt, and technology transfer resists in the outcome texts of the Conference:

> To fulfill their responsibilities [with regard to climate debt] we call on the developed countries to: ... remove barriers such as intellectual property rights ... and to ... provide additional unconditional financial resources to enable technology transfer, capacity building and adaptation in developing countries.
>
> *(see Final Conclusions working group 8: Climate Debt, http://pwccc.wordpress.com/2010/04/30/final-conclusions-working-group-n%C2%BA-8-climate-debt/, accessed 10/15/2012)*

56 And legitimacy is a crucial element of Law *qua* Law (Fuller 1969, 1977; Postema 1994; Brunnée and Toope 2010).

57 Carbon markets have been plagued by two main problems: oversupply of credits (Allan 2012; Harvey 2012), and cheating (Neslen 2012).

References

Abramsky, K. (2009) Energy, Crisis and World-wide Production Relations, in Brand, U., Bullard, N., Lander, E., and Müller, Tazio (eds.) *Contours of Climate Justice Ideas for Shaping New Climate and Energy Politics*, Critical Currents No. 6, October 2009.

Ahmed, Haseed and Cutrone, Chris (2012) The Occupy Movement, a Renascent Left, and Marxism Today: An Interview with Slavoj Žižek, *Platypus Review* 42, December 2011–January 2012. Accessed 10/22/2012 from http://platypus1917.org/2011/12/01/occupy-movement-interview-with-slavoj-zizek/.

Allan, A. (2012) CERs Case Below a Euro as Sellers Dump Credits, PointCarbon Carbon Market News Service, October 22.

Austen, M. and Bedall, P. (2010) Climate Justice – Bezugspunkt einer Gegenhegemonie oder wolkige Leerformel?, *analyse & kritik – Zeitung für linke Debatte und Praxis*, 549, April 16, 2010.

Austen, M. and Bedall, P. (2012) Climate Justice – Point of Reference for a Counter Hegemony or Nebulous Empty Phrase? Accessed 10/22/2012 from http://notesfrombelow.wordpress.com/2010/07/29/climate-justice-point-of-reference-for-a-counter-hegemony-or-nebulous-empty-phrase/#_edn1.

Benney, M. (1983) Gramsci on Law, Morality, and Power, *International Journal of the Sociology of Law*, 11: 191–208.

Brand, U. (2007) The Internationalization of the State as the Reconstitution of Hegemony, IPW Working Papers No. 1/2007.

Brand, U., Bullard, N., Lander, E., and Müller, T. (2009) Radical Climate Change Politics in Copenhagen and Beyond: From Criticism to Action, in Brand, U., Bullard, N., Lander, E., and Müller, T. (eds.) *Contours of Climate Justice: Ideas for Shaping New Climate and Energy Politics*, Critical Currents No. 6, October 2009.

Brunnée, J. and Toope, S. J. (2010) *Legitimacy and Legality in International Law: An Interactional Account*, Cambridge: Cambridge University Press.
Bruno, K., Karliner, J., and Brotsky, C. (1999) Greenhouse Gangster vs. Climate Justice, CorpWatch, November 1, 1999.
Cox, R. (1981) Social Forces, States and World Orders: Beyond International Relations Theory, *Millennium, Journal of International Studies*, 10(2): 126–155.
Cox, R. (1983) Gramsci, Hegemony and International Relations, *Millennium, Journal of International Studies*, 12(2): 162–175.
Cox, R. (1993) Gramsci, Hegemony and International Relations: An Essay in Method, in Gill, S. (ed.) *Gramsci: Historical Materialism and International Relations*, Cambridge: Cambridge University Press.
De Lucia, V. (2009) Hegemony and Climate Justice: A Critical Analysis, in Böhm, S. and Dabhi, S. (eds.) *Upsetting the Offset: The Political Economy of Carbon Markets*, London: MayFlyBooks.
Dutta, Soumya (2009) Offsetting Lives and Livelihoods: Atmospheric Brown Cloud and the Targeting of Asia's Rural Poor, in Böhm, S. and Dabhi, S. (eds.) *Upsetting the Offset: The Political Economy of Carbon Markets*, London: MayFlyBooks.
Ellul, J. (1964) *The Technological Society*, New York: Vintage Books.
ETCGroup (2012) The Case for Technology Assessment. Bad Timing. Accessed 10/22/2012 from www.etcgroup.org/sites/www.etcgroup.org/files/TAF5_Bad%20Timing_042612.pdf.
Feenberg, A. (1999) *Questioning Technology*, London and New York: Routledge.
Feenberg, A. (2005) Critical Theory of Technology: An Overview, *Tailoring Biotechnologies*, 1(1): 47–64.
Femia, J. (1981) *Gramsci's Political Thought: Hegemony, Consciousness, and the Revolutionary Process*, Oxford: Clarendon Press.
Friends of the Earth (2009) A Dangerous Obsession: The Evidence Against Carbon Trading and for Real Solutions to Avoid a Climate Crunch. Summary Report. Accessed 10/22/2012 from www.foe.co.uk/resource/reports/dangerous_obsession_sum.pdf.
Fuller, L. (1969) *The Morality of Law, Revised Edition*, Yale: Yale University Press.
Fuller, L. (1977) *The Anatomy of the Law*, Westport: Praeger.
Gill, S. (ed.) (1993) *Gramsci, Historical Materialism and International Relations*, Cambridge: Cambridge University Press.
Gramsci, A. (2007a) *Selections from the Prison Notebooks*, translated by Hoare, Q. and Nowell-Smith, G., London: Harbour Pub. Co. Ltd.
Gramsci, A. (2007b) *Quaderni del Carcere, Edizione Critica dell'Istituto Gramsci*, ed. Valentino Gerratana, Turin: Einaudi.
Harvey, F. (2012) Global Carbon Trading System Has "Essentially Collapsed," *Guardian*, September 10. Accessed 10/22/2012 from www.guardian.co.uk/environment/2012/sep/10/global-carbon-trading-system.
Higgins, P. (2010) *Eradicating Ecocide: Laws and Governance to Stop the Destruction of the Planet*, London: Shepheard-Walwyn.
Hornborg, A. (2001) *The Power of the Machine: Global Inequalities of Economy, Technology, and Environment*, Walnut Creek: Altamira Press.
Hultman, N. (2011) The Durban Climate Talks: Making the Technology Mechanism Useful, Brookings, opinion, December 9. Accessed 10/22/2012 from www.brookings.edu/research/opinions/2011/12/09-technology-mechanism-hultman.
ICHRP (International Council on Human Rights Policy) (2008) Climate Change and Human Rights. A Rough Guide, International Council on Human Rights Policy, Versoix, Switzerland. Accessed 10/22/2012 from www.ichrp.org/files/reports/45/136_report.pdf.
IPCC (Intergovernmental Panel on Climate Change) (2000) Methodological and Technological Issues in Technology Transfer. Summary for Policymakers. A Special Report of IPCC Working Group III. Accessed 10/22/2012 from www.ipcc.ch/pdf/special-reports/spm/srtt-en.pdf.
Khor, M. (2012) The Equitable Sharing of Atmospheric and Development Space: Summary, South Centre Paper for UNFCCC AWG-LCA workshop on Equity, Bonn, May 16.
Laclau, E. (1989) Politics and the Limits of Modernity, *Social Text*, 21 (Universal Abandon? The Politics of Postmodernism): 63–82.
Laclau, E. and Mouffe, C. (2001) *Hegemony and Socialist Strategy: Towards a Radical Democratic Politics*, 2nd edition, London and New York: Verso.
Law, A. (2009) The Callous Credit Nexus: Ideology and Compulsion in the Crisis of Neoliberalism, *Sociological Research Online*, 14(4): 5.

Litowitz, Douglas (2000) Gramsci, Hegemony and the Law, *Brigham Young University Law Review*, 2: 515–551.
Lohmann, L. (2006) Carbon Trading: A Critical Conversation on Climate Change, Privatisation and Power, Development Dialogue, 48, September 2006. Dag Hammarskjöld Center.
Lohmann, L. (2008a) Carbon Trading, Climate Justice and the Production of Ignorance: Ten Examples, *Development*, 51(3): 359–365.
Lohmann, L. (2008b) *Six Arguments against Carbon Trading, Climate and Capitalism*. Accessed 10/22/2012 from http://climateandcapitalism.com/2008/09/29/carbon-trading-the-wrong-way-to-deal-with-global-warming/.
Marcuse, H. (2006) *One Dimensional Man: Studies in the Ideology of Advanced Industrial Society*, London and New York: Routledge Classics.
Morton, A. D. (2007) *Unravelling Gramsci: Hegemony and Passive Revolution in the Global Economy*, London: Pluto Press.
Mouffe, C. (1979) Hegemony and Ideology in Gramsci, in Mouffe, C. (ed.) *Gramsci and Marxist Theory*, London: Routledge & Kegan Paul Ltd.
Mumford, L. (1974) *Pentagon of Power: The Myth of The Machine*, Vol. II, New York: Harcourt Brace Jovanovich.
Neslen, A. (2012) Polish "Ghost" Coal Plants Ignite Emissions Trading Outrage, *Guardian*, July 11. Accessed 10/22/2012 from www.guardian.co.uk/environment/2012/jul/11/coal-energy.
Parry, S. and Douglas, E. (2011) In China, the True Cost of Britain's Clean, Green Wind Power Experiment: Pollution on a Disastrous Scale, Mail Online, January 26. Accessed 10/22/2012 from www.dailymail.co.uk/home/moslive/article-1350811/In-China-true-cost-Britains-clean-green-wind-power-experiment-Pollution-disastrous-scale.html.
Paterson, M. (2010) Legitimation and Accumulation in Climate Change Governance, *New Political Economy*, 15(3): 345–368.
Paul, J. and Campbell, G. (2011) Investigating Rare Earth Element Mine Development in EPA Region 8 and Potential Environmental Impacts, EPA Document-908R11003, August 15.
Postema, G. (1994) Implicit Law, *Law and Philosophy*, 13(3): 361–387.
Poulantzas, Nicos (1980) *State, Power, Socialism*, London: Verso.
Pusey, A. and Russell, B. (2010) The Climate Crisis or the Crisis of Climate Politics? *Perspectives* 2010. Accessed 10/22/2012 from www.anarchiststudies.org/node/423.
PushEurope (2012) *A Planetary Emergency*. Global Campaign to Demand Climate Justice. Accessed 10/22/2012 from http://pusheurope.org/wp-content/uploads/2013/05/PlanetaryEmergency-AnOverview-online-circulation.pdf.
Rappaport, R. A. (2010) *Ritual and Religion in the Making of Humanity*, Cambridge: Cambridge University Press.
Sauter, R. and Watson, J. (2008) *Technology Leapfrogging: A Review of the Evidence*, Tyndal Center for Climate Change Report.
Sheppard, K. (2009) Poor Countries Reject "Suicide Pact," Mother Jones, December 9. Accessed 10/22/2012 from www.motherjones.com/environment/2009/12/poor-countries-g77-suicide-pact-copenhagen.
Sweney, M. (2009) Copenhagen Climate Change Treaty Backed by "Hopenhagen" Campaign, *Guardian*, June 23. Accessed 10/22/2012 from www.guardian.co.uk/media/2009/jun/23/hopenhagen-climate-change-campaign.
Third World Network (2008) Note on Access to Technology, IPR and Climate Change. Briefing Paper No. 1. Climate Change Subsidiary Bodies' Sessions. June 2–13, 2008, Bonn.
Third World Network (2009a) TWN Info Service on Intellectual Property Issues. Accessed 10/22/2012 from www.twnside.org.sg/title2/intellectual_property/info.service/2009/twn.ipr.info.090609.htm
Third World Network (2009b) Developing Countries Proposed Relaxing IPR Rules for Climate Technologies. Bonn News Update, April 4, 2009.
Tucker, V. (1999) The Myth of Development: A Critique of Eurocentric Discourse, in Munck, R. and O'Hearne, D. (eds.) *Critical Development Theory: Contributions to a New Paradigm*, London: Zed Books.
Walker, C. (2010) Copenhagen in Context, *Dissent*, Autumn/Winter: 17–19.
Watts, J., van der Zee, B., and Vaughan, A. (2009) Copenhagen Day of Mass Protest Passes without Major Incident, *Guardian*, December 16. Accessed 10/22/2012 from www.guardian.co.uk/environment/2009/dec/16/copenhagen-protest.
Winner, L. (1978) *Autonomous Technology: Technics-out-of-Control as a Theme in Political Thought*, Cambridge, MA: MIT Press.

6

CLIMATE JUSTICE FROM THE PERSPECTIVE OF PHILOSOPHY

Andreas Niederberger

1 Climate justice between theories of justice and applied ethics

The salience of philosophical concepts and themes is certainly related to new insights and developments within philosophy. But it also always reflects political and social changes. In this light one can explain the preeminent role of theories of justice and as a consequence also the preeminent role of political philosophy today. The popularity and importance of these two theories in the last 40 years can be explained by the end of the confrontation between Marxist and liberal approaches and the emergence of new political actors, like, for instance, new social movements. And the current success of applied ethics is (at least also) due to the transformation of the universities and the growing importance of criteria of applicability to scientific research (and its funding). If one considers the urgency which many attribute to climate change and its impacts, it is, given the preceding diagnosis, not surprising that philosophy reacts to this issue. Furthermore, this preoccupation is, since the beginning of the new Millennium,[1] growing in importance, increasingly international, and it transcends established schools. Philosophy offers itself as an interlocutor in the political debate where different political actors already refer to the notion of climate justice.[2] One could observe such direct references to principles of climate justice during the UN conferences on climate change, when spokespersons for the Global South claimed greater negotiation power and when they asked for acknowledgment of or even compensation (for instance in the form of financial or technology transfers) for the advantage states in the Northern hemisphere had through their previous ability to use resources which contribute or contributed to climate change.[3]

For authors in the philosophical debate climate change reminds of important problems which are discussed in research on global justice with its focus on famine and poverty in many parts of the world. They point out unequal impacts of climate change at different points of the globe where the poor and those who are disfavored anyway are hit hardest by the effects of changing weather and temperature. This distribution of effects is, on the one hand, caused by geographic situations (maybe in contrast to famine and poverty), but, on the other hand, it is also related to different infrastructures which depend on respective positions in the global economy and the unequal distribution of access to resources. The expected effects of anthropogenic climate change not only reproduce existing inequalities and injustices, they rather reinforce the latter and contribute to a situation in which measures against injustices fail.[4] Climate change is, therefore, also an important subject in the philosophical debate because it expounds the problems of usual strategies intended to improve domestic or global justice.[5] Climate change raises questions especially in non-ideal theory which, so far, has mostly concentrated on problems related to missing motivations to comply with moral duties. Competing moral obligations or mutually reinforcing morally 'corrosive' processes are a new challenge for this theory.[6]

One of the interesting facets of philosophical research on anthropogenic climate change is its position between theories of justice and applied ethics. This position results from the assumption of many authors that we can neither neglect the question of principles for a just reaction to climate change, nor the necessity of developing applicable and realizable solutions. Theories of justice, as they have been presented since the publication of John Rawls' *Theory of Justice* in the 1970s, are primarily concerned with the issue of deciding which goods we should distribute according to which criteria. They search for a just 'basic structure' of society which means an institutionally guaranteed distribution of resources, chances, or status and which reflects all in their equality or particular needs.[7] In this context authors present the idea that it is only the basic structure which constitutes the framework within which people (at least in the most important respects) cooperate. The question is thus how to distribute the resources, etc. which directly depend on cooperation or which are available in cooperation. The distribution does not concern share in clean drinking water as such, but it only relates to drinking water which could in principle be exploited and which could be available for and in joint actions. The primary question in theories of justice is not how much clean drinking water should be available for each and everyone. It rather concerns the issue who can oblige whom to which degree to make drinking water accessible, to share it or not to prevent others from using available drinking water. Compared to the debate on climate justice and under the condition that this debate locates itself in the field of theories of justice, it is important to mention that these general theories of justice conceive of goods in terms of a definite, scarce amount which could in principle be subject to some distribution in the form of claims to actions and omissions of actions. Goods, which cannot be related to such actions and omissions, are not at stake in theories of justice.

Applied ethics, on the other hand, is characterized by developing principles and virtues for the responses to important moral and often specific challenges (problems which one cannot not react to without moral difficulties). The realization of the principles or virtues aims at dealing adequately with the challenges. Applied ethics searches for solutions for single issues which can be normatively justified, for instance, whether we should make use of a technology, if we cannot exclude that it will have problematic effects, or how to distribute scarce medical goods in a group of sick people who cannot all be healed. Often, these solutions do not pretend to cohere in a single ethical system and sometimes applied ethics calls the principles it recommends principles of justice. But most solutions choose a teleological or consequentialist perspective with a focus on the problem that must be solved (see, for example, Nida-Rümelin 1996: 57–63). The discussion thus never only considers a mode of distribution or reaction. It always (also) aims at really responding to a given challenge in the given circumstances.

By using climate justice as a key concept the philosophical debate on climate change positions itself within the field of theories of justice and this implies within political philosophy. At the same time since the point of departure is a specific topic, namely the adequate response to the effects of global warming, which results from anthropogenic greenhouse gas emissions, the debate is also closely related to research and methods in applied ethics. In this sense the protagonists of the debate understand the effects of climate change to structurally affect humanity as a whole and to cause the urgent need for comprehensive and global measures.[8] Climate change is not only a basic moral problem which affects some significantly stronger than others or probably even only some (which is the case of hunger as one of the key issues in the debate on global justice).[9] But often approaches conceive of the effects of climate change in terms of goods.[10] *Adaptation* to climate change or changes in behavior to stop or limit emissions (*mitigation*) is considered in the form of costs or limits in access to goods.[11] If one starts with such a goods-centered perspective, it seems natural to see climate change and its effects as a question of the just distribution of the respective costs, advantages, or claims. One can find similar goods-centered definitions of problems and following suggestions for solutions from the perspective of theories of justice also in other domains of applied ethics, most prominently in the field of the allocation of scarce medical resources (like medication, medical equipment, availability of personnel, etc.). The particular difficulty of research on climate change is that for principled reasons one cannot define the precise amount of the costs and advantages that must be distributed. There might be an external limit to climate change beyond which we must assume that human existence on Earth becomes impossible or that irreversible processes will begin which severely curtail human lives.[12] But, first, there is the problem that before reaching this limit the scientific evidence with regard to certain or possible developments and consequences is not definite (especially concerning the question which consequences are to be expected once the average global temperature increases by more than 3°C). And second, there are multiple scenarios with different combinations of adaptation, mitigation, and climate engineering. Given this, the costs and advantages, which must be distributed, are closely related to the question which scenario with its model of living and natural conditions we (or imagined future human beings) choose or consider being acceptable. We must first decide which conditions should be preserved or secured or how future circumstances should be offset against current costs and advantages, before we have an idea of the amount of costs and advantages that must be distributed.

The philosophical debate on climate change is, thus, confronted with the methodological challenge how to combine a teleological perspective (with the question which goal we should and could strive for) with the aim of finding a just distribution of the costs which we will incur while trying to reach the goal. This combination presents a major challenge because there is no clear primacy of one of the two elements which must be combined.[13] There are, on the one hand, arguments against a certain goal, such as the limitation of global warming to an average of 2°C because it would cause too high costs and losses in socio-economic developments. The position would, thus, refer to unacceptable distributions of costs and advantages or point out the absolute weight of costs, for example with regard to economic development or in the form of restraints which would have to be accepted. On the other hand, we can only find a just distribution of costs if all those who are affected can rationally accept the resulting perspective regarding the distribution or reduction of emissions and possibilities to react to climate change. It might be just to distribute costs and burdens such that global warming is limited to 5°C. But since such a distribution would, with very high probability, have the consequence that some or possibly even many islands and coastal regions were flooded, it would not be acceptable for those living in the flooded regions without further qualifications.[14]

The philosophical debate on climate justice is not necessarily concerned with the general teleological questions how life will succeed and which should be the form of good coexistence. But it must develop a precise idea of the goods and amounts of goods which should generally be available.[15] Developing this idea will depend on conceptions of the good life or on hierarchies of interests. And these must be conceptions or hierarchies which are not too formal because otherwise one could not comprehend the weight of goods in question in the perspective of those who are affected.[16] With these difficulties the debate on climate justice is located between applied ethics and theories of justice and the philosophical disciplines and methodologies they represent: On the one hand, approaches consider emissions and costs for the adaptation to climate change as advantages and burdens which we should distribute according to principles of justice. On the other hand, the acceptability of limitations to emissions or the unacceptability of costs depends on the way they make valuable forms of life and coexistence impossible. Philosophically seen, we can, thus, not separate climate justice from defining the specific challenge; this means from defining the state which we should achieve, preserve, or not trespass.[17]

This position of the philosophical debate on climate change between theories of justice and applied ethics makes it quite complicated[18] – and it has some consequences for the issues that are discussed. One of the foremost consequences is that the issues, which the theories must address, are themselves objects of the debate. In the following I will demonstrate this in an exemplary way in some key fields of the philosophical debate on climate change. I will end by offering some perspectives on the future of the debate and ask how important the debate is within philosophy and which developments we can expect or hope for.

2 Dimensions of the philosophical debate on climate justice

2.1 *Climate justice vs. social justice*

Against the background of what has just been said, it is not surprising that one of the most important controversies in the debate on climate justice turns around the question whether a just distribution should only concern costs related to climate change or whether it should relate to the broader perspective of social justice in general. In a certain sense the answers in other parts of the debate also depend on the position authors take with regard to this key issue. Anthropogenic climate change is, most of all, the result of the usage of fossil fuels since the beginning of the industrial era. Industrialization was (and is) a major factor for the development of societies and their wealth, which is why one can detect in (most) societies a direct connection between the current level of well-being and the earlier and partially also current use of fossil fuels. In addition to this, there is the widespread belief that the future wealth of developing countries and emerging market societies decisively depends on their ability to make use of significant amounts of such fuels. In the philosophical debate on climate change – just like in the political discussion and during the international negotiations for an agreement succeeding the Kyoto Protocol[19] – this brought approaches to the forefront which deny that we can detach the question of how to distribute the costs of climate change from considerations on the general conditions for social or even global justice. If current wealth relies on the earlier use of fossil fuels, the argument goes, we must acknowledge this earlier use (and this also means the ensuing tightening of emission possibilities in the present) when we talk about the distribution of current or future costs.[20] The aim should be comparable living standards and not only an acceptable distribution of the costs resulting from necessary reactions to climate change.[21]

These thoughts have led some to expect that climate justice will be the way to achieve global justice in general, for instance through a just distribution of emissions rights and their global trade or probably even through 'back payments' of developed countries.[22] Others object that this inclusion of climate justice into a general theory of justice is implausible because it would confront climate justice with the difficulties which approaches to global justice could not overcome so far – and in the case of climate change this failure would constitute itself a problem of justice. In this sense Eric Posner and David Weisbach argue that many approaches to climate justice cannot explain that and why rich states and their populations, but also emerging market societies should (and this means under which conditions they would indeed agree to) accept high costs or losses in their living standards or in their pace of development. And since progressive climate change affects and will affect poor states and populations much more directly than rich societies, any delay in responding to climate change will aggravate the unjust situation of poor countries. In face of this it is unjust if we do not take into account the conditions which would make it likely that rich states and emerging market societies agreed to join a common climate agreement (Posner and Weisbach 2010: 11–13). These states will only join such an agreement – at least given the current motivations – according to Posner and Weisbach, if they do not perceive the distribution of the costs of climate change as a general social redistribution program. The authors, therefore, suggest searching for an "International Paretianism," this means for a distribution of the costs, which is advantageous for all and does not impose high costs on some, while others only profit from it (see also Posner and Sunstein 2008a, 2008b, 2009; Posner and Weisbach 2013).

Many have raised objections against this view. They mostly concern the economic and utilitarian core of the argument, namely the direct correlation of the willingness to join international agreements with their economic advantages and disadvantages (see, for instance, Jamieson 2013). And one can also contest many of the assumptions concerning the demonstrability and attributability of emissions which would allow for answers to the complicated questions on the resources, which are at the base of the situations of different states or persons (see, among other, Meyer 2013). At the same time, Posner and Weisbach are certainly right in stressing the fact that we should not delay dealing with the consequences of climate change. Already now the costs we incur as well as the risks of future costs are increasing. And, moreover, we cannot exclude that we will reach the 'tipping points' mentioned before beyond which cost might be irreversible and unpayable (see among others Barnosky et al. 2012). It might, thus, be right that it is not necessary and normatively seen very likely impossible to detach the distribution of the costs of climate change from historic advantages which the past use of fossil fuels allowed. But in difference to other goods, which theories of justice discuss, climate change has the property of directly affecting the conditions for the applicability of distributive principles. Given this, we cannot wait with the realization of such principles until all preconditions are fulfilled for the achievement of global justice (see also Shue 2010).

This brings the debate back to the tension between theories of justice and applied ethics: Posner and Weisbach argue teleologically – the aim is to prevent the poorest from further harm – and given this aim they conceive of their theory of justice in a consequentialist way. They conclude from their consequentialist check of other theories that it would be wrong and unjust to require a (just) distribution which will lead to no significant reaction to climate change because those who are incurring the costs will refuse to bear these costs.[23] From the perspective of most theories of justice one can respond that this raises the question of the normative standard used in such a consequentialist check of theories of justice. There are possible ways of justifying obligations to emissions reductions or the bearing of costs in a consequentialist perspective – and many authors in the debate argue exactly in this way. But this is not necessary because one could also infer such obligations from the immediate duty to establish just or legitimate relationships to others. According to this view it would be unjust if rich states or emerging market societies simply kept to their claims, advantages, or privileges. Considered now in a consequentialist perspective it could be just to coerce these states or those in the states who would have to bear the biggest share of the burdens if necessary into a different behavior.[24]

2.2 *Climate justice as an individual virtue or responsibility?*

Another important controversy within the philosophical debate concerns the issue of individual obligations resulting from climate justice. This controversy is related to the first because some view the orientation towards individual responsibility or duties as a way out of the confrontation between general and particular claims of justice. The confrontation mentioned before considers groups or whole states as the relevant actors for bringing about justice or implementing the principles of justice. But there are also two alternative forms of argumentation with regard to the question of individual responsibility: For some positions the key point is that an adequate response to climate change requires changing habits and assuming responsibility for the well-being of the entire world population. In this perspective the debate on climate justice focuses on individual behavior and attitudes, maybe even in the form of the global civic virtue of being disposed to act in line with a global common good.[25] At the same time it is clear that modified individual behavior will not be sufficient to cope with the effects of climate change. And moreover, one could even argue that individual behavior is a negligible factor: First, below a certain threshold individual actions are largely irrelevant for climatic transformations – and in this sense, it is very difficult to prove that individual actions harm others (see Sinnott-Armstrong 2005). Second, individual actions will only have significant 'positive' effects with respect to climate change (this means effects concerning the amelioration or at least stabilization of climatic developments) if others also change their behavior.[26] And finally, third, a perspective on individual actions might more generally be insufficient since it cannot address the important issues of regulations, redistributions, and, most of all, solutions which are complex in a way that they cannot be achieved by interaction alone. They require institutional actors like states, big companies, and/or international organizations including their steering capacities.

In light of these objections to a focus on individual changes of behavior one can understand why there are only a few approaches to climate justice referring to virtue ethics or conceiving of it in terms of calls for individual responsibility. This is a major difference to the importance of this dimension in the ecological movement and in environmental ethics. Nevertheless, these few approaches offer valid arguments which support the idea of individual duties despite the objections just mentioned. The most general argument underlines that for most parts of the First and Second World changes in behavior will become necessary to cope with the effects of climate change.[27] Given this, it makes a lot of sense to start making these changes immediately because this stresses that the changes are feasible and this creates favorable conditions for further changes. Others again point out that individual behavior is not as unimportant as some suggest. Habits of consumption such as eating meat or the use of computers and other electronic devices contribute significantly to global carbon dioxide emissions.[28] Changes in these habits are, thus, an important dimension in the adequate reaction to climate change.[29]

Philosophically even more central for the defense of individual responsibility is to argue in terms of individual rights and duties. Such an argumentation also offers the ground for a more convincing perspective to address climate justice as part of a more general theory of social and global justice which can respond to objections like those brought forward by Posner and Weisbach. Many theories of climate justice start with individual entitlements to emissions or to goods which persons need to pursue their respective life plans (this is the case in equal per capita approaches, but also in theories, which favor an emissions tax). If such an allocation of emission rights is plausible, then any action is wrong in which somebody uses more than his or her allotted amount of emissions. In these cases persons would make use of emissions or goods which they do not own. According to this view one could not say that most actions entailing emissions are unobjectionable because they would not cause damages to others. Instead many actions (or series of actions performed by one person) harm others, because they imply an inadmissible appropriation of the emission rights of others.[30] Some authors draw the conclusion from this that emissions and their effects can be handled according to principles of strict liability as they are known from liability law. Such an approach has the special advantage of securing responsibility also in those cases, when there is no intentional, conscious, or even negligent harm of others.[31]

2.3 Polluter pays principle vs. beneficiary pays principle

The equal distribution of entitlements on all individuals sounds easy and plausible. But there are many problems which immediately come to mind: For instance, one will have to say how one could determine the respective entitlements or, most of all, the limits to emissions. As was already said before, there is no 'objective' scarcity of emission options. They will only result from decisions to limit the amount of available emissions to preserve certain options and living conditions for the future.[32] But even if we could reach an agreement on the amount of emissions[33] which should be distributed, there would be a further problem for an equal per capita distribution of emission rights. And this problem is an important reason why many authors in the philosophical debate on climate justice think it is wrong to give center stage to such entitlements or their correlated duties and rights.

The principle of an equal distribution of entitlements and duties and the correlated principle of individual responsibility essentially depend on the idea that everyone bears the responsibility for his own actions. And this means that everyone is obliged not to exceed his or her entitlements or to compensate others in the case of such an excessive use. Everyone would, thus, have the responsibility to live a life which does not require more emissions than have been allotted to him or her. If the person needs more emissions, he or she must deliver something else to compensate for the encroachments upon the other's entitlements. Persons would, thus, have to bear the responsibility for the harm of others because the distribution assumes a limited amount of emission options with the consequence that any overuse will require further limitations elsewhere. Such a foundation for climate justice seems elegant and narrow. It is, as was mentioned before, not confronted with the difficulties raised by Posner and Weisbach: It starts from an exact equal distribution of entitlements and, thus, any deviation from the distribution is easy to measure and to evaluate according to further clear criteria (possibly even on a market where those possessing entitlements to emissions could sell them – the so-called cap-and-trade model).[34] Those who require more emissions due to their lifestyle will have to bear the costs for it and this means that they will have to transfer other resources or make payments to those who (can) renounce[35] to use their own entitlements. This distribution of emission entitlements amounts to a polluter-pays-principle according to which everyone bears the responsibility for the (further) limitation of amounts of admissible emissions caused by one's actions. Such a polluter-pays-principle seems to be the best expression of an egalitarian theory of climate justice.[36]

But many contest that the polluter-pays-principle serves this purpose well. They point out that the polluter-pays-principle together with an equal per capita distribution of emission rights could only offer a convincing solution if emissions were largely optional goods; this means if persons could freely choose to perform actions requiring emissions or not. But given the current circumstances we cannot assume that this condition is fulfilled. The possibilities of living a low-emissions life are very unequally distributed. The application of the equal per capita principle would, thus, give an advantage to those who are living in places where due to natural conditions or historical achievements it is much easier to renounce emissions than in other places.[37] But this is not all: As the example of Norway shows, the application of the principle could even lead to double injustices. Norway earns a lot of money from selling oil which comes from its own territory. At the same time because of its location it can satisfy its own energy needs to a large degree with hydroelectric power or with other low-emission forms of energy. The polluter-pays-principle together with the equal per capita principle would thus put each Norwegian citizen – without any change of behavior – in a much better position than most other persons on Earth.[38]

In response to the difficulties the polluter-pays-principle encounters some have brought forward the alternative beneficiary-pays-principle. According to this principle what must be distributed (equally) are the advantages and the utility (eventually also) resulting from emission possibilities (see, among others, Gosseries 2004). In the view of this approach making use of emissions will lead only to duties and responsibilities if these emissions provide those using them with an advantage over others. Poor persons living in a country with few resources and who must buy oil, for example, from Norway to heat their homes will very likely not have an advantage over their Norwegian counterparts by the use of the emissions in question. At the same time the Norwegians have an advantage over others because others must buy the oil. Given this, the Norwegians are at least also beneficiaries of the emissions and these benefits must also be considered in the distribution of costs and duties. Similarly one could (or probably even should) take past emissions into consideration if they provide currently living persons with advantages.

But the difficulties of applying such a principle are also obvious – and this is why the elegance of the polluter-pays-principle has been highlighted before: How should we determine when there is a relevant kind of advantage and how to weigh these advantages? In which respects must the opportunities or living conditions of persons be equal or comparable such that they could serve as a measure for advantages and disadvantages? And how to weigh advantages which combine unmerited elements with proper accomplishments? This indicates the difficulty of evaluating the advantages of Norwegians or the inhabitants of other countries who use their profits from selling oil or more generally from using fossil fuel to create research institutions developing low-emission forms of energy, while a poor country makes no use of building more energy-efficient housing or power lines. Such proper accomplishments are, thus, often also due to the renunciation to use the resources or revenues in a different way which is much more directly linked to individual advantages. Should such accomplishments completely be disregarded and if not to which degree and in which form should they be acknowledged?[39]

2.4 *Individual claims vs. collective claims*

The defense of individual responsibility is not only confronted with the difficulty of justifying that individual actions are relevant at all. The distribution of emission shares with corresponding rights and duties is also less than clear if we reject the perspective of a general and equal per capita distribution of emissions. In addition, there is the problem of how to assess previous decisions with the consequence of reducing the future need of emissions. For many authors these difficulties show that relating to individual responsibility and individual claims is not convincing. These approaches confront challenges like those presented by Posner and Weisbach and concentrate on collective entities as the point of reference for requirements of climate justice. According to these positions both the advantages and disadvantages of respective emissions or climate change as well as the duties and responsibilities, which must be acknowledged, should be determined with regard to collective entities. In their view such an approach is particularly salient if one considers the obvious inequality in emissions and advantages resulting from these emissions with regard to nations or states[40] and not with respect to individuals.[41]

In this perspective authors conceive of collective entities such that we can trace back their way and circumstances of life to a common economic development and to shared or at least partially similar ideas of the good life. As a consequence we should not assume that the occurring emissions will be contrary to the interest and perspectives of a significant part of the population, even if there is only some tacit consent. Members of such collective entities can live their lives within a context which the entity in its preceding history has created and further developed. At the same time the life courses of the different members of the collective mutually depend on one another. Some paths of life are much easier to realize than others and some goods are available in sufficient quantities, while this is not true for others.[42] Even if individuals cannot take over the direct responsibility for decisions because they do not dispose of alternative ways of acting, this is not valid for collectives. That is why collectives are an important point of reference because they are to be understood as instances that have nearly all the conditions of their members acting at their disposal (without having additional different resources, but those, which the society's members are providing).

In this way a solution is to be found for some complicated questions of distribution that were described above because domestic injustices do not necessarily lead to global problems or limitations of third-party demands. If it can be shown that the organization of a collective was designed in a way that some have lived off others, then there is a primary duty to care for more just conditions within the collective. The global perspective that comprises the duty of all other human beings comes only subordinately into play.

Such a primacy of collective responsibility compared with individual claims and duties however also leads to significant difficulties. First, the continued existence of socio-political contexts and the unproblematic applicability of the so-called grandfathering principle are not given unequivocally in all cases. There are even situations in which the disintegration of such contexts leaves some people with advantages and others with disadvantages. Second, philosophical disputes especially arose about the question whether persons can be obliged to take over responsibility for actions and conditions that they were not involved in and that they could not decide on in principle. Can somebody be punished or held responsible (in the form of being put in a significantly worse-off situation compared to others) for conditions that he cannot personally be blamed for? And possibly it can even be assumed that the punished person would have massively fought against these conditions (see Caney 2005: 758–760). Furthermore, can actions carry a punishment for those, who could not know during the performing that they were causing damages?

2.5 Sufficientarianism – prioritarianism

Consequently it is difficult to apply the register of responsibility or even culpability for individual or collective actions[43] and to incorporate such a consideration of emissions into an egalitarian theory of distributive justice. The attempt to identify individual claims to emissions or to goods and opportunities that possibly require the use of emissions creates serious problems. The same is true for the consideration of reference points for a current distribution of claims and duties in the light of historical collective decisions and developments. With these problems one cannot exclude that climate justice itself is in considerable respects unjust. A distribution of the utility and the costs that are the consequence of reduced emissions or adaptation to climate change, according to absolute entitlements or options to pursue own goals, could lead to a situation where other important principles of justice are violated. In this category belong the principles that one should not be punished for actions that one was not involved in or that those who are the worst off in a situation should take the biggest advantage from a just structure.

These objections are not so fundamental that all representatives of egalitarian positions would give up the idea of justifying an equal distribution of entitlements to emissions or opportunities to live a good life – with respect to the current situation alone or also in historical perspective.[44] However, there are many contributions to the philosophical debate that suggest to leave out the question about the historical origins of factors relevant for climate change and for the moment go without the constructivist relational idea of justice, which egalitarian models favor. The basic idea is not that equality is not any longer a point of reference, but rather that equality is moved from the starting position to the aims of the theory of justice. In the face of real severe injustices we encounter today it is more plausible for the authors to think about ways to deal with these injustices so that more equality is achieved instead of choosing a fictitious equality as a starting point and using it to differentiate between those actors who have already exhausted their claims and those who have not done so yet.

The two central suggestions for such a consequentialist theory of justice, which is often designed in a utilitarian style, are sufficientarianism (or some kind of threshold theory) and its further development into prioritarianism. Sufficientarianism presumes that justice primarily relates to the conditions below the threshold beyond which there is an unproblematic plurality of heterogeneous ways of life, etc. Problematic are from this perspective those conditions or states that make it impossible for the affected people to reach the threshold. It is therefore just to provide those below the threshold with all they need to reach the threshold so that they dispose of sufficient goods, abilities, and options in order to live in dignity.[45]

Sufficientarianism distributes responsibilities and costs with regard to the current status of actors and the expected effects of climate change or mitigation policies on those who are below the threshold. As long as there are persons who are below the threshold, those who are above it are obliged to transfer goods so that they do not fall under the threshold and the others come closer to the threshold. Critics of sufficientarian theories of justice have brought up especially two arguments against them: First, many consider the determination of the threshold as difficult if it is not thought to express a purely natural and therefore rather low level of goods, etc., and takes into account diverse ways of life and their requirements. Second, the threshold is also limited regarding the conditions below its reaching. It is possible that especially those benefit from transfers who are located just below the threshold, while the situation of the poorest hardly changes.

In reaction to these objections sufficientarianism was further developed into prioritarianism. In some of its forms this approach still refers the idea of a threshold, but it insists on the fact that the distribution of responsibilities, duties, and claims is in general assessed on those who are worst situated. As long as there are persons who are living in worse conditions than others, the privileged have to give up their resources and options if this contributes to ameliorating the situation of the worst-off (see Meyer and Roser 2009: 236–238).

Sufficientarianism and prioritarianism both initially occurred in the general discussion on theories of justice. However, they are particularly prominent in the climate justice debate because they allow leaving out the complicated problems of a basic structure referring to the principle of an equal distribution of claims and duties and the construction of current entitlements to which it leads. At the same time they allow answering rather precisely questions about which political actions we are to take in light of given climate developments. Regardless whether the current situation of actors results from some earlier usage of fossil fuels or not, they are obliged to help those who live in poverty and cannot improve their situation by themselves. According to sufficientarianism and prioritarianism the distribution of emission options therefore has to pursue the goal that those who are below the threshold or are absolutely worst situated benefit the most. That does not necessarily mean that they will be entitled to the most emissions, for example if it can be shown that the relevant goods and performances can be produced somewhere else with less emissions and if there is a transfer mechanism for the corresponding goods and performances.

Sufficientarianism and prioritarianism, of course, are also criticized within the philosophical debate on climate justice. All the major objections can be brought up that were advanced in other contexts against consequentialist approaches as well – such as the blindness to relations between actors (and the corresponding claim to compensatory justice for past violations of duties or injustices) and especially the paternalism this blindness permits.[46] It is not the actors themselves who determine that and in which form they have to or can do without emissions. In contrast theorists of justice decide in the face of the needs and the corresponding resources which is the ideal solution in a given situation. But one can still hold that the two versions of a consequentialist theory of justice have high plausibility regarding the distribution of duties that do not stem from direct harms to others. They convince on the field of *remedial responsibility*, that is to say the responsibility that leaves out the conditions that have led to a given situation and asks who must pay the costs and actions in order to solve the situation.[47] In the face of the previously mentioned fact that it could be unjust not to react to climate change – if and because the right principles of justice cannot be applied – more and more authors in the philosophical discussion concentrate on the question of a permissible solution. All in all sufficientarianism and prioritarianism consequently seem to be attractive options[48] – however, only on the premise that no procedures must be found that include all affected persons in the common decision making. For questions about procedural justice sufficientarianism and prioritarianism are not interesting.

2.6 Intergenerational justice

To conclude there is a last field of the philosophical debate on climate justice, which must be mentioned, namely intergenerational justice. It is nearly undisputed that we can already observe the first consequences of anthropogenic climate change. Periods of drought and heat are getting longer, violent storms are occurring more often or more intensively, and some coastal regions are acutely threatened by rising sea levels. Nevertheless, these developments are (still) in general at a manageable size so that climate change is not yet a decisive factor for social and political positions of persons.[49] If we can believe the prognoses of climate research, this will seriously change during the next 50 years if we do not manage to stop or at least limit climate change to a certain low level. Climate change could affect severely the use and usability of certain areas and regions. Rising sea levels could make whole countries uninhabitable and violent storms could lead to high costs and thus massive economic and social impacts for the affected – not to mention the general changes in lifestyle, which higher temperatures or changed weather conditions might render necessary, and military conflicts that are to be feared in face of scarce raw materials and highly coveted agricultural land.[50] Against this background it is obvious that climate justice also connects the currently living generation with future generations. Every decision about distributive principles for emission claims or burdens to mitigate or adapt to climate change implies choosing aims, which cannot only be justified to those who are currently affected by costs and burdens. It is also necessary to acknowledge that future generations have to live under the conditions of the chosen aims. In the end there are two axes of justice that have to be combined with each other.

The definition of justice for future generations raises further important philosophical questions: How can future generations claim something from today's generation if the whole existence of following generations is dependent on decisions of today's generation?[51] The answer of egalitarian approaches is that the equal per capita-principle of emissions will be expanded to the whole of mankind including future generations. Today's amount of emissions therefore does not comprise all possible available future emissions, but future generations will have the same entitlements. Or in the future the same chances of leading a successful life will have to exist for all. Today's decisions and actions should not take away from future persons the same chances to similar decisions and actions, which we can take. The extension of the equal per capita-principle to all future generations, however, is not entirely plausible because there will obviously be some technical progress that is to be taken into account: If, e.g., it was possible today to secure a future emission-free energy supply by using more emissions today, it would surely be wrong not to do this because this would mean to violate future claims to emissions.[52] But also the perspective of the same options in the future that could react to the previously mentioned problem does not convince. Historical experience shows that ways of life change over time and in reaction to new technical and cultural opportunities so that it would be even more complicated to determine which should be the same options. In the face of these problems one can assume that both mentioned axes of justice – distribution of costs/advantages between the living *and* distribution of costs/advantages between now and future living – cannot simply be integrated into a common comprehensive distributive principle.

Sufficientarian and prioritarian approaches therefore suggest proceeding in two stages. In this two-stage model the first question is how it can be secured that future generations do not fall below the threshold or are not situated worse than the currently living. On this basis one investigates how the advantages and burdens or duties are to be distributed today in order to reach the aim of intergenerational justice and justice between the currently living.[53] The question is, however, whether the definition of the minimal threshold does not bear similar problems as the equal opportunity approach. There might be some elements of the threshold that are not dependent on historical and cultural developments. But the threshold cannot be reduced to these elements without being set too low in order to adequately take into account the interests of future generations.

To summarize one can say that the question about intergenerational justice is considered to be relevant and partly even as the central issue by nearly everybody who is involved in the debate on climate justice. At the same time there still is no fully convincing answer to this question. Especially the controversy between egalitarian and sufficientarian approaches touches upon important basic problems. In the end the decisive question of what the same chances or the threshold for future generations should look like is mostly not answered. Thus this field of the discussion cannot yet fulfill its important general function for the debate on climate justice, namely to clarify the aim by answering the question to which extent climate change must be limited.

3 Climate justice and its future role in philosophy

In the philosophical debate on climate justice especially two approaches are facing each other: On the one hand, there are egalitarian constructivist theories that search for a basis on which to build the distribution of claims to emissions and of responsibilities or duties. In the field of these theories there are some versions that concentrate on the equal distribution of emission claims, while others rather choose equality of opportunities or the equality of living conditions as points of reference. The two versions are both defended with respect to the present alone and to the time since the industrialization. On the other hand, there are consequentialist or outcome-oriented theories whose central question is how the burdens are to be distributed that are necessary today and in the future in order to stop climate change or limit it. These theories are interested in balancings that imply that for all affected only advantages are to be expected. Or they aim at goals that enable at least in some areas more equality or generally better and humane living conditions to be achieved.

Both approaches articulate important ideas for a theory of climate justice and encounter criticism that is also present in other discussions of theories of justice.[54] But they also come across strong objections because they do not succeed in integrating the convincing parts of the respective other approaches. The egalitarian approaches continue the considerations of general theories of justice and can – by dealing with the consequences of climate change – even add essential new aspects to the general theory, where the question of 'natural' components of the basic structure of society has up to now been largely absent. Seen from the perspective of the general theory of justice these egalitarian approaches are certainly more convincing than consequentialist or outcome-oriented theories. It would be difficult to find a just solution for coping with the effects of climate change, which would not reflect on the reasons for existing inequalities.

However, these egalitarian approaches do not have a good answer to the problem how to deal with the worsening of the situation that results (also) from the fact that up to now general principles of justice have not been realized or implemented. In this respect outcome-oriented theories that are partly methodologically close to applied ethics seem to make more plausible suggestions. A threshold theory or a combination of sufficientarianism and prioritarianism might be suitable as a normative point of reference for pragmatic compromises. But in all probability they will not calm down the controversy about the 'really' just solution. The reason is that it makes a huge difference whether someone gets a higher amount of emissions or compensations for the adaptation to climate change because he or she is badly or worse situated than others or whether he or she has emissions at his/her disposal because it is his/her just share of the natural resources. Despite these limits both approaches influence the political discourse and become stooges in the public debate. This happens not least because some protagonists of the philosophical dispute are directly involved in the writing of the next IPCC report[55] and because the criteria that are discussed in the report are relevant for the procedure which will serve to distribute future emission rights all over the globe.

At the same time despite the range of approaches important perspectives keep being unconsidered in the philosophical debate on climate justice. This is true at both ends of the spectrum of available positions if the existing debate is depicted as such a spectrum. At the beginning of this article it was pointed out as a central characteristic feature that theories of justice concentrate on the distribution of resources, goods, and entitlements that are subject to human disposal. In the end theories of justice focus on relations between human beings and on the question how the relations between persons should look like, for instance, with respect to the access to certain resources or chances. This means for the debate on climate justice that a central disputed point is how claims to emissions and responsibilities or opportunities are to be distributed in order to limit or stop anthropogenic climate change. Two important aspects are left out in the debate so far: First, there is no convincing position regarding the demands of 'nature' itself, that is to say climate change is reduced to its effect on human existence. It might be right that a theory of justice itself cannot give answers to the question how and to which extent nature and other creatures are entitled to moral claims. But this does not mean that this question could be left out completely. The theory of climate justice must, thus, be complemented with considerations on the question how nature can be preserved or at least how we could achieve sustainability in the relationship between human beings and nature.[56]

Second, theories of justice and political philosophy as a whole have taken an institutional turn during the last couple of years, that is to say that distributive principles are no any longer the only focus of attention. The institutions and forms that enable or achieve the binding character of principles are also considered now. The question becomes important who can and in which way influence or control the operation of institutions. Thus the previously dominant and nearly ethical perspective of theories of justice is politicized because power, politics, and institutional operations are discussed (see also Niederberger 2013). Such a change in the debate on climate justice is yet to come since at the moment central topics are still the principles that should be applied. Institutions and structures, in contrast, that could carry out or put into practice principles are rarely discussed.[57] It is obvious that these institutions would have to be very powerful if they achieved validity for the far-reaching principles that theories of climate justice develop.

The debate on climate justice has to become reflexive and it must discuss its own function for global, international, and national politics. This reflective turn also includes the investigation of links between capitalism and climate change or the limits for a just coping with the consequences of climate change that exist because of the global capitalist economic order (see Storm 2009). To distribute claims to emissions or to ascribe responsibilities on the basis of an argumentation from the theory of justice is highly controversial and thus an empowerment due to its argumentative context. Such an empowerment can itself be unjust if it contributes to justifying globally or domestically unjust conditions by referring to supposed constraints from climate policy. In this sense a comprehensive theory of climate justice is still to be developed – and this will only be convincingly possible by talking and arguing with all citizens worldwide and especially with all the important actors in the conflict about the global strategy for dealing with climate change.

Notes

1 Some first philosophical articles on global warming and related moral questions have been published since the mid 1990s, for example, Jamieson (1992, 1996), Coward and Hurka (1993), Shue (1993), or Grubb (1995). But a broad philosophical discussion of the moral dimensions of climate change has only occurred since the early 2000s and especially since the publication of the last IPCC-Report in 2007 (see as an important summarizing and starting point for this new debate Gardiner 2004). Since then there has been a multitude of books, articles, and conferences dedicated to climate justice or other philosophically relevant aspects of anthropogenic climate change.

2 See among others the website www.klimagerechtigkeit.de (accessed 05/25/2013) or the websites of the networks *Climate Justice Action* and *Climate Justice Now*: www.climate-justice-action.org, www.climate-justice-now.org (both accessed 05/25/2013).

3 See on the question, if emerging market countries rightly present themselves as representing the Global South and the difficulties this raises for just negotiation procedures, Hurrell and Sengupta (2012).

4 Examples are inundations in the last couple of years in Colombia where budgetary funds for the development of social institutions and rights had to be used to cope with damages caused by flooding.

5 This mostly affects strategies concentrating on just domestic constitutions that aim at local redistributions between the rich and the poor.

6 Moreover Dipesh Chakrabarty pointed out that anthropogenic climate change might require us to revise or at least supplement usual models of human agency (which implies for him as a postcolonial theorist that the opposition of the Western Enlightenment model of autonomy and the postcolonial – poststructuralist difference-model loses its poignancy and meaning). The idea of an 'Anthropocene,' i.e., a geological era defined by human actions, refers to an understanding of human agency which cannot be reduced to the single actions of specific persons because of its extension in time. In this sense philosophy returns to the question how the different temporal, individual, and collective dimensions of human agency are interrelated. See Chakrabarty (2012).

7

> By the basic structure I mean a society's main political, social, and economic institutions, and how they fit together into one unified system of social cooperation from one generation to the next. The initial focus, then, of a political conception of justice is the framework of basic institutions and the principles, standards, and precepts that apply to it, as well as how those norms are to be expressed in the character and attitudes of the members of society who realize its ideals.
>
> *(Rawls 1993: 11–12)*

8 See on the dangers of climate change including the differentiation between risk and danger Moellendorf (2011).

9 For a general critique on the too philosophical and too little policy-oriented character of the debate on climate justice see Light (2011).

10 On the other hand, there are also approaches arguing against a theory of climate *justice* and opting for a more general climate or environmental *ethics*, for instance, Gesang (2011).

11 One could add costs for possible *climate engineering*, this means the attempt to steer the developing of the climate. See on this topic with more references to the discussion Gardiner (2010).

12 See in this respect, for instance, the discussion of knowledge about "tipping points" which lead to the irreversible processes mentioned in the text above, in Lenton et al. (2008).

13 Consider in this respect also the epistemological, decision theoretical, and normative discussion of a just response to climate change, if it is a so-called 'wicked' or even 'super wicked problem.' This term relates to problems where it is difficult to find clear solutions because every solution raises so many new and complicated problems that one cannot simply recommend or realize one solution. See for an overview of this discussion including a suggestion how to deal with the difficulties Levin et al. (2012).

14 One can observe this, for instance, in the rejection of even the 2°C-criterion by the people living in the Southern pacific region. See on the discussion of the effects of current developments Ferris et al. (2011).

15 Similar to general theories of justice there is no discussion on resources which are in principle unavailable, but only on those resources which are (possibly) disposable for human beings. Given the uncertainties of scientific studies on future developments, it is unclear to a certain degree what exactly will be disposable.

16 See in this context, for instance, the primacy of the interests and needs of the poor which Caney (2005: 770) supports.

17 Within theories of justice some authors also discuss the question what the aim of a just distribution of goods or a just establishment of institutions is. And this question has even led to some of the key controversies within philosophical theories of justice, such as, for instance, the controversy between communitarians and liberals (see among others Honneth 1993) or to the controversy about the importance of equality for justice (see among others Krebs 2000). But within these discussions the aim of the theory of justice is itself considered to be a primarily philosophical problem (namely to view different aims as different responses to the question which the good for human beings is). It is rather not assumed that the aim of the theory is to solve a puzzle – which is what characterizes the debate on climate justice.

18 In the perspective of some authors the mentioned complexity (the character of a 'wicked problem') has the consequence that no just solution will be found because it always offers the option to point out factors which have not been considered so far – options which themselves facilitate manipulative or self-deceiving behavior. See on such a diagnosis Gardiner (2006).

19 See, for instance, the reconstruction of China's position during the Copenhagen conference 2009 in light of philosophical approaches to climate justice in Harris et al. (2012).

20 In addition, there are also approaches which do not assume that the necessity of a more general perspective on social or global justice depends on past advantages in the use of greenhouse gas emissions. The mere fact of existing severe inequalities or of hunger and poverty is sufficient for these approaches to distribute the costs of climate change in accordance with general standards of justice or to relativize climate duties in light of claims related to the overcoming of poverty. See, for example, Caney (2005: 763) or Shue (1993).

21 See on this point for the different 'schools' in the theory of justice Shue (1999), Singer (2002: 14–50), and Caney (2010).

22 The main point of reference for such positions is the 'equal per capita' distribution of emissions rights; this means the exact same disposal of each and every world citizen of emissions rights. See, for instance, Ekardt (2012: 47–52).

23 A parallel, but different 'doubling' of justice is the separation of justice and fairness where fairness is understood as the "subjective experience of division that respects individuals' specific situations"; this means as perceived justice in contrast to its real claims or to certain principles of justice. See Klinsky and Dowlatabadi (2009: 90).

24 This again could bring up the consequentialist question under which conditions such coercion is likely to occur, etc. On the justification of the right to coerce others into complying with duties of climate justice see Meyer and Sanklecha (2011).

25 See in this respect the discussion of potentially relevant virtues in Thompson and Bendik-Keymer (2012).

26 See in this respect the thesis that changes in the behavior of some with the aim of adequate reactions to climate challenges (for instance the reduction of the use of fossil fuels) might entail the consequence that climatically harmful behavior could become 'less expensive' for others which would make it more likely to occur. Sinn (2008) defends such a thesis.

27 In this context Sonja Klinsky and Hadi Dowlabati point out that in view of global history such radical changes in behavior have led in many cases to the demise or to the loss of importance of civilizations (Klinsky and Dowlabati 2009: 104). Beyond the question of how necessary such developments are this

mostly raises the difficulty of deciding whether those who are the addressees of a more just climate policy will also be those who will really exist in the future and profit from some current climate policy. If there were good reasons to assume that future developments (for instance as a result of a specific climate policy) will have the effect that some persons will not be born or pursue very different paths of life, it would be very difficult to make uncomplicated assumptions regarding future generations or prospective persons who are supposed to benefit from climate policy.

28 See also the recent research of the Research Program on Climate Change, Agriculture and Food Security (CCAFS). This research indicates that the global agriculture and production of food contributes a fifth to a third of climatically relevant emissions (see http://ccafs.cgiar.org/news/press-releases/agriculture-and-food-production-contribute-29-percent-global-greenhouse-gas, accessed 05/31/2013).

29 One could add that many structural or regulatory measures only make sense if individual actions are not adversely affecting them. Some assume, for instance, that measures intended to increase energy efficiencies in houses and vehicles will not lead to savings in energy consumption if people use the saved money for long-distance journeys or to buy new devices which will entail additional energy use.

30 See on such an argumentation, for instance, Ekardt and Hövel (2009). Schüssler (2011), in contrast, makes a suggestion how to conceive of emissions and more particularly historical emissions on the basis of a Lockean theory of property and appropriation (including the Lockean proviso that enough must be left for others).

31 The central contribution to the debate is Neumayer (2000).

32 There are obviously scarce resources, such as oil or coal, which lead to emissions when they are used – and one will have to decide on their distribution (see, for instance, the emerging debate on energy poverty). But since the use of renewable resources, such as wood, also releases carbon dioxide into the atmosphere, the full amount of possible emissions is largely open ended.

33 See on this also the discussion in Section 2.6 of the question to which degree and in which form we should take into consideration the interests of future generations.

34 See on a conception of achieving global justice through a global emissions trade Page (2011).

35 It is important in this perspective (therefore also the inclusion of the 'can') that those who renounce to use their share of emissions can do so more or less voluntarily. There could not be some who decide paternalistically to use more and pay others for this with the mere assumption that these other persons will anyway renounce to make this use because given their level of development, they will not be able to use more. A convincing version of the equal per capita distribution, therefore, presupposes that, in principle, all are really entitled to use their share of emissions. Another problem with this 'can' will be discussed in the next paragraph.

36 In addition to the polluter-pays-principle there is also a more general version of the causative principle that obliges persons to answer for damages caused by one's actions. But this more general version has the problem that there are only very few occasions where we can show that individual actions are directly causing climatic harm. Mostly individual actions harm the environment or other human beings, only if there are further individual actions entailing emissions. Therefore, we cannot apply the more general causative principle of individual duties if we have not distributed individual entitlements beforehand – which is why the more general principle is usually brought forward by theories of collective responsibility. See on these approaches the following Section 2.4.

37 This inequality is particularly important, when one compares developed countries and emerging market societies. Given the unequal access to technological progress it will, in the foreseeable future, be much easier for persons in developed countries to drive cars with lower fuel consumption, to build factories which emit less pollutants, and in general to utilize new energy-efficient technologies. Some authors, therefore, highlight the important role of technology transfer. Such transfer would not only allow for reactions to new climatic conditions, it would also enable poorer states to reduce the effects of mitigation. The UN Clean Development Mechanism already includes such technology transfers.

38 Another difficulty of the equal per-capita distribution of emission rights is that it could conflict with other measures of global development policy. The possibility for poor states to gain income by selling emission rights could be an incentive for population growth or to stop or limit measures against such increases. To prevent this, there must be limits on cap-and-trade markets and/or emission shares should be tied to the population size at a certain point.

39 See for a criticism of the beneficiary pays principle along these lines Schüssler (2009, 2011) and Gesang (2011: 65–66).

40 In parts of the debate on climate policy there are also considerations on other collective agents, for

instance multinational corporations like Shell or other oil companies which in some cases have used or are using much higher amounts of emissions than states. In the philosophical debate such economic actors are not very important so far. There are different reasons for this. The most important reason can be found in the methodological assumptions of political philosophy and applied ethics according to which there are no collective agents sui generis. Ultimately we can and should reduce any collective agent to the individual persons who have created the seemingly collective agents and have an interest in their existence and operation. Emissions of oil companies could, thus, be accounted for in the shares of emissions of those who own the companies or have an advantage from their operation. See on the difficulties of a convincing account of business ethics (with only a limited role of environmental protection and climate duties) Neuhäuser (2011).

41 The important number in the research literature is that 90 percent of all emissions which have occurred so far can be attributed to 25 nations. See Klinsky and Dowlatabadi (2009: 90).

42 For a general theory of justice which ties the production and availability of goods to the expectations of a political and social community see Walzer (1983).

43 See for critical comments on applying the idea of culpability (in the sense of the categories of American tort law) also Posner and Sunstein (2008b).

44 See for a different distinction between egalitarian and nonegalitarian positions Grasso (2012).

45 See regarding this topic the presentation of sufficientarianism in the context of the debate on climate justice in Meyer and Roser (2009: 235–236).

46 Therefore it is not surprising that authors defending sufficientarian or prioritarian positions more often than others demand a 'climate avant-garde' that makes use of existing power structures in international organizations (and also, if necessary, avoids democratic processes) in order to act.

47 The differentiation and connection between 'responsibility' and 'capability' used in legal documents should also be considered in this context. See Moellendorf (2009: 250).

48 Grasso argues that even if sufficientarianism is not convincing per se, it seems to be an attractive compromise in international negotiations (Grasso 2012: 681–683).

49 See in contrast the debate about the question whether there are already the first climate refugees (e.g., Biermann and Boas 2010 or Farbotko and Lazrus 2012).

50 See regarding this topic for example the various scenarios for paradigmatic cases in Cullen (2010).

51 In the research literature this is called the nonidentity problem, that is to say the identity of future generations essentially depends on decisions of previously living people. Those who are referred to as future generations are possibly not those who will in fact be the later generations. It could be that renouncing certain improvements of today's conditions in order to preserve living conditions for future generations could lead to a situation where currently living persons will not procreate. As a consequence the paradox situation is that the previous generation refrains from using certain goods or options in favor of a generation that will not exist after all – and therefore they will also have no advantage from this sacrifice, which is why it would not be necessary to abstain from anything with regard to this future generation.

52 This is partly discussed as the possibility of a discount of the interests of future generations. The question is which cuts of interests are permissible if one can presume that, for instance, technical progress leads to the fact that future living conditions will structurally be better than the current ones. See for a depiction and critique of this economically inspired idea Caney (2008).

53 Prioritarian approaches may have counterintuitive implications if they are applied to intergenerational justice. Thus it is possible that they do not only not exclude future poverty, but that they even require actively striving for it. This can result from the focus on those who are worst situated and who can claim help even from those who are also not well (but better) situated. Therefore the duty could be to father children who will not be well situated as long as it is known that there will always be some persons who are worse situated. In addition, the generally unlimited amount of future persons who are worse situated than the currently living would have the consequence that those now living have to pay extremely high costs in order to care for the persons situated worse in the future. These implications let authors who argue for prioritarianism change their perspective to sufficientarianism if it comes to intergenerational questions. See Meyer and Rosner (2009: 231–232, 234–235).

54 Among these objections is, e.g., the critique of the neglect of communitarian conditions for the constitution of socio-political collectives, the critique of the assumed risk aversion, or the insufficient consideration of principles of desert.

55 John Broome (Oxford) and Lukas H. Meyer (Graz) are members of the Working Group III 'Mitigation of Climate Change' which takes part in the preparation of the next IPCC report for 2013–2014.

56 Some preliminary thoughts on this issue can be found in Schlosberg (2012).

57 By looking at claims to emissions and development as human rights that climate change endangers we come closer to a more 'institutional' perspective. However, existing approaches define 'human rights' mostly as extremely deeply rooted moral claims. See for example the contributions in Humphreys (2010).

References

Barnosky, A. D., Hadly, E., Bascompte, J. et al. (2012) Approaching a State Shift in Earth's Biosphere, *Nature*, 486: 52–58.

Biermann, F. and Boas, I. (2010) Preparing for a Warmer World: Towards a Global Governance System to Protect Climate Refugees, *Global Environmental Politics*, 10(1): 60–88.

Caney, S. (2005) Cosmopolitan Justice, Responsibility, and Global Climate Change, *Leiden Journal of International Law*, 18(4): 747–775.

Caney, S. (2008) Human Rights, Climate Change, and Discounting, *Environmental Politics*, 17(4): 536–555.

Caney, S. (2010) Climate Change and the Duties of the Advantaged, *Critical Review of International Social and Political Philosophy*, 13(1): 203–228.

Chakrabarty, D. (2012) Postcolonial Studies and the Challenge of Climate Change, *New Literary History*, 43(1): 1–18.

Coward, H. and Hurka, T. (eds.) (1993) *Ethics and Climate Change. The Greenhouse Effect*, Waterloo: Wilfried Laurier University Press.

Cullen, H. (2010) *The Weather of the Future. Heat-Waves, Extreme Storms, and Other Scenes from a Climate Changed Planet*, New York: HarperCollins.

Ekardt, F. (2012) Grundlagen einer geisteswissenschaftlichen Klimadebatte, in Ekardt, F. (ed.) *Klimagerechtigkeit. Ethische, rechtliche, ökonomische und transdisziplinäre Zugänge*, Marburg: Metropolis.

Ekardt, F. and Hövel, A. v. (2009) Distributive Justice, Competitiveness, and Transnational Climate Protection: 'One Human – One Emission Right,' *Carbon & Climate Law Review*, 3(1): 102–113.

Farbotko, C. and Lazrus, H. (2012) The First Climate Refugees? Contesting Global Narratives of Climate Change in Tuvalu, *Global Environmental Change*, 22(2): 382–390.

Ferris, E., Cernea, M. M., and Petz, D. (2011) 'On the Front Line of Climate Change and Displacement.' Accessed 05/30/2013 from www.brookings.edu/~/media/research/files/reports/2011/9/idp%20climate%20change/09_idp_climate_change.pdf.

Gardiner, S. (2004) Ethics and Global Climate Change, *Ethics*, 114(3): 555–600.

Gardiner, S. (2006) A Perfect Moral Storm: Climate Change, Intergenerational Ethics, and the Problem of Moral Corruption, *Environmental Values*, 15: 397–413.

Gardiner, S. (2010) Is 'Arming the Future' with Geoengineering Really the Lesser Evil?, in Gardiner, S., Caney, S., Jamieson, D., and Shue, H. (eds.) *Climate Ethics. Essential Readings*, Oxford: Oxford University Press.

Gesang, B. (2011) *Klimaethik*, Berlin: Suhrkamp.

Gosseries, A. (2004) Historical Emissions and Free-Riding, *Ethical Perspectives*, 11(1): 36–60.

Grasso, M. (2012) Sharing the Emission Budget, *Political Studies*, 60(3): 668–686.

Grubb, M. (1995) Seeking Fair Weather: Ethics and the International Debate on Climate Change, *International Affairs*, 71(3): 463–496.

Harris, P., Chow, A., and Karlsson, R. (2012) China and Climate Justice: Moving Beyond Statism, *International Environmental Agreements*, online first publication. Accessed 11/03/2012 from www.springerlink.com/content/hml507u521w8201k/.

Honneth, A. (ed.) (1993) *Kommunitarismus. Eine Debatte über die moralischen Grundlagen moderner Gesellschaften*, Frankfurt/Main, New York: Campus.

Humphreys, S. (ed.) (2010) *Human Rights and Climate Change*, Cambridge: Cambridge University Press.

Hurrell, A. and Sengupta, S. (2012) Emerging Powers, North – South Relations and Global Climate Politics, *International Affairs*, 88(3): 463–484.

Jamieson, D. (1992) Ethics, Public Policy, and Global Warming, *Science, Technology, Human Values*, 17(2): 139–153.

Jamieson, D. (1996) Ethics and Intentional Climate Change, *Climatic Change*, 33(3): 323–336.

Jamieson, D. (2013) Consequentialism, Climate Change, and the Road Ahead, *Chicago Journal of*

International Law, 13. Accessed 09/26/2012 from www.law.uchicago.edu/files/files/Jamieson%20 paper_0.pdf.

Klinsky, S. and Dowlatabadi, H. (2009) Conceptualization of Justice in Climate Policy, *Climate Policy*, 9: 88–108.

Krebs, A. (ed.) (2000) *Gleichheit oder Gerechtigkeit. Texte der neuen Egalitarismuskritik*, Frankfurt/Main: Suhrkamp.

Lenton, T., Held, H., Kriegler, E., Hall, J., Lucht, W., Rahmstorf, S., and Schellhuber, H. J. (2008) Tipping Elements in the Earth's Climate System, *Proceedings of the National Academies of Sciences*, 105: 1786–1793.

Levin, K., Cashore, B., Bernstein, S., and Auld, G. (2012) Overcoming the Tragedy of Super Wicked Problems: Constraining Our Future Selves to Ameliorate Global Climate Change, *Policy Sciences*, 45(2): 123–152.

Light, A. (2011) Climate Ethics for Climate Action, in Schmidtz, D. and Willot, E. (eds.) *Environmental Ethics: What Really Matters? What Really Works?*, Oxford: Oxford University Press.

Meyer, L. (2013) Why Historical Emissions Should Count, *Chicago Journal of International Law*, 13(2). Accessed 09/26/2012 from www.law.uchicago.edu/files/files/Meyer%20paper.pdf.

Meyer, L. and Roser, D. (2006) Distributive Justice and Climate Change: The Allocation of Emission Rights, *Analyse & Kritik*, 28(2): 223–249.

Meyer, L. and Roser, D. (2009) Enough for the Future, in Meyer, L. and Gosseries, A. (eds.) *Intergenerational Justice*, Oxford: Oxford University Press.

Meyer, L. and Sanklecha, P. (2011) Individual Expectations and Climate Justice, *Analyse & Kritik*, 33(2): 449–471.

Moellendorf, D. (2009) Treaty Norms and Climate Change Mitigation, *Ethics & International Affairs*, 23(3): 247–265.

Moellendorf, D. (2011) A Normative Account of Dangerous Climate Change, *Climatic Change*, 108(1): 57–72.

Neuhäuser, C. (2011) *Unternehmen als moralische Akteure*, Berlin: Suhrkamp.

Neumayer, E. (2000) In Defence of Historical Accountability for Greenhouse Gas Emissions, *Ecological Economics*, 33: 185–192.

Nida-Rümelin, J. (1996) Theoretische und Angewandte Ethik, in Nida-Rümelin, J. (ed.) *Angewandte Ethik*, Stuttgart: Kröner.

Niederberger, A. (2013) Gerechtigkeit und die Grenzen von moralischem Handeln und Demokratie, in Haardt, A. (ed.) *Pluralität und Interkulturalität von Gerechtigkeitskonzeptionen*, München: Fink.

Page, E. (2011) Cosmopolitanism, Climate Change, and Greenhouse Emissions Trading, *International Theory*, 3(1): 37–69.

Posner, E. and Sunstein, C. (2008a) Justice and Climate Change, The Harvard Project on International Climate Agreements Discussion Paper 08–04 2008. Accessed 09/26/2012 from http://belfercenter.ksg.harvard.edu/files/SunsteinWeb6.pdf.

Posner, E. and Sunstein, C. (2008b) Climate Change Justice, *Georgetown Law Journal*, 96: 1565–1612.

Posner, E. and Sunstein, C. (2009) Justice and Climate Change: The Unpersuasive Case for Per Capita Allocations of Emissions Rights, in Aldy, J. (ed.) *Post-Kyoto International Climate Policy. Implementing Architectures for Agreement*, Cambridge: Cambridge University Press.

Posner, E. and Weisbach, D. (2010) *Climate Change Justice*, Princeton: Princeton University Press.

Posner, E. and Weisbach, D. (2013) International Paretianism: A Defense, *Chicago Journal of International Law* 13. Accessed 26/09/2012 from www.law.uchicago.edu/files/files/posner%20weisbach%20International%20Paretianism%20defense.pdf.

Rawls, J. (1993) *Political Liberalism*, New York: Columbia.

Schlosberg, D. (2012) Justice, Ecological Integrity, and Climate Change, in Thompson, A. and Bendik-Keymer, J. (eds.) *Ethical Adaptation to Climate Change: Human Virtues of the Future*, Cambridge: MIT Press.

Schüssler, R. (2009) Global Governance and Climate Change: A Question of Historical Justice?, *Filosofski Godišnjak*, 21: 51–66.

Schüssler, R. (2011) Climate Justice: A Question of Historic Responsibility?, *Journal of Global Ethics*, 7(3): 261–278.

Shue, H. (1993) Subsistence Emissions and Luxury Emissions, *Law & Policy*, 15(1): 39–59.

Shue, H. (1999) Global Environment and International Inequality, *International Affairs*, 75(3): 531–545.

Shue, H. (2010) Deadly Delays, Saving Opportunities: Creating a More Dangerous World?, in Gardiner,

S., Caney, S., Jamieson, D., and Shue, H. (eds.) *Climate Ethics. Essential Readings*, Oxford: Oxford University Press.

Singer, P. (2002) *One World: The Ethics of Globalization*, New Haven: Yale University Press.

Sinn, H.-W. (2008) *Das Grüne Paradoxon. Plädoyer für eine illusionsfreie Klimapolitik*, Berlin: Econ.

Sinnott-Armstrong, W. (2005) It's Not my Fault: Global Warming and Individual Moral Obligations, in Sinnott-Armstrong, W. and Howarth, R. (eds.) *Perspectives on Climate Change: Science, Economics, Politics, Ethics – Advances in the Economics of Environmental Resources*, Amsterdam: Elsevier.

Storm, S. (2009) Capitalism and Climate Change: Can the Invisible Hand Adjust the Natural Thermostat?, *Development and Change*, 40(6): 1011–1038.

Thompson, A. and Bendik-Keymer, J. (eds.) (2012) *Ethical Adaptation to Climate Change: Human Virtues of the Future*, Cambridge: MIT Press.

Walzer, M. (1983) *Spheres of Justice. A Defense of Pluralism and Equality*, New York: Basic Books.

PART II

Components of the climate movement

7
THE GREEN MOVEMENT IN BRITAIN

Elaine Graham-Leigh

In the planning stages, the demonstration at the climate talks in Copenhagen in December 2009, and the associated protests in cities around the world, were often discussed as if they were the culmination of all previous years of campaigning against climate change. Indeed, from the perspective of 2012, it is easy to look back on them as the apogee of the climate movement, meaning that from then on it has been all downhill.

1 The apparent rise and fall of the green movement

The environmental movement in Britain has a long history,[1] but it is possible to see the origins of the modern movement in the late 1960s and early 1970s, a period which saw the foundation of organizations such as Friends of the Earth (1970) and the Green Party. The latter was founded as People in 1973, and subsequently became the Ecology Party and then the Green Party in 1985 (Rüdig and Lowe 1986; Wall 1994). The first major electoral breakthrough came in 1989, when the Green Party achieved 15 percent of the vote in England and Wales in the European elections. Although this success was not repeated in the following election, this can be seen as a landmark for developing awareness of climate change in Britain.

The period 2000–2009 saw the emergence of a number of different groups, reflecting both increased concern about climate change in the light of George Bush's refusal to sign the Kyoto Protocol and arguably the upturn in the movement in the widest sense in the same period. The growth of the anticapitalist movement from the protests at the conference of the World Trade Organization in Seattle in 1999 on had a general fertilizing effect on the environmental movement. It is also reasonable to speculate, for example, that the demonstrations on climate change especially from 2002 onwards would have been smaller had it not been for the effect of the Stop the War Coalition in energizing and radicalizing a new generation of activists.

There have been several different strands within the climate movement in Britain. Groups like Campaign against Climate Change (founded in 2001) and Stop Climate Chaos (a coalition of NGOs which organized the iCount demonstration and the Wave demonstration taking place in London in 2006 and 2009) concentrate on pressure for national and international governmental action on climate change. Furthermore, there are also direct action groups like Rising Tide, Plane Stupid, and most prominently the Camp for Climate Action, which held its first camp at Drax power station in August 2006. In addition, there are groups focusing not on demonstrations, protests, and so on but on encouraging actions to reduce individual or corporate emissions by engaging with people or organizations directly. Examples are the 10:10 campaign, where people and companies can sign up to reduce their emissions by 10 percent in a year and Transition Towns, which aims to create sustainable societies on a local basis.

There have been some notable successes in these years: the iCount and Wave demonstrations were both the largest climate demonstrations in Britain at the time, showing that it was possible to engage a wide range of people with the necessary to address climate change. When the jury in the case against the Greenpeace activists who shut down the Kingsnorth power station for a day in October 2007 found that their actions had been justified because they were trying to prevent worse harm, this was an indication of how far the movement had succeeded in winning the argument with ordinary people that climate change mattered. Campaigns against particular projects have also been successful: in 2009, the energy company E.on announced that they would not be building a new power station at Kingsnorth, while the third runway at Heathrow, the subject of a well-organized and committed local campaign, was shelved in 2010 (although at time of writing it appears to be back on the agenda).

However, after the protests at the Copenhagen summit, the picture seems very different. Indeed, the comparison between the climate movement in Britain before and after Copenhagen is as instructive as it is depressing. The protest at Copenhagen itself was made up of at least 100,000 people, while Stop Climate Chaos' The Wave demonstration in London culminated in 50,000 people encircling the Houses of Parliament. This was the largest ever climate protest in the UK. During and immediately after Copenhagen, there was talk of new alliances between environmental groups across Europe, and the possibility of creating a wider, stronger movement which could take the energizing experience of the Copenhagen protests and use it to take the campaigns against climate change forward. While there have been some excellent attempts to turn this promise into reality, such as the Climate Alliance, the position in 2012 is not what anyone would have hoped for in January 2010.

The London climate change demo in December 2010 was only 1,000 people strong, and the demonstration in December 2011 was not significantly larger. Climate Camp, arguably one of the most visible parts of the movement, decided to wind itself up to allow its members to concentrate on other things. This is not to say that there have not been successes as well. The Green Party saw its first Member of Parliament (MP), Caroline Lucas, elected in 2010, and the Green candidate in the London mayoral elections, Jenny Jones, beat the Liberal Democrats for third place.[2] However, as time slips by without the necessary reductions in emissions – according to the Committee on Climate Change, UK emissions fell by 9 percent in 2009 as a result of the recession, but the cold winter in 2010 caused a 2.9 percent rise[3] – and possible responses fail to get off the ground,[4] there is a distinct impression that the climate change movement in general is in trouble.

2 Positions and debates in the movement

In one view, the movement is simply the victim of unfavorable objective circumstances. The furore over 'Climategate,' the hacked emails from climate researchers at the University of East Anglia, along with the contemporaneous accusation of mistakes in IPCC reports, empowered a new wave of climate change denial from late 2009 on, while the failure of world leaders to come up with a meaningful deal at Copenhagen dealt a significant blow to the idea of climate change as a solvable problem. Despite efforts by Chris Huhne MP and Lord Stern to present the 2011 Durban climate summit as a success, the agreement to have an agreement for climate change reductions is hardly sufficient to reverse this Copenhagen effect. Even the weather played a part in the creation of these difficult times for the green movement. Popular confusion between weather and climate may have helped keep climate change in the news when 'the hottest ... since records began' headlines gave journalists an opportunity to report on it. But the particularly cold winters, as the UK experienced in 2009/2010, from some were also taken as evidence that climate change is not happening.[5] Alongside these difficulties, it also could be argued that the economic crisis has pushed climate change down the political agenda, although this does not explain why the antiwar movement had the opposite effect.

That Climategate and Copenhagen were a difficult double whammy for the green movement is fairly uncontroversial, but there is another argument which sees climate change campaigns as the authors of their own misfortunes, actually damaging the agenda they are seeking to promote. This is not solely a post-Copenhagen contention: Ted Nordhaus and Michael Shellenberger of The Breakthrough Institute put their case that the green movement was neither right for nor up to the task of combating climate change. In 2007 they published their book *Break Through. Why We Can't Leave Saving The Planet To Environmentalists*. It has however been given extra impetus by the sense of crisis after the failure of the Copenhagen summit. Since early 2010, criticisms have been leveled at climate change campaigners by figures who were generally seen as having come from within the movement, and who were certainly widely respected in it, such as Mark Lynas and George Monbiot. This has been of particular importance because of the standing such figures have in the movement. While it would be an exaggeration to term either Lynas or Monbiot as universally respected by green activists, it is nevertheless the case that criticisms from them are inevitably felt as more serious than those from other quarters. It is therefore worth examining their critiques of the environmental movement since 2010 in some detail.

The most obvious point of contention between Lynas and Monbiot on the one hand and many greens on the other is the issue of nuclear power. Opposition to nuclear power is a given for most green groups, whereas for Lynas, nuclear power is a clear necessity, so much so that he states in his latest book, *The God Species*, that "anyone who still marches against nuclear today ... is in my view just as bad for the climate as textbook eco-villains like the big oil companies" (Lynas 2011: 10). Monbiot is less unequivocal than Lynas in his nuclear position, but he increasingly regards nuclear power as a least-worst, necessary replacement for coal.[6] Like Lynas, he appears to see green opposition to nuclear as not only mistaken but in rather bad faith. For example, he commented in a reply to Jonathon Porritt in August 2011 that we should ask ourselves if the aim is to stop climate change or to get maximum takeup of renewables, as it sometimes seems that greens prioritize the latter over the former (Monbiot and Goodall 2011). Both Lynas and Monbiot therefore present a picture of green opposition to nuclear power which is ideological rather than arising from an understanding of the nature of nuclear power.

For both Lynas and Monbiot, the nuclear argument became more urgent to win as a result of the decision of the German government among others to abandon nuclear power following the explosion at the Fukushima nuclear plant in Japan in Spring 2011. They both argue that Fukushima should not be seen as a confirmation of antinuclear arguments but the final contradiction of them. It shows, they think, that even in the worst circumstances imaginable, the consequences of the explosion were actually fairly limited, particularly, as Monbiot points out, in comparison to the likely death toll from unchecked climate change (Monbiot and Goodall 2011). Given the significant levels of radiation discovered months after the disaster as far away as Tokyo, this confidence in its limited effects may be rather premature. However, the point is not Lynas and Monbiot's support for nuclear power per se but what it shows about their views of the problems of the green movement.

Lynas and Monbiot are clearly coming from different places politically, and both differ again from the position of the Breakthrough Institute, but they are united by what seems to be an appeal to pragmatism. In different ways, all these arguments present the idea that one of the major problems with the green movement is that it has become too ideological, and that this ideological nature is preventing it from simply embracing what might work to deal with climate change. This is not restricted to the nuclear issue: for both Lynas and the Breakthrough Institute, the root of the problem lies in the fundamentals of what they characterize as green thinking.

The central opposition for these authors is between green desires for untouched nature, secured from human influence, and those who, like them, appreciate the primary importance of human society. It is true that the idea of a pristine natural world is both impossible, resulting from our alienation from nature under capitalism, and a failure to appreciate the dialectic between human society and the natural world, shaping and being shaped by each other. Lynas and the Breakthrough Institute are not however asserting the importance of understanding this dialectic, but making the extremely undialectical argument that what matters is the irreversible status of man's domination over nature.

That this is real domination is shown by the metaphors used to describe it: for Lynas, we need to embrace our role as the conquerors of the world and realize that "the first responsibility of a conquering army is always to govern" (Lynas 2011: 12). For Nordhaus and Shellenberger, our position vis-à-vis nature is even more lofty, as they quote US environmentalist Stewart Brand, "We are as gods and might as well get used to it" (Nordhaus and Shellenberger 2007: 271). Lynas' expression of human dominance of nature in military terms is telling in the context of his argument that greens have allowed a straightforward message about climate change to become contaminated by other political agendas. Given the involvement of many greens in campaigning against imperialist wars in Afghanistan and Iraq in recent years, the characterization of humans in general as a conquering army comes across as a very definite statement of position. It is a challenge to the totality of what the green movement might be thought to stand for.

The idea that all environmentalists pursue the separation of nature from humans is of course highly debatable. Reviews of the first edition of *Break Through* pointed out that that the authors had set up a straw-man environmental movement against which to argue. Nordhaus and Shellenberger responded in the second edition with the unanswerable claim that any objections only prove them more correct:

> Perhaps the reason these environment exemplars [cited in the book] seem so alien is because modern green readers have experienced a shift in their own values, one that has occurred gradually over the past decade. It may be that we are all now, to one degree or another, post-environmentalists.
>
> *(Nordhaus and Shellenberger 2007: viii)*

Clearly, these authors do believe their characterization of the green movement as technophobic fundamentalists is a fair one, but it's also the case that framing the argument in these terms enables them to start from the implication that the green position is to be opposed to any technical innovation to deal with climate change.

The debate is thus portrayed as between on one side a green view which sees energy-saving austerity and cuts in living standards as the only way to reduce emissions, and on the other the sane, balanced thinkers who perceive that modern technology can do the job on emissions while enabling us to keep the lights on. Lynas sets out the position effectively:

> Global warming is not about overconsumption, morality, ideology or capitalism. It is largely the result of human beings generating energy by burning hydrocarbons and

> coal. It is, in other words, a technical problem, and it is therefore amenable to a largely technical solution, albeit one driven by politics.
>
> (*Lynas 2011: 66*)

It is an eloquent statement, but is only achieved by conflating green suspicion of technofixes with technology, a conflation which reveals rather neatly the ideology beneath Lynas' claims to a nonideological position.

The notion that greens in general are opposed to using technology to reduce emissions is at the very least difficult to sustain. One would be hard put to find green opposition to the concept of renewable energy, for example, or to the development and improvement of these sorts of technologies. The problem with the technofix is not the use of technology itself, as opposed to the return to the pretechnology age for which greens are supposedly nostalgic, but the attempt to use technology as an alternative to making changes in the way the current system works.

Technofixes, in other words, are technological measures designed to permit the continuance of the system which created the problem in the first place. They maintain the status quo and Lynas appears to embrace them for precisely that reason. Lynas' argument is essentially that solution to climate change can only be found within the current system, and they can therefore only be technical ones. This is presented as a nonideological position but it is of course intensely ideological: arguing that no solutions for climate change which call for system change are legitimate is either an argument that the current system is the best possible, or at the very least, that attempts to change it are doomed to failure.

Given this basic assumption that only solutions within the current system are acceptable, it is not surprising that Lynas views greens who are critical of the system as actively unhelpful and misguided. They drive away those who have a stake in the current system, who need only to be convinced that dealing with climate change does not mean consorting with 'hairy lefties.' This is perilously close to the argument put by Anthony Giddens in his *Politics of Climate Change* (1992) which was that governments and market leaders have failed to act on climate change until now simply because to do so would entail associating themselves with such 'dreadful people' as greens.

This is an important argument because Lynas and Giddens have to explain why, for all the eminent solvability of climate change within the current system it has not, in fact, been solved. The technofixes may be easily available, but evidence for their large-scale imminent deployment is sadly lacking. If this failure cannot be laid at the door of counterproductive green campaigners, the only option is to conclude that the system is not able to respond to climate change by introducing measures to combat it directly, despite persuasive arguments that preventing catastrophic, irreversible warming will be far less costly than allowing it to happen and accepting the costs of attempting to live with it.

This has been the conclusion for example of a number of climate scientists who produced the 2010 Hartwell Paper. They effectively argued that actions to address climate change for the sake of addressing climate change are impossible, and that the only way forward is to encourage changes which provide direct economic benefits, and climate benefits as a side effect.[7] The Hartwell Paper was written explicitly as a response to the 'crash of 2009': the twin effects of Climategate and Copenhagen, which for the authors meant that existing approaches to climate change were completely discredited. While Lynas presents his arguments within a more positive framework, it is difficult not to see *The God Species* in a similar light, as a product of the perceived defeat of the green movement in 2009.

George Monbiot, while writing from a more generally left-wing perspective than Lynas, seems to have a similar take on the movement, in that the perspective now is one following a significant setback. Monbiot's adoption of a broadly pronuclear power position can be seen as a study in his move towards a view of the green movement which holds that it has been damaging its own cause by not adopting a pragmatic view on what is possible now. This is because, while it is possible to trace in Monbiot's recent writing a move towards more enthusiastic support for nuclear power, it is clear that he remains far from the wholehearted nuclear enthusiast that Lynas appears to be. For Monbiot, nuclear power remains a far from ideal solution, but is the solution which is both available now, and is significantly better in climate terms than fossil fuel power stations.

That this is a capitulation to the status quo is demonstrated by the fact that Monbiot does not dispute that it would be theoretically possible to decarbonize energy generation, noting that Zero Carbon Britain and others have shown how this could be done in principle through a combination of renewable energy generation and energy efficiency (Monbiot 2009). The barriers to a large-scale shift to renewable energy are not physical but political: the additional costs of managing the grid with a significant proportion of electricity from renewable sources, the time it would take to construct the new infrastructure, and local objections to new renewable installations, particularly to wind farms and to the new power lines they would require (Monbiot 2011).

It is interesting that for Monbiot, protestors against wind farms are a political reality whose views have to be accommodated, even to the extent of abandoning the idea of a decarbonized energy system derived from renewable sources. In contrast, antinuclear protestors simply have to accept the error of their ways and stop their opposition to this best available answer to climate change. Why both groups of protestors do not have equal weight is not clear. Even if any suggestion of a preference for the antiwind farm over the antinuclear protestors were to be dismissed, an argument that antiwind farm protestors have more political weight than antinuclear ones is difficult to sustain. While wind farms are a country rather than an urban issue, this does not automatically make them more influential even under a conservative coalition government, and it is worth noting that the areas with the most wind farm campaigns are not the traditional Tory shires of South East England, but Scotland, Wales, Northern England, and Devon and Cornwall.[8] Compared with fears expressed by both Monbiot and Lynas that in the wake of Fukushima, antinuclear campaigns could derail a new generation of nuclear power stations, this does not appear to add up to a uniquely powerful antiwind lobby, so why they should be accommodated while antinuclear protestors are dismissed remains unexplained.

The other arguments in favor of nuclear as opposed to renewable energy also work only with the assumption that the system, and therefore the rules within which the different options for low-carbon energy generation are considered, cannot be changed. This is the assumption that increased generation costs could not be met with increased government subsidies, and would therefore always end up with higher bills for consumers. It similarly underlies calculations of how long the creation of a renewable infrastructure would take. Statements of the time needed for large-scale construction projects can appear to be describing immutable natural laws but, of course, they aren't; they are estimates of how long projects would take with a given level of labor and investment. This level could be changed, and could be changed more markedly if the profit requirements of private companies did not have to be taken into account. Not everything has to take several years. Even in a capitalist economy, changes can be made remarkably quickly if there is sufficient motivation, as the US economy's 'turn on a dime' when they entered World War II showed.

However, since Monbiot effectively rules out any such dramatic shifts in the current economy, the task of shifting energy generation to renewables appears an impossible one. Unlike Lynas or the Breakthrough Institute, Monbiot remains unconvinced by the possibility of responding to climate change within the current system, yet for him no solution which looks beyond the current system is possible. It is small wonder that he sees the green movement as lacking in answers, given the unanswerable conundrum this reasoning sets up, nor that his writing on energy generation from 2010 on has become steadily more pessimistic. This sense of pessimism is very much in keeping with the idea that Copenhagen and Climategate added up to a defeat for the movement. Indeed, Monbiot wrote in June 2012 that following the Earth20 summit in Rio, there was now no hope of achieving any international agreement on emissions reductions and any movement to try to achieve them was hopeless (Monbiot 2012). It is not made explicit in this article, but it seems clear that this conclusion about the future for the green movement is related to the assumption that a practical response to climate change means not looking beyond the status quo. If the climate movement's death knell was sounded by Rio, the process was started by Copenhagen. Therefore it is worth examining why Copenhagen is now assumed to have been such a defeat for the climate movement.

3 Copenhagen and the crisis in the movement

That the Copenhagen summit ended without the agreement of a follow-up deal to the Kyoto Protocol was of course a serious setback to hopes that governments could get together and agree a rational response to climate change. This is not the same however as saying that it was also necessarily a serious setback for the climate change movement. The immediate run-up to Copenhagen was presented by the media, with their short news cycles and attention spans, as the crucial period for putting pressure on world leaders, but by that time it was realistically almost a year too late. The crucial time was actually the climate summit of 2008, which could, but failed, to prepare for a genuine agreement at Copenhagen. The media attention on climate change during the Copenhagen summit made it a perfect time for climate change campaign groups to air their message to a wider than usual audience, but this is unlikely to have been in the widespread belief that the chances of a meaningful deal were very high.

Concluding that the climate movement was defeated because the Copenhagen summit did not result in a deal is rather like the argument that the Stop the War coalition failed in 2003 because it didn't prevent the war in Iraq: an argument leveled at it by opponents which completely ignores how the activities of the antiwar movement shifted general opinion against the wars in both Iraq and Afghanistan and eventually also led to the resignation of Tony Blair. However, it is true that there appeared to be considerable opportunities arising from the demonstrations around the world and in Copenhagen, along with the Copenhagen alternative summit, for creating a wider, more united movement against climate change, which have largely failed to come to fruition.

In part, capitalizing on the links made by activists at Copenhagen would have required both energy and optimism, and it's possible to speculate that the very discourse of failure in the media coverage of the summit played its part in the sense that campaigns against climate change had suffered a serious setback. There is a sense in which this perception of failure could have become self-fulfilling: the green movement was defeated at Copenhagen because after the fact it decided that it had been.

It is important not to underestimate this retrospective sense of failure, but it is equally important not to take it as the whole story of Copenhagen and its aftermath. This would, after all, be to see the movement as divorced from its objective circumstances. Understanding an event or a period as a success or failure can indeed help to make it so, but the progress or otherwise of a movement does not happen solely in the heads of its participants. Aside from the discourse of failure, Copenhagen did objectively herald a change for the movement. Even if groups had been able to grasp the opportunities for working together, it would have had to have been in a very different way from campaigning in the lead-up to the summit.

The demonstrations around the annual climate summits have tended to work as protests calling for government action to reduce greenhouse gas emissions, with the specific slogans comparatively muted by this general message. That they were focused around the summits went some way to creating this effect: since the question for the summits has been whether or not an international agreement on emissions targets could be agreed, it was difficult to see the demonstrations as not pushing for such targets, whatever the placards said. It was also partly a holdover from the early days of the movement, before claims to take climate change seriously were a part of every political party's manifesto.

It has now been some years since we got their attention, but although the issues around climate change have moved on, the idea of calling for government action, pretty much any government action, on climate change, has remained in some ways a meeting point for the movement. It has been the point at which those who believe in market solutions, lifestyle changes, or revolution have come together to demonstrate for action, with the precise action in the small print.

The idea that governments could be expected to take some action which would, however partially and slowly, go some way to reducing the likelihood of catastrophic climate change was not on the face of it an unreasonable one. The effects of climate change, if unchecked, would after all prove costly for capitalism to live with, while the development of new, green technologies could open up a whole new area for potential profits. That the Kyoto Protocol will go down in history not as a weak and limited first step but as the high point of intergovernmental efforts on climate change could be seen as a testament to the irrationality of capitalism, that it is unable to promote sensible action even when the long-term interests of capitalists are at stake. The less charitable view would point out that there is no such thing as an unprofitable disaster. Either way, without the hope of concerted government action within the current system, the green movement is left between two poles – system change or the profitable technofix – without the point at which it was able to cohere in the middle. It is easy to see why this might be disorientating for the movement, but more than that, it could call into question the need for a mass movement specifically and solely on climate change at all.

4 A future for campaigning against climate change

The strategy of building a mass movement on climate change contains the assumption that climate change can usefully be seen as a discrete problem for which there are solutions available within the current system, regardless of how difficult a political task it would be to compel the government to adopt them. In the wake of Copenhagen, however, this view of climate change no longer seems realistic. The governmental failure to agree even minor reductions in emissions emphasizes that it is not a separate problem but an issue arising from the workings of the system itself, and so not amenable to amelioration by statute. If we therefore accept that the action we need to fight for is to change the capitalist system, rather than for emissions reductions within it, then maintaining campaigning against climate change as a separate activity appears as a tactic which can be questioned.

This is not to argue that climate change campaigners should simply widen their remit. This would be unlikely to be a successful strategy, and in practice would be difficult to do. It has been suggested that part of the reason for the Camp for Climate Action's decision to release its participants for other projects was the difficultly in moving from an understanding of themselves as activists against climate change specifically to more general anticapitalism within the Climate

Camp framework. Climate Camp could not itself take on the entire system, but there seems to have been an increasing sense that concentrating only on the climate change aspect of the wider problem was not enough.[9]

Climate change campaigns may not be able to bring down the system on their own, but what we can do is place ourselves at the center of the movements which are taking on capitalism at the sharp end – campaigns against austerity, against cuts, against unemployment, against the war. If we understand climate change as not a technical problem but an integral part of capitalism then joining these campaigns *is* fighting for climate action, just as much as marching on a climate change demo and it is concentrating the fight where we have the greatest chance of building a genuinely mass movement.

Where climate change campaigners have been able to make the connections between environmental and austerity issues, as for example in the climate jobs campaigns, they have been markedly successful. The One Million Climate Jobs campaign set up by the Campaign against Climate Change trade union groups, has gained support from a number of trade unions and was able to put on a Climate Jobs Caravan, taking the message that creating a green infrastructure could answer both austerity and climate change, around the country in June 2012.[10] There is clearly considerable potential for integrating this campaigning with the movement against cuts and austerity. Indeed there was a climate jobs contingent on the Trades Union Congress' march on 20 October 2012, which saw 300,000 people on the streets of London to protest against the government's cuts agenda.

Lynas and Monbiot in many ways are correct: under the current system, the only response to climate change is to develop and use commercially profitable technofixes, and if you don't think that sounds like a good idea, then there are no answers at all. But the current system is not immutable, and if we don't see privatized sky mirrors as the answer for global warming, the movement against the Conservative–Liberal Democratic coalition government's cuts may actually be the best place to continue the fight.

The period after the demonstrations around the Copenhagen climate summit in late 2009 has been in many ways a difficult one for the British climate movement. The falling off in attendance at the annual climate demonstrations, in the face of increasing evidence of government indifference if not hostility to green concerns, and the winding up of groups such as the Camp for Climate Action, can seem to be evidence of a downturn in the movement. At the same time, green campaigners have come in for sustained criticism from figures who are often regarded as leading authorities within the movement. It is not necessary to conclude from this, however, that there is no future for campaigning for action on climate change in Britain. The work done by some groups around green jobs, and links with antiausterity campaigning, points to a different way forward for the movement beyond concerns about a post-Copenhagen slump. As the anti-austerity movement develops, it remains to be seen how green groups will respond to the possibilities it represents for climate campaigning, but the initial signs are good.

Notes

1 Note that in this discussion I use the green, environmental, and climate change campaigns synonymously. The term 'green' is a useful catchall to signify both environmental and climate change campaigns, and it is reasonable in the British context to see different campaigns as part of one green movement. It is difficult to argue for the existence of an environmental movement separate from climate change campaigning; because, for example, environmental groups such as the Royal Society for the Protection of Birds and the World Wildlife Fund are also members of the Stop Climate Chaos campaign. There are of course campaigns which oppose renewable energy projects like wind turbines or tidal electricity generation on the grounds of environmental harm, but given the disagreements

discussed below on energy within the climate change movement, it would seem overly pedantic to insist on a formal separation between the environmental and the climate change movement on these grounds.

2 The achievement of Jenny Jones may however say as much about Londoners' view of the Liberal Democrats' participation in the coalition government as it does about the Greens.

3 See www.theccc.org.uk/reports/3rd-progress-report, accessed 11/12/2012.

4 See for example the announcement by Scottish Power in October 2011 that the plan to convert their coal-fired power station at Longannet in Fife to carbon capture and storage had been scrapped.

5 This tendency is not restricted to the UK: the snowstorm which hit the eastern US in October 2011 was similarly treated by climate change skeptics as evidence that global warming theories were flawed. See www.bishop-hill.net/blog/2011/10/31/snow-in-new-england.html, accessed 11/12/2012.

6 See the range of articles on nuclear power reposted on www.monbiot.com, accessed 12/15/2012.

7 See http://eprints.lse.ac.uk/27939/, accessed 11/10/2012.

8 This observation is based on a list of 271 local wind farm campaigns. See www.countryguardian.net/Campaign%20Windfarm%20Action%20Groups.htm, accessed 11/14/2011.

9 My thoughts on the wind-up of Climate Camp owe much to an excellent manuscript from Sophie Lewis (Lewis 2011). However, my conclusions and any misapprehensions are my own. Also interesting on this point are comments made by some of the interviewees in the 2011 film *Just Do It* on their thoughts on the climate change movement post Copenhagen. See http://justdoitfilm.com/, accessed 11/10/2012.

10 See www.climate-change-jobs.org, accessed 11/10/2012.

References

Giddens, A. (2009) *The Politics of Climate Change*, Cambridge: Weley & Sons.

Lewis, S. (2011) The Rise and Fall (and Rise) of the Camp For Climate Action UK: Notes on Interventions in Challenging Carbon Democracy, unpublished manuscript, Oxford University.

Lynas, M. (2011) *The God Species. How the Planet Can Survive the Age of Humans*, London: Fourth Estate.

Monbiot, G. (2009) Nuked by Friend and Foe. Accessed 10/15/2012 from www.monbiot.com/2009/02/20/nuked-by-friend-and-foe.

Monbiot, G. (2011) The Lost World. Accessed 10/15/2012 from www.monbiot.com/2011/05/02/the-lost-world.

Monbiot, G. (2012) End of an Era. Accessed 10/15/2012 from www.monbiot.com/2012/06/25/end-of-an-era.

Monbiot, G. and Goodall, C. (2011) The Moral Case for Nuclear Power. Accessed 10/15/2012 from www.monbiot.com/2011/08/08/the-moral-case-for-nuclear-power/.

Nordhaus, T. and Shellenberger, M. (2007) *Break Through. Why We Can't Leave Saving The Planet to Environmentalists*, second edn., New York: Mariner Books.

Rüdig, W. and Lowe, Philip D. (1986) The Withered 'Greening' of British Politics: A Study of the Ecology Party, *Political Studies*, XXXIV: 262–284.

Wall, D. (1994) *Weaving a Bower Against Endless Night. An Illustrated History of the Green Party*, London: Green Party.

8
THE CLIMATE MOVEMENT IN GERMANY[1]

Georg Kössler

1 To be or not to be as a central question

This would be a sad text if it was starting with the message that a climate movement in Germany does not exist. Nevertheless, activists and scientists are permanently discussing this question. This chapter holds the founded assumption that there is a climate movement in Germany. The character, activities, problems, and challenges of the movement shall be described. Furthermore, the different approaches of acting existing in the movement are analyzed. The chapter finally calls on the movement to become more aware of itself.

A short look at scientific definitions of social movements is sufficient to justify the classification of climate activism in Germany as movement. In literature, social movements are described as collective actors (Raschke 1985: 76) which consist of persons, groups, and organizations related in networks (Rucht and Neidhardt 2007: 634) and engaged in (symbolical) interaction (Raschke 1985: 78). The members of a movement are characterized by a collective identity (Hellmann 1998: 19–20) and their participation is of certain continuity (Raschke 1985: 77–78). As regards the classification of the German case it is also useful to look at established movements such as the antiwar movement, the women's movement, or the antinuclear movement. These examples show that movements do not have to be coherent and permanently active collectives, although these are popular requirements for movements. People from the German left, who some years ago started to engage in climate politics, have contributed to the foundation of a phenomenon that indeed can be termed as climate movement.

Actors of civil society often seem to mix up doubts about the existence of a climate movement with doubts about the effects and successes of the movement. Besides, young activists often do not have experience with the foundation of movements. These processes can be accompanied by difficulties and conflicts. Scientists seem to be skeptical about the existence of the climate movement because their empirical indicators are not able to measure it. As a consequence, different perspectives on the climate movement and definitions of it are existing. According to Descartes' "Cogito ergo sum," this chapter will not question the existence of a climate movement in Germany anymore. In contrast, it will describe and analyze it.[2]

Nevertheless, it is important to mention that different groups use the label climate movement in different contexts. In parts the term is used outside the classical social-emancipatory context. For example, Daniel Boese describes the climate movement as a huge societal phenomenon. For him, the climate groups are the youth movement of the twenty-first century and stand in the tradition of 'movements' such as Rock'n'Roll, hippies, or punks (Boese 2012). It is questionable whether they will succeed in their mission[3] or – as their historical predecessors – will become part of the mainstream of society. From these considerations it is just a small step to the eco-warriors which are portrayed by another author.[4] A problem of these perspectives describing the climate movement as "green, clever and sexy" (advertising slogan for Boese's book) or as young and ecological oriented metropolitans is that they underestimate or even ignore members of the climate movement holding more critical positions. These perspectives on the climate movement are in parts too narrow and sometimes also too broad. For example it is problematic that for Boese the activities of the movement are ranging from demo-tourism to the damage of property:

> Climate activists block airports, hijack trains transporting coal and shut down power stations. They are present on the streets of 193 countries. They install solar panels, wind miles and plant crops in the cities. They save human kind from climate change. They plant millions of trees and have made the climate movement an international political force.
>
> *(Boese 2012)*

At first glance, the climate alliance (Klima-Allianz), a network of NGOs, church groups, and others, could be a central actor for the coordination of the climate movement in Germany. For example, its anticoal campaign was quite popular. Nevertheless, the climate alliance was not able to realize its potential. Today it is an elite project with limited scope and impact. This development is based on several reasons: First, members of the network got into conflict with each other. Second, the network does not act like a social movement because it focuses on consulting politics instead of engaging in demonstrations and civil disobedience. Several members of the climate alliance aim at cooperation with politics and economic actors and reject criticizing the current system for not being effective. In contrast, many other climate groups are convinced that established forms of political activism are not sufficient to achieve radical change and therefore oppose current politics (McCarthy and Mayer 2006: 537). Reacting on this cooperative approach, which is represented by the climate alliance and others, Marxist groups started to speak of a climate justice movement in order to distinguish themselves from the cooperative groups. Besides, they oppose the dominant interpretation of climate change sent out by the established NGOs.[5] The emancipatory[6] and capitalism critical left, for which a solution of the climate crisis within the existing system is not possible, is an decisive and activating part of the climate movement in Germany. Nevertheless, this chapter uses the term climate movement and not climate justice movement to name the object of investigation.

2 Development and current state of the climate movement in Germany

The appearance of the climate movement in Germany resulted from the convergence of two political groupings, which felt that they were not represented by mainstream organizations dealing with climate change such as Germanwatch or the green party. At the end of the first decade of the twenty-first century, civil society actors moved to the terrain of climate and energy politics (Bedall et al. 2011). Activists from Attac, Antifa, the antinuclear, and the antiwar movement were attracted by climate politics. One reason for this was the intensification of the discourse about climate change due to the Stern report from 2006 and the IPCC report from 2007. The engagement of these groups in climate politics was characterized by their radical and

Marxist political background. A plural mixture of autonomous people, members of the movement of 1968, and young and ecologically motivated activists came together. The latter ones were passionately engaged for the planet, but less critical in their political attitudes. Bricke describes the merging of these two different groupings into the German climate movement (Bricke 2008). From a skeptical point of view, activists from the youth organizations of environmental NGOs and the green party were allowed to listen to older activists and finally could discover their own radicalism. This process led to a spread of critical thoughts such as the theory of societal relations with nature. For many left activists, this was an important secondary effect of the appearance and development of the climate movement.

Despite their aim of climate justice, the German climate movement rather creates the impression of being engaged against something than for something. A popular critique of the movement tackles the neoliberal character of the discourse on climate politics. The publication '20 Theses Against Green Capitalism' by Tadzio Müller and Alex Passadakis was one of the central documents of the debates on climate justice before the UN Climate Change Conference in Copenhagen in 2009. The text dealt with members of the environmental movement, which were frustrated about the inability of politics (in particular of the green party) to cope with climate change. It outlined a new field of engagement for the global justice movement and its critique of capitalism. These critical approaches of the German climate groups have similarities to the international debates of the movement. Some years ago, in 2004, the Durban Declaration against Carbon Trading criticized radically the flexible mechanisms of the Kyoto Protocol. Other documents and contributions followed.[7]

In addition to these theoretical relations to the international level, activists from Germany participated in the climate camps in Great Britain (see chapter by Graham-Leigh in this book). Due to their successes in motivating the participants, these protest events were transferred to the German context. For example, the newspaper *taz* from Berlin described Tadzio Müller as the "importer of the climate camps" (Schumacher 2008). But of course it was not only him who brought the camps to Germany.

Climate Camps are meetings oriented at the grassroots level of the climate movement. They intend to create a platform for learning about a world respecting the principles of climate justice (theory). Furthermore, they aim at strengthening the climate movement (strategy) and are used for the organization and realization of direct actions against huge polluters (practice) (Bedall et al. 2011: 42). These three aims created high expectations and put pressure on the climate camps. For a long time, the camps were not able to lower this pressure. Bedall et al. (2011: 42) observed the following processes during the appearance of the climate movement in Britain in 2005: The euphoria about successful protest actions was accompanied by the frustration that the government tried to generate new legitimacy from their engagement in climate protection. As a consequence, the publications of the group of 8 (G8) on climate change were intensively criticized by members of civil society. These attempts to abuse climate change for political purposes sensitized the German left for the topic of climate change (Bedall et al. 2011: 41). This sensitization is still working today. The successful blockade of the G8 conference in Heiligendamm in 2007[8] contributed to the rising popularity of the topic among left groups. This popularity made it easier for the radical climate groups to express their differences with classical environmental and climate NGOs.

Neither for the public, nor for climate politics was the UN Climate Change Conference in Bali in 2007 of huge importance. The environmental minister of Germany wore a colored summer shirt and the representatives of the US held neutral positions in the negotiations. More important for the climate movement was the splitting of the climate groups which happened during the conference. Several NGOs left the Climate Action Network (CAN) and founded the network Climate Justice Now! (CJN!). In particular climate groups from the Global South were engaged in the foundation of the new network. In the following years, many groups joined CJN!, but the network never reached the size of CAN. One reason for the splitting was that the climate conferences of the UN were mostly dominated by mainstream NGOs supporting the negotiations. This resulted in internal conflicts of the climate movement, because radical NGOs rejected this cooperative approach (for the different approaches in the climate movement see the chapter by Görg and Bedall in this book). In Germany, no splitting of the climate movement took place, although there are some practically and cooperatively oriented actors. For these groups, effective climate protection and the economic system do not stand in contradiction to each other. Or at least they do not articulate their doubts about a possible contradiction. Mainstream NGOs like Oxfam are part of the climate movement in Germany. As a consequence, radical and moderate climate groups are in contact and communicate with each other. The German environmental NGO BUND can be characterized as a rather radical actor. Nevertheless, BUND is a member of CAN as well as of CJN!. The organization acts as a kind of moderator between the two wings of the climate movement. On the whole, the fragmentation of the German movement is less intensive than the fragmentation of the international climate movement. Furthermore, the media does not report about this topic.

In Summer 2008, two events, which were of importance for the climate movement in Germany, happened in Hamburg. First, a conservative-green government was established under mayor Ole van Beust. In the same year, this government authorized the construction of a coal plant in Hamburg-Moorburg. This decision stood in contradiction to the election promises of the green party. For many activists of the climate and environmental movement, the permission for the construction of the power plant in Moorburg led to the conclusion that the Green Party has moved to the right and should be treated as a political enemy. During the discussions about the construction permission for the power plant in Moorburg, the first German climate camp took place in Hamburg (August 15th to August 24th). After long debates regarding the planning of the event, finally a combined camp of climate activists and antiracism activists was organized. The latter ones were integrated due to their advanced mobilization and organization skills. The preparation of the camp was a difficult process, because the organizers got into conflict, but later cooperated again. Different groups joined the camp: Youth organizations from the left and green party participated as well as activists from Avanti and Attac. Public attention was raised by the attempt to occupy the construction side of the power plant in Moorburg. Climate activists judged the camp in an ambivalent way: "Many media attendance was raised, but only few people from outside of the movement were activated" (Bricke and Müller 2011).

While the year 2008 did not offer much potential for mobilization – for example because of the unspectacular climate summit in Poland – 2009 was very important for the climate movement. It was the peak of the activity and public presence of the climate movement so far. In April 2009, the weblog of the climate movement network (Klima!Bewegungsnetzwerk, K!BN) was launched. The first published message was the following one:

> Welcome to the blog of the climate movement network! We are a collective stemming from the preparation of the climate camp in Hamburg in 2008. Currently, we are engaged in the preparation of the COP15 conference in Copenhagen in December 2009. Besides, we are active in local energy battles. More information will be published soon!

Although the opening text of the blog refers to local conflicts, it was written in a time when many climate groups were intensively oriented towards protests at the climate summits. In 2012 the blog was still active, but mostly contained appeals for protest events. It became a kind of event calendar. Debates sometimes take place through the mailing list of the blog, but mostly face to face communication is preferred.

A meeting of climate groups in March 2009 for the preparation of the Climate Summit in Copenhagen can be described as an important step in the development of the international climate movement, if not as the birth of the movement. A special characteristic of this meeting was that many different and highly motivated actors took part. Among others, many groups from Germany (environmental groups, autonomous groups, youth organizations of parties) joined the meeting. Academic members of the climate justice movement such as Nicola Bullard or Patrick Bond participated as well as the US organization 350.org, which was founded just a short time before. During the meeting conflicts over the forms and messages of the planned protests were raised by members of the emancipatory left. They recognized parallels between the UN Conferences and the meetings of the G8 and WTO. Since then such conflicts appeared at almost every meeting of the climate movement. A certain left habit is expressed in these conflicts over forms, intensity, and content of planned protests and publications. A result of the described meeting for the preparation of the Climate Change Conference in Copenhagen was the foundation of the network Climate Justice Action (CJA). It defined itself as a radical grassroots version of Climate Justice Now!. Despite this radical character, only a short time after the foundation, members of CJA split off and founded the collective Never Trust a COP (NTAC). Most of the members of NTAC were German groups. The international mobilization for the climate summit in Copenhagen helped the German climate movement to define its identity and to develop internal structures. A central question of these processes of self-discovery dealt with the legitimacy of inside activism (participation in the negotiations, lobbying) versus outside activism (demonstrations and other protest events outside of the conference area). Important also was the question how both approaches could be connected. While NTAC rejected all forms of inside activism, CJA suggested the creation of a common space at the fence of the conference center, where progressive activists from the inside of the negotiations could meet with activists engaged outside the negotiations.

In this period, the public attention for the climate groups was rather weak, despite some actors trying to create publicity. The media concentrated on single eloquent activists. This made it difficult for the movement to present and organize itself without hierarchies:

> The personalisation of the reports on the movement and its actions as well as the resulting conflicts can be exemplified by the public relations work of the Climate Justice Action Network and the reports about its activity (Anon. 2010: 10f.). This example shows the necessity of a conscious dealing with the mechanisms of mass media and the threat of internalising their hierarchical interpretation of the movement's structure.
>
> *(Korte 2011: 205)*

At the McPlanet conference in Berlin in April 2009, in which different left groups took part, an appeal to disturb the climate summit massively was spread. Established NGOs felt threatened by the appeal. The mainstream environmental news service Wir-Klimaretter.de assumed that paid lobbyists were responsible for it. At that time, radical critique of the system or at least doubts regarding the ecological engagement of economic actors existed only rarely in the climate movement and therefore provoked skeptical reactions. Nevertheless, the failure of Copenhagen proved that these radical approaches were more realistic than the moderate approaches of mainstream NGOs. They prevented the movement from being publicly judged as 'slobs' with unrealistic positions. The interested public showed sympathy for the radical rejection of the climate negotiations. Despite this late 'success' of prognosticating correctly the results of the summit, the climate movement was not able to use it as a catalyst for its mobilizations. While the following climate summits in Cancun (2010) and Durban (2011) in large parts were ignored by the movement, there was some interest for the alternative climate summit in Cochabamba in 2010 (see chapter by Kruse in this book).

The period after the Copenhagen summit was very difficult for the climate movement. On optimism "as plentiful as carbon emissions in the US" (Lawton 2009) a deep internal crisis followed. Reasons for this crisis were the failure of the summit, the prevention of racial protest actions by the repressive intervention of the Danish police, as well as the sudden decline of public attention and activists after the end of the summit. The loss of activists after Copenhagen shows that a part of the climate movement is strongly related to the political and societal mainstream. Besides, there was a criminalization of climate activists. The popular German activist Tadzio Müller was arrested at the beginning of the summit. Mobile phones and laptops were spied on by the police without judicial permission. These repressions have effects up until today. At some meetings of the German climate movement, mobile phones have to be turned off. As another consequence of the police activity in Copenhagen, the climate movement discovered how radical it really is. Perhaps this experience threatened some activists and made them leave the movement. More radical activists were disappointed with the implementation of the planned protest events in Copenhagen. An interesting conclusion regarding the impact of critical actors in Copenhagen can be drawn from these observations: Friends of the Earth supported a counterhegemonic discourse against mainstream NGOs and their inside activism more effectively than NTAC whose activists were arrested precociously. Friends of the Earth organized a blue climate wave across Copenhagen. It criticized the climate negotiations by using classical protest instruments such as demonstrations and party. "The space for discussion on the ways of solving the climate crisis has become more open and plural due to the failure of Copenhagen. Also solutions outside of the hegemonic order are now considered" (Korte 2011: 205). After Copenhagen the importance of the network CJA declined, while the German environmental organization BUND intensified its cooperation with climate NGOs.

According to Bedall et al. (2011), three important developments emerged during the preparation of the climate summit in Copenhagen in 2009: First, there were changes in the protest behavior of the climate NGOs. Second, the climate camps became more popular. Third, actors from the global justice movement discovered the issue of climate change. This constellation provokes the question whether Copenhagen was the coming out or just a limited rise of the climate movement. Shortly after Copenhagen, there was some optimism regarding this question. Chris Methmann from Attac said: "On the streets of Copenhagen a new movement was born, but it still has to learn how to walk" (Götze 2010). Others observed alliances between the climate movement and environmental NGOs and demanded that these should be continued. Regarding the limited scope of the protest activity in Copenhagen and the weak resistance against the arrests of activists, Mona Bricke and Tadzio Müller (2011) concluded that "it seems not to have been a coming out."

It is not possible to measure the number of German activists who traveled to Copenhagen. A huge part of the 100,000 people demonstrating in Copenhagen on December 16th were Germans. Nevertheless, just some hundred activists participated in the radical protest events on December 18th. Many grassroots groups focusing on climate change were founded in 2009 in Berlin, Hamburg, or the Rhineland. This led to a relevant growth of the German climate movement. In 2008 only some groups and activists joined the protests against the construction of a new coal-fired power plant in Hamburg-Moorburg. In 2010 the new capacities of the movement declined. Activists had burnouts, climate groups stopped to hold meetings, and the debates of the movement were full of doubts. There was only some hope due to the new sensitivity and tolerance for radical critique. This period was characterized by intensive discussions about reconfiguring the strategy of the climate movement: "Due to its limited resources, the movement should reduce international demo-hopping and concentrate on the local level" (Kössler 2010).

Gegenstromberlin (gsb), an organization founded in the aftermath of the Climate Camp 2008, and other groups organized some small protest events in 2010. For example, gsb protested during the marathon in Berlin, which was sponsored by the energy company Vattenfall. Gsb also interrupted the Vattenfall Contemporary Art Show[9] and demonstrated against the production of energy from coal. By organizing several meetings of climate groups, KBN! contributed to a change of the strategy of the German Climate Movement. At these meetings, the climate groups decided to focus on local actions such as protests against planned coal plants. The idea of NTAC – to orient to the basis of climate change by tackling the port of Copenhagen – was picked up and decentralized. Despite this shift in the strategy of the movement, there were still voices of climate activists demanding an orientation towards the concept of climate justice. For example, the group Climate Justice Leipzig (Klimagerechtigkeit Leipzig) organized a demonstration with 200 participants in front of the emission trading market in Leipzig.[10] Nevertheless, the second half of 2010 was not characterized by this strategy shift, but by another issue. Many climate activists took part in the protests against the transport of nuclear waste across Germany. They joined the antinuclear movement because in Fall 2010 the German government cancelled the decision to phase out nuclear energy. Due to the high popularity and mobilization potential of this topic, the climate movement got engaged in the antinuclear protests. For example, many climate activists participated in the campaign 'Castor? Schottern!' Nevertheless, this cooperation with the anti-nuclear movement reduced the capacities and resources of the climate movement for months. Furthermore, the nuclear protests achieved high popularity and a broad participation of the population. Climate activists were interested in transferring this dynamic to the issue of global warming, but the opposite happened. The broad protests against nuclear energy created the impression that the climate movement was comparatively small and weak. Not only did more people participate in the antinuclear protests, but also the cooperation among civil initiatives, members of the movement, and left actors worked better than in the climate movement.

As a consequence of these two post-Copenhagen processes – orientation change towards decentralized activity and the cooperation with the antinuclear movement – a special conference was organized in February 2011 in Berlin. Climate groups and civil initiatives engaged against the extraction of coal and the storage of carbon below the ground met with environmental NGOs (for example BUND) and youth organizations of the green and left party. The organizers of the conference explicitly invited representatives of the antinuclear movement. Their intention was to learn about the cooperation experiences of the antinuclear movement. The participants of the conference found out that their opponents were the same: Energy companies, nuclear, and coal plants as well as the extraction of coal and the storage of carbon below the ground. The strategies and slogans used by the different actors engaged against energy companies are still quite different. Some call their activities 'energy struggles' (Energiekämpfe), others 'energy autonomy.' Nevertheless, the protests against the transport of nuclear waste in Germany have proven that residents, hippies, autonomous and ecological activists can work together. At the conference in Berlin, these groups decided to organize a common climate camp. The idea was to merge the aims of the diverse actors. These aims consist of radical action, intensification of political pressure, and enhancement of practical solidarity. In 2011 this special climate camp was realized in Brandenburg. It was called Lausitzcamp and about 250 people participated. Simultaneously the youth organization of the environmental NGO BUND organized a climate camp in a traditional coal region in the Rhineland. In 2012 there was the attempt to coordinate the different climate camps in Germany. But this did not really work. For example, a common website[11] of the camps was launched just some weeks before the start of the camps. Despite missing coordination, the camps organized several protest events. During the climate camps in the Rhineland in 2011 and 2012, a train route was blocked in order to prevent the transport of coal from the sites of extraction to the coal plants of the region. In Brandenburg participants of a climate camp protested during a speech of the head of the local government.[12] Besides they demonstrated in front of a coal plant in Jänschwalde. In Brandenburg a continuous football match between a local team and the 'FC Climate Camp' was established. The energy company Vattenfall felt bothered by these matches and tried to stop them.[13] In contrast, due to broad media coverage and the positive mobilization effects on the local population, the majority of the climate activists were satisfied with the climate camps.

The shift of the orientation of the climate movement after Copenhagen towards local activity (see Methmann 2011) was shared and implemented by most of the grassroots groups in 2010 and 2011. In the debates of the movement, the relevance of local struggles against projects of the fossil fuel industry were framed as a part of larger "energy battles" (Müller 2011). The idea behind this concept is to bring together all activists who are engaged in energy politics. A powerful and solidly united movement should be created with the aim to build up an energy democracy. To achieve this aim strategic opponents should be tackled together. Despite this concept, the temporary alliance between the climate movement and the antinuclear movement had, as described, no positive effects on the former one. The popularity of the nuclear topic led to a loss of attention for climate change. One reason for this could be that the issues and movements were too diverse or that it was not possible to show the relations among them.

Despite these problems after Copenhagen, several German climate groups kept active. In cooperation with a critical collective of artists, gegenstromberlin organized protest events in Berlin, Leipzig, and Cancun.[14] The climate section of the alliance for a left movement (Bündnis für eine linke Strömung, FeIS) started to participate in the round table for energy in Berlin. This round table consists of groups such as BUND Berlin, NOYA, and Powershift and aims at buying the local energy system and bringing it under public control. This is a quite concrete and practical demand for the German climate movement. Furthermore, the diverse actors participating at the round table make it necessary to use careful and sensitive procedures. Even gegenstromberlin, which for a long time favored the (theoretically based and ambitious) concept of climate justice, started to support the round table in late 2011. At this time another climate camp in Brandenburg was organized. Its participants tried to block a local coal plant and occupied the office owned by a party of the local government. In 2012 some of the groups taking part in the round table in Berlin demanded protest actions instead of just collecting signatures. In that year there was again a climate camp in the Rhineland. Its participants occupied a forest in order to protect it from destruction by the energy company RWE, which planned to extract coal from this area. Besides, in 2012 a network of the climate camps in Brandenburg, the Rhineland, Hohenmölsen, Attaching, and Zürich was founded. Due to these activities, one can conclude that the regions Brandenburg/Berlin as well as the Rhineland have become important cores of the German climate movement. Nevertheless, also climate groups from South Germany organized successful activities in 2012. A popular example is the prevention of the construction of a third runway of the airport in Munich by an alliance of civil initiatives and the British climate group Plane Stupid.

3 Aims and programmatic ideas of the German climate movement

The central demands of the climate movement in Germany are:

- stop energy production from fossil fuels;
- change of consumption behavior;
- global redistribution; and
- a rights-based policy approach.

One reason for the limited influence of the concept of climate justice on the German climate movement is the low visibility of climate change in Germany. There are no population groups which are affected by global warming and therefore articulate demands for climate justice. While liberalizing policies in the last decades resulted in unemployment and provoked huge demonstrations, there are no victims of carbon emissions in mid-European spheres. People from the German middle class are engaged in organizations such as NABU or BUND and express their wish for enhanced climate protection in moderate and concrete ways. More radical climate groups are discussing a lot and risk getting lost in their own debates. It was a huge success to unite these diverse groups and people under the common banner of the climate movement. Since the beginning of the movement, there were always conflicts about the adequate proportion of protest action and theoretical or ideological background. While some actors criticized the weak theoretical basis of the movement, others rejected an ideological framing of their activities. For example, some climate groups presented anticoal protests as an attack on capitalism. Existing hopes to use the climate question as a catalyzer for a world revolution (Müller 2011) were disappointed because they ended in long debates on concepts such as energy democracy and energy sovereignty.[15]

In 2009 the agenda of the German climate movement was dominated by emission trading, CCS, and agro fuels. The debates of the movement dealt with the thesis that the politicians participating in the climate negotiations in Copenhagen rather were a part of the problem of climate change than a part of the solution of the problem (Passadakis 2009). According to Bedall et al. (2011), the aim of the climate movement is the creation of a project which stands in antagonistic opposition against the hegemony of neoliberal climate politics. Groups such as gegenstromberlin were founded with the explicit aim to raise the importance of the environmental and climate crisis in the discourses of the political left. Furthermore, the aim is to confront the mainstream interpretation of climate change as a natural problem solvable by technology with an interpretation of climate change considering the social dimensions of the problem and its solution. For many members of the climate movement, mainstream NGOs such as Germanwatch and WWF have lost their sensitivity for questions of international justice and solidarity due to their involvement in the UN negotiations. In the eyes of their critics, they have either dropped these aims completely, or they subordinated them under progresses in the climate negotiations. In contrast, the radical group FeIS aims at identifying opportunities for intervention in the mainstream discourse on climate change and wants to create a vision of an ecological and solidly united society. By doing this, they also focus on the deficits of the left regarding these points.[16] Until today one of the main tasks of the climate movement is the unification of the diverse groups engaged in the movement. This can be exemplified by the slogan of the Blockupy protests in May 2012 in Frankfurt/Main: "It's not about polar bears – unite social and ecological battles."[17]

The anticapitalistic left in Germany interpreted climate change as a new battlefield against the neoliberal discourse. Due to the rising popularity of the issue in Spring 2007 resulting from the

IPCC report, many activists hoped to have discovered a weak point of the neoliberal project: "[A]s much there's a movement without a story, there's also a story without a movement: Climate change" (Müller 2008: 246). Nevertheless, an intensified strategical and ideological framing of protest activity seems to correlate with a reduced intensity of these protests (indicators from Dietz 2010: 13). One activist characterized the German climate movement in the following way: "The Germans, when they see climate change, they just want to throw copies of Adorno and Foucault at it and hope that it goes away" (Interview D, in Bricke 2008).

Apart from these debates, the members of the climate movement discuss several economic questions. In 2009 they criticized the idea of green capitalism and a green new deal. After that they discussed the question of economic growth which led to new alliances and conflicts in the movement. In the context of the Rio+20 summit in 2012 many activists argued against the idea of a green economy. Nevertheless, only some members of the climate movement participated in this neoliberal discourse on green economy because the majority of the movement was occupied with local protests and campaigns. Activists justified their preference for local actions instead of huge discourses with the argument that "[i]t's not useful, if you always want to storm the winter palace" (Bricke and Müller 2011).

While the mainstream discourse on climate politics tried to integrate confrontative positions by developing win-win scenarios,[18] radical members of the climate movement are in a state of doubled confrontation. First, they engage against actors like energy companies which they assume to be responsible for climate change. Second, they engage against mainstream groups of the climate movement who they claim to support false solutions and a too cooperative approach of acting. On the opposite, the mainstream groups criticize the radical actors due to the moral character of their appeals. This critique hit the radical actors and resulted in a weakening of the climate movement because activists felt they were being treated as ecological dreamers. Their reaction was to widen their focus and hint at supposed relations of the climate crisis to other current crises:

> More and more people can't pay their energy bills because current politics and their cuttings make them poor. The same mechanisms of the system prevent that people are able to determine their lives. These mechanisms result in walls. They create wars aiming at the domination of people, land and resources.
>
> *(appeal of the campaign ausgeco$_2$hlt)*[19]

4 Actors, structures, and strategies of the German climate movement

An experienced activist described the foundation of the German climate movement as a "decided move amongst long-term activists from quite different backgrounds" (Bricke 2008). Several of these activists felt personally guilty when the popularity and importance of the climate movement began to decline in 2010.

The climate groups in Germany are not based on formal membership comparable to established NGOs or parties. Nevertheless, their structures and contacts consist of more than loose mailing lists and digital signatures. The climate activists have built informal networks, which are mostly working covertly in order to prevent observation[20] and influence by actors from the outside of the movement. Social relations are very important for keeping the movement together. The German climate movement would almost be similar to a circle of friends if actors were not disappearing and new ones were joining the movement. Activists have left the German climate movement due to frustration, burnouts, or the beginning of their working life. An important task resulting from this fluctuation is the transfer of knowledge from experienced activists to new ones. This issue consciously gets discussed and treated in the German movement.

In order to achieve its aim to influence public discourses, the German climate movement has always tried to start alliances with other actors of civil society. These processes were accompanied by conflicts. Members of the climate movement had difficulties in deciding between radical, left-wing, and clandestine actions on the one hand and broad, civil actions on the other hand. The diversity of the groups engaged in the climate movement result in opportunities and problems at the same time. According to Bricke (2008), in the first years of the movement, there were conflicts between eco-anarchist libertarian political approaches and Marxist-inspired approaches of intervening in hegemonic discourses. Up until today, the approaches of acting and the forms of communication vary heavily among climate activists, members of the anti-nuclear movement, leftwing organizations, and antifascistic groups (see chapter by Brunnengräber in this book).

An important current conflict in the German climate movement deals with the problematic relation of the movement to the green party. According to Bricke, many climate activists hold the opinion that the greens and their participation in local and national governments are responsible for the weakness of eco-activism in Germany. The critics argue that people lose their motivation to engage in the environment if there is an ecological party in government, which could – but in fact does not consistently do this – implement ecological politics. Despite this criticism, members of the green party sympathize with the climate movement. For example, their youth organization cooperates with the climate movement and criticizes its own mother party. Debates on the relationship with the green party will probably also in future be an important issue for the climate movement.

As regards the theory and debate oriented character of the climate movement in Germany, even the critique of imperial lifestyles (Brand, in Müller 2011) was picked up and spread by climate activists. This issue was raised, although the media attention for the movement and its activity declined in the aftermath of Copenhagen. Furthermore, as Korte (2011: 187) observed, media reports on the German climate movement rather focus on the forms of their activity than on their messages and contents. This discrepancy is increased by the following asymmetry:

> The deficit of real power and the logic of acting lead to movements dependent on the media. . . . Movements have a contradictive relation to the media, because they need their attention. In order to raise the interest of the media, they have to satisfy their need for simple messages and coloured pictures.
>
> *(Brunnengräber 2005: 353)*

In reaction to this media logic, the climate movement organizes mass demonstrations and produces background materials such as a newspaper about the COP 15. Articles are written by activists in order to intervene in hegemonic discourses. But these texts are often just published in left magazines and not in the mainstream press. Furthermore, even climate camps and direct actions consciously intend to create media attention, for example if they fight against symbols of injustice.

When the climate groups realized that their movement will not achieve the size of the global justice or antinuclear movement, they started to engage in multi-issue protests such as those of the Bloccupy movement. Nevertheless, the climate movement was also able to succeed in its original field. The Yasuni-ITT campaign[21] created pressure on development minister Niebel. Several climate camps were organized by intensively engaged activists. Finally, also the engagement of climate groups and civil initiatives in Munich was successful because they stopped the planned construction of a third runway for the airport in Munich.

5 Challenges for the future

The popularity of climate camps in Germany brings the climate movement back to its roots. Critical voices question the effects of the camps: Do they achieve their aims of creating a collective identity and spreading the dynamical feeling of a movement (Bedall et al. 2011: 43)? These doubts are based on the thesis that "the [ ] camps need to be part of a wider project that gives them some political meaning beyond their highly localized intervention" (Müller 2008). Not all members of the climate movement in Germany share such theoretical demands.[22] Therefore the theoretical and discursive aims of the radical part of the climate movement create problems in the cooperation with the mainstream actors of the movement. Despite these doubts and conflicts, the reorientation towards eco-anarchic protest forms and civil disobedience for example in the context of the climate camps has positive effects on the movement. These protest forms strengthen the feeling and motivation of the movement as well as its public visibility. Apart from solving the question of how much discourse is necessary, it is also important that the climate movement starts to use the Internet in an adequate way. It has huge potential for the enhancement of their networking and protest activity.

Parts of the climate movement are currently building strategical coalitions with other actors. According to Foucault, broad alliances can be successful, even if this requires that their members limit their own demands. Korte (2011: 200) observes the use of this strategy in several social movements. Despite the potential of such broad coalitions, it is important not to forget the traditional allies of the climate movement such as the anticapitalism and global justice movement. Their interest for the issue of climate change as well as their relations to the climate movement must be kept alive because they support it with a lot of resources and energies. In contrast, alliances with the Occupy movement could have negative effects: If a movement always wants to become broader, more inclusive, and more coherent (see Müller 2011), it will risk losing its individual profile and messages which are necessary to mobilize the public and to change discourses.

Alliances of the climate camp in Brandenburg with Sorbian activists were initiated with positive intentions, but resulted in negative effects. They limited the scope of the activities of the climate camp. The climate movement is already characterized by diverse members. Alliances with activists from other backgrounds create further heterogeneity and conflicts. As a consequence, even the radical wing of the movement could get into internal discussions and contradictions. If the climate movement defines itself as part of broad energy conflicts, it will lose its label, activists, and orientation. If these thoughts are considered, if dynamic and influential actors (Bricke and Müller 2011) know their roles and positively influence young activists and if the German climate movement is able to relate local resistance with global theoretical discussions, its short history will go on.

Notes

1 Many thanks for their critical advice and enduring engagement in the climate movement go to the commentators of this text, in particular to Philip Bedall.
2 Not all members of the climate movement in Germany are mentioned in this text. Some activists were left out due to their wishes, others could not be described because of the limited length of the text. The author is responsible for the selection of the presented actors and therefore also for resulting deficits. He lives in Berlin and is engaged in the climate movement.
3 According to the publishing company of Boese's book, this mission consists of revolution and salvation of the planet.
4 See Hunter (2012).
5 The limited focus on climate change held by mainstream NGOs is problematic because as a consequence

carbon is often treated as the central cause of climate change. Other roots such as injustice and unequal distribution of wealth are often not considered. German NGOs show diverse attitudes towards the question of justice. During an internal meeting in 2011, a member of Germanwatch rejected the integration of justice demands into the agenda of the climate movement. He feared that the claim for justice could overburden activists and the public.

6 Korte defines emancipatory social movements as "movements, whose aims and instruments intend to reduce hegemonic structures in society and want to enhance the ability of subjects to act. Such movements have concrete political aims, but nevertheless want to create societal change." By using the term social movements in this article, the emancipatory character of these movements will always be considered.

7 For an introduction see Lohmann (2006) and Klein (2011).

8 As a consequence of the blockade, the politicians were forced to enter the area of the conference by helicopters.

9 A video of the protest action against Vattenfall can be watched at http://youtu.be/btA5IxG4pDE (accessed 10/15/2012).

10 See http://youtu.be/Nr_DH3v8v1A (accessed 10/15/2012).

11 See www.klimacamp.org/ (accessed 10/15/2012).

12 In front of the building of the political event, climate activists demonstrated together with activists engaged for education. Three activists dressed as ice bears entered the building with a protest banner and interrupted the speech of the head of the local government.

13 There are reports that staff of Vattenfall called people of the village and told them that the football matches could have a negative effect on the support of the village by Vattenfall.

14 The Berlin Eclectic Electric Collective used a ten meter high hammer as a symbol of their rage against the ineffective negotiation process. The hammer should also mobilize climate activists for other protest actions.

15 Energy sovereignty points at the emancipatory potential of energy supply and is therefore an attractive political concept.

16 See http://fels.nadir.org/de/whois (accessed 10/15/2012).

17 This slogan was part of the appeal for the Blockupy protests in Frankfurt in May 2012 (see http://blockupy-frankfurt.org/de/blockade/aktionsthemen/oekologie). The slogan was also used by the climate camp in Rhineland in 2012.

18 This perspective underlines the potential of climate politics to generate growth and green jobs. Meanwhile, this perspective negates the necessity of a social reconstruction of society for solving the climate crisis.

19 See www.ausgeco2hlt.de/klimacamp/hintergrund/kohle-kapitalismus-klima/ (accessed 10/15/2012).

20 During the Climate Summit in Copenhagen, the mobile phones of some activists of the climate movement were spied on by the police. In trials activists were confronted with passages of their own phone calls. These illegal observations were not researched judicially and had no consequences for the responsible police officers.

21 This campaign dealt with a certain area of rainforest in Ecuador. The Ecuadorian government demanded €100 million from the international community as compensation for the protection of the forest. Otherwise, it would permit the exploitation of the forest for example by the extraction of oil (Klimaretter 2011).

22 In the debates of the climate movement, there are several references to Marx (class struggle), Gramsci (role of hegemony), and Foucault (discourse intervention).

References

Bedall, P., Bricke, M., and Tzschiesche, S. (2011) Klima- und Energiekämpfe, in Bois, M. and Hüttner, B. (ed.) *Beiträge zur Geschichte einer pluralen Linken*, Berlin: Rosa Luxemburg Stiftung. Accessed 10/15/2012 from www.rosalux.de/fileadmin/rls_uploads/pdfs/RosaLux/Rosalux_3–2010_web.pdf.

Boese, D. (2012) [Personal Website]. Accessed 11/15/2012 from www.danielboese.de/wordpress/.

Bricke, M. (2008) What Role Do Radical Climate Change Campaigns and the Social Movements That Generate Them Play in the Securitisation of Climate Change, and How Are They Affected by the Process of Securitisation? – A Case Study of Climate Camps in the UK and Germany, Master Thesis.

Bricke, M. and Müller, T. (2011) Kurze Geschichte einer kurzen Geschichte, *arranca*, 44. Accessed 07/11/2012 from http://arranca.org/ausgabe/44/kurze-geschichte-einer-kurzen-geschichte.

Brunnengräber, A. (2005) Gipfelstürmer und Straßenkämpfer. NGOs und globale Protestbewegungen in der Weltpolitik, in Brunnengräber, A., Klein, A., and Walk, H. (eds.) *NGOs im Prozess der Globalisierung. Mächtige Zwerge – umstrittene Riesen*, Schriftenreihe der Bundeszentrale für politische Bildung 400, Bonn: BpB.

Dietz, M. (2010) Gibt es eine internationale Klimabewegung?, *Forschungsjournal Neue Soziale Bewegungen*, 23(4): 6–18.

Götze, S. (2010) Wie weiter nach Kopenhagen?, *Neues Deutschland*, January 25. Accessed 07/21/2010 from www.neues-deutschland.de/artikel/163661.wie-weiter-nach-kopenhagen.html.

Hellmann, K.-U. (1998) Paradigmen der Bewegungsforschung- Forschungs- und Erklärungsansätze – ein Überblick, in Hellmann, K.-U. and Koopmans, R. (eds.) *Paradigmen der Bewegungsforschung. Entstehung und Entwicklung von Neuen sozialen Bewegungen und Rechtsextremismus*, Opladen and others: Westdeutscher Verlag.

Hunter, E. (2012) *Öko-Krieger*, Frankfurt a. M.: Fischer.

Klein, N. (2011) Capitalism vs. the Climate, *The Nation*, November 28. Accessed 10/15/2012 from www.thenation.com/article/164497/capitalism-vs-climate/.

Klimaretter (2011) Die Klimabewegung entdeckt Yasuní, November 9. Accessed 10/15/2012 from www.klimaretter.info/umwelt/hintergrund/9839-die-klimabewegung-entdeckt-yasuni.

Korte, M. (2011) Der Köder muss dem Fisch schmecken und nicht dem Angler – Vom Umgang sozialer Bewegungen mit den Medien, in Brunnengräber, A. (ed.) *Zivilisierung des Klimaregimes. NGOs und soziale Bewegungen als Akteure der nationalen, europäischen und internationalen Politik*, Wiesbaden: VS Verlag.

Kössler, G. (2010) Wie weiter mit der Klimabewegung? Accessed 07/21/2012 from http://klima-der-gerechtigkeit.boellblog.org/2010/02/22/wie-weiter-mit-der-klimabewegung/.

Lawton, C. (2009) Climate Change Summit Becomes a Target for Protest, *SpiegelOnline*, November 25. Accessed 08/05/2012 from www.spiegel.de/international/europe/from-seattle-to-copenhagen-climate-change-summit-becomes-a-target-for-protest-a-663142.html.

Lohmann, L. (2006) Carbon Trading. A Critical Conversation on Climate Change, Privatisation and Power. Accessed 10/15/2012 from www.thecornerhouse.org.uk/resource/carbon-trading-0.

McCarthy, J. and Mayer, Z. (2006) Resource Mobilization Theory: Vigorous or Outmoded?, in Turner, J. (ed.) *Handbook of Sociological Theory*, New York: Springer Science.

Methmann, C. (2011) NGOs in der staatlichen Klimapolitik – Zwischen diskursiver Intervention und hegemonialer Verstrickung, in Brunnengräber, A. (ed.) *Zivilisierung des Klimaregimes. NGOs und soziale Bewegungen als Akteure der nationalen, europäischen und internationalen Politik*, Wiesbaden: VS Verlag.

Müller, T. (2008) The Movement Is Dead, Long Live the Movement, in *Turbulence*, July. Accessed 05/25/2012 from http://turbulence.org.uk/turbulence-4/the-movement-is-dead-long-live-the-movement/.

Müller, T. (2011) Von der Klimagerechtigkeit zur Energiedemokratie – für's Klima kämpfen, ohne vom Klima zu reden, Forum Wissenschaft. Accessed 06/11/2012 from www.bdwi.de/forum/archiv/themen/oekologie/5415876.html.

Müller, T. and Passadakis, A. (2008) 20 Thesen gegen den grünen Kapitalismus, prager frühling. Accessed 08/05/2012 from www.prager-fruehling-magazin.de/article/192.20_thesen_gegen_den_gruenen_kapitalismus.html.

Passadakis, A. (2009) Auf dem Weg nach Kopenhagen, *taz*, November 10. Accessed 08/05/2012 from www.taz.de/1/archiv/print-archiv/printressorts/digi-artikel/?ressort=tp&dig=2009%2F11%2F10%2Fa0206&cHash=fda2561c19.

Raschke, J. (1985) *Soziale Bewegungen. Ein historisch-systematischer Grundriß*, Frankfurt am Main and New York: Campus.

Rucht, D. and Neidhardt, F. (2007) Soziale Bewegungen und kollektive Aktionen, in Joas, H. (ed.) *Lehrbuch der Soziologie*, Frankfurt am Main and New York: Campus.

Schumacher, J. (2008) Der Importeur des Klimacamps, *taz*, August 22. Accessed 08/05/2012 from www.taz.de/!21955/.

9

MOVEMENTS FOR CLIMATE JUSTICE IN THE US AND WORLDWIDE

Brian Tokar

1 Introduction

Just a few years into the current millennium, popular movements to address the emerging global climate crisis appeared to be at an impasse. Although scientific evidence was rapidly converging with respect to the nature and consequences of human-derived climate changes, public attention to the issue lagged far behind, especially in the United States. Efforts to engage the public were most often limited to explaining the science of global warming and emphatically making the case that the phenomenon was real. This limitation was most apparent in Al Gore's popular documentary film, *An Inconvenient Truth*, but appeared endemic to the movement as a whole.

Even in the aftermath of the 2007 report of the Intergovernmental Panel on Climate Change (IPCC), which declared that the evidence for human-caused climate change was now "unequivocal," global warming was most often viewed as a somewhat esoteric scientific concern, with consequences that would be felt at some indefinite future time and most dramatically affect the inhabitants of remote and uniquely endangered ecosystems. The most iconic symbol of this earlier wave of climate activism was the polar bear, struggling to stand its ground amidst shrinking ice flows in the Arctic north.

When tens of thousands of people converged on the streets of Copenhagen during the landmark UN climate conference in December of 2009, a different perspective was clearly emerging (see Figure 9.1).[1] Some activists still dressed as polar bears, while others aimed to focus the world's attention on the need to reduce the atmospheric concentration of carbon dioxide to a maximum of 350 parts per million (Hansen et al. 2008). But a far more urgent outlook on the climate crisis was also beginning to capture the world's attention. This view, known as 'climate justice,' is rooted in vulnerable communities around the world that have experienced severe and destabilizing climate-related disruptions to their lives and livelihoods. This chapter will examine the perspective of climate justice, its origins and scope, as well as some of its challenges in moving forward.

2 Origins of climate justice

The first acknowledged reference to the concept of climate justice appeared in a 1999 report titled *Greenhouse Gangsters vs. Climate Justice* by the San Francisco-based Corporate Watch group

Figure 9.1 Indigenous activists at the UN climate change conference in Copenhagen in 2009 (*source:* Orin Langelle).

(Bruno et al. 1999). The report was mainly an examination of the petroleum industry and its disproportionate political influence, but it also made an initial attempt to define a multifaceted approach to climate justice, including:

- addressing the root causes of global warming and holding corporations accountable;
- opposing the destructive impacts of oil development, and supporting impacted communities, including those most affected by the increasing incidence of weather-related disasters;
- looking to environmental justice communities and organized labor for strategies to support a just transition away from fossil fuels; and
- challenging corporate-led globalization and the disproportionate influence of international financial institutions such as the World Bank and World Trade Organization.

The report's conclusions were highlighted at a 1999 rally at Chevron Oil's headquarters in San Francisco, as well as at international conferences held in the Netherlands in 2000 and on the Indonesian island of Bali in 2002.

The Corpwatch authors were active participants in the US movement for environmental justice, which began in the 1980s and had become a focus for inner city, indigenous, and poor rural communities confronting their disproportionate exposure to a variety of environmental hazards. The movement was galvanized by several successful local campaigns, as well as a landmark, church-sponsored report, *Toxic Wastes and Race*, which revealed that the racial composition of communities is the largest factor in the siting of hazardous waste facilities in the US. The report documented that three out of five African Americans nationwide live in close proximity to abandoned toxic sites (United Church of Christ 1987).

News of the *Toxic Wastes and Race* report helped to unite a variety of currents that had been challenging this reality on the local level for many years, and helped empower African American, Native American, and Latino activists to demand a greater voice within the Euro-American-dominated environmental movement (Tokar 1997: chapter 6). In 1991, a National People of Color Environmental Leadership Summit issued a comprehensive public statement against environmental racism and for environmental justice.[2] By the mid-1990s, leaders such as Tom Goldtooth of the Indigenous Environmental Network (IEN) were articulating the need to bring the deepening climate crisis into this framework, understanding that people of color would be as disproportionately impacted by climate disruptions as by chemical toxins. The movement's second Leadership Summit in 2002 issued a document titled '10 Principles for Just Climate Change Policies in the US.'[3]

Also throughout the 1990s, international NGOs such as the World Rainforest Movement, Friends of the Earth International, and the Third World Network drew public attention to local struggles of indigenous and other land-based peoples in the global South against the rising levels of resource extraction that accompanied neoliberal economic policies. They joined with Corpwatch, Indigenous Environmental Network, and others at a meeting in Bali in 2002 to develop the Bali Principles of Climate Justice, a comprehensive, 27-point program aimed to "begin to build an international movement of all peoples for Climate Justice."[4] Campaigns to highlight indigenous land struggles helped shape the international movement against corporate-driven globalization in the late 1990s and early 2000s, and became a significant focus for organizations engaged in international climate justice activities today, including IEN, the Global Forest Coalition, and the Global Justice Ecology Project.[5]

Several years earlier, at the 1997 UN conference that produced the Kyoto Protocol, Vice-President Al Gore had offered that the US would endorse an agreement only if mandated greenhouse gas reductions were implemented through a system of market-based, tradable emission permits. The protocol eventually came into effect without US ratification, and policy measures in the EU and other countries increasingly come to rely on such 'cap-and-trade' measures to nominally reduce greenhouse pollution. Market skeptics, concerned about the injustices inherent in this approach, convened a meeting in Durban, South Africa in the fall of 2004 that included representatives of social movements and indigenous peoples' organizations based in Brazil, India, Samoa, the US, and UK, as well as South Africa. That gathering drafted the Durban Declaration on Carbon Trading, which has gained over 300 endorsements worldwide.[6]

When the UN's annual climate conference was held in Bali in 2007, the Durban Group for Climate Justice and numerous allies from around the world gathered in significant numbers. Representatives of communities disproportionately affected by global inaction on climate presented a strong and unified showing both inside and outside the official proceedings, and a more formal worldwide network emerged under the slogan, 'Climate Justice Now!' At a series of side events, press conferences, and protests throughout the Bali conference, representatives of affected communities, Indigenous Peoples, women, peasant farmers, and their allies articulated their demands for (CJN! 2007):

- reduced consumption in the global North;
- huge financial transfers from North to South based on historical responsibility and ecological debt, paid for by redirecting military budgets, innovative taxes, and debt cancellation;
- leaving fossil fuels in the ground and investing in energy-efficiency and community-led renewable energy;

- rights based resource conservation that enforces indigenous land rights and promotes peoples' sovereignty over energy, forests, land, and water;
- sustainable family farming and food sovereignty.

A more detailed statement of principles for Climate Justice Now! (CJN!), developed the following year, begins in part:

> From the perspective of climate justice, it is imperative that responsibility for reducing emissions and financing systemic transformation is taken by those who have benefited most from the past 250 years of economic development. Furthermore, any solutions to climate change must protect the most vulnerable, compensate those who are displaced, guarantee individual and collective rights, and respect peoples' right to participate in decisions that impact on their lives.
>
> *(CJN! 2008)*

By 2010, some 750 international organizations had joined the CJN! network, including numerous grassroots groups throughout the global South, and had become a clearinghouse for information and the continuing involvement of many groups seeking to further these goals.[7] At the annual UN climate conferences, the network offers an inclusive meetingplace for critical perspectives on the unfolding international climate negotiations.

Over the past several years, climate justice has come to embody several distinct but largely complementary currents from various parts of the world. In the global South, demands for climate justice unite an impressive diversity of indigenous and other land-based people's movements. They include rainforest dwellers opposing new mega-dams and palm oil plantations, African communities resisting land appropriations for industrial agriculture and agrofuel production, Pacific Islanders facing the loss of their homes due to rising seas, and peasant farmers fighting for food sovereignty and basic land rights. A statement to the 2009 Copenhagen climate conference from the worldwide confederation of peasant movements, La Vía Campesina, stated in part:

> Climate change is already seriously impacting us. It brings floods, droughts and the outbreak of pests that are all causing harvest failures. I must point out that these harvest failures are something that the farmers did not create. Instead, it is the polluters who caused the emissions who destroy the natural cycles.... [W]e will not pay for their mistakes.[8]

In the US, environmental justice activists continue to be the leading voices for climate justice – mainly representatives from African American, Latino, and Native American communities that have been resisting daily exposure to chemical toxins and other environmental hazards for more than 25 years. A conference in New York City in early 2009, organized by West Harlem Environmental Action (WEACT), brought together inner city activists, community and youth organizers, indigenous representatives and farmworker advocates with students, environmental lawyers, scientists, public health advocates, and government officials to discuss the relevance of the climate justice framework for communities of color and their allies across the US.[9] Many speakers described the emerging climate justice movement as a continuation of the US civil rights legacy, and of their communities' continuing "quest for fairness, equity and justice."[10] Others explained how, in recent years, the environmental justice movement has broadened its scope to areas of food justice, housing justice, and transportation justice, as well as opposition to the commodification of the atmosphere through global carbon markets. The national Grassroots Global Justice Alliance continues to bring delegations of US environmental justice activists to the annual UN climate conferences, while the Labor Network for Sustainability and allied groups work to raise support for climate justice among the ranks of organized labor in the US.[11]

In much of Europe, climate justice emerged as a further evolution of the global justice and anticapitalist movements that arose in opposition to the World Trade Organization and annual G8 economic summits during the late 1990s and early 2000s. A March 2010 discussion paper from the European Climate Justice Action network (CJA) explained that "Climate Justice means linking all struggles together that reject neoliberal markets and working towards a world that puts autonomous decision making power in the hands of communities." The paper concluded: "Fundamentally, we believe that we cannot prevent further global warming without addressing the way our societies are organized – the fight for climate justice and the fight for social justice are one and the same" (Climate Justice Action 2010). While Climate Justice Action proved to be relatively short-lived, this approach is also expressed through ongoing networks such as Rising Tide – which was formed in the Netherlands and boasts chapters in the UK, US, Mexico, Ecuador, and Australia – as well as the UK Climate Camp movement, which organized high profile actions between 2006 and 2010 at major power plant sites, Heathrow Airport, London's financial district, and the Edinburgh headquarters of the Royal Bank of Scotland.[12]

While various organizational expressions have proved difficult to sustain, the outlook of climate justice continues to have significant appeal in many parts of the world, and the informal Climate Justice Now! network serves as a consistent point of contact among these disparate currents, especially around the ongoing UN climate negotiations. Between the UN conferences, people and groups collaborate through a variety of online forums to share news, debate perspectives and strategies, and further the scope of climate justice organizing. Demands for climate justice have recently been voiced by representatives of waste pickers in Durban, South Africa, migrant farmworkers in the hills of Vermont, and Rising Tide activists challenging the practice of "mountaintop removal" coal mining in the US state of West Virginia, among others.[13] As the consequences of climate destabilization continue to be felt by people around the world, the disproportionate effects on those least responsible for excess greenhouse gas emissions continue to be a powerful motivator for engaged civil society actions.

3 The case for climate justice

The disproportionate impacts of a changing climate are most clearly illustrated by the marked increase in weather-related disasters over the past half-century. The stories of people affected by unprecedented droughts, floods, wildfires, and other such incidents have most shaped public perceptions of the case for climate justice. From Hurricane Katrina in the US to the flooding of the Indus River valley in Pakistan, uncontrollable wildfires in Russia, and years of protracted drought in the Horn of Africa, the stories of vulnerable and often helpless people facing increasingly dangerous and unanticipated weather events have stirred the conscience of people around the world.

While scientists may disagree to what extent particular events are attributable to climatic changes, three things are clear: first, that the rate of weather-related disasters is increasing rapidly, upsetting even the authoritative projections of the global insurance industry; second, that these incidents are wholly consistent with the predictions of climate models for the behavior of a warmer, more turbulent atmosphere; and third, that when the climate contribution to particular weather events can be measured, the signal of global warming consistently stands out as an essential strong contributing factor.[14]

The 2007 UN Human Development Report revealed that one out of every 19 people in the so-called developing world was affected by a climate-related disaster between 2000 and 2004, compared to one out of every 1,500 people in the OECD countries (Human Development Report 2007/2008: 16). A 2009 Oxfam study found that of nearly 250 million people who are now affected by natural disasters every year, 98 percent of them are falling victim to climate-related events such as floods and droughts. They predict that this could increase to over 375 million people per year as soon as 2015 (Oxfam 2009). Columbia University's International Earth Science Information Network predicts that by 2050 the world will see as many as 700 million climate refugees (cited in Parenti 2011: 7). The IPCC's ongoing assessment of current climate research has confirmed these realities, while summarizing numerous studies of the potential impacts of observed trends. Their 2007 report predicted a worldwide decrease in crop productivity if global temperatures rise more than three degrees Celsius, and that crop yields from rain-fed agriculture could be reduced by half as early as 2020. Current climate trends would expose between 75 million and 250 million people to "increased water stress" in Africa alone (IPCC 2007).

Some of the grimmest consequences for human rights are contained in the IPCC's assessment of the health consequences of climate changes, which predicts

> increases in malnutrition and consequent disorders ...; increased deaths, disease and injury due to heatwaves, floods, storms, fires and droughts; the increased burden of diarrheal disease; the increased frequency of cardio-respiratory diseases due to higher concentrations of ground-level ozone ...; and, the altered spatial distribution of some infectious disease vectors.
>
> *(IPCC 2007)*

This will also include malaria. It is clear that those populations with "high exposure, high sensitivity and/or low adaptive capacity" will bear the greatest burdens; those who contribute the least to the problem of global warming will continue to face the most severe consequences (IPCC 2007: 393).

A study by Rafael Reuveny of Indiana University examined 38 cases over seven decades where populations were forced to migrate due to a combination of environmental and other factors. Half of these cases led to violent conflict between the migrating populations and those in the host areas. Reuveny points out that those who depend the most on the environment to sustain their livelihood, especially in regions where arable land and fresh water are scarce, are most likely to be forced to migrate when faced with rapid and unplanned-for climate changes (Reuveny 2007). Journalist Christian Parenti's book, *Tropic of Chaos*, offers a particularly graphic, on-the-ground tour of climate-exacerbated conflicts throughout Africa, Asia, and Latin America, illustrating what he describes as "the catastrophic convergence of poverty, violence and climate change" (Parenti 2011: 5).

"There could be no clearer demonstration than climate," explained the 2007 Human Development Report, "that economic wealth creation is not the same as human progress" (Human Development Report 2007/2008: 27). Those who benefited least from the unsustainable pace of economic growth and capitalist expansion between 1950 and the early 2000s are facing a future of suffering and dislocation that may be unprecedented in world history (see Figure 9.2). To prevent this will require a thoroughgoing reversal of patterns of exploitation and maldevelopment that many in the global North have come to take for granted. This understanding lies at the very core of the case for climate justice.

Figure 9.2 Demonstration during the UN climate change conference in Bali in 2007 (*source:* Orin Langelle).

4 Challenging fossil fuels and the false solutions

While movements for climate justice do not claim a unified political strategy, their praxis to date clearly fall into three main arenas: strategic interventions at the annual negotiations carried out under the UN's Framework Convention on Climate Change (UNFCCC); challenges to the expanding extraction of fossil fuels; and a variety of efforts to expose corporate-driven 'false solutions' to the climate crisis. Even as fossil fuel companies and other transnational corporate interests underwrite efforts to deny the reality of global climate disruptions, they have simultaneously aimed to influence the debate over climate solutions. Several sectors of the global energy industry have proposed technological and policy measures purported to address the climate situation while assuring their continued hegemonic role in the global economy. Campaigns to challenge the false solutions to climate change focus mainly on these corporate initiatives. The false solutions framework has engaged those who identify with the messages of climate justice in opposition to a new generation of energy megaprojects, as well as the global proliferation of carbon markets and a variety of proposed geoengineering schemes.[15]

The petroleum industry's most aggressive measures, however, are more closely linked to their support for climate change denial. With oil prices rising and conventional reserves rapidly dwindling, the industry is pursuing new technologies to extract oil and gas from increasingly inaccessible and hazardous locations, including deep beneath the oceans, once-impervious geological formations, and tar sands deposits from Alberta, Canada to the Congo Basin (Wykes 2009; Birger 2012). These projects have sparked widespread opposition from the people most affected by these projects, including Inuit communities near the shores of the Arctic Ocean, Cree people in central Alberta, and people living in the vicinity of major shale formations that have shown to be lucrative sources of oil or gas when subjected to the high-pressure injections of water and chemicals known as hydrofracturing, or "fracking" (for example, see Federman 2010; Raoul 2010; Broder and Krauss 2012). Campaigns in solidarity with those most affected by these most extreme forms of energy development have engaged people throughout the world in recent years.

The technology of fracking for natural gas links opposition to conventional fossil fuels with the emerging opposition to false solutions to the climate crisis. Influential environmental NGOs in the US and beyond continue to view natural gas, with roughly 30 percent less carbon dioxide emitted than from burning oil, as a 'bridge fuel' to help facilitate the future expansion of renewable energy. But the hazards of modern gas extraction methods have sparked opposition in many regions, including some of the most politically conservative parts of the United States. Amid numerous reports of contaminated water, dying livestock, and hazards to human health, landowners are organizing to resist the pressure to sign drilling contracts, while towns and entire counties are enacting bans. Vermont passed a statewide ban on fracking in 2012, and New York's governor proposed to prohibit the practice in most of the state (Cantarow 2012).

False solutions to the climate crisis include a variety of technologies that are claimed to expand energy supplies without increasing greenhouse gas emissions, as well as market-oriented policy measures such as carbon markets and offsets that aim to substitute for regulations against pollution.[16] Technological false solutions include the expansion of nuclear power, which was widely advocated before Fukushima and will likely be considered again in the near future. Indigenous inhabitants of areas known to be rich in uranium, including the US Southwest and central Canada, are organizing against the expansion of mining in their territories. The false promise of 'carbon capture and storage' from new coal-fired power plants is another important focus, as a new generation of purportedly 'capture-ready' coal plants has been proposed in many countries, despite a scientific consensus that reliable underground carbon capture is many decades away, at best.[17] Local activists tied to the Sierra Club have reportedly halted the construction of 132 new coal plants in the US in recent years (Hertsgaard 2012). While utilities claim that new coal plants will replace older, less efficient ones and thus reduce the climate impact of US electricity production, activists argue that new construction will sustain the economy's dependence on coal far into the future.

In many countries, opposition to industrial-scale plantations for biofuel – more appropriately 'agrofuel' – production is a central focus of resistance. Expansion of soya, maize, and sugarcane fields for fuel production encroaches on traditional land uses and contributes to ongoing food shortages, while scientists increasingly question whether there is any climate advantage to burning agrofuels instead of gasoline (Tokar 2010a). The challenge to land rights is most acute in South America – where native forests and grasslands, as well as traditional polycultures, are being plowed under to facilitate a massive expansion of soy production – as well as in Africa and Southeast Asia, where rainforests and pasture lands are overrun by expanding palm oil plantations (Smolker et al. 2008). The Global Forest Coalition and World Rainforest Movement, among others, have brought international attention to people's movements challenging these rapidly expanding agrofuel monocultures. The same regions of the world also face a new generation of massive hydroelectric dams, including projects in uniquely endangered areas such as the Amazon basin. For example, indigenous and fishing peoples have repeatedly attempted to block the construction of a new dam on Brazil's legendary Xingu River; when developers built an initial barrier to begin obstructing the river, opponents blocked roads and hand-dug a trench to free the river once again (Lyons 2012).

On the policy front, campaigners for climate justice have led the environmental opposition to the expansion of carbon markets, along with several newer schemes that further commodify nature. Climate justice activists tend to view tradable emissions permits as a boon to corporations seeking to postpone investments in energy-saving technologies, while simultaneously creating a new property right to pollute the atmosphere.[18] The consistent over-allocation of permits has allowed politically powerful corporations to exceed legislated 'caps' on emissions with ease. One significant impetus for the creation of Climate Justice Now! was the advocacy in favor of carbon markets by several larger environmental NGOs, represented at the UN climate conferences as the Climate Action Network (CAN). CAN describes itself as a broadly inclusive alliance working toward "the coordinated development of NGO strategy on international, regional, and national climate issues," and boasts a network of 700 international organizations, both large and small.[19]

The British columnist George Monbiot has aptly described global carbon trading as "an exuberant market in fake emissions cuts" (Monbiot 2007). Carbon offsets, which allow corporations to invest in nominally carbon-reducing projects around the world in exchange for higher emissions at home, are frequently described as a gaping 'hole' in any mandated emissions cap. A German study of UN-approved carbon offset projects reported that at least 40 percent, and as many as 86 percent of all offset-funded projects (hydroelectric and wind power developments, industrial energy efficiency, etc.) would likely have been carried out anyway (Schneider 2007). This runs counter to the Kyoto Protocol's guideline requiring that projects granted emissions offsets must be "additional," i.e., not already under way.

In California, environmental justice groups went to court in 2011 to challenge a new state cap-and-trade law. They argued that its provisions for international offsets would encourage corporations to continue increasing in-state CO_2 emissions, together with other forms of pollution that more directly affect public health (Conant 2011). A decision in the plaintiffs' favor was overturned on appeal, and advocates went on to appeal to the governor, and then to the federal Environmental Protection Agency (EPA), on grounds that the scheme violates the civil rights of communities of color that are disproportionately exposed to unsafe levels of air pollutants (Center on Race, Poverty & the Environment 2012; Egelko 2012).

Despite widespread opposition, and mirroring the predictions of many market skeptics, UN negotiators continue to expand the scope of pollution markets beyond the trading of permits to emit carbon dioxide. The Indigenous Environmental Network has been in the forefront of organizing global opposition to an enhanced offset scheme aimed to address deforestation, which is responsible for up to 25 percent of global greenhouse gas emissions (Boas 2011). This plan, known as REDD (Reducing Emissions from Deforestation and Forest Degradation), has encouraged new proposals for tradable biodiversity credits, clean water credits, and other novel market-oriented measures (see chapters of Hein and Garrelts as well as of Segebart and König in this book).

5 To the summits and beyond

While the principles of climate justice have the potential to unite a wide variety of people's movements around the world, many groups engaged in local struggles against new energy developments still identify rather loosely with the climate justice movement. This is especially true in the US and Europe, where climate justice campaigns began to coalesce in the lead-up to the 2009 Copenhagen climate summit, only to subside in the summit's aftermath.

Indeed, some active participants have concluded that Copenhagen was the inadvertent peak of global efforts to create a unified climate justice movement. Some 50,000 people attended the alternative 'Klimaforum' people's summit in Copenhagen, and over 100,000 marched in the streets during the conference, many conveying demands for climate justice and for 'System Change, Not Climate Change.' However, this proved far from sufficient to prevent a diplomatic meltdown inside the conference center. The Copenhagen effort, according to CJA activists Nicola Bullard and Tadzio Müller,

> failed to establish an anti-capitalist CJ-discourse that was visible and understandable beyond the subcultures of activists and policy-wonks, and thus failed to provide a visible alternative to despair; failed to establish a new 'pole of attraction' that would substantially reconfigure the political field around climate change; and failed to do anything to significantly advance the fight for climate justice. In some sense, the *global* CJM [Climate Justice Movement; emphasis in original] remained something more of a potential than a reality.
> *(Bullard and Müller 2012: 57)*

In the US as well, the most visible manifestations of climate justice occurred in the lead-up to Copenhagen. North American activists created a Mobilization for Climate Justice network and organized a series of actions across the continent on November 30, 2009, the ten-year anniversary of the landmark demonstrations in Seattle against the World Trade Organization.[20] In the San Francisco Bay Area, the Mobilization for Climate Justice-West (MCJ-West) organized seven high-profile demonstrations and actions during the five months prior to Copenhagen, often acting in solidarity with community activists in Richmond, California in opposition to a major Chevron oil refinery located in that North Bay city (see Figure 9.3).

MCJ-West proved unable to sustain its high public visibility into 2010 and beyond, despite a principled and politically significant effort to restructure the group to better reflect the priorities of organized local communities. The broader mobilization ceased to operate following a similar internal discussion. Although participants agreed that 'frontline' environmental justice communities should ideally be in the forefront of the climate justice movement, organizations struggling with the daily impacts of political and economic marginalization did not appear to have the capacity – nor perhaps the inclination – to sustain a unified national climate justice coalition. The lessons of these experiences, however, have inspired some insightful new approaches to political alliance-building across barriers of race and class, initiated in part by a San Francisco Bay Area group called the Movement Generation Justice and Ecology Project.[21]

Efforts to reunite various climate justice tendencies at the 2010 US Social Forum in Detroit proved largely symbolic and limited in scope. However, Movement Generation and several allied groups, including the Indigenous Environmental Network, Grassroots Global Justice, and others, continued to meet and develop strategies. A September 2010 position paper proposed uniting around four themes: root cause remedies; human rights and antiracism; reparations for historic injustices; and directly democratic control by people over the decisions that affect their lives.[22] In 2012, nearly 30 groups organized as the Climate Justice Alignment proposed a nationwide campaign for a 'just transition' away from fossil fuel dependence, including the creation of ten million new jobs in renewable energy, public transportation, local food, waste reduction, and related areas.[23]

Internationally, a significant effort to coalesce climate justice-oriented movements was realized in Cochabamba, Bolivia during April 2010. At the end of the Copenhagen climate conference, Bolivian president Evo Morales invited participants to Bolivia for a People's Conference on Climate Change and the Rights of Mother Earth. Some 30,000 representatives of civil society, indigenous peoples, and social movement actors from around the world, along with a delegation of public officials from Bolivia and allied countries, met in Cochabamba to develop a 'People's Agreement,' assembled from the products of 17 onsite working groups.

In marked contrast to the rather empty 'accord' that came out of the UNFCCC negotiations in Copenhagen, the People's Agreement was rooted in indigenous views of harmony, complementarity, and anticolonialism, and proposed a Universal Declaration on the Rights of Mother Earth. The

Figure 9.3 Demonstration of MCJ in Richmond, California, in August 2009 (*source:* Brooke Anderson).

document condemned carbon markets and the commodification of forests for carbon offsets; asserted the rights of climate migrants; and proposed an International Climate and Environmental Justice Tribunal to judge and punish activities that promote further climate change (World People's Conference 2010). However, the agreement fell short of endorsing the full climate justice agenda, most notably the call to keep fossil fuels in the ground. Popular movements critical of Bolivia's economic dependence on resource extraction formed a separate working group, which was compelled to operate outside of the official proceedings (Müller 2012).

Further public expressions of climate justice continued to manifest at the annual conferences of the UNFCCC in Cancún, Mexico at the end of 2010 (see Figure 9.4) and Durban, South Africa in 2011.[24] In both instances, international NGOs affiliated with Climate Justice Now! brought representatives of popular movements from throughout the global South to the UN summits to testify at official side events and demonstrate outside of the proceedings. La Vía Campesina and its affiliated peasant farmer movements were in the forefront of public events in Cancún, challenging the limitations of the official proceedings. In Durban, conflicts between civil society groups participating in the UN conference and those who remained outside came to a head on the very last day during an Occupy-style demonstration just outside the conference hall. While representatives of organizations such as Greenpeace and 350.org urged cooperation with UN security in clearing the building of protesters, several activists refused to leave and some were forcibly removed (Petermann and Langelle 2012). Whereas many groups affiliated with Climate Justice Now! have had an increasingly difficult time airing their issues within the UN process – pointing to a concerted effort by officials to marginalize civil society voices – others remain hopeful about the potential for a coordinated inside/outside strategy around these annual events.

6 Into the future

Where is the movement for climate justice today? Many groups continue to intervene under the banner of climate justice at the UN level, despite what some groups perceive to be diminishing returns on these efforts. But perhaps more important, local activists around the world continue to bring climate justice messages into their campaigns. Some activists have voiced concerns that the message of climate justice has been appropriated by large international NGOs, governments, and development agencies, while others celebrate the emergence of official climate justice initiatives, such as a £3 million fund announced by the Scottish government in 2012 to support water projects in eastern and central Africa (TckTckTck 2012).[25] As awareness grows of

Figure 9.4 Solidarity action of MCJ-West in the US during the UN climate summit in Cancún in December 2010 (*source:* Mobilization for Climate Justice West).

increasingly erratic and extreme weather patterns throughout the world, the social justice dimensions of the climate crisis have proved increasingly difficult to overlook.

One promising development is the increasingly unified opposition to the increasing pace of oil extraction from western Canada's Alberta tar sands, which has widely been described as the most climate-damaging project in the world (Fairley 2011; Benzak and Droitsch 2012). In 2011, organizers with the traditionally cautious 350.org network overcame an initial reluctance and actively supported a civil disobedience campaign in Washington, D.C. Over 1,200 people were arrested outside the White House in opposition to a new US pipeline for tar sands oil, and representatives of the Indigenous Environmental Network were prominent participants in those actions.[26] A year later, '350' (see organization profile in this book) was actively supporting a pipeline blockade in Texas, an organizing effort in northern New England against the reconfiguration of a 60-year-old pipeline to transport tar sands oil, and a civil disobedience campaign at Montana's state capitol over the issue of coal exports. The tar sands pipeline – proposed to run from the Canadian border to oil refineries on the Gulf of Mexico – even became an issue in the 2012 US presidential campaign. These efforts have significantly raised the stakes for climate activism in general, and for furthering the understanding that marginalized peoples – starting with the First Nations communities in central Alberta – remain at the frontlines of destructive new energy developments in North America and beyond. Despite concerns about the potential dilution of climate justice principles, the movement's core messages continue to motivate concerted actions for climate protection and for the rights of affected communities around the world.

Movements for climate justice continue to be multifaceted, and draw upon a perhaps unprecedented diversity of perspectives and strategies. In many ways, their diversity is their greatest strength, given the multiplicity of peoples affected by extreme weather events and increasing climate chaos, as well as the need to develop appropriate strategies for a wide variety of political contexts. But the climate crisis is also inherently global in scope, and the lack of progress toward global reductions in greenhouse pollution speaks to the need for ever greater coordination, determination, and commonality of vision.

Climate justice activists' involvement at the UN level has helped to forge unique and encouraging alliances among people and organizations throughout the world, but the 2011 Durban Platform's deferral of new mitigation measures until 2020 at the earliest helped further a lingering crisis of confidence in the entire process. The delay could spell a "death sentence for Africa, small island states, and the poor and vulnerable worldwide," in the words of Friends of the Earth International chair Nnimmo Bassey, and increasing "climate racism, ecocide, and genocide," according to Tom Goldtooth from the Indigenous Environmental Network (quoted in Petermann and Langelle 2012). Perhaps the climate movement's best hope lies in the combination of rising climate militancy in the North and the increasing international visibility of struggles in the South. Short of a comprehensive strategy to overturn the stalemate in global negotiations – in turn a product of the political hegemony of fossil fuel interests – South African analyst and activist Patrick Bond suggests that: "we should remind ourselves of the most important features of a future climate justice politics: in thinking locally, nationally and globally, and also acting in each sphere with the appropriate analysis, strategies, tactics and alliances" (Bond 2012a). Humanity's future may rest on that somewhat tentative but undoubtedly essential prospect.

Notes

1 For a comprehensive review of climate justice perspectives in the lead-up to Copenhagen, see Brand et al. (2009).
2 See www.ejnet.org/ej/principles.html (accessed 06/14/2012).
3 See www.ejnet.org/ej/climatejustice.pdf (accessed 06/14/2012).
4 See www.ejnet.org/ej/bali.pdf (accessed 06/14/2012).
5 The climate-centered activities of the Global Justice Ecology Project are highlighted on their blog (see http://climate-connections.org and IEN's at http://ienearth.org/climatejustice.html, accessed 06/14/2012).
6 See www.durbanclimatejustice.org/durban-declaration/english.html (accessed 06/14/2012).
7 A full listing of the members of CJN! is available at www.climate-justice-now.org/category/climate-justice-movement/cjn-members (accessed 06/14/2012).
8 Statement of Henry Saragih, general coordinator of La Vía Campesina, to the Klimaforum alternative summit, December 7, 2009, via CommonDreams.org, see La Vie Campesina (2009).
9 WEACT's origins are discussed in Dawson (2010). Also see WEACT (2009).

10 Comments of Robert Bullard at 'Advancing Climate Justice: Transforming the Economy, Public Health and Our Environment' conference (New York, January 30, 2009).
11 The activities of the Grassroots Global Justice Alliance at the 'Rio+20' environmental summit in Brazil in 2012 are outlined at http://ggjalliance.org/node/982, and the Labor Network for Sustainability's perspectives are described at www.labor4sustainability.org (both accessed 06/14/2012).
12 See www.risingtidenorthamerica.org/about-rising-tide-north-america/our-history/ (accessed 06/18/2012) and Shifter (2011).
13 A further listing of climate justice successes by communities in the US and south Asia can be found in Bond (2012b: 188–194).
14 For example, see World Resources Institute (2005), Reichenmiller et al. (2010), Min et al. (2011), Pall et al. (2011), and Coumou and Rahmstorf (2012).
15 Comprehensive reviews of the false solutions framework include *Hoodwinked in the Hothouse: False Solutions to Climate Change* (Rising Tide North America and Carbon Watch 2009) and *Indigenous Peoples' Guide: False Solutions to Climate Change* (Indigenous Environmental Network 2009).
16 For a more complete discussion of these issues, see Tokar (2010b: 76–86).
17 The scientific case against carbon capture and storage (CCS) is described in Rochon et al. (2008). In April 2011, several prominent scientists signed on to an environmental justice letter to the US EPA Administrator outlining the case against CCS; available from www.precaution.org/lib/ej_and_science_ccs_letter_final.110411.pdf (accessed 06/14/2012).
18 For detailed reviews of these issues, see Lohmann (2006, 2009).
19 See www.climatenetwork.org/about/about-can; a full listing of affiliated groups is available at www.climatenetwork.org/about/members (accessed 06/14/2012).
20 See http://actforclimatejustice.org; see also Russell (2009) and Rising Tide North America (2010).
21 Movement Generation's outlook and activities are described at www.movementgeneration.org (accessed 06/14/2012). Their distinct approach to climate justice organizing, developed in collaboration with the Ruckus Society and other groups, is most fully explored in Moore and Russell (2011).
22 Available at www.climate-justice-now.org/cj-in-the-usa-root-cause-remedies-rights-reparations-and-representation (accessed 06/14/2012).
23 Personal communication.
24 For samples of some of the exemplary alternative media coverage of these events, see www.democracynow.org/topics/climate_change (accessed 06/14/2012).
25 The Global Campaign for Climate Action – known as the 'TckTckTck' campaign after the sound of a ticking clock – claims to represent a broad alliance of environmental, youth, and antipoverty groups; it is one group that has promoted a climate justice frame since Copenhagen but has not embraced the comprehensive political critique supported by most groups in Climate Justice Now!. The Mary Robinson Foundation – Climate Justice, based in Dublin, has also been a focus of criticism for its ties to corporate-linked foundations (see www.mrfcj.org/partners_and_links/partners/funding_partners.html, accessed 12/15/2012).
26 A more skeptical view of current tar sands campaigns, and the often co-optive role of large ENGOs appears in Stainsby (2011).

References

Benzak, J. and Droitsch, D. (2012) *Going in Reverse: The Tar Sands Threat to Central Canada and New England*, New York: Natural Resources Defense Council.

Birger, J. (2012) Why Shell is Betting Billions to Drill for Oil in Alaska, *Fortune*, June 11. Accessed 06/27/2012 from http://features.blogs.fortune.cnn.com/2012/05/24/oil-shell-alaska-drilling.

Boas, H. (ed.) (2011) *No REDD Papers*, Volume One, Portland: Indigenous Environmental Network and Carbon Trade Watch.

Bond, P. (2012a) Durban's Conference of Polluters, Market Failure and Critic Failure, *Ephemera*, 12(1/2): 42–69.

Bond, P. (2012b) *The Politics of Climate Justice: Paralysis Above, Movement Below*, Durban: University of KwaZulu-Natal Press.

Brand, U., Lander, E., Bullard, N., and Müller, T. (eds.) (2009) *Contours of Climate Justice. Ideas for Shaping New Climate and Energy Politics*, Critical currents no. 6, Occasional Papers Series, Uppsala: Dag Hammerskjöld Foundation.

Broder, J. and Krauss, C. (2012) New and Frozen Frontier Awaits Offshore Oil Drilling, *New York Times*, May 23.

Bruno, K., Karliner J., and Brotsky C. (1999) *Greenhouse Gangsters vs. Climate Justice*, San Francisco: Transnational Resource & Action Center. Accessed 10/15/2012 from www.corpwatch.org/article.php?id=1048.

Bullard, N. and Müller, T. (2012) Beyond the 'Green Economy': System Change, Not Climate Change? *Development*, 55(1): 54–62.

Cantarow, E. (2012) Fracking's 'Little Revolution': How a People-Powered Movement Fights Big Business for Clean Water. Accessed 01/23/2012 from www.alternet.org/story/153853/fracking%27s_%22little_revolution%22%3A_how_a_people-powered_movement_fights_big_business_for_clean_water.

Center on Race, Poverty & the Environment (2012) Civil Rights Complaint Filed with the US. Accessed 06/27/2012 from http://climate-connections.org/2012/06/11/groups-claim-cap-and-trade-violates-civil-rights-2/.

Climate Justice Action (2010) What Does Climate Justice Mean in Europe? Discussion Paper, Amsterdam, February 2010. Accessed 08/23/2013 from http://www.climate-justice-action.org/resources/documents/what-does-climate-justice-mean-in-europe/.

Climate Justice Now! (2007) Climate Justice Now! Principles of Unity, draft version, May, via Climate Justice Now! email list.

Conant, J. (2011) New Ruling Puts California's Cap and Trade on Permanent Hold. Accessed 05/25/2011 from http://climate-connections.org/2011/05/23/new-ruling-puts-california%E2%80%99s-cap-and-trade-on-permanent-hold/.

Coumou, D. and Rahmstorf, S. (2012) A Decade of Weather Extremes, *Nature Climate Change*, 2(7): 491–496.

Dawson, A. (2010) Climate Justice: The Emerging Movement against Green Capitalism, *South Atlantic Quarterly*, 109(2): 325–326.

Egelko, B. (2012) Cap-and-trade Plan Upheld by Calif. Court. Accessed 06/21/2012 from www.sfgate.com/cgi-bin/article.cgi?f=/c/a/2012/06/19/BASU1P4MFB.DTL&type=science.

Fairley, P. (2011) Alberta's Oil Sands Heat Up, in *Technology Review*, November/December. Accessed 10/10/2012 from www.technologyreview.com/energy/38888/.

Federman, A. (2010) To Drill or Not to Drill: Natural Gas Rush Forces Farmers to Choose Between Income and Land, *Earth Island Journal*, March. Accessed 10/15/2012 from www.alternet.org/story/145856/to_drill_or_not_to_drill%3A_natural_gas_rush_forces_farmers_to_choose_between_income_and_land.

Hansen, J., Sato, M., Kharecha, P., Beerling, D., Masson-Delmotte, V., Pagani, M., Raymo, M., Royer, D., and Zachos, J. (2008) Target Atmospheric CO2: Where Should Humanity Aim? *Open Atmospheric Science Journal*, 2: 217–231.

Hertsgaard, M. (2012) How a Grassroots Rebellion Won the Nation's Biggest Climate Victory. Accessed 04/03/2012 from http://motherjones.com/environment/2012/04/beyond-coal-plant-activism.

Human Development Report (2007/2008) *Fighting Climate Change: Human Solidarity in a Divided World*, United Nations Development Program.

Indigenous Environmental Network (2009) Indigenous Peoples' Guide: False Solutions to Climate Change. Accessed 10/15/2012 from www.ienearth.org/docs/Indigenous_Peoples_Guide-E.pdf.

IPCC (Intergovernmental Panel on Climate Change) (2007) Report of Working Group II 'Impacts, Adaptation and Vulnerability,' Geneva: Intergovernmental Panel on Climate Change.

La Via Campesina (2009) Why we left our farms to Copenhagen. Accessed 21/11/2013 from http://viacampesina.org/en/index.php/actions-and-events-mainmenu-26/-climate-change-and-agrofuels-mainmenu-75/807-why-we-left-our-farms-to-come-to-copenhagen.

Lohmann, L. (2006) Carbon Trading: A Critical Conversation on Climate Change, Privatization and Power, *Development Dialogue*, 48.

Lohmann, L. (2009) Neoliberalism and the Calculable World: The Rise of Carbon Trading, in Böhm, S. and Dabhi, S. (eds.) *Upsetting the Offset: The Political Economy of Carbon Markets*, London: MayFlyBooks.

Lyons, C. (2012) The Dam Boom in the Amazon, *New York Times*, July 1.

Min, S., Zhang, X., Zwiers, F., and Hegerl, G. (2011) Human Contribution to More-intense Precipitation Extremes, *Nature*, 470: 378–381.

Monbiot, G. (2007) Hurray! We're Going Backwards!, *Guardian*, December 17.

Moore, H. and Russell, J. (2011) *Organizing Cools the Planet: Tools and Reflections on Navigating the Climate Crisis*, Oakland: PM Press.

Müller, T. (2012) The People's Climate Summit in Cochabamba: A Tragedy in Three Acts, *Ephemera*, 12(1/2): 73–74.

Oxfam (2009) *The Right to Survive: The Humanitarian Challenge for the Twenty-First Century*, London: Oxfam International.

Pall, P., Aina, T., Stone, D., Stott, P., Nozawa, T., Hilberts, A., Lohmann, D., and Allen, M. (2011) Anthropogenic Greenhouse Gas Contribution to Flood Risk in England and Wales in Autumn 2000, *Nature*, 470: 382–386.

Parenti, C. (2011) *Tropic of Chaos: Climate Change and the New Geography of Violence*, New York: Nation Books.

Petermann, A. and Langelle, O. (2012) The Durban Disaster, *Z Magazine*, February. Accessed 10/15/2012 from http://globaljusticeecology.org/stopgetrees_news.php?ID=602.

Raoul, E. (2010) Petropolitics versus Democracy: Canada's Bitter Black Sands, *Le Monde Diplomatique*, English version, May.

Reichenmiller, P., Spiegel, A., Bresch, D., and Schnarwiler, R. (2010) *Weathering Climate Change: Insurance Solutions for More Resilient Communities*, Zürich: Swiss Reinsurance Company Ltd. Accessed 10/15/2012 from http://europa.eu/epc/pdf/workshop/2–3_pub_climate_adaption_en.pdf.

Reuveny, R. (2007) Climate Change-induced Migration and Violent Conflict, *Political Geography*, 26(6): 656–673.

Rising Tide North America (2010) The Climate Movement Is Dead. Long Live the Climate Movement. Accessed 03/11/2010 from www.risingtidenorthamerica.org/wordpress/wpcontent/uploads/2010/03/climatemovement_longlive.pdf.

Rising Tide North America and Carbon Trade Watch (2009) Hoodwinked in the Hothouse: False Solutions to Climate Change. Accessed 10/15/2012 from risingtidenorthamerica.org/special/hoodwinkedv2_WEB.pdf.

Rochon, E., Kuper, J., Bjureby, E., Johnston, P., Oakley, R., Santilo, D., Schulz, N., and Goeme, G. (2008) *False Hope: Why Carbon Capture and Storage Won't Save the Climate*, Amsterdam: Greenpeace International.

Russell, J. (2009) Actions Speak Louder Than Words: Climate Justice Actions Sweep the US before Copenhagen Talks. Accessed 10/10/2012 from www.grist.org/article/actions-speak-louder-than-words-climate-justice-activists-across-u.s.-mobil.

Schneider, L. (2007) Is the CDM Fulfilling its Environmental and Sustainable Development Objectives? An Evaluation of the CDM and Options for Improvement. Report prepared for WWF by Öko-Institut E.V. Berlin, November 5, 2007.

Shifter, A. (2011) Remember, Remember: Climate Camp, *Shift Magazine*, 12: 24–26. Accessed 06/18/2012 from http://libcom.org/files/Issue%2012_Shift%20magazine.pdf.

Smolker, R., Tokar, B., and Petermann, A. (2008) *The Real Cost of Agrofuels: Impacts on Food, Forests, Peoples and the Climate*, Asunción, Paraguay: Global Forest Coalition.

Stainsby, M. (2011) A Tar Sands Partnership Agreement in the Making?, *Canadian Dimension*, August 1. Accessed 10/15/2012 from http://oilsandstruth.org/tar-sands-partnership-agreement-making.

TckTckTck (2012) Scotland Reveals Details of World's First Climate Justice Fund. Accessed 12/15/2012 from http://tcktcktck.org/2012/06/scotland-reveals-details-of-worlds-first-climate-justice-fund/27831.

Tokar, B. (1997) Environmental Justice, in Tokar, B. (ed.) *Earth for Sale: Reclaiming Ecology in the Age of Corporate Greenwash*, Boston: South End Press.

Tokar, B. (2010a) Biofuels and the Global Food Crisis, in Magdoff, F. and Tokar, B. (eds.) *Agriculture and Food in Crisis: Conflict, Resistance, and Renewal*, New York: Monthly Review Press.

Tokar, B. (2010b) *Toward Climate Justice: Perspectives on the Climate Crisis and Social Change*, Porsgrunn, Norway: Communalism Press.

United Church of Christ (1987) *Toxic Wastes and Race in the United States: A National Report on the Racial and Socioeconomic Characteristics of Communities with Hazardous Waste Sites*, New York: United Church of Christ.

WEACT (2009) *Advancing Climate Justice: Transforming the Economy, Public Health and Our Environment*, Conference Agenda and Resource Guide, New York: WEACT.

World People's Conference (2010) *People's Agreement*, Cochabamba, Bolivia. Accessed 06/27/2012 from http://pwccc.wordpress.com/2010/04/24/peoples-agreement/.

World Resources Institute (2005) *Synthesis: Ecosystems and Human Well-Being: A Report of the Millennium Ecosystem Assessment*, Washington DC: Island Press.

Wykes, S. (2009) *Energy Futures: Eni's Investments in Tar Sands and Palm Oil in the Congo Basin*, Berlin: Heinrich Böll Foundation.

10

THE AUSTRALIAN CLIMATE MOVEMENT

A disparate response to climate change and climate politics in a not so 'lucky country'

Hans Baer

1 Introduction

In frustration at the lack of action on the part of the Liberal-National government under Prime Minister John Howard from 1996 to 2007, around the beginning of the present century a growing number of Australians began to form and join local grass-roots climate action groups (CAGs). These groups have since grown exponentially and now constitute the most significant new social movement in Australian society. Many of these CAGs are linked in complex ways to other organizations, such as established environmental nongovernment organizations (ENGOs), the Australian Greens (a growing political force in parliament), and extra-parliamentary political parties and groups. Many of them have also become affiliated, beginning in early 2008 with various regional climate action networks and in early 2009 with a national climate action network. The Australian climate movement is a dynamic but still disparate movement which consists of two broad layers. The first of these functions primarily within mainstream civil society, though occasionally in parliaments and local councils, includes the Southern Cross Climate Coalition, peak ENGOs (such as Greenpeace, Friends of the Earth, the Wilderness Society, Climate Action Network Australia), state-based nature conservation societies, and labor unions. The other layer acts primarily at the grass-roots level and generally below the radar of mainstream media and includes regional and local climate action groups, socialist groups, and various special climate groups. Some climate activists have become so angered by the Rudd and Gillard governments' failures to adopt adequate climate change mitigation strategies that they have embarked upon direct actions at the Newcastle coal port, coal mines, and coal-fired power plants.

2 The topsy turvy world of Australian climate politics

Australian climate politics has come to encompass numerous actors, including corporations, the Rudd and Gillard (Australian Labor Party) governments and various state governments, the Opposition, the Greens, think tanks, the Commonwealth Science, Industrial and Research

Organisation, the Australian Bureau of Meteorology, academic research centers, public intellectuals, labor unions, and last but not least a burgeoning and disparate climate movement. Despite assertions of moving beyond a business-as-usual approach and acceptance of the climate science regarding the seriousness of climate change, including for Australia, the Rudd and Gillard governments in reality adopted rather conservative approaches. The Rudd government proposed the implementation of a Carbon Pollution Reduction Scheme (CPRS) in 2008. It committed itself by 2020 to "reduce Australia's carbon pollution by up to 15 percent below 2000 levels in the context of a global agreement where major economies agree to substantially restrain carbon pollution and advanced economies take on reductions comparable to Australia" and "an unconditional five percent in carbon pollution below 2000 levels by 2020" (Australian Government 2008: iv).

The initial CPRS called for a cap and trade scheme that would set a limit on total annual emissions and stipulated a carbon price of ten dollars a tonne in the first year. It proposed to cover some 1,000 companies, emissions from stationary energy, transport, industrial processes, waste and 'fugitive emissions,' such as waste emissions resulting from oil and natural gas production. The initial CPRS stipulated that emissions-intensive trade-exposed industries and producers of goods for transport would be granted 90 percent of their permits free and other businesses would get 60 percent of their permits free. Despite many concessions to industry, various corporate groups, such as the Minerals Council of Australia and the Australian Industry Group, either opposed the scheme or wanted its introduction delayed. In order to win over the Opposition under its leader Malcolm Turnbull, the Rudd government announced a modified proposal for an emissions trading scheme (ETS) known as CPRS II on May 4, 2009. Apart from delaying implementation of the scheme by a year, which the Coalition had always favored, the new scheme proposed a 5–25 percent reduction band, a new ten dollar cap on carbon pricing, and more free trading permits for emissions-intensive trade-exposed industries. However, in large part because a substantial number of Coalition parliamentarians opposed even CPRS II, Tony Abbott managed to depose Turnbull as Opposition leader. Abbott repeatedly asserted that the CPRS constituted a "great big new tax." He offered an "environmental policy" rather than a "tax policy." Abbott's plan proposed to subsidize companies for voluntarily cutting greenhouse gas emissions and thus reduce 140 million tonnes of CO_2 by 2020. In achieving such a cut, the Coalition also proposed "direct action on soil carbons" and tree planting. The CPRS was defeated in the Senate on August 13, 2009.

Beginning in late 2010, much of Australia shifted from a land of drought and heat waves to one of cyclones, heavy rains, and floods. Indeed, world-wide 2010 turned out to be the wettest year on record but also a very hot year. The World Meteorological Organisation maintained that 2010 was the hottest year since records began in 1850. NASA and the National Ocean and Atmospheric Administration also reported that 2010 was the wettest year on record as well as a very hot year, tying with 2005 as the hottest year on record. Particularly Queensland and to a lesser degree the states of New South Wales and Victoria experienced heavy rains and floods during late 2010 and into early 2013. Australia experienced its hottest month on record with a nation-wide heat wave in January 2013, with an average mean maximum temperature of 36.92°C (Australian Bureau of Meteorology 2013). In that month, Sydney hit a record high of 45.8°C on January 18 and Hobart a record high of 41.8°C. Moona in South Australia had the highest temperature (49.6°C on January 12 recorded during the heat wave). While parts of interior Australia were experiencing ravaging bushfires, coastal Queensland and New South Wales experienced heavy rainfall and flooding.

The Rudd government ran into difficulties when the mining sector mounted a media campaign to counter a proposed Resource Super Profits Tax. Julia Gillard, the former Deputy

Prime Minister, assumed the Prime Ministership in a set of circumstances that may have appeared to the rest of the world as a *coup d'état*. One of the first actions that she took as Prime Minster was to whittle down the Resource Super Profits Tax from 40 percent to 22 percent and rename it the Minerals Resource Rent Tax – one which is to be applied only to iron ore and coal companies and only those who earn profits above 50 million dollars (McAuley 2010: 20). In arriving at this new arrangement, Gillard entered into negotiations with the three largest multinational corporations in the mining industry, namely BHP Billiton, Rio Tinto, and Xstrata.

Both the ALP and the Coalition downplayed climate politics during the 2010 election which eventually permitted Gillard to continue on as Prime Minster in a minority government in which the Greens and the independents came to hold the 'balance of power.' Particularly with the demise of the Australian Democrats in the Senate, the Greens have become more and more of a presence in parliamentary politics, both in the federal government and various state governments. The Australian Greens have generally made climate change one of their various environmental concerns. While not speaking officially on behalf of their party, Bob Brown and Peter Singer already expressed concern about the 'greenhouse effect' in the mid-1990s and proposed a carbon tax as a climate change mitigation strategy (Brown and Singer 1996). More recently, the Greens asserted that the Rudd government's CPRS overlooked the risks associated with catastrophic climate change, essentially proposed to hand billions of dollars to the 'big polluters,' and undermined international efforts to avert global climatic and environmental disasters. As part of the debate on the CPRS, the Greens submitted a Safe Climate Bill which called for a more stringent emissions trading scheme without free permits, compensation, and a higher carbon price. Components of the bill proposed that if an international agreement were achieved on reducing greenhouse gas emissions, Australia's emissions would fall to 40 percent below 1990 levels by 2020. It stipulated provisions based on the following principles: "Making polluter pay, not paying polluters," "repowering Australia with renewable energy," "save energy, save money, save the planet," "travelling with a light footprint," and "protecting our green carbon."[1]

On January 21, 2010 the Greens announced an 'interim carbon price proposal' whereby CO_2 would be taxed within the framework of the CPRS. The Greens proposed taxing carbon at a rate of 23 dollars per tonne in the first year, starting in July 2010, and 24 dollars in the second year. Green politicians, particularly Bob Brown and Christine Milne, have been speakers at climate change conferences and symposia in universities and climate action conferences and rallies, including the 2009 Climate Action Summit in Canberra. Both of them became staunch proponents of the Gillard government's Carbon Price Mechanism as did Adam Bandt, a Green Member of Parliament from the seat of Melbourne. The Greens served as an important force in keeping climate politics alive during the 2010 election. The Multi-Party Climate Committee consisting of ALP and Greens representatives proposed that the 'Carbon Price Mechanism' could begin as early as July 1, 2012 if it were to pass in both the House and Senate and would operate for three to five years, after which time it would convert into a flexible cap-and-trade emissions trading scheme.

While the ALP and Greens joined forces in pushing for a Carbon Price Mechanism, the opposition accused the Gillard government of trying to impose 'a great big new tax' on the Australian people. On July 10, 2011, dubbed Carbon Sunday, the Gillard government announced the details of its planned Carbon Price Mechanism. It proposed the implementation of a carbon tax on July 1, 2012, which will transition into an ETS on July 1, 2015. It is designed to result in a 5 percent cut in Australia's CO_2 emissions, based on 2000 levels, by 2020, and an 80 percent cut on those levels by 2050. Some 500 big polluters will pay a tax to be set at 23 dollars for every tonne of CO_2 emissions, rising to 24.15 dollars in 2013–2014 and 25.40 dollars in 2014–2015. Agricultural emissions are exempted. It is an extremely complex scheme that includes numerous subsidies for both industry and low- and middle-income households, and free permits for some industries. The scheme compensates low-income households by lifting the tax-free threshold from 6,000 to 18,000 dollars in 2012 and then to 19,400 dollars in 2015–2016 and grants a tax cut to all households below 80,000 dollars. The government projects that two out of three households will have their costs fully offset and that nine out of ten households will receive some compensation.

Compensation to industry was exceptionally generous. The scheme allocates 9.2 billion dollars over three years to support EITE (emissions exposed, trade intensive) industries; heavy polluting industries such as steel, aluminium, zinc, pulp, and paper will get 'free permits' covering up to 94.5 percent of their emissions; 1.2 billion dollars is allocated for a 'clean technology program' to help manufacturers; 1.3 billion dollars will assist affected coal miners find alternative employment. To encourage renewable energy, ten billion dollars is set aside to create a Clean Energy Finance Corporation, which will invest in renewable energy projects; and 3.2 billion dollars is allocated for research and development of renewable energy. The scheme maintains that by 2020, 20 percent of Australia's energy shall come from renewable sources, such as wind, solar, and geothermal. Large-scale renewable energy shall provide 40 percent of electricity by 2050. The dirtiest generators will be paid to close, cutting 2,000 megawatts of coal-fired power by 2020. Gas-fired electricity shall increase by more than 200 percent by 2050. Other funding initiatives include: 429 million dollars for a Carbon Farming Initiative for reforestation and revegetation schemes that would allow farmers to earn money by selling carbon credits to big polluters; 947 million dollars over six years for a biodiversity fund to protect Australian native species from climate change; and 40 million dollars over the next four years to promote energy efficiency measures among small businesses and community groups.

Tony Abbott referred to it as a scheme for redistributing wealth. He reiterated that the Coalition had proposed that emissions should be cut by planting numerous trees, shifting from chemical to organic fertilizer, and improving energy efficiency, but the Coalition would continue to support the coal-fired power industry. Legislation for the Carbon Price Mechanism passed the House of Representatives 74–72, assisted by votes from one Green member and four Independents, on October 12, 2011. On November 8, the Senate passed the Carbon Price Mechanism or the 'carbon tax' as it has come to be known in everyday parlance. Only time will tell whether this legislation will be a step forward, one on which climate change mitigation can move forward in Australia, as the Greens often and most environmental NGOs and at least some, if not many, climate activists assert, or whether it will merely create the appearance that effective action is being taken, thus circumventing more radical forms of action on creating a safe climate not only for Australia but the globe.

In 2012 the Australian Government released its *Energy White Paper* (EWP) which boasted that Australia is the leading exporter of coal in the world and is projected to become the second-largest liquefied natural gas (LNG) exporter, both part and parcel of a 'clean-energy transformation.' The document discusses an increasing reliance on natural gas, the development of carbon capture and sequestration as a way of reducing emissions from coal-fired power plants, and a qualified support for renewable energy sources. In their critique of the EWP, Edwards and Brait assert:

> The EWP ... shows that there is a contradiction at the heart of the Gillard government's climate and energy policies. On the one hand, it fought to get the carbon price through parliament in 2011, while on the other hand Australia's planned fossil fuel export products will generate at least eleven times as many annual emissions as will be saved by the Clean Energy Future package.
>
> *(Edwards and Brait 2012: 6)*

3 Typologies of actors in the Australian climate movement

Mark Diesendorf (2009: 7), an environmental scientist and a climate activist, delineates several types of *climate movement organizations* (CMOs) in Australia: (1) grass-roots groups; (2) "large environmental and social justice NGOs for which climate action is one program"; and (3) "some student, faith, trade union, business and professional groups." He observes that CMOs may operate within coalitions and alliances with "influential organisations, such as large businesses, trade unions and professional organisations" (Diesendorf 2009: 140). While Diesendorf's scheme is quite comprehensive, it overlooks a number of key players in the Australian climate movement. Bearing this thought in mind, I propose an alternative scheme depicted in Box 10.1 which distinguishes between actors in the climate movement that tend to function in two broad layers. This scheme illustrates the evolutionary process that social movements often go through if they achieve some level of institutional success. While some might argue as to which groups in the climate movement operate within mainstream civil society and which operate at the grass-roots level, some actors in the climate movement tend to be more bureaucratic and procedural whereas others tend to be more spontaneous, egalitarian, and participatory. Nevertheless, even groups at the grass-roots level may exhibit tendencies toward bureaucratization and compromise with the powers-that-be.

Box 10.1 Layers and actors in the Australian climate movement

- Mainstream civil society
  - Southern Cross Coalition
    - Climate Institute
    - Australian Conservation Foundation
    - Australian Council of Trade Unions
    - Australian Coalition of Social Services
  - Peak environmental NGOs
    - Greenpeace
    - Friends of the Earth
    - Climate Action Network Australia
    - World Wildlife Federation Australia
    - Wilderness Society
    - Safe Climate Australia
    - Brotherhood of St. Laurence
  - State Nature Conservation Councils
- Grass-roots climate movement
  - Regional climate action networks
  - Local climate action groups (CAGs)
  - Special climate action groups
  - Socialist groups
  - Student environmental groups

3.1 Mainstream civil society

Various groups and organizations operating within the corridors of mainstream civil society as well as occasionally in parliaments and local councils, and within the view media of mainstream media, have come to address climate change. These include the Southern Cross Climate Coalition, peak ENGOs, state-based nature conservation societies, and various labor unions, either in consultation with the Australian Council of Trade Unions or on their own. In the interests of space I discuss the Southern Cross Climate Coalition and four ENGOs involved in the Australian climate movement, namely the Climate Action Network Australia, Friends of the Earth Australia, Environment Victoria, and Safe Climate Australia.

The Southern Cross Climate Coalition

The Southern Cross Climate Coalition is an alliance of the Climate Institute, the Australian Conservation Foundation (ACF), the Australian Council of Social Services, and the Australian Council of Trade Unions (ACTU). Essentially it operates as a lobby group as well as an information-disseminating entity. Historically, the Australian Conservation Foundation has been a moderate ENGO, including on the issue of climate change. It called on the Australian government to set legally binding national targets to cut greenhouse pollution by at least 30 percent by 2020 (from 1990 levels). ACF has consistently supported the implementation of an emissions trading scheme that purportedly would drive down greenhouse gas emissions, avoid loopholes and giveaways, and generate revenues to invest in a transition to 'clean economy' (ACF 2008). It also advocated stabilizing greenhouse gases at 350 ppm CO_2 and cutting CO_2 emissions by at least 40 percent by 2020. It accepted the Rudd government's CPRS but withdrew its support on November 25, 2009 after the Rudd government agreed to compromise with Malcolm Turnbull by whittling it down. Conversely, the ACTU continued to support the CPRS despite these developments. The Southern Cross Climate Coalition proved to be a strong supporter of the Carbon Price Mechanism.

Climate Action Network Australia

Climate Action Network Australia is a broad-based network which has numerous members, including Friends of the Earth, Greenpeace Australia, WWF-Australia, the Climate Institute, the Wilderness Society, the Australia Institute, the Australian Youth Climate Coalition, GetUp, Oxfam Australia, the Australian Religious Response to Climate Change, the Liquor, Hospitality and Miscellaneous Union, Environment Victoria, the Climate Emergency Network, Rising Tide, and several local climate action groups.[2] The Climate Action Network Australia had recommended 40 percent emissions cuts by 2020. Its policy paper on emissions reductions called for Australia's greenhouse gas emissions to peak by 2010, drop by at least 40 percent by 2020, and by at least 95 percent by 2050 (with a baseline of 1990 levels). CANA criticized the Rudd government for low emissions targets and its willingness to allow the coal industry to continue functioning and its advocacy of CCS. It came out in favor of the Gillard government's Carbon Price Mechanism.

Friends of the Earth Australia

Burgmann (2003: 192–194) characterizes Friends of the Earth, an international NGO, as a "radical, egalitarian, unbureaucratic organisation, committed to sustainable activism on the ground" and the "most radical of the principal [Australian] green organisations, and participated enthusiastically" in demonstrations against the World Economic Forum conference in Melbourne. Friends of the Earth Australia is a federation of approximately ten local branches. It has maintained a vigorous Climate Justice Campaign based on the notion that "climate justice means providing particular assistance to those with the least financial and economic means to adjust." Friends of the Earth has maintained a strong campaign to address the problems related to climate change faced by in particular the people of low-lying islands in the South Pacific. It has demanded "immediate deep cuts," arguing that "Australia should set national target of reducing our emissions by at least 90–95 percent (from 1990 levels) by 2050." Friends of the Earth has called for a moratorium on biofuels, a carbon accountability alternative to carbon offsets, and changes in the transportation infrastructure. The organization expressed skepticism as to whether the adoption of an ETS "will be able to withstand political pressures of fossil fuel industry and other polluters who will attempt to reduce liabilities under such scheme." Conversely, Friends of the Earth admitted that an ETS potentially serves as "part of the solution to growth of Australia's carbon emissions but only if it avoids problems of previous schemes and only if it is tightly structured and managed." Such an ETS would require that most sources of emissions would be covered, the exclusion of forest and soil sinks from the scheme, the auctioning of all permits, no granting of free permits to purportedly emissions-intensive trade-exposed industries, eschewal of CDM and other offsetting schemes, and no borrowing of permits. Friends of the Earth also recommended stabilization of CO_2 concentrations at below 350 ppm (Friends of the Earth Australia 2008). In early 2009, Friends of the Earth initiated a Climate Displacement Coalition which seeks to lobby the Australian government to assist climate refugees to relocate. In responding to the proposed Carbon Price Mechanism, Friends of the Earth rejected the plan because of an anticipated low price on carbon and its belief that carbon trading "does not work – across the world cap-and-trade has failed to reduce emissions, harmed local communities and delayed real action, whilst delivering huge profits to polluters and financial speculators" (Friends of the Earth Australia 2011).

Environment Victoria

Environment Victoria, along with the Nature Conservation Council of New South Wales, is a leading nature state conservation society involved in the climate movement. It released a report titled *Turning It Around: Climate Solutions for Victoria* which seeks to demonstrate how Victoria can cut its greenhouse gas emissions 54 percent by 2020. It identifies four areas in which emissions could be cut: (1) the promotion of sustainable products and consumption; (2) increased energy efficiency; (3) the creation of a low carbon energy supply; and (4) an increased shift to public transportation. In a more recent report, Environment Victoria (2009) makes numerous recommendations to the Victoria government, including cutting emissions by at least 50 percent by 2020 from 1990 levels, switching from coal to natural gas as a 'transition fuel,' and creating new jobs in the natural gas and renewable energy industries. Environment Victoria has been the principal organizer of the 'Walks Against Warming' in Melbourne.

Safe Climate Australia

Al Gore launched Safe Climate Australia at a Business Breakfast for a Safe Climate in Melbourne on July 13, 2009 that brought together some 1,000 "community leaders," including those representing "some of the largest and most influential corporations in Australia." Safe Climate Australia defines itself as a NGO "formed and steered by a foundation group of concerned scientists, community and business leaders with a shared understanding of the scientific imperatives for emergency action to restore a safe climate." It acknowledges that climate change threatens the Australian environment and economy. Safe Climate Australia seeks to promote technological and social changes needed to achieve a safe climate for Australia and the planet. Its climate change mitigation strategies clearly fall very much within the parameters of ecological modernization as is exemplified by its *Safe Climate Transition Plan* which claims to "encompass an all-sector approach including stationary energy, energy efficiency, housing and commercial buildings, transport, industrial and extraction processes, land use and agriculture" which are intended to maintain a "high standard of living, security of energy and food supply, access to mobility and comfort and well-being for all." Safe Climate Australia seeks to develop a comprehensive transition plan to create a "zero-carbon economy." It supports a cut in CO_2 levels to between 280 and 325 ppm in order to create a "safe climate."[3]

3.2 The grass-roots climate movement

The grass-roots climate movement, which for the most part operates below the radar of mainstream media, includes numerous local climate action groups (CAGs), a still evolving national climate action network, two regional CAGs, various socialist groups, and student environmental collectives. The great majority of grass-roots groups involved in the climate movement belong to a green social democratic wing which urges lobbying politicians and persuading business people to embrace a regulated green capitalism which would result in a reduction of greenhouse gas emissions. They favor a program of ecological modernization which would entail a strong shift to the use of renewable energy sources (particularly solar, wind, thermal, and wave power), energy efficiency, and public transport. While generally critical of the CPRS which the Rudd government tabled, most of the green social democratic groups have been open to an ETS scheme and supported with some reservations the Gillard government's Carbon Price Mechanism.

A much smaller democratic eco-socialist wing maintains that meaningful climate change mitigation ultimately requires the transcendence of global capitalism which fosters a treadmill of production and consumption and economic expansion heavily reliant on fossil fuels. Actors in this wing include members of the Socialist Alliance, Solidarity, eco-socialists within the Green Party and perhaps the ALP, and an assortment of independent eco-socialists. Eco-socialists, however, recognize that the mitigation process cannot await the collapse of global capitalism, although some point to the recent global financial crisis as a possible sign of the unraveling of this system. In the interim period, rather than simply awaiting the 'revolution,' the eco-socialists call for various transitional reforms, such as public ownership of utilities, the expansion of public mass transport, minimization of car use, renewable energy sources, and redistributive mechanisms. Whereas the Socialist Alliance toyed for a while with the implementation of steep carbon tax at the site of production, with rebates for low-income people, Solidarity vehemently opposed any sort of carbon tax on the grounds that it would hurt low-income people.

3.3 Local climate action groups and the drive for a national community climate network

The origins of local climate action groups in Australia remain obscure. Figures on the number of these groups vary. The climate network website lists over 100 CAGs but others have claimed the existence of some 280 CAGs. In reality, some CAGs have come and gone and others are dormant. As Box 10.2 below illustrates, the Australian grass-roots climate movement has engaged in a wide array of activities, ranging from conferences to rallies and climate camps to lobbying politicians and business people.

The Climate Action Summits of 2009 and 2010 were part and parcel of an effort to create a national Community Climate Action Network. The Climate Action Summit held at the Australia National University from January 31 to February 2, 2009 was organized by a committee of activists from the Australian Student Environment Network, Rising Tide (Newcastle), Greenpeace, Friends of the Earth, the Canberra Climate Action Network along with various individuals. In part the Climate Action Summit was designed to coordinate the numerous local climate action groups, many of which have focused on individual solutions, such as promoting solar panels. In commenting on the diversity of people present at the summit, White and Nicholls (2009: 2) observe that

> [f]rom vegan to vegetarian groups to local sustainable technology businesses, from professors of environmental science to high school students, from local Climate Action Group activists to organic farmers, from trade unionists to environmentalists active around water and forests, from Greens to socialists and people who 'hate politics,' the summit was a representative pot pourri of the growing army of Australians who are alarmed about global warming and determined to act against it.

Box 10.2 Activities of the climate movement

- Educational
  - Meetings
  - Conferences
    - Climate Justice Seminar at Melbourne University on August 2, 2008
    - Climate Action Summits 2009–2013
  - Online information
- Rallies
  - Climate Action Rally at Parliament House on February 3, 2009
  - National Climate Emergency Rally on June 13, 2009 in capital cities and other cities
- Climate camps followed by rallies
  - Newcastle (New South Wales), Hazelwood (Victoria), Port Augusta (South Australia), Helensburgh (New South Wales)
- Lobbying politicians and business people

The summit consisted of several plenary sessions and numerous concurrent workshops. Eighty-one recommendations were made at the summit. Of these, 72 received majority support. Nine recommendations, including ones on nuclear power, carbon capture and storage, and biofuels, were rejected. The summit passed a resolution demanding that the Australian government shall participate in an international effort to reduce global levels of carbon dioxide to 300 ppm no later than 2050, and that Australia's CO_2 emissions should be reduced by at least 60 percent by 2020 and 90 percent by 2030 (from 1990 levels). It also demanded that the government enact a policy of 100 percent renewable energy by 2020 and establish a moratorium on all new coal- and gas-fired power plants immediately, revolutionize energy efficiency, promote mass transportation, foster agricultural biological resilience, and create a moratorium on native forest logging.

Participants at the Climate Action Summit had different views on the merits or flaws of emissions trading schemes in general, but agreed that the CPRS is a flawed proposal, with its free permits and compensation to big polluters and low emission reduction targets. The summit called for the creation of 'green jobs.' No consensus was achieved on the structure for a national climate network. While the Green Party was not one of the sponsors of the Summit, Christine Milne, a Green Senator from Tasmania, spoke at the summit and supported the actions of its participants. Bob Brown, also a Green Senator from Tasmania and the national spokesperson for the Green Party, addressed the Climate Action Summit's rally at Parliament House on February 3. While the summit's organizers claimed some 2,500 attendees at the rally, the *Canberra Times* reported on February 4 the presence of 1,250 attendees.

The Community Climate Network emerged following much organizing when over 100 CAGs endorsed the effort in late 2009. The network organized the second Climate Action Summit which occurred on March 13–15, 2010 at the Australian National University. The second Climate Action Summit only drew some 300 attendees and consisted of six campaign streams, namely ones on the national climate action network, 100 percent renewables, 'quit coal,' trade unions and green jobs, and the climate emergency mode. It retained the previous summit's rejection of the CPRS, call for 100 percent renewables by 2020, and a target of 300 ppm CO_2 globally. The summit urged CAGs to campaign for candidates who support its climate policies. The summit adopted World Environment Day on June 5, 2010 as a national day of climate action revolving around the themes of renewable energy sources, cease coal production, a safe climate transition, and green jobs.

Climate Action Summit 2011 occurred on April 9–10 at the University of Melbourne. Some 300 people representing over 100 climate action groups attended the conference to hear various climate activists, scholars, and climate change experts discuss and debate climate action strategies. Much of the discussion in the panels and in the halls revolved around the proposed Climate Price Mechanism, whether climate activists should support a defective scheme or oppose it because of its inadequacies. In a communiqué released to the media, the Summit declared that "current climate policy options in Australia are inadequate and bear little relationship to actions demanded by the science of climate change." The Summit called on the government to:

> develop and implement a comprehensive national climate policy framework that includes a national plan for transitioning to a zero emissions economy, with substantial reductions in energy use ... make polluters responsible by ceasing subsidies for fossil fuels use immediately, and not compensating polluting industries ... redirect subsidies from animal agriculture to sustainable plant-based farming and to just transition programs for affected workers and communities ... provide generous support providing a just transition for affected workers and communities via income redistribution and/or direct assistance, or other measures ... [and] place a much greater focus on so-called 'complementary' measures.
>
> *(Australian Climate Action Summit 2011)*

3.4 Regional climate action networks

Climate action regional networks are a relatively recent development and may have started on February 9, 2008 at the Climate Movement Convergence that met at Northcote High School in an inner Northern Melbourne suburb. While most speakers and workshop organizers at the Climate Movement Convergence proposed strategies of adaptation and mitigation that clearly seek to address climate change within the parameters of 'green capitalism,' and writing letters to and lobbying politicians and business leaders, some conferees seemed to be committed to mass actions and moving beyond 'business-as-usual.' A highlight of the conference was the launching of a report titled *Climate Code Red: The Case for a Sustainability Emergency* authored by David Spratt (CarbonEquity) and Philip Sutton (Greenleap Strategic Institute) and sponsored by Friends of the Earth Australia (Spratt and Sutton 2008a). Spratt and Sutton proposed a framework for climate change campaigning based on the argument that the time to address climate change is urgent and requires emergency measures on those adopted by the Allied Powers against the Axis powers during World War II. The launch of the report led to the formation of the Climate Emergency Network, an organization based in Victoria. Spratt and Sutton (2008b) expanded their monograph into a book which achieved widespread reception in Australia but also outside of Australia to some extent.

Regional climate emergency networks exist in only two states, namely Victoria and South Australia. Climate Emergency Network based in Victoria has sought with mixed success to serve as an organization linking local climate action groups.

CEN has provided financial support for the group Beyond Zero Emissions (BZE) which has been involved in designing a near-zero emissions/ten-year transition plan for Australia. The work of BZE has focused on technological innovations in the stationary energy, transport, land use, housing and commercial buildings, industrial processes, and export revenue sectors of the economy. CEN in collaboration with Friends of the Earth Australia launched in 2009 an ambitious Transition Decade campaign committed to creating a 'safe climate outcome' which would reduce the Earth's temperature by between 0.3 and 0.8°C and reduce the CO_2 in the atmosphere in the range of 280–320 ppm. The first two years of the Transition Decade proposed to get whole communities to say 'yes' to the campaign and the subsequent eight years to obtain an initial phase of government commitment to the plan and several years of functioning in 'emergency transition mode' during which 'everyone gears up for physical rebuilding' and then engages in this process on a massive scale. The Transition Decade was launched at Public Meeting at the Melbourne Town Hall on February 14, 2010 before an audience of some 700 people.

The Climate Emergency Network (CLEAN) largely originated out of a conference held in conjunction with the Australian Education Union on 'Climate emergency – no more business as usual' in Adelaide on October 10–11, 2008. Members of the Eco-Socialist Network in Adelaide played a key role in the formation of CLEAN. While like CEN, CLEAN emphasizes the adoption of renewable energy sources, energy efficiency, and other technological innovations, largely as a result of a strong eco-socialist and eco-anarchist presence in its membership, it touches more upon social justice issues than does its Victorian sister organization. In its positions statement drafted on November 18, 2009, the following three items strongly addressed social justice issues:

- "As the wealthy nations of the world have very largely caused the climate emergency and have unfairly exploited many of the poorest nations, people, and natural resources, the wealthy nations owe a very substantial duty to compensate materially poorer nations by providing them with financial aid to develop green energy infrastructure and to enable them to raise their standards of living to reasonable level."
- "Wealthy nations, due to their privilege and economic power, must play a leading role in mitigating climate change and be willing to make much more substantial cuts in their greenhouse gas (GHG) emissions, in absolute and relative terms, than poorer nations."
- "In dealing with the climate emergency, all solutions to the emergency must be based on the principles of both ecological sustainability and social and economic justice between all nations and between individuals within nations."

Like most grass-roots climate action groups, CLEAN supports the goal of 100 percent renewable energy infrastructure by 2020 and opposes the CPRS and nuclear power plants. It also favors the implementation of an "effective carbon pricing mechanism," particularly a carbon tax.[4]

3.5 Socialist parties and groups

Various socialist parties or groups, particularly the Socialist Alliance and Solidarity, have become key players in the climate movement. Members of socialist groups are often visible at climate rallies and conferences and have been active particularly in the Climate Emergency Network (CEN) in Victoria, the Climate Emergency Action Network (CLEAN) in South Australia, and the Climate Action Summit Network. Socialists have also played a role in the efforts to create a national climate action network.

The Socialist Alliance in its newspaper *Green Left Weekly* regularly publishes articles on the impact of climate change on human societies, including Australia, and climate politics at both the international level and in Australia. The Socialist Alliance has developed a Climate Charter which calls for an emergency plan to mitigate climate change. Its 10-Point Climate Action Plan proposes the following measures:

1. The introduction of emission reduction targets of at least 5 percent a year and achieving "100% renewable energy by 2020";
2. The implementation of an international agreement that aims to "get all countries to agree on a global target of at least 90 percent emissions cuts on 1990 levels by 2030";
3. A shift to a "zero-waste economy";
4. The requirement that existing houses and commercial buildings as well as all new buildings become energy efficient;
5. The implementation of public ownership of utility industries, immediate phasing out of coalmining and coal-fired power plants, providing jobs and retraining to workers in affected industries, and operating the "maximum possible base-load power from existing natural gas and/or hydro power stations instead of coal until renewable energy is available";
6. Nationalizing the car industry and retooling it to "manufacture wind turbines, public transport vehicles and infrastructure, solar hot water, solar photovoltaic cells";
7. The construction of solar thermal plants and wind farms and the upgrading of the "national grid to make it compatible with 100% renewable energy";
8. The banning of old-growth forest logging and the creation of a program of "reforestation, carbon farming and biodiversity protection";
9. The phasing out of industrial farms reliant on oil-resources fertilizers, pesticides, and fuel; and
10. Provision of free public transport, upgrading of existing public transport services, nationalizing and upgrading interstate and ferry services, and facilitating cycling.

The Socialist Alliance opposed passage of both the CPRS and the Carbon Price Mechanism. It released a statement on March 5, 2011 titled 'A carbon price does not equal climate action.' The statement asserts that the "highest prices now being discussed will simply stimulate a mass rollout of gas, extending Australia's commitment to fossil fuels at the expense of renewable energy." Although it argued that a "carbon tax can be a secondary or additional aspect of the government's climate policy," the statement goes on to note that the

> Socialist Alliance stands for large-scale public investment in a publicly-owned and run renewable energy sector, a rollout of public transport, a shift to sustainable farming and other carbon abatement programs, as well as government regulation to phase out dirty industries.

It also maintains that the "the government needs to stop fiddling with the market to cut carbon emissions and instead set emissions cuts targets based on the science."

While not as large as Socialist Alliance, Solidarity has made climate change one of its principal campaign streams. Solidarity formed in February 2008 when three social groups, namely Solidarity per se, the Socialist Action Group, and the International Socialist Organisation merged. Solidarity regularly publishes articles on climate change and climate politics in its monthly magazine *Solidarity*. In early 2010 Solidarity stated its opposition to the ALP/Green Carbon Price Mechanism in part because the polluters will pass on the costs incurred by a carbon tax to consumers. It maintained that the "climate movement will not be able to win the mass support needed for serious action by telling workers to take a cut in their living standards" (Solidarity 2011).

3.6 *Special climate groups*

Particularly prominent among the special climate groups are Rise Ride and the Australian Youth Climate Coalition. On its website, Rising Tide Australia states that it is a

> grassroots Newcastle group taking action against the causes of anthropogenic climate change and for equitable, just, effective, and sustainable solutions to the crisis. We are committed to the principles of Non-violent Direct Action. We are part of the global Rising Tide climate justice movement. We live in the biggest coal port in the cosmos.
>
> *(Rising Tide Australia 2012)*

The Australian Youth Climate Coalition was created by lawyers Amanda McKenzie and Anna Rose in November 2006. It consists of some 27 youth organizations, including the Australian Medical Student Association, the Australian Student Environment Network, Engineers without Borders, National Indigenous Youth Movement of Australia, and the Sai Youth Hindu Youth Network. AYCC favored passage of the Carbon Price Mechanism.

4 An overview of the climate movement in Australia

The Australian climate movement has been divided on whether or not virtually any type of ETS, such as the CPRS or the Carbon Price Mechanism, would be better than none at all. In 2011 an intense debate has erupted within the climate movement over the proposed Carbon Price Mechanism. While some climate activists accept the dilemmas of an ETS, they maintain that the movement should support the proposed legislation in order to get a carbon tax and then campaign against it rolling over into an ETS. Whereas the ACTU and various environmental NGOs such as the Australian Conservation Foundation, the Climate Action Network Australia, and the Brotherhood of St. Laurence joined the Greens in supporting the Carbon Price Mechanism, many grass-roots climate activists, such as ones in the Socialist Alliance and Solidarity opposed it. While most participants at the Climate Action Summit which met at the University of Melbourne on April 9–10, 2011 expressed qualified support for a carbon price, they regarded the ALP/Green Carbon Price Mechanism as grossly inadequate as a climate change mitigation policy, although many attendees favored it as a way of moving forward on climate action.

Beyond Zero Emissions and other groups in the climate movement have expressed concern that the Gillard government is not highly committed to renewable sources of energy but to a shift from coal power to natural gas power. Indeed, on March 9, 2011, Greg Combet stated: "For baseload electricity generation it will be a gas-fired electricity that we see emerge, and for that investment to be committed, we need a carbon price in the economy" (quoted in Moore 2011). Indeed, the government has approved exploration for coal seam gas sites in Queensland, Western Australia, Victoria, South Australia, Tasmania, and the Northern Territory. In a history-making, path-breaking alliance, farmers and graziers joined with climate activists in the Lock the Gate movement. The Lock the Gate Alliance is a national coalition of 122 community groups and hundreds of individuals concerned, including environmentalists, farmers, and graziers, about CSG. According to Munro (2012: 2), government-supported coal seam gas exploration has been "creating alliances that transcend social and political strata and state borders, and the most unlikely people have been forced into defensive action."

The strong anti-CSG sentiment in rural communities has been assisted by the US documentary film *Gasland*. Local rallies against CSG have proliferated along with the wells themselves. With Victoria facing a wave of exploration licenses for CSG, huge shows of community opposition have in particular occurred in New South Wales and Victoria. Friends of the Earth Melbourne called on the Victorian government in September 2011 to ban CSG.

Aside of efforts such as the anti-coal seam gas campaign in Queensland and New South Wales and a campaign on the part of Friends of the Earth to halt plans to build a natural gas power plant in the La Trobe Valley east of Melbourne and its efforts to halt coal production in the same area, Greenpeace's campaign to 'leave coal in the ground,' the Australian climate movement hit a lull of sorts beginning in late 2011 in the aftermath of the passage of the Carbon Price Mechanism. According to Tietze and Humphrys,

> Rather than being a 'historic first step,' the carbon price has been a distraction from serious action. The belief that the political class will rush out to drive more change after the political turmoil the current policy has caused – bringing down an opposition leader [Liberal Malcolm Turnbull] and a prime minister [ALP Kevin Rudd], contributing to near defeat for a first-term federal government and playing a part in ALP support – verges on delusional. The lack of momentum for further action was on display in the government's 2011 energy white paper, which left climate action to 'market mechanisms' while signalling a commitment to expanding fossil fuel industries, and showing minimal interest in renewables.
>
> *(Tietze and Humphrys 2012: 12)*

Some social activists shifted their energies to the Occupy movement in Melbourne, Sydney, and other capital cities and time will only tell whether the Occupy and climate movements will find a common ground in recognizing global capitalism as the source of both social inequities and degradation of the environment. A national Climate Action Summit did not occur in 2012 and there does not appear to be a concrete plan to convene a Climate Action Summit in 2013. The Climate Emergency Network in Victoria has gone into abeyance for the moment. On the positive side, however, Climate Action Network in South Australia has joined forces with Beyond Zero Emissions, the Socialist Alliance, and community activists to pass a referendum to replace the two coal-fired power plants in Power Augusta with a solar energy plant.

Australian climate politics has taken on a new tenor with the election of Tony Abbott as Liberal-National Prime Minister in September 2013. He has promised to abolish the Carbon Price Mechanism and already dissolved the Climate Commission and the Department of Climate Change. The Abbott government claims to reduce CO_2 emissions through its Direct Action Plan which entails planting trees, improving soil carbon by organic fertilizer, and improving energy efficiency.

The disparate nature of the Australian climate movement, even at the grass-roots levels, raises certain questions about the social scientific portrayal of new social movements. For example, Faulks (1999: 88) maintains that the

> novelty of NSMs can be seen in their disillusionment with the statist politics of the socialist left and the neo-liberal right, and their explicit rejection of the state as a tool that can be utilised to create social justice and ensure democratic accountability.

In the case of the Australian climate movement, many climate activists adhere to the notion that they can persuade politicians and even corporate elites to reform the political economy in such a way to prevent dangerous climate change. Furthermore, there are socialists involved in the movement who while eschewing efforts to appeal to the corporate elites to act in a socially responsible manner on climate change, have engaged in electoral politics. Aside of the socialist groups and some groups such as Friends of the Earth and a few others, a major weakness of the Australian climate movement is muted critique of global capitalism and a tendency to deemphasize social justice issues. This is quite in contrast with various other climate action groups around the world, particularly in developing countries, many of which might be more aptly termed *climate justice groups* rather than simply *climate action groups*.

Despite certain weaknesses in the Australian climate movement as a whole, I remain committed to struggling within it. Along with others, particularly those who take a strong social justice stance and critique of global capitalism, I have the hope to transform it from a climate action movement to a climate justice movement. Stilwell poignantly portrays the difficulties of working within social movements but also the possibilities for meaningful social change that they may provide:

> No doubt there are recurrent frustrations for activists working within social movements – it often seems like 'one step forward, two steps back.' Precisely what are the common elements for cooperation is unclear – some movements have a reformist character while others have a more radical agenda. But the breath and diversity are also sources of strength in a politics of opposition. This is a distinctively dialectical element in the process. It is because the prevailing forces of globalisation, neo-liberalism and corporate managerialism recurrently violate numerous social concerns that these movements are continuously reactivated.
>
> *(Stilwell 2000: 163)*

Notes

1 See www.safeclimatebill.org.au, accessed 02/20/2012.
2 See www.cana.net.au, accessed 05/20/2013.
3 See www.safeclimateaustralia.org, accessed 05/20/2013.
4 See www.cleansa.org.au, accessed 05/20/2013.

References

Australian Bureau of Meteorology (2013) *Hottest Month on Record, Final Figures Highlight Climate Extremes*, Significant Weather Media Release, February 1. Accessed 02/20/2012 from www.bom.gov.au/climate.

Australian Climate Action Summit (2011) *Not Good Enough!* Accessed 10/10/2012 from http://climatesummit.org.au/previous-summits/communique/.

ACF (Australian Conservation Foundation) (2008) *What Is an Emissions Trading Scheme?* Accessed 10/10/2012 from www.acf.org.au.

Australian Government (2008) *Carbon Pollution Reduction Scheme: Australia's Low Pollution Future*, Canberra: Australian Capital Territory.

Brown, B. and Singer, P. (1996) *The Greens*, Melbourne: Text Publishing Company.

Burgmann, V. (2003) *Power, Profit and Protest: Australian Social Movements and Globalisation*, 2nd edn., Crows Nest, NSW: Allen & Unwin.

Diesendorf, M. (2009) *Climate Action: A Campaign Manual for Greenhouse Solutions*, Sydney: UNSW Press.

Edwards, T. and Brait, P. (2012) Entrenching Energy Interests, *Arena: The Australian Magazine of Left Political, Social and Cultural Commentary*, 116: 5–6.

Environment Victoria (2009) *The People's Climate White Paper: A Community Action Plan for our Climate*, September 2009.

Faulks, K. (1999) *Political Sociology: A Critical Introduction*, Edinburgh: Edinburgh University Press.

Friends of the Earth Australia (2008) *Open Submission to Garnaut on the Emissions Trading Scheme*, April 17.

Friends of the Earth Australia (2011) Hard-wired Path to Carbon Trading Goes Wrong Direction, Press Release, February 24.

McAuley, I. (2010) Taxing the Miners' Uncommonly Large Profit, *Dissent*, 33: 20–25.

Moore, J. (2011) Carbon Price: The Gas Agenda, *Green Left Weekly*, June 29.

Munro, S. (2012) *Richland, Wasteland: How Coal Is Killing Australia*, Sydney: Macmillan.

Rising Tide Australia (2012) *About Us*. Accessed 10/15/2012 from www.risingtide.org.au/.

Solidarity (2011) Backing Tax on Workers No Way to Stop Abott, *Solidarity*, 33. Accessed 10/15/2012 from www.solidarity.net.au/33/backing-tax-on-workers-no-way-to-stop-abbott/.

Spratt, D. and Sutton, P. (2008a) *Climate Code Red: The Case for a Sustainability Emergency*, Fitzroy: Friends of the Earth Australia.

Spratt, D. and Sutton, P. (2008b) *Climate Code Red: The Case for Emergency Action*, Melbourne: Scribe.

Stilwell, F. (2000) *Changing Track: A New Political Economic Direction for Australia*, Sydney: Pluto Press.

Tietze, T. and Humphrys, E. (2012) The Science Cannot Save Us, in Loewenstein, A. and Sparrow, J. (eds.) *Lefttum. Political Essays for the New Left*, Melbourne: Melbourne University Press.

White, D. and Nichols, D. (2009) Socialist Alliance and the Movement Against Global Warming. *Alliance Voices*, Socialist Alliance Discussion Bulletin and National Newsletter, 9(2). Accessed 10/15/2012 from http://socialist-alliance.wikispaces.com/file/view/%0920090325+Vol+9+No+2+March+2009.pdf.

11

OUT OF THE FOREST – THE CLIMATE MOVEMENT IN BRAZIL

Dörte Segebart and Claudia König

1 Introduction

In 2012 the Brazilian government reported the lowest annual deforestation rate since monitoring began 24 years ago (Brasil.gov.br 2012). The deforestation of the Brazilian Amazon has decreased steadily over recent years, not least due to pressure from a growing environmental and climate movement within civil society. However, Brazil simultaneously fell from rank 7 to rank 33 in the Germanwatch Climate Change Performance Index 2013 (Germanwatch and CAN Europe 2012); primarily because emissions due to deforestation were included in the index for the first time.

Both reports indicate Brazil's unique position within the climate change debate: the country achieved a significant reduction in the deforestation rate but this still remains a massive problem in the Amazon region. Emissions from the deforestation of the Amazon region play a central part in international climate debates. Deforestation also takes center stage in Brazilian climate policy and concerns the majority of the country's climate movement.

The downgrading of Brazil on the Climate Change Performance Index was in part also due to a decreased rating of Brazilian climate policy (Germanwatch and CAN Europe 2012). This is reflected in complaints from Brazilian climate groups who see the previously close and successful cooperation between politics and parts of civil society jeopardized by the new government of President Dilma Rousseff.

Large sections of the Brazilian climate movement are highly professionalized and internally differentiated. Many actors focus on the Amazon forest but due to the geographical size and great expanse of the country there are many regionally specific interests, target groups, and strategies. In the Amazon region the discussion mainly focuses on the reduction of greenhouse gas emissions, especially through implementing measures to reduce emissions from deforestation (REDD).[1] This approach is seen as controversial within the Brazilian climate movement. In the Northeast of Brazil, questions of adaptation to increasing drought periods are a major concern to environmental and climate groups. The vulnerability of urban environments on the other hand, especially in areas of extreme agglomeration like São Paulo or Rio de Janeiro, or the dangers for certain ecosystems in coastal regions[2] or other ecosystems (for example Cerrado) are often neglected in national debates on climate policy and in climate movements.

This chapter presents the aims and strategies of different actors in the Brazilian climate movement and exemplifies their positions and international relations concerning economic instruments for climate protection.

2 Brazilian climate policy

Hosting the Rio Conference on Environment and Development in 1992 marked Brazil's debut as an actor in the field of international environmental and climate policy. The United Nations Framework Convention on Climate Change was agreed at the conference, and subsequently also signed and ratified by Brazil. As a BRICS state[3] and a country with one of the largest primary forest areas worldwide, Brazil's stature in international climate politics has steadily increased. Brazil's importance for climate negotiations is mainly based on its high emissions and the subsequent potential to reduce emissions by decreasing deforestation. Over 70 percent of the country's emissions are still caused by changes in land use (Germanwatch and CAN Europe 2012), mainly in the Amazon region.

Internationally, Brazil champions the Clean Development Mechanisms (CDM) with Brazilian diplomats having been closely involved in its development (Zilla 2009). However, for a long time the country resisted attempts to include forest- and land-use-related emissions in the international climate regime. After Marina de Silva, who had a background in the environmental movement, took office as Minister for the Environment in 2003 under President Lula, the position of the Brazilian government changed, mainly because of the activities of Brazilian NGOs. At the COP 2003 in Milan, for example, members of the NGO IPAM (*Instituto de Pesquisa Ambiental da Amazônia*) presented one of the first draft proposals for the inclusion of deforestation emissions into the international climate regime, forcing the Brazilian government to position itself and take action (Moutinho et al. 2009; see Section 4).

In the following years, a string of climate policy measures were enacted under the Lula administration. Whereas Brazil was previously a strict adherent to the 'common but differentiated responsibilities' approach in international climate policy and resisted voluntary emissions goals for developing and emerging countries, at the COP 15 in Copenhagen in 2009 Lula for the first time presented voluntary emissions reduction goals: within the framework of the *National Climate Change Plan* (Governo Federal do Brasil 2008) Brazil announced it would reduce its emissions by 36–39 percent by the year 2020. This was to be achieved primarily by reducing deforestation in the Amazon region by 80 percent compared to the average of the years 1996–2005. Following the COP 16, the National Climate Change Law[4] was passed, with specific but voluntary goals for emissions reduction. For the implementation of the law, a *National Climate Change Plan* was drawn up, specifying emissions goals, reduction, and adaptation measures for all policy sectors. These goals and measures are ambitious in international comparison. They were drafted in cooperation with NGOs (Couto 2012) and welcomed positively by the Brazilian climate movement. Internationally, Brazilian climate policy has also been positively received. However, there are increasingly dissenting voices correlating with Brazil's drop to rank 33 on the Germanwatch Climate Change Performance Index 2013. The inadequate implementation of national climate policies and the incompatibility of the goals with watered-down environmental regulations and infrastructure projects in the Amazon region have attracted much criticism (Zellhuber 2012).

Deforestation control is a key aspect of climate policy in Brazil. Since 1988, the Brazilian Environmental Ministry (MMA), in cooperation with INPE (Instituto Nacional de Pesquisas Espaciais), has been monitoring deforestation of the Amazon region annually with satellite data. In addition, the NGO *Instituto do Homem e Meio Ambiente da Amazônia* (IMAZON) has specialized in deforestation monitoring. Since 2008, in addition to the Amazon region, other ecologically relevant areas are also monitored, namely Cerrado, Caatinga, Pantanal, Pampas, and the Atlantic forest. Furthermore, the 'National Action Plan for the Prevention and Control of Deforestation of the Amazon region' (PPCDAm) has since 2004 been considered the most important instrument for implementing the climate and environmental agenda and for achieving the deforestation reduction goals. Today the plan is also valid for the above-mentioned regions. In addition to regulatory policies on regional planning, land regulation, environmental monitoring and control, the promotion of sustainable farming systems also forms part of the plan. Since 2007, several decrees have been issued aiming at a local approach to fighting deforestation. These include the blocking of loans for illegal logging operations. At the same time, the entire value chain involved in illegal deforestation was held responsible. In 2008, with support from industry and NGOs, the Brazilian government signed the 'soy moratorium' (Moratório da Soja). The participants agreed to stop trading soybeans grown on previously deforested areas. From 2009, headed by the Federal Public Prosecutor, the 'Legal Meat' (Carne Legal) initiative was set up; consumers were encouraged to use the new means of tracing the origin of beef.

A well-publicized result of these measures and PPCDAm's campaigns was the publication in 2007 of a 'red list' of 43 municipalities with a very high rate of deforestation. The list is still updated annually. Consequently, some municipalities have tried to get off of this embarrassing list, for example by participating in the program 'Green Municipalities' Municipios Verdes. In Pará, the program developed into a government program involving several state ministries (Governo do Estado do Pará 2011). In multistakeholder processes, a municipal 'pact to reduce deforestation' was agreed upon and measures to change the economic structure towards a sustainable use of natural resources were developed. These processes are supported by IMAZON, The Nature Conservancy (TNC), the Vale Foundation, and the public prosecutor. Guidelines for the creation of green municipalities were published in 2011 (Guimarães et al. 2011). Currently, 89 municipalities are involved in this program; some have received money to support their transformation efforts from the Amazon Fund.

The Amazon Fund (Fundo Amazônia) was established in 2008. It was designed as a pilot program for the fund solution preferred by Brazil during the discussions about REDD. The idea of a REDD fund is seen as a countermodel to the integration of emissions from deforestation into international carbon trading. Nation states or other donors are supposed to pay into these funds so that countries with large prime forests can finance emission-reduction projects. The main donor to the Amazon Fund is Norway, which contributed 450 million of the total 500 million US dollars which will be paid out by 2015. The donors receive certificates recording the emissions saved by their contribution. The money is used to fund projects by NGOs, foundations, municipalities, federal states, and universities.[5]

Another environmental initiative, which has topped the agenda of the Brazilian environmental and climate movement in recent years, is the reform of the Brazilian Forest Law (Código Florestal). The reform affects forests in private ownership and provides, among other things, a weakening of safeguards for certain areas such as riverbanks and hilltops. It relieves the small-scale producers of the legal obligation to protect a specific area (depending on the biome between 20 and 80 percent) by nonuse or to reforest it after deforestation. Small-scale producers are those who own between one and four fiscal modules of land. The size of a fiscal module varies between 20 and 400 hectares depending on the municipality. Furthermore, illegal logging prior to 2008 will not be prosecuted if the owners register the land and illegal land seizures prior to 2008 can be legalized. Under strong protest from the environmental movement, the reform proposal was finally adopted in October 2012 after years of debate. In its protests against the retrospective legalization of deforestation and other aspects of the reform, the Brazilian environmental and climate movement was united as hardly before. The larger organizations in particular repeatedly introduced analyses and alternative proposals into the debate, but, despite their close relations to politics, they could not stop the reform.

3 Origins and profile of the Brazilian climate movement

The closeness to politics of many Brazilian organizations is also reflected in their participation in the Climate Forum (*Forum Brasileiro de Mudanças Climáticas*, FBMC) convened in 2004. Through the establishment of the Forum, the Brazilian government has expressed its willingness to engage in a process of dialogue with civil society in an institutionalized form. Today, the forum is one of the most important places for representatives of the climate movement to be critically involved in the government's climate policy.

A precise definition of the Brazilian climate movement is not easy, because it does not refer to itself as such. Furthermore, policy in the fields of energy, agriculture, natural resources, or forestry is also always linked to climate issues. Today, many NGOs engaged in the fields of climate change and protection have recently adopted the topic but treat it as just one aspect among many. They continue to mainly focus on their traditional issues such as social inequality and the critique of existing production and consumption patterns, seeing here the cause of climate change and its consequences. Few of the organizations that are active in the Brazilian environmental movement began with a specific focus on climate change and its causes and consequences. For these organizations, fighting deforestation in the Amazon region was at the core of their activities right from the start. Existing organizations also united in forums and networks where the issues of climate change and, increasingly, climate justice are being met with new strategies. The actors build on social movements that have already been involved in conflicts over natural resources and the fight against environmental crime for decades.

The history of the environmental movement in Brazil is often described as that of a protest movement that has become a professionalized, well-informed, science-driven social movement (see Alonso et al. 2005; Hochstetler and Keck 2007; Alonso and Marciel 2010). This enables cooperative and synergistic relationships between activists, political authorities, and renowned research groups. Consequently, the Brazilian environmental movement today is highly professionalized. However, in some areas there are divergent views and ideological divisions. The main point of divergence of the competing factions consists in their position towards the concepts of ecological economics, in particular towards payments for environmental services (see Sections 3.2 and 4).

3.1 From environmental to climate justice and back – origins of the environmental movement in Brazil

The environmental movement in Brazil originated in a conservation movement, which campaigned for the preservation of certain endangered biomes (e.g., Mata Atlântica or Pantanal). The conservation movement also denounced the use of environmental toxins in agriculture and various forms of industrial environmental crime. It was initially carried mainly by an urban, educated middle class in Southern Brazil. In 1971, for example, the influential *Associação Gaúcha de Proteção ao Natural Ambience* (Agapan) was founded in the Southern state of Rio Grande do Sul by the environmental activist, later Minister for the Environment and winner of the Alternative Nobel Prize, José Lutzemberger. The 1970s and 1980s saw a time of creative and effective demonstration and protest (see Alonso et al. 2005).

In the Amazon region, however, many groups that can be counted as part of the environmental and climate movement were not set up to protect nature but as initiatives campaigning for the rights of traditional user groups. The most prominent example is the rubber tappers' movement, formed in the federal state of Acre around the activist Chico Mendes. The former Minister for the Environment Marina Silva was also originally part of this movement. In 1977 Chico Mendes founded the union of rubber tappers, which later became the National Council of Rubber Tappers (CNS), renamed in 2011 as the National Council of Extractivists (Conselho Nacional das Populações extrativistas).[6] Some large environmental NGOs on the other hand, like IMAZON or the Instituto de Pesquisa Ambiental da Amazônia (IPAM), which are very prominent actors in Brazil's environmental movement today, did not originate from initiatives for the rights of local user groups, but emerged in the context of international discussions about the Amazon region and in close connection to US institutions (Buclet 2009; Couto 2012).[7] Since the 1970s, many international NGOs have started working in Brazil in a similar context (see Section 4).

Despite these relatively early roots, it is only since the 'Rio+10' conference in Johannesburg in 2002 and the Kyoto Protocol taking effect in 2005 that some real impetus has come into climate policy and the environmental movement in Brazil (see Section 4).[8]

From environmental to climate justice

In addition to the established environmental agenda on conservation (see, for example, key areas of activity of the organization SOS Mata Atlântica), in Brazil and other Latin American countries, the concept of environmental justice became part of the debate in social movements in the 1990s. Even though the original American concept was used as a frame of reference (see Caruthers 2008), a unique interpretation of environmental justice developed in Brazil (see Acselrad 2008).

The creation of the Environmental Justice Network (*Rede Brasileira de Justiça Ambiental*)[9] in 2001 was a pivotal moment in Brazil. The network is characterized by the systematic study of the issues, inter- and transdisciplinary organizational processes (see De Freitas et al. 2009), and special participatory procedures (Porto and Milanez 2009). Today, the network consists of almost 100 NGOs, scientific, and grassroots organizations. Interestingly, these organizations are more closely associated with urban social movements, especially in the Southeast and South of Brazil. The larger and well-known environmental organizations that are vocal on climate change and especially REDD tend to be more active in Northern Brazil and are not represented in this network.

In addition to the concept of environmental justice, the concept of climate justice in a relatively broad sense has been established in Brazil (see, for example, FASE 2009). In particular, the organizations of the *Rede Brasileira de Justiça Ambiental* have integrated the demand for climate justice into their agendas. The relatively broad interpretation of climate justice in Brazil in comparison to debates in other countries and at the international level has the effect that the concept of climate justice is hardly distinguished from the term environmental justice. Due to this broad focus, the general criticism of the debate on climate change as being too focused on the issues of reduction (or sequestration) of greenhouse gases and adaptation to climate change and thereby neglecting issues such as sustainable development and social justice, is not valid for Brazil.

Since the UN environmental conference in Rio in 1992, there has been a dialogue in Brazil between the environmental movement and the movements of small farmers and traditional and/or indigenous users of natural resources. Consequently, these groups do not hold opposing views on the sustainable use of natural resources, but are able to express common positions on socio-ecological conflicts. The discourse on climate change and climate justice is anchored in various social movements in Brazil. Furthermore, the issues of climate justice and environmental justice are mostly discussed in relation to issues of social justice and human rights.

3.2 The Brazilian climate movement today: strategic alliances and ideological differences

The climate movement is perhaps the most scientifically oriented and qualified part of today's environmental movement in Brazil. Many of the movement's actors are professionalized NGOs who critically monitor and constructively advise national and international climate policy debates, often formulating their own conceptual proposals.

Research institutions or research-oriented NGOs (such as IPAM, INPE, IMAZON, the *Instituto Nacional de Pesquisas da Amazônia* – INPA), research institutes at universities or individual scientists (e.g., Carlos Nobre from INPE) play a central role in both climate policy and the environmental movement in Brazil. Their studies – in particular, their models and maps – are used by civil society organizations to support their arguments and demands. However, they also conduct specific studies commissioned by civil society groups or in cooperation with them. At the least in an advisory position, research institutions are nearly always involved in studies or other activities of environmental and climate groups. Civil society debates on climate change thus have the same high scientific and technical standards as at the political level or in government. For grassroots organizations, rural communities, or traditional groups, it is increasingly challenging to follow and understand these scientific debates. Consequently, in recent years many NGOs have invested heavily in educational projects. Specific seminars on topics such as the effects of climate change, emissions reduction and adaptation, payments for environmental services, and specifically on REDD are on offer.

As part of national climate policy, climate researchers and civil society groups are invited by the government to join special advisory councils; for example, in the technical steering committee of the Amazon Fund and in the federal forums on climate change. Scientists could thus be considered new, non-traditional actors in the environmental and climate movement in Brazil.

Positions

The Brazilian climate movement links the issue of climate justice to social justice. All actors in the movement share the opinion that these issues belong together. One focus of the discussions is on the analysis and criticism of payments for environmental services – in particular REDD – and in this context of the idea of a green economy. Views diverge in this point and, to simplify somewhat, form two different currents within the Brazilian environmental movement.

The main argument of the one group can be viewed as a criticism of concepts of ecological economics. From a position of criticism of capitalism, it is argued that an internalization of ecological costs in a market-oriented economy is not capable of reducing the negative environmental and social consequences of the economic system. The market economy is seen as a fundamentally destructive force; social redistribution and the realization of ecological sustainability are impossible within its framework (see Grupo Carta de Belém 2011; Terra de Direitos 2012). On the basis of this logic, several organizations also reject the certification of wood and other forest products.[10]

On the other hand, at a practical-political level there are fears that by integrating economic instruments into climate policy, especially through REDD, the pressure on industrialized countries for far-reaching reforms of their power supply will subside, patterns of production and consumption will not change, and there will be no system change or global social redistribution.

This criticism was already leveled over the inclusion of forests in the international climate regime at the start of the 2000s (Belém Manifesto 2000; see Section 4). Similarly, there are fears about the privatization of some previously public and common-use spaces, especially habitats of indigenous people and traditional groups (Grupo Carta de Belém 2011; Terra de Direitos 2012; CIMI 2012).

This criticism is connected with demands for social justice and recognition of traditional rights and traditional knowledge. Special emphasis is placed on the discussion of Payments for Environmental Services (PES) and TEEB (The Economics of Ecosystems and Biodiversity). The commodification of nature, especially the forest, is viewed very critically. In contrast to other Latin American countries though, there is no mention of the spiritual side of nature that must not be sold, in the sense of the 'Madre Tierra' law following from the concept of Buen Vivir that was added to the Bolivian constitution (see Fatheuer 2011). Rather, the privatization of common goods (*Commons*) is feared since it would take away or reduce the habitat of the traditional forest user groups. Private actors are assumed to have a reduced and functional perspective on the forest resources, which does not cover the holistic importance of forests to society. Consequently, the forest per se would not be protected, only its economically viable functions (especially CO_2 storage). This would relieve state actors of their political and financial responsibility (Grupo Carta de Belém 2011; Terra de Direitos 2012).

The organizations in this wing of the Brazilian climate movement differ in general from the other organizations discussed below with respect to their modest financial resources, lower political influence, and weaker media coverage on climate change issues. The driving force in this wing of the movement is established civil society organizations, who in addition to their primary issue of social justice now also work on climate issues. Since 2009 they are better organized, mainly in the group 'Belém Letter' (*Grupo Carta de Belém*)[11] and the Climate Change and Social Justice Forum (*Fórum Mudanças Climáticas e Justiça Social*) which has close links with organizations from the Catholic Church. They include, for example, FASE, the *Instituto de Estudos Socioeconômicos* (INESC), *Terra de Direitos* (TDD), the *Fórum da Amazônia Oriental* (FAOR), *La Via Campesina Brazil*, *Conselho Indigenista Missionário* (CIMI), and some international organizations such as *Friends of the Earth Brazil* and *Amazon Watch*.

The policy documents of this grouping usually do not directly mention climate justice, but most of the organizations have integrated climate justice into their demands. Their positions are consistent with those of other groups of the Latin American and international environmental and climate movements. An example within the Latin American context is Observatório Latinamericano de Conflitos Ambientales (OLCA). Internationally, the demands and criticisms of this faction can be found in the *Durban Declaration on Carbon Trading* (Climate Justice Now! 2009). FASE is the most prominent Brazilian signatory of the Declaration.[12]

The other grouping within the Brazilian environmental movement does not form a united entity but consists of organizations that do not generally reject the idea of payments for environmental services. On the contrary, they see the internalization of ecological costs into economic calculations as necessary. This wing of the environmental movement consists primarily of environmental and climate groups and organizations that deal with socio-ecological issues. The concept of environmental and climate justice features less prominently on the agendas of these groups.[13] The actors are characterized by their closeness to politics. Former members can often be found in public institutions or government bodies. The origin of this grouping can be localized in the years before the Kyoto Protocol came into effect and in their joint support of the inclusion of forests in the international climate regime. For these organizations, climate change has been an important part of their core business for many years. The topic became a vehicle for many to siphon off international climate change funds to Brazil.

The best opportunity to do so was perceived in potential compensation payments for avoiding deforestation (see Section 4).

Organizations in this grouping include, for example, IPAM, IMAZON, the *Instituto Socioambiental* (ISA), *SOS Mata Atlântica*, the *Instituto Centro de Vida* (ICV), larger organizations representing the social movement of the Amazon region like the *Grupo de Trabalho Amazonico* (GTA) and the CNS, and international conservation organizations such as *WWF*, *The Nature Conservancy* (TNC), and *Conservation International* (CI). Most of these organizations are part of the Observatory of the Climate (*Observatório do Clima*) network. The Observatory was founded in 2002 and currently none of its 29 members is part of the aforementioned *Grupo Carta de Belém*. Many of the organizations express their positions systematically in publications. They welcome the potential that could lie in payments for environmental services. However, they certainly criticize the lack of efficiency of existing instruments and the negative social consequences that result from current practices. They also make suggestions for better practices (see ISA 2009; CGEE et al. 2011).

Organizations such as IPAM and IMAZON, who are among the major architects and advocates of REDD in the Brazilian environmental movement, are explicitly against the integration of REDD into international emissions trading and instead support a fund solution. The organizations share this position with the Brazilian government and participate as important actors in the international debate on this issue. Some NGOs are also directly involved in REDD pilot projects in the Amazon region (e.g., IPAM, IMAZON, GTA, TNC) and use their practical experience as a basis for developing new concepts.

The ideological differences between the two groupings recently became more acute over the issue of payments for environmental services and REDD with indigenous groups (see König and Segebart 2013). The discussion centers on the new program for environmental services (Sistema de Incentivos para Serviços Ambientais, SISA) in the federal state of Acre and the protests by indigenous groups and activists against plans for carbon trading in California (US) that propose including emissions reductions from avoiding deforestation in the tropics. Conselho Indigenista Missionário (CIMI), which is associated with the International Conference of Bishops and is also a member of the Grupo Carta de Belém, is a prominent actor vehemently arguing against economic instruments. In 2011, the organizations co-authored a declaration (Carta do Acre 2011) with the World Rainforest Movement. The indigenous organizations themselves have adopted a moderate position in the conflict. This is shown by the fact that the Belém Letter published in 2009 was not signed by any indigenous organization, although the document mentions the rights and visions of the 'povos originários' (indigenous and traditional peoples).

Of the organizations in the Grupo Carta de Belém, FASE is the one that cooperates most with organizations belonging to the other wing. This is also confirmed by the fact that FASE is involved in a project funded by the Fundo Amazônia, despite their critical position towards the REDD fund. FASE is also one of the very few organizations that sit together with organizations from other factions as civil society representatives in the climate forum of the Brazilian government.

The Brazilian Network on Multilateral Financial Institutions (Rede Brasil sobre Instituições Financeiras Multilaterais) and several working groups of the Brazilian Forum of NGOs and Social Movements for the Environment and Development[14] are still places where a few ideologically divergent organizations are active together. Outside of these structures, organizations of the Brazilian climate movement which differ ideologically have little contact with each other.

Strategies

Many organizations in the climate movement in Brazil are active as policy advisors due to their scientifically supported positions. Almost all use the Internet to a large extent and some organizations are strongly involved in international climate movement networks.

Regional networking with organizations in Latin America is still quite rare compared to international networking and compared to other Latin American countries. Very few Brazilian organizations are, for example, a member of the *Climate Action Network* (CAN) Latin America, which, however, is not yet very active on the regional or international level. Similarly, only a few organizations are members of CAN International,[15] several more belong to 350.org, the majority of which have international roots (WWF Brazil, Greenpeace Brazil, Amazon Watch, Amigos da Terra Brazil).[16] Most organizations are members of Climate Justice Now!, primarily organizations and networks considered social movements that can be associated with the market critical grouping.

Whereas organizations based in social movements often try to use their network membership to reach an international audience, the larger environmental NGOs are characterized by their close links to Brazilian politics and used this route in the past to communicate their ideas on the international level. This is especially true for organizations that were established specifically to address international environmental issues and climate change (IPAM and IMAZON for example) and some international NGOs such as WWF and TNC.

Cooperation with private sector actors is one of the strategies of some organizations. For example, the Amazon Sustentável Forum founded in 2009 operates nationwide and unites organizations from civil society, government, academia, and the private sector. A prominent example of the cooperation between the government and the private sector with the participation of NGOs is the Juma REDD project in the state of Amazonas. Funded by the Fundo Amazônia and with money from the Bradesco Bank and the Marriott Hotel Group monthly payments to the residents of an Extractivists Reserve are made in order to protect the area from deforestation. However, the project is controversial in the Brazilian climate movement.[17]

Not least because of the comparatively high interest among European and North-American publics in the Amazon region can Brazilian organizations create an extraordinary political audience for their local issues. One example is the protest movement against the Belo Monte mega-dam project in the federal state of Pará. The protest is internationally known and combines environmental and climate issues with social issues. Similar to the protest against the reform of the Forest Act, and unlike the discussion of REDD and other market instruments, all organizations of the climate movement agree on central criticisms like forced relocation of indigenous people, negative environmental impacts, and indirect deforestation.

In addition to energy policy, social movements in Brazil also focus on economic, agricultural, and infrastructure policy – and this also applies for the climate movement. There is unity and strength here too: for example, on issues such as deforestation for the expansion of agricultural areas, often associated with marginalization and displacement of the local population, or in the protests against the negative effects of transport infrastructure (e.g., BR 163)[18] or agrofuels.

The Internet plays a central role for both the international and the Brazilian environmental movement. Every major civil society actor has a website, runs a blog,[19] a mailing list, or newsletter. Many Brazilian organizations report regularly on their websites on national and international climate policy (especially on the COPs) and publish critical and in-depth analyses.

The Brazilian government has recognized the highly professionalized and scientifically well-founded organizations of the climate movement as a constructive dialogue partner and acknowledges their potential for giving qualified policy advice. The Climate Forum is a good example.

It consists of members of government, including the Brazilian President Dilma Rousseff, representatives of civil society and academia. The members of the Climate Forum meet regularly and together are shaping Brazilian climate policy at national and international level. Recently, however, dissatisfaction and criticism of the government from civil society organizations has increased. The main issues are the mega-dams in the Amazon region and the cuts in climate protection funds, such as the National Climate Fund.

Observation platforms, such as the Observatory on REDD (Observatório do REDD), with more than 30 participating organizations, are also part of critical policy monitoring by Brazilian civil society. The Amazon Fund is also carefully monitored by NGOs – as civil society representatives in the Technical Committee on the one hand and as 'watchdog' organizations via the blog 'De olho no fundo' (Keep an eye on the fund) on the other. The climate and environmental organizations, however, apply for and receive money from the Amazon Fund themselves.

The institutionalized and professionalized parts of the environmental movement are walking a fine line: they actively shape Brazilian climate policy and demand transparent and fair policy in the process. This creates the risk that they legitimize any potential negative effects of state climate policy, and of being accused rightly or wrongly of being co-opted.

4 The Brazilian climate movement on the international level: the role of BINGOs and the conflict about forests in the international climate regime

The major international environmental and climate organizations ('BINGOs' – Big International NGOs) began their activities in the 1970s. Among the BINGOs most active today in Brazil are Conservation International (CI), Greenpeace, The Nature Conservancy (TNC), and the WWF. In the years before and after the Rio conference, CI, Greenpeace, and WWF set up local organizations registered in Brazil. In the first half of the 1990s, various large Brazilian NGOs were founded, like IMAZON, the Instituto Centro de Vida (ICV), ISA, or IPAM. For the development of IPAM and IMAZON, in particular, international connections, especially with the US, were of direct relevance (see Section 3.1). The time around the Rio Conference in 1992 can therefore be seen as the beginning of a greater internationalization of the Brazilian environmental movement.

4.1 *A formative conflict*

International NGOs have played an important part in the formation of the Brazilian climate movement and are now an integral part of it. At the beginning, however, they pursued an agenda largely independent of the Brazilian organizations. This is especially evident in the conflict over the inclusion of forests in the international climate regime that occurred in the years between the passing of the Kyoto Protocol in 1997 and its taking effect in 2005. There was disagreement about whether economic actors from industrialized countries should be able to buy emissions permits by means of reforestation activities, agroforestry projects, or avoiding deforestation under the Clean Development Mechanism (CDM). The debate took place among politicians and members of civil society organizations. However, the main result was the split in the civil society climate movement (Fearnside 2001).

The majority of international NGOs working in Brazil, especially Greenpeace Brazil and WWF Brazil, but also, e.g., Friends of the Earth, were decidedly opposed to the inclusion of forests in emissions trading under the Kyoto Protocol. Their view was shared with their international headquarters (WWF International and Greenpeace International). TNC and CI were in favor of including forests, but stayed out of the conflict.[20] In the view of WWF and Greenpeace, only projects for direct emissions reductions from the use of fossil fuels should be considered for the CDM, in order to not absolve the industrial countries from their historic responsibility and to achieve long-term reductions in emissions (Fearnside 2001; Lamy et al. 2008). This was also the position of the Brazilian government. Only a few organizations from Brazilian civil society followed this argument (Vitae Civilis et al. 2000).

The opponents faced a large majority of Brazilian NGOs in favor of inclusion. These were led by IPAM and ISA, arguing in particular for the inclusion of avoided deforestation (REDD) into the international climate regime. In addition to IPAM, ISA, and other major NGOs, organizations of local and traditional communities, such as GTA and the umbrella organization of the Indigenous Peoples of the Brazilian Amazon (COIAB), positioned themselves in the camp of the supporters. This group of supporters is broadly in line with the current grouping of REDD-friendly organizations. The proponents justified their position with the extremely high rate of deforestation in Brazil and the resulting important contribution that could be made by the country through the reduction of deforestation within the international climate regime, with the help of its international funds (Belém Manifesto 2000). Through the close connection between scientific work and policy advice, they managed to convince important elements of the Brazilian government of their concept of 'compensated emissions reduction' which they had been presenting since COP 6 in Den Haag (Santilli et al. 2000). At the COP 9 meeting, IPAM and ISA representatives presented the concept officially as an initial negotiation proposal. Today it is part of the basis of the internationally discussed REDD concept.

Furthermore, WWF Brazil and Greenpeace Brazil changed their position and thus presented, at least initially, a different view to their international headquarters. Although Greenpeace Brazil has a critical view on REDD, it cannot be assigned to the group that speaks out against REDD. WWF Brazil is now a key player in the debate on REDD. It was, for example, involved in the development of socio-ecological standards for REDD measures (Manceira Bonfante et al. 2010).

The conflict over the role of forests in the international climate regime can, for several reasons, be described as being formative for the Brazilian environmental movement:

i Within the framework of this conflict, many Brazilian organizations positioned themselves for the first time on international climate issues, which is why it can be seen as the birth of the Brazilian climate movement.

ii It created a division within the Brazilian environmental movement between organizations who saw climate change as an opportunity due to the inflow of international funds and those who considered climate change and related international policy instruments primarily as a threat to Brazil and its local population.[21]

iii In the context of overcoming the conflict, the international organizations gained greater independence from their headquarters, because in the wake of the conflict they increasingly formulated their own Brazilian positions.

Today's relative autonomy of the Brazilian branches of BINGOs was also due to the strong Brazilian organizations who were already pursuing and implementing their own climate change discourse at the start of the movement. In some cases, the Brazilian branches of BINGOs even initiated changes in their headquarters. For example, Greenpeace Brazil has played a crucial role in the now global (and not uncontroversial) involvement of Greenpeace International in the Forest Stewardship Council (FSC) and the certification of forest management by logging companies.[22]

After overcoming the initial conflict between national and international organizations, today's situation is characterized by cooperation between them. Brazilian organizations often receive project funding from international organizations and both Brazilian and international organizations advise the Brazilian government directly or indirectly. Greenpeace and the WWF, for example, are members of the above-mentioned Fórum Brasileiro de Mudanças Climáticas; TNC was one of the first organizations to receive funding from the Fundo Amazônia and plays a central role in the implementation of the Municipios Verdes program (see Section 2). Today's closeness between national and international organizations is also shown by the strong fluctuation of personnel between them. It is often the case that employees of national organizations change to the international one and vice versa.[23]

5 Conclusion

The Brazilian climate movement presents itself as a strong, well-organized, professionalized, and financially sound actor. It can build on a decades-long civil society organization culture that is supported by a politically active middle class and growing and strong grassroots organizations primarily from rural areas. Furthermore, it is based on a professionalized NGO sector, which has grown significantly since the early 1990s and has well-trained professionals, a high degree of political influence, and years of experience.

The climate movement has also profited in recent years from some additional factors: (i) the huge ecological relevance of the Amazon region within the context of climate change – as a source of CO_2 emissions through deforestation, as CO_2 storage, and as a tipping point – and the ensuing international grants; (ii) the relatively 'simple' and inexpensive way to reduce emissions by curbing deforestation; (iii) from Brazil's political profile abroad; and (iv) by a relative openness of the Brazilian government for ecological issues. Although the last point is not necessarily reflected in the current economic, energy, and infrastructure policy in Brazil, nevertheless, the country has made its mark internationally with the organization of the UNCED Conference in 1992 and the 'Rio+20' conference in 2012. Of particular note, however, is that the question of adaptation is hardly covered in the considerable wealth of Brazilian publications on climate change. In particular, the majority of institutionalized and professionalized NGOs focus on mitigation. This might also be a reflection of the lack of international attention to this subject.

Scientific actors play an important role in the Brazilian climate movement; both individuals and entire research institutions. They often operate in a zone between policy advice and lobbying. Scientifically oriented NGOs from Brazil have refined political articulation skills at the national and international level. Their studies, which are often translated into English or Spanish, their forecasts, models, and maps on climate change and specifically on payments for environmental services are often used strategically in politics. Scientifically oriented NGOs use the Internet competently and successfully to increase the visibility and impact of their research.

A variety of NGOs are active in the Brazilian climate movement. There are strong differences concerning their abilities to articulate their goals and capabilities and capacities to influence national and international policy. Through the Internet, and a relatively low digital gap in Brazil with respect to NGOs, a democratization of the debate is possible. Nevertheless, there is a growing danger that due to increasing scientific differentiation and internationalization of the climate debate, smaller, less well-funded organizations might be marginalized or left behind. They often cannot afford to build up scientific expertise or participate in international conferences. This also applies to indigenous and other traditional groups, who are less organized in Brazil in comparison to other Latin American countries. In the current conflicts involving indigenous peoples and REDD, the impression is that much is spoken about these groups but too little with them.

The vast majority of Brazilian NGOs active in the climate field today are still partly or even mainly financed by international donors. The national funds, however, have increased significantly in recent years. Therefore, organizations today increasingly receive financial support from private Brazilian foundations or state institutions. Since 2009, the Fundo Amazônia has been an important source of financing. The majority of international donors are private foundations in the US, followed by European organizations and foundations, although funds from the European Union have declined in recent years. A tendency can be observed that organizations which are critical of capitalism are often financially weaker and more likely to get funding from individual European foundations, political and religious organizations, and regional and international networks. In general, a trend towards greater diversification of funding sources can be seen. The Brazilian organizations in particular seem to have a more privileged access to international project funds. However, in order to function at a relatively high standard, these funding sources are essential. This is also why all Brazilian NGOs have professionalized their fundraising. Nevertheless, larger and internationally well-known Brazilian organizations such as IPAM, IMAZON, or ISA usually have greater financial resources.

The Brazilian environmental movement does not seem to be blocked by the ideological polarization described above, which is especially apparent in the context of payments for environmental services. The polarization rather emphasizes the diversity and openness of the democratic debate that exists within Brazilian society. Both groupings of the Brazilian environmental movement have the potential to provide impulses to the global climate movement and global climate policy.

On the one hand, the demand to not only take emissions into account in the climate debate but also issues of social justice and global production and consumption patterns is not sufficiently reflected in international climate negotiations. The Brazilian organizations with decades of experience, close ties to local realities, and their scientific analysis can make an important contribution to this change.

On the other hand, the conditions for the development of innovative approaches to REDD in Brazil are favorable: Brazil has a rule of law that is more stable than in many other countries with tropical forests, good preliminary conceptual work has been done by civil society, and in international comparison there is an open political climate forming the basis for government dealings with the proposals of civil society. As a result, the Brazilian government could, together with civil society, take bold steps to develop compelling and socially just concepts for payments for environmental services that could become a model for the global level.

Notes

1 REDD stands for 'Reducing Emissions from Deforestation and Forest Degradation.' The frequently used term REDD+ indicates the inclusion of forest management measures to raise the carbon levels of forests. In the following, 'REDD' is used to signify these and other current developments.

2 Brazil has about 7500 km of coastline. A third of the Brazilian population lives in coastal regions.

3 The BRICS countries comprise Brazil, Russia, India, China, and South Africa. All five states are characterized by high economic growth. There are regular meetings between the members of this group.

4 See www.planalto.gov.br/ccivil_03/_Ato2007–2010/2009/Lei/L12114.htm (accessed 12/08/2012). Under the law a national climate change fund was set up. At the end of 2012 its funding was reduced, a move that was criticized by the climate movement.

5 Comprehensive information about donors, projects, and project applications, also in English and Spanish, can be found at www.amazonfund.gov.br (accessed 04/11/2013).

6 In Brazil – and in this paper – extractivism means the collection of nontimber forest products, i.e., of forest products not based on the removal of trees (wood) (e.g., nuts, seeds, fruits, resins). In the context of a traditional way of living and working, this might include hunting and fishing.

7 Two US scientists living in Brazil played a prominent role in founding IMAZON. The organization was initially funded by the US-based MacArthur Foundation. IPAM was also founded with the strong participation of Americans (some of the same people as in IMAZON) and in close connection to the Woods Hole Research Center and the Environmental Defense Fund. This connection still exists today (Buclet 2009; Couto 2012).

8 This is also supported by figures from the Brazilian Institute of Geography and Statistics (IBGE): the number of environmental NGOs in Brazil increased between 2002 and 2005 by over 60 percent. The total number of Brazilian NGOs in this period rose by only 22 percent. The study was conducted by the IBGE together with the Instituto de Pesquisa Economica Aplicada (IPEA) and published in 2008. See www.mudancasclimaticas.andi.org.br/node/734?page=0, 2 (accessed 11/22/2012).

9 For the history of *Rede Brasileira de Justiça Ambiental* see Acselrad et al. (2004) and Acselrad (2008).

10 As was related to the authors in an interview.

11 The group's name is based on a letter written in 2009 in the run-up to COP 15 in which the Brazilian government is asked to reject REDD and other economic climate protection instruments. The group currently consists of 53 organizations, many of whom are from the state of Pará, where the first meeting took place. Increasingly, the group has developed into a nationwide network. See http://grupocarta-debelem.wordpress.com/ (accessed 11/18/2012).

12 In addition to FASE the other Brazilian signatories are mainly small, local organizations considered part of the social movement.

13 The organizations forming this grouping are usually not members of Rede Brasileira de Justiça Ambiental.

14 Fórum Brasileiro de ONGs e Movimentos Sociais para Meio Ambiente e Desenvolvimento (FBOMS). See www.fboms.org.br/ (accessed 04/08/2013). Interview partners told us that there is very little interaction within the forum between the ideologically diverse actors.

15 IPAM, Vitae Civilis, and Fundação Grupo Esquel Brasil are members of CAN International.

16 As well as these four international NGOs, Vitae Civilis and some local organizations are members of 350.org.

17 See the debate on the REDD-critical website 'REDD Monitor': www.redd-monitor.org/2010/08/02/juma-reserve-project-in-brazil-fundacao-amazonas-sustentavel-responds-to-criticism/ (accessed05/06/2013). Representatives of various organizations criticized the project in interviews with the authors.

18 The BR 163 runs nearly 1800 km from Cuiabá in Mato Grosso to Santarém in Pará, where there is a port facility primarily for soy. For a long time the road was not properly surfaced. The plans for surfacing the road, mainly to improve the marketing of soy, led to a broad civil society protest, which denounced the increasing deforestation and the displacement of peasants by large soy producers. As early as 2003, civil society organized the Forum of the Social Movements of BR163 (*Fórum dos Movimentos Sociais da BR-163*). After long conflicts, the Forum Sustainable BR 163 (Forum BR 163 Sustentável) was established in 2007, comprising various ministries, federal and municipal representatives as well as representatives of civil society and indigenous groups. In 2009 a regional development plan (Plano de Desenvolvimento Regional Sustentável para a Área de Influência da Rodovia BR-163 – Cuiabá-Santarém) was jointly developed and ratified. Although the mobilization of civil society for and within this plan was strong, the lack of implementation has been strongly criticized and many regard the plan as a failure (see Klingler in press). Today, a large part of the BR 163 has been surfaced. The conflicts though are still ongoing and their resolution is part of the work of many NGOs and social movements in the area.

19 An example is *Vozes do Clima* from FASE, at http://vozesdoclimabrasil.org (accessed 04/11/2013).

20 Neither organization signed the proponents' manifesto (Belém Manifesto 2000).

21 In an interview with the authors, Adriana Ramos, a senior staff member at ISA, differentiated Brazilian climate movements organizations into those who saw climate change as an opportunity and those who perceived it primarily as a threat.

22 This was said to the authors in an interview. For positions of the organizations and their history also see: www.fsc-watch.org/archives/2007/07/02/Greenpeace_attacked_for_supporting_FSC__call_for__suppressed__report_to_be_published (accessed 04/11/2013) and www.greenpeace.org/international/en/campaigns/forests/amazon/amazon-and-sustainable-forestr/ (accessed 04/08/2013).

23 One example is the current Greenpeace Brazil director (Sérgio Leitão), who previously worked for ISA.

References

Acselrad, H. (2008) Grassroots Reframing of Environmental Struggles in Brazil, in Caruthers, D. (ed.) *Environmental Justice in Latin America: Problems, Promise, and Practices*, Cambridge; London: MIT Press.

Acselrad, H., Herculano, S., and Pádua, J. A. (eds.) (2004) *Justiça Ambiental e Cidadania*, Rio de Janeiro: Relume Dumará.

Alonso, A. and Marciel, D. (2010) From Protest to Professionalization: Brazilian Environmental Activism After Rio-92, *The Journal of Environment Development*, 19(3): 300–317.

Alonso, A., Costa, V., and Maciel, D. (2005) The Formation of the Brazilian Environmental Movement, IDS Working Paper 256: Brighton, Sussex.

Belém Manifesto (2000) Manifestação da Sociedade Civil Brasileira sobre as Relações entre Florestas e Mudanças Climáticas e as Expectativas para a COP-6, Belém.

Brasil.gov.br (2012) Amazônia registra menor taxa de desmatamento, November 27. Accessed 12/10/2012 from www.brasil.gov.br/noticias/arquivos/2012/11/27/amazonia-registra-menortaxa-de-desmatamento.

Buclet, B. (2009) Os peritos não governamentais da biodiversidade amazônica e seus financiadores internacionais: uma parceria desigual em torno de interesses comuns, *Revista Pós Ciências Sociais*, 6(12): 89–113.

Carta do Acre (2011) *Carta do Acre. Em defesa da vida, da integridade dos povos e de seus territórios e contra o REDD e a mercantilização da natureza*, Rio Branco.

Caruthers, D. (ed.) (2008) *Environmental Justice in Latin America: Problems, Promise, and Practices*, Cambridge; London: MIT Press.

Center for Strategic Studies and Management (CGEE), Amazon Environmental Research Institute (IPAM), and Secretariat for Strategic Affairs of the Presidency of Brazil (eds.) (2011) *REDD in Brazil : A Focus on the Amazon: Principles, Criteria, and Institutional Structures for a National Program for Reducing Emissions from Deforestation and Forest Degradation – REDD*, Brasília: CGEE.

Climate Justice Now! (2009) The Durban Declaration on Carbon Trading. Accessed 04/25/2013 from www.fern.org/sites/fern.org/files/media/documents/document_3614_3622.pdf.

Conselho Indigenista Missionário (CIMI) (2012) *A Natureza não tem preço: capitalismo verde é neocolonialismo. Diga NÃO ao REDD!* Luziâna.

De Freitas, C., Tambellini, A., Schultz, G., Bertolini, V., and Abreu Franco Netto, F. (2009) Quem é quem na saúde ambiental brasileira? Identificacao e caracterizacao de grupos de pesquisas e organizacoes da sociedade civil, *Ciência & Saúde Coletiva*, 14(6): 2071–2082.

Fatheuer, Th. (2011) *Buen Vivir. Eine kurze Einführung in Lateinamerikas neue Konzepte zum guten Leben und zu den Rechten der Natur*, Berlin: Heinrich Böll Stiftung.

Fearnside, P. M. (2001) Saving Tropical Forests as a Global Warming Countermeasure: An Issue that Divides the Environmental Movement, *Ecological Economics*, 39: 167–184.

Federação de Órgãos para Assistência Social e Educacional (FASE) (2009) Compromisso da FASE Contra a Mercantilização e pela Justiça Climática. Accessed 12/10/2012 from www.asc-hsa.org/node/793.

Germanwatch and Climate Action Network (CAN) Europe (2012) *The Climate Change Performance Index. Results 2013*, Bonn: Germanwatch and CAN Europe.

Governo do Estado do Pará (2011) Municípios Verdes, Belém. Accessed 12/10/2012 from www.sema.pa.gov.br/download/SEMA-Programa-%20Municipios-Verdes.pdf.

Governo Federal do Brasil (2008) Plano Nacional Sobre Mudança do Clima (PNMC), Brasília. Accessed 12/08/2012 from www.mma.gov.br/estruturas/smcq_climaticas/_arquivos/plano_nacional_mudanca_clima.pdf.

Grupo Carta de Belém (2011) Quém ganha e quém perde com o REDD e Pagamento por Serviços Ambientais? Documento de Sistematização das Convergências do Grupo Carta de Belém extraídas do. seminário sobre REDD+ e Pagamento por Serviços Ambientais X Bens Comuns, Brasilia.

Guimarães, J., Veríssimo, A., Amaral, P., and Demacki, A. (2011) *Municípios Verdes: Caminhos Para a Sustentabilidade*, Belém: Imazon.

Hochstetler, K. and Keck, M. E. (2007) *Greening Brazil: Environmental Activism in State and Society*, Durham; London: Duke University Press.

ISA (Instituto Socioambiental) (2009) *É pagando que se preserva? Subsídios para Políticas Públicas de Compensação por Serviços Ambientais*, Série Documentos ISA: São Paulo.

Klingler, M. (in press) *Novo Progresso. Der Südwesten Parás im Spannungsfeld zwischen Inwertsetzungsstrategien und nachhaltiger Regionalentwicklung*, Doctoral Thesis: University of Innsbruck.

König, C. and Segebart, D. (2013) Zukunftsfähigkeit oder Ausverkauf traditioneller Lebensweisen? Emissionshandel in indigenen Gebieten Brasiliens, in Tschorn, L., Schmitt, T., and Husseini de Araújo, S. (eds.) *Widerständigkeiten im Land der Zukunft. Ein anderes Brasilienbuch*, Münster: Unrast.

Lamy, A. C. M., Mertens, F. A. G., and Moutinho, P. R. (2008) Florestas Tropicais nas Negociações Internacionais sobre a Mudança do Clima Global: Barreiras Históricas e Perspectivas Futuras, IV Encontro Nacional da ANPPAS. Accessed 04/25/2013 from www.scielo.br/scielo.php?script=sci_arttext&pid=S0034-73292012000300009&nrm=iso.

Manceira Bonfante, T., Voivodic, M., and Meneses Filho, L. (2010) *Desenvolvendo Salvaguardas Socioambientais de REDD+. Um guia para processos de construção coletiva*, Piracicaba, São Paulo: Imaflora.

Moutinho, P., Cenamo, M. C., and Moreira, P. F. (2009) Reducing Carbon Emissions by Slowing Deforestation: REDD Initiatives in Brazil, in Palmer, Ch. and Engel, S. (eds.) *Avoided Deforestation: Prospects for Mitigating Climate Change*, London; New York: Routledge.

Porto, M. F. and Milanez, B. (2009) Eixos de desenvolvimento econômico e geração de conflitos socioambientais no Brasil: desafios para a sustentabilidade e a justiça ambiental, *Ciência & Saúde Coletiva*, 14(6): 1983–1994.

Santilli, M., Moutinho, P., Schwartzmann, S., Nepstad, D., Curran, L., and Nobre, Carlos (2000) *Tropical Deforestation and the Kyoto Protocol: An Editorial Essay*, Belém: IPAM.

Terra de Direitos (2012) Pagamento por 'Servicos Ambientais' e flexibilização do código para um capitalismo 'verde.' Accessed 12/09/2012 from http://terradedireitos.org.br/wp-content/uploads/2011/08/Analise-PSA-CODIGO-Florestal-e-TEEB-_Terra-de-direitos.pdf.

Vitae Civilis (2000) *Declaração die ONGs Brasileiras sobre Florestas e Mudanças de Clima no Âmbito do Mecanimso de Desenvolvimento Limpo do Protocolo de Quioto*. São Lourenço da Serra.

Zellhuber, A. (2012) Umweltpolitik in Brasilien. Spannungsfeld zwischen Umweltschutz und Wachstumsideologie, in Fontaine de la, D. and Stehnken, Th. (eds.) *Das politische System Brasiliens*, Wiesbaden: VS Verlag.

Zilla, C. (2009) Brasilien und die Klimapolitik: Ein kreativer Partner mit Potential, in Dröge, S. (ed.) *Die internationale Klimapolitik. Prioritäten wichtiger Verhandlungsmächte*, Berlin: Stiftung Wissenschaft und Politik.

12

INDIA – THE LONG MARCH TO A CLIMATE MOVEMENT

Arne Harms and Oliver Powalla

1 Introduction

Hardly any nation state is affected in so diverse ways by the dynamics of anthropogenic climate change as India. Committed to the ideology and practice of rapid economic growth, contemporary India experiences an almost unrestricted exploitation of resources, the establishment of polluting industries, and the lure of new consumerist lifestyles. Therefore India belongs to the group of nations aggravating climate change. At the same time, however, there is hardly any disagreement that the accelerated climate change will have diverse, yet dramatic consequences for India. In fact dislocations have already begun to make themselves felt. They range from irregularities or shiftings in rainfall patterns over the melting of Himalayan glaciers and the endangerment of vast coastal landscapes to the increase of 'natural disasters' (see Toman et al. 2003; Dash 2007; Indian Network for Climate Change Assessment 2010). The immense diversity of likely consequences rests partly, of course, on the geographic and climatic diversity of the gigantic state. But the consequences of climate change in India can hardly be constricted to the mentioned effects. Large parts of the population will become victims of shifting meteorological regimes or disasters precisely due to their social vulnerability and poverty. The increasing contributions of emerging middle classes to accelerated climate change stand therefore in a sharp contrast to the continuously deepening vulnerabilities in other strata of Indian society.

Precisely in its paradoxical dimensions this constellation has been addressed for quite some time in India – particularly by proponents of the Indian environmental movement (Agarwal and Narain 1991). However, the Indian environmental movement itself is characterized by extreme heterogeneity and, at times, contradicting interests. In a recent paper that reflected on the state of the movement, the environmental scientist and activist Sharachandra Lele identified three broad currents. These are (i) 'wildlife conservationists' emphasizing rather narrowly the conservation of particular biospheres; (ii) the so-called 'livelihood movements' focusing on the habitats of poor populations and addressing their structural exploitation and marginalization; and (iii) the 'urban environmentalists' who are mainly committed to more sustainable patterns of resource usage (Lele 2012: 213). In order to unite these diverse environmentalist currents into a social movement, according to Lele, two conditions have to be met. What is needed is, on the one hand, a multidimensional analysis of the roots and consequences of climate change complemented, on the other hand, by the acceptance and linking of plural engagements (Lele 2012:

214–215). Both principles – an elaborated identification of the issue and the correlation of various forms of action – certainly mark the fundamental conditions for social movements that are to evolve. The Indian climate movement, however, is still in its infancy and thus characterized by isolated criticism and dispersed protests. In this chapter we will introduce possible footholds of such a movement and illuminate their stance on climate change. Approaching the problem thus we follow Lele's justified demand. We root our analysis in the interrogation of both discursive and practical dimensions, and ask how particular actors correlate with diverse dimensions of climate change and which alliance and coalitions are fostered through their practical engagements. On such an interaction of diverse layers and actors rests the potential of a broad political mobilization and intervention.

2 Environmentalism in India: trajectories

The fragments of an Indian climate movement are intimately related to Indian environmentalism, yet partly transgress the latter. Surely this development is based on the conceptional, ethical, and practical difficulties to engage in climate change – difficulties that emerge in particular ways under postcolonial conditions (Chakrabarty 2009; Gardiner 2011). Through a brief historical contextualization we will now outline fundamental themes and conflicts of both spectra.

One theme of the debates is the problem of economic growth. The rapid industrialization that India has seen after Independence contributed as a non-intended consequence to accelerated climate change, but is similarly understood to be a crucial, if not irreplaceable instrument to mitigate it (see, e.g., Government of India 2008; Ramesh 2012). Both the highly charged character of this paradox as well as the persuasiveness of the dictum of a rapid development are rooted in the postcolonial predicament of India. From this perspective previous generations not only caused climate change, but they have done so as *colonizing* agents. The problem of intergenerational equity entwines, therefore, with that of historical injustice.

To escape from the consequences of colonialism, Gandhi argued for a return to village life and therefore to economic practices that exploited natural resources somewhat less. Certainly Gandhi's approach relied heavily on the ultimately imagined autarchy of *the* village (Srinivas 2004). Regardless of its faults, this construction played an important role for the success of the Independence movement and continues to pervade, as will be shown later, parts of Indian environmentalism. Overall more influential in this regard proved to be India's first Prime Minister, Jawaharlal Nehru. He inscribed the idea of rapid growth into the postcolonial state (Escobar 1995). Rapid industrialization was meant to guarantee the autonomy of the nation state, to lead out of poverty, and, finally, to strengthen the processes of nation building (Guha 2007: 204–218). This basic approach continued to influence the political and economic agenda when Nehru's gargantuan projects and centralized industries long had lost their appeal and were displaced, partly by a liberalized economy after the 1990s. Among large parts of the climate-interested public, further industrializations are thus understood to be crucial to reduce vulnerabilities to climate change.

Nehru, the third Prime Minister of India, and his daughter Indira Gandhi were outspoken nature lovers. Both continued colonial practices of conservation. Steeped in bourgeois approaches to nature, specific animals, habitats, or landscapes were identified with the nation and put under the control of state administrations (Greenough 2003). The institutionalization of wilderness rested, as often has been noted, on the exclusion of local resource users. Conflicts that emerged against this background point to the second theme of fault line within discourses on climate change in India. Here, more than elsewhere, opposed interests of different strata of society come to the fore. The intention to save local environments or to oppose their devastation is, however, not equally distributed along socio-economic categories. The latter are rather enfolded in social imaginations and practical engagements that have to be taken into consideration. In other words, the prejudice of environmental destructions by and through the poor is (often enough) misleading. Postcolonial India, for instance, has seen substantial and successful environmentalist mobilizations of poor populations that were rooted in localized imageries. Most prominent among these is the Chipko movement opposing the deforestation in the western Himalayas. Taking a cue from this successful, grassroots movement Ramachandra Guha and Juan Martinez-Alier coined the notion of an 'environmentalism of the poor' (Guha and Martinez-Alier 1997).

Through critical engagements with the notion of nature and an increasing inclusion of social dimensions into environmentalist approaches, some of these conflictive dynamics have been modified. In many ways they still pervade environmentalist approaches and are reintroduced in climate change discourses. After all, climate change is envisioned and experienced as a worsening of conditions for both humans and large parts of nonhuman biota. The changes are likely to be interpreted as deep antagonisms. The conflicts surrounding India's endangered tiger population are a case in point: Dominant cosmopolitan groups enforced national parks especially for tigers and excluded other (social) interests (Jalais 2008). Furthermore, the present and future of climate change might, as it is discussed, transform these sanctuaries into densely populated areas.

The third theme pervading Indian environmentalism emerges from the deep roots of environmental initiatives in religious traditions. In spite of the assumed antiquity of relevant practices and ideas, such initiatives gained momentum only in the last few decades. With respect to climate change, two moments of these activities are immediately relevant. That is, on the one hand, the return to village economies mentioned above. Gandhi and his followers called for this radical shift as a break with modernity and a return to a spiritual unity with nature. Gandhi himself saw the foundation for this practice in one of the most important texts of the Hindu tradition, the *Bhagavadgita*. His exegesis of this text then influenced the idea of deep ecology (see for instance Naess 1995) and other radical environmentalist frameworks. Beyond this reformulation of environmental relations that in turn had a deep effect on Indian environmentalist thought, concrete initiatives rest on the ideas of sacral nature. Prominent examples are the conservation of rivers considered holy by large parts of India's population (e.g., Alley 2003) and particular patches of greenery as 'sacred groves' by traditional or indigenous populations (Apffel-Marglin and Parajuli 2000; Karanth 2007). The latter processes are intimately related to identity politics and include therefore powerful, yet highly contested imageries of Indian environmentalism.

3 India's climate policy: between the extension of renewable energy and a strategic lack of commitment

One of the basic ideas of international climate policy, the so-called 'common, but differentiated responsibility' makes India a permanent but inassessable partner in the climate negotiations. Although India is pretty much interested in a reliable climate protection, due to increasing environmental problems and rising energy consumption (Wagner 2009: 61), the country is still more concerned about ensuring its national sovereignty. Agency to solve the climate crisis is still regarded as a duty of developed countries.[1] This very defensive attitude, which is related to the historical asymmetries of economic development, is more and more losing its credibility in face of the rapid growth of the Indian economy. Updated prospects show that India's carbon dioxide emissions will have a share of 9.6 percent of global emissions in 2035 if the current economic development sustains. Right now India is already the world's third biggest emitter of carbon dioxide (IEA 2012).

If one observes the Indian climate policy more closely, some progress becomes visible irrespective of the ongoing blockade of a post-Kyoto agreement. The Indian government does not totally renunciate long-term goals for climate protection. Manmohan Singh at least accepted the two degree limit of global warming as the corridor for future emissions. Furthermore he declared that India's emissions per head will never overshoot those of developed countries (Wagner 2009: 63). At the same time it is clear that such assertions leave enough space for the real dynamic of effective emissions. Instead of determining absolute budgets and by doing so affirming the criteria of the Kyoto protocol, India rather tends to agree on an orientation of climate policies by the energy intensity of the GDP. On this dimension just the efficiency of the Indian economy gets regulated, but not its proper environmental impact. Nevertheless, fundamental breakthroughs have appeared in the history of India's climate diplomacy, but have been withdrawn at once. At the climate summit 2010 in Cancun the former environmental minister Ramesh declared that his country was on the verge of making legally binding commitments in climate protection. He was heavily criticized for that political turnover in the national public – foremost from pro-development NGOs – to such an extent that he had to recede from his position (Betz 2012: 3).

This very wobbly political stance in discourses about global warming has already affected India's internal affairs. In 2008 the National Action Plan on Climate Change (NAPCC) was approved. According to that programmatic document the goal for climate and environmental politics is "to enhance the ecological sustainability of India's development path" (Government of India 2008: 1). The main idea is to give a sustainable direction to the rapid growth of the Indian economy. It includes protection of natural resources and the use of economic profits for social benefits. To cover at least the ecological part of that claim, eight 'missions' were defined and future working packages were sketched out. Several actions are supposed to be realized: Among them are the adaptation of livelihoods to environmental change (National Mission on Sustainable Habitat), the extension of renewable energy (National Solar Mission), and the improvement of environmental education (National Mission on Strategic Knowledge for Climate Change). There are also some concrete goals mentioned such as a 1,000 megawatt increase of installed solar energy capacity per year (Government of India 2008: 11–13). The NAPCC was accompanied by an institutional backing: The relatively new *Advisory Council on Climate Change* subordinated to the prime minister is responsible for the implementation of the program.

At the same time it should not be forgotten that the production and implementation of green technology is also part of the dynamic development of the Indian economy. Climate protection is not exclusively a political task. Enterprises like First Solar or BP-Tata are strongly capitalized project bodies that have discovered renewable energy as an emerging market.[2] Not surprisingly the share of photovoltaic in India's energy supply is steadily increasing. In 2012 it had already reached 11.7 percent.[3] In the same year the symbolic line of one gigawatt of installed capacity was crossed.[4]

Finally the Indian government can't be accused of staying totally passive in the face of dangerous climate change, although the hesitant manner of Indian climate policy is a target of critique from civil society actors. They focus mainly on the doubtful implementation of the NAPCC as well as on the general direction of climate policy which doesn't seem to be in compliance with the two degree limit of climate change. Furthermore, although the field of adaptation may not be totally ignored by Indian politics, it is still standing in the shadow of the thriving combination out of economic growth and green technology.

4 Between fragmented initiatives and substantial overlaps: scales and actors of climate politics in the Indian civil society

4.1 Public protest and scenarios of ecological conversion: the political anatomy of Greenpeace India

With its study *Hiding Behind the Poor* Greenpeace India has caught international attention and has become one of the most important players in the climate politics of the Indian civil society. The global scope of the study contrasts stunningly with its national focus of research. Its object is composed out of statistical insights about carbon dioxide emissions of different layers of the Indian population. Greenpeace has analyzed the energy consumption of 819 households and concluded that the consumption habits of higher income groups already exceeded the threshold of sustainability. Consequently the lifestyle of societal elites can't be regarded as a role model from the perspective of climate policy. On the opposite side the low emissions of the 823 million Indian poor hide the augmented ones of the Indian elite. The low numbers of emissions per head which are cited by Indian politics to reject climate obligations on an international level are founded on the economic exclusion of the majority of the population and prevent a more differentiated discussion about India's responsibility for climate change and alternative future development paths (Greenpeace 2007). That insight forms the basis for the demand of Greenpeace to renew the energetic infrastructure of the country which seems to be necessary to gain space for a more inclusive development of the society: "A major revision of the future of the power sector is needed, shifting investments from coal and nuclear to renewables and energy efficiency, to create the carbon space for the poor to develop" (Greenpeace 2007: 2).

Around the release of *Hiding Behind the Poor* the whole spectrum of actions and contents becomes visible which creates the identity of Greenpeace India: It connects the offensive intervention in India's climate policy with a critical support for the economic growth of the country and a strong lobbying for the poor subaltern. The report was released before COP 13 and was supposed to sensitize the Indian public for the upcoming negotiations about a post-Kyoto agreement.[5]

In contrast to the Government of India there are clear announcements from Greenpeace how carbon dioxide emission should develop in the future. Thereby the two degree limit is accepted as the critical border of global warming.[6] India's contribution to stabilize the carbon dioxide concentration in the atmosphere is seen in freezing its emissions at one tonne per head and year (Greenpeace 2009a: 44). To achieve this Greenpeace is mobilizing for a so-called energy revolution which should contain five elements: (i) the widespread introduction of renewable energy; (ii) respecting the natural limits of the environment; (iii) shutdown of non-sustainable energy resources such as nuclear energy; (iv) social justice in the access to natural commodities; and (v) a decoupling of economic growth from a raise of carbon dioxide emissions (Greenpeace 2009b).

Greenpeace India acts like most of Western NGOs – which is not surprising due to the global range of the organization. What's rather unique is that ecological conversion shall be transferred to a country which had been rejecting an ambitious climate policy. On a principle level the goals and instruments of Greenpeace could be implemented in any other part of the world as well. For example Greenpeace advertises the European concept of emission trading. In this context it becomes obvious that Greenpeace is not wishing for a fundamental opposition against the Indian government, but is rather inclined to support the first steps of a dedicated climate and energy policy. The NAPCC was commented on in a very positive way:

> India's solar mission plan if implemented would benefit millions of Kalavatis across the country by creating clean jobs, triggering high technology diffusion, and help with poverty alleviation in the country. It would also contribute to the fight against climate change and its impacts.[7]

The Indian engagement of Greenpeace is not limited to the role of being just a moderate agency of implementation. Their study *Hiding Behind the Poor* was already characterized by the will to enfold harsh social criticism. That interest is also reflected in the efforts of Greenpeace to electrify rural areas. Low carbon dioxide emissions of low income groups are not necessarily based on a much elaborated environmental consciousness; rather they are symptomatic of the bad energy supply in rural India. In this context, Greenpeace speaks about a cleavage between cities and countryside. While power cuts in urban areas normally do not last longer than two hours, people from the countryside have to live for hours without electricity (Greenpeace 2009a: 4). Although the capacities for energy production are increasing, rural households are the last to be supplied. To highlight these deficiencies Greenpeace is talking about 'energy injustice.' This term is hard to separate from the concept of 'climate injustice' which refers to the disconnection between the responsibility for climate change and the vulnerability towards its consequences. According to this concept, energy supply is unjust if a small, privileged minority has a reliable access to it, while the majority is forced to live in a maintained insecurity about the functioning of important household devices. As a possible solution for that problem Greenpeace is emphasizing the potentials of decentral, renewable energy. Therefore the already existing grid infrastructure which is fed by centralized power plants shall be supplemented by local energy production consisting of solar power and biomass. In such a system the surplus of the individual producers can be passed to other village inhabitants. For the politics of Greenpeace this approach is so far emblematic, as the NGO is concerned about arranging coalitions of stakeholders which play a crucial role in the decentral proliferation of renewable energy. The activities to introduce that model concentrate on the federal state of Bihar. Greenpeace is in close contact to the local population and has documented their livelihood during 'social audits.' The audits were supposed to examine the electrification program which was launched by the federal government in 2005. Beside this rather informative and publicity related exchange Greenpeace itself tries to take action and to bring relevant actors together that have the capacity to accelerate the conversion to rural energy autonomy. At the 'International Renewable Energy Business Conference,' which was also called 'Bihar Yatra on energy revolution,'[8] people affected by energy poverty were represented by delegates of organized civil society. The focus was on a higher ranked patronage. Representatives of financial institutions and potential investors from industrial corporations were invited to the conference as well as developers of renewable energy and members of think tanks.

Around the issue of climate change the actions of Greenpeace are rather orientated to global events. One of these was the visit of Hillary Clinton in Mumbai in 2009. For that occasion Greenpeace brought children from the city together with the US foreign affairs minister whereby they had the opportunity to report about the flooding of their districts. On that day another action of Greenpeace also pointed out the consequences of climate change. In front of the Red Fort in Delhi – an emblem of the city – portraits and citations of inhabitants of the Ganges Delta were presented. Greenpeace doesn't count solely on the political performance of human destinies, but is also taking action in protests. Some of these protests have a global framework like the international 'Polar Bear Day' or 'Global Action Day on Climate Change.' With this continuous engagement Greenpeace is emphasizing its core competence which always has entailed the mobilization of a broad public for environmental problems.

Greenpeace thereby acts predominantly in an autonomous way. Although the organization is somehow connected to other actors by global days of action and common conferences, it would never participate in alliances which could weaken its political independency. There is one exception where the NGO commits itself permanently to a bigger network of exchange and discussion: Like many other organizations Greenpeace belongs to the supporters of 350.org which is an international NGO network with different regional branches. It is founded on the very ambitious demand to downsize the carbon dioxide concentration in the atmosphere to the level of 350 ppm. The membership of Greenpeace India of 350.org is of special interest in the context of this chapter because Vandana Shiva, the founder of the Indian organization Navdanya, also advocates it.

4.2 The globalization of organic farming: Navdanya's combination of international star attraction and local participation

According to Navdanya climate change is a massive societal challenge which is not only converting, but also threatening and destroying the livelihood of the majority of the 1.2 billion inhabitants of the Indian subcontinent.[9] This very drastic appraisal has also to do with the geographical position of the organization. Navdanya's range of action crosses one of the evidence zones of climate change: the Himalayas.[10] It is the location of the water basin on which the majority of the population of Northern India depends. In its publications Navdanya cites a lot of studies which detected an increase in temperature of about 0.6 degrees in the Himalayan area in the last 30 years and diagnosed an annual glacier melt of 10–60 meters. In some areas the glaciers have already receded around 21 percent over the last 42 years.[11] Once glaciers disappear and stop regulation of the water supply in North India, there will be the threat of total dryness in times of heavy drought (Shiva and Bhatt 2009: 6–7).

Peasants in the region between Ladakh and Dehradun had already experienced the effects of climate change in the summers from 2009 to 2011. Foremost precipitation, a crucial factor in agriculture, varied heavily. After the extreme drought in 2009 with massive losses in harvest an extraordinary precipitation with dangerous cloudbursts followed. In 2010 dozens of people died due to a spring tide in Ladakh. In contrast to that the monsoon in the upcoming year was more or less normal. Rain started at the right time and with a sufficient volume.

For Navdanya and its members thus climate change is foremost a source of big insecurity. It destabilizes established expectations about the seasonal climate and normalizes the unnormal.[12] The work of Navdanya is therefore characterized by an idiosyncratic epistemology. In its core stand the so-called "barefoot scientists" (Shiva and Bhatt 2009: xii). These testimonies of climate change are integrated in the scientific work of the organization. As a result climate change is not just an abstract or prospective reality for Navdanya. Instead it already gains space in rural everyday life. Consequently the "barefoot scientists" document ongoing changes of the climate or remember the already observed. Navdanya's climate knowledge gets another special note by the fact that it is transcending pure physical causalities, whereby an ethical stigma gets attached to climate change. The overconsumption of resources appears as a violation of the duties of everyone towards the constitutive conditions of life, which become the object of prudence as they represent more than pure material matter. From this perspective, climate change becomes more than a danger for human life. The degradation of fossil resources counts as a cut in the nutrimental body of mother earth which disturbs a ritualized symbiosis, an assumed harmonic balance. The use of scientific results from other institutions (like IPCC or WWF) delivers therefore an adjust instrument to contextualize these local impressions and to apply them on a regional or global level. By the use of that external knowledge the very diffuse diagnosis of a new unpredictability of the weather becomes enriched with precise numbers and causal chains.

The production of climate knowledge by Navdanya is led by a method which is also used by Greenpeace: the 'Yatra.' It refers to a traditional pilgrimage where peasant communities are consulted. From 2008 until 2009 Navdanya performed a climate change campaign in Garwhal, the region at the foot of Himalaya. Two purposes were behind it: on the one hand a research on the current effects of the phenomena, on the other hand informing the resident population about present and future changes. The campaign consisted of a row of different elements and actions: In interviews so called "climate witnesses" were asked for their environmental experiences (Rawat 2010: 49), thousands of leaflets were distributed and several workshops and conferences conducted. The activities culminated in the already mentioned 'climate Yatra' which lasted for six days. As a part of it numerous presentations were organized to support the exchange with local residents. The participative peak followed a few months later. During the interviews with peasants and people from the villages "veteran intellectuals" were selected (Rawat 2010: 36). These are acknowledged persons from the rural communities who were supposed to input their expertise into a common plan about the mitigation of and adaptation to climate change. After several meetings a common document was given to the Minister of Forest and Environment of Uttarakhand in September 2009.[13]

Regardless of Navdanya's basic democratic approach, the international star cult of Vandana Shiva is constitutive for the organization. She founded Navdanya, developed a philosophy of sustainable, organic agriculture, and cultivated the first fields according to her own method. Her presentations on several conferences made her a global icon of alternative movements. Since the 1970s she has been struggling together with the peasants of the Himalaya region whereby she had also supported the 'chipko movement.' If Vandana Shiva is the face and the heart of Navdanya, then Bija Vidyapeeth has to be called its organism and fertile soil. This so called 'earth university' became a political and agricultural nodal point which forms the base of Navdanya beside the offices in Delhi and Dehradun. On its own soils the organization cares about more than just the biodiversity of farming. Frequently there is also training for peasants from the region at Bija Vidyapeeth which is supposed to be a tool of distributing the benefits and techniques of organic agriculture.

Navdanya is not a welfare organization, but a political project. That becomes evident around the conflict between nature and economy which triggered its foundation. The genetical modification of seeds by companies like Monsanto is regarded as a destruction of traditional peasant habits. It becomes more serious through the fact that this specific agricultural practice is already disappearing since the 'green revolution' and statal programs started to support industrial agriculture (Shiva 1993). In contrast to that, the pursued transformation of Navdanya shall address three dimensions of problems: (i) the conversation of biodiversity; (ii) the challenges of climate change; and (iii) the eradication of poverty. Organic farming is bound with the promise to increase the harvest of Indian peasants and to prepare them for a soaring unstable weather. Besides, the seeds of transnational enterprises would augment the peasants' vulnerability because they get dependent on very expensive and very inflexible agricultural practices. In this context the adaptation to climate change is just another justification to quit chemical farming.[14] Finally it has to be concluded that Navdanya is working on a collective return to more original forms of rural life. The destructive effects of the capitalist market are to be counteracted with the approved traditions of organic farming.

Taking all this together, the work of the NGO is very much characterized by the charismatic leadership of Vandana Shiva, but is absorbing its societal power from a deep embeddedness in local communities. Regarding an Indian climate movement it has to be noticed that Navdanya works foremost in an autonomous way. Sometimes the organization participates in international networks or collaborates with other Indian activists. Nonetheless Navdanya doesn't rely on permanent political partnerships. Vandana Shiva and her work force rather try to be an independent pol of action and initiatives than becoming the institutional arm of peasants' fights or mobilizations.

4.3 Activism in the Indian Ganges delta and the climatological interpretation of environmental hazards

In the south of India's metropolis Kolkata, in East India (the former Calcutta), the fissured and not easily accessible deltaic landscape of the Sundarbans is situated. Because of its peripheral position this archipelago became an arena and subject for heterogeneous environmental initiatives. In recent years, fragments of a regional climate movement appeared. Various civil society actors move about in a complex dynamic marked by the hazardous socio-ecological context, on one hand, and, on the other hand, by the collision of local and regional interests. Distant and isolated, the Sundarbans yet emerge as an important node in entangled environmental discourses (Greenough 1998; Sarkar 2010). Within the framework of socio-ecological analysis the Sundarbans are an extremely low-lying, yet densely populated coastal landscape that is marked by chronic environmental hazards as well as by poverty and deep governmental neglect. In the same time, large parts of the Sundarbans are understood to be untouched by human activity. The latter tracts are subjected to direct regulation by India and Bangladesh, form together the largest continuous mangrove forest of the world, and are habituated by a considerable population of Royal Bengal Tigers. Intimately linked to each other, mangroves and tigers enjoy a crucial role in global environmental narratives and through the identification of national elites with the tigers (Chakrabarti 2009; Jalais 2010). Official conservation efforts, then, lead to highly restrictive regulations, exclusions, and violent encounters – and result, therefore, in the articulation of environmental conflicts.

Actors, ideas, and activities of a localized climate movement are bound up in these complex dynamics. Among the web of actors and institutions we will highlight two organizations: the globalized WWF and the deeply rooted Tagore Society for Rural Development (TSRD). Through this approach we do not intend to emphasize the faulty dichotomy of global process and local actors. We will rather look at two of the most important organizations contemporarily devoting considerable efforts towards climate change, condense interpretations, and try to give a voice to local experiences.

The WWF, for one, has been present in the Sundarbans for several decades. Until quite recently focusing entirely on the tiger population, the regional emphasis of the organization has shifted considerably and included progressively the human presence in the delta. This mirrors both fundamental reorientations within the WWF, as well as in the wider environmental movement. Locally the WWF has taken up the task of giving climate change a human face. Beyond that, it aims at developing strategies to mitigate climate change and to adapt to its destructive consequences. Needless to say, these activities are tied to the perspectives of Kolkata's urbanized middle classes. For the latter the Sundarbans are primarily relevant as a tiger habitat, and only recently as a *symbol* of vulnerabilities to climate change. In contrast to the WWF, which approaches climate change as a powerful presence, urban circles continue to be pervaded by doubts of its actuality or its harmful character.

The TSRD followed a different trajectory. Its focus lays clearly on long-term interventions in localized politics, pedagogic approaches and economic measures to reduce multifaceted vulnerabilities of the marginalized populations. The TSRD, therefore, is among those initiatives that work for a certain degree of development and welfare precisely in those areas that are neglected by the state (Kanjilal 2000). As an actor within, what Lele called "livelihood movements," the TSRD demands a more just right to natural resources very much in line with social reformist principles expressed by the Nobel laureate Rabindranath Tagore and Mahatma Gandhi.[15] Now, in the Sundarbans 'natural' disasters played an important role from the beginning; and possibilities to mitigate and to adapt to these events have not ceased to be so since. Recently climate change was included into the array of environmental hazards. In contrast to the WWF, however, TSRD hardly addresses a transregional public.

Summarizing, the WWF emerges as particularly investing into the conservation of the biosphere reserve, while the TSRD focuses heavily on social issues such as poverty and vulnerability. Nevertheless, both approaches intermingle in conceptional and practical engagements since the mutual dependence of man and environment is undisputed. Both organizations, however, stop short of treating climate change as, ultimately, an entanglement of social and ecological spheres – and frame it as a further environmental hazard, a disaster, or a catalyst of pre-existing hazards. Beyond these theoretical dimensions we are concerned with the usage of the notion of climate change to produce a broad public. In which relation stand the efforts of both organizations with a possible mobilization and, hence, the construction of a future climate movement?

Both organizations are situated at the interface of differing audiences – and not only in an intermediating role. Rather, they translate concepts and experiences, include them in respective debates, and shape the latter thereby. The production of an interested and informed public is a noted aim and emphasis of the climate change related efforts made by both the WWF and TSRD. However, the direction of the interventions – that is, the intended public – differs fundamentally.

In its climate program, the WWF primarily aims at a regional or global public and, indeed, works to foster it. The strengthening of local capacities or a local public clearly is in the shadow of the prominent and highly visible *Climate Witness* program. The latter involved repeated interventions in the media landscape and the circulations of various publications. The interventions were efforts to give a human face to the abstract problem of climate change in order to confront doubts about its existence and to prevent political inactivity. However, the publication of 'faces' and the production of 'actuality' lead directly towards the volatile politics of representation.

On the website of WWF India it is noted: "Climate Witnesses are not victims, they are leaders."[16] Their testimonies would allow for the development of adaptation measures and the recognition of the global causes of climate change. Furthermore, these witnesses would allow for the identification or, at least, the imagination of climate change in other parts of the world. In a publication WWF India relates the narratives from the Sundarbans with others from the Indian Himalayas:

> what follow [*sic*] are voices. Voices of individuals who have witnessed change and are living these changes, of those who are finding ways to survive and move on in the hope of a better future by adapting to climate vulnerabilities and alternative livelihood options.
>
> *(WWF India 2010: 1)*

Beyond voice and agency, what is stressed here is witness.[17] As persons who witnessed global warming, they are included in the publication in order to prove the actuality of the deteriorations and disruptions brought about by climate change. To be sure, the witnesses are awarded a sense of hope and resilience, but this is ultimately not the reason why they matter. They do so rather because of their witness, their direct experience of something that is (still) hidden to the direct experience of the readers: vulnerability, degradation, and disaster. Not unproblematic, of course, is the reduction of hazardous experiences on matters of 'climate change.'[18] But what is more: These voices are included into global discourses, so much has to be critically noted, in order to generate attention. In the present they are already a witness to what will threaten other parts of the world in later years. Within the ethical-moral dilemmas of climate change these voices, therefore, are awarded political weight. The relevance of the witnesses is mainly based on the effect that they allow the readers of their reports to imagine possible vulnerabilities: "Their lives remind us of how precarious our existence is" (WWF-India 2010: 1). The dynamics of witnesses as an act and as a dialogical potential (Das 2006) therefore is muted – in favor of what might be termed a 'proxy witness' of the recipients in Indian urban centers and the Global North.

To be sure, the focus on social dimensions of environmental changes, as much as the intention to make local voices be heard, is rooted in important theoretical shifts, and marks an important development. But the very positionality of these interventions, so far, does not contribute to a *regional* climate movement in India. As a correction of earlier, exclusively conservationist approaches the *Climate Witness* initiative certainly can be understood as enabling a climate movement, but the focus of the latter is ultimately beyond the social present of the Sundarbans themselves.

TSRD, in contrast, devotes its activities entirely to the local habitat and tries to counter rural poverty and widespread development deficits. Its focus rests on marginalized populations and the potentials of agriculture. Different from Gandhi's thoughts, however, nature appears less as a nourishing base of life, but instead as a presence that is considered foremost as a threat. Against the background of the multifaceted environmental hazards in the Sundarbans (cyclones, floods, erosions, predators) as well as the overall fragility of human presence on these unstable islands, this approach is hardly surprising. Nevertheless, this does not mean that environmental relations are of an antagonistic or violent nature. The measures and strategies of the TSRD are rather permeated by the above mentioned environmentalism of the poor. As elsewhere, the environment here is not perceived as lost or endangered paradise, but is sought to be cared for in its double role as everyday hazard and central resource. Both features are transformed under conditions of climate change: hazardous relations increase and resources are modified in ambivalent ways.[19] Against the backdrop of these shifting and still highly uncertain patterns of localized hazards, communication marks presently the cornerstone of climate-related activities by TSRD. Through informal and formal interactions awareness and knowledge for wider changes are to be established; both as precondition and means of mitigation, strengthening of resilience, and pragmatic adaptations.

With regards to the possibilities of a regional climate movement the nature of these interventions entails that the political dimensions and practical negotiations of rights and entitlements rooted therein are contemporarily beyond the scope. Although the empowerment of localized capacities is important, political contestations or the prioritization of climate change are only marginally detectable among local populations. If and to what degree the aware public envisioned by TSRD could translate into a political movement is currently uncertain. Situated in highly vulnerable and already affected coasts, the public is rarely addressed as a political subject in terms of a social movement.

5 Climate change as a complex scenario and the futures of an Indian climate movement

Among the actors presented in this chapter, climate change oscillates between the global and the local, between far-reaching scenarios and concrete experiences. The shared assumption of profound changes in climate are woven into various political approaches and hence differentiated as well as negotiated. From the wide array of articulations, we chose relevant case studies that allow reflection about the general condition of Indian climate activism. To begin with, the following constellation has to be noted: Greenpeace and WWF aim to actively influence national and global discourses. In doing so, they emerge as climate lobbyists somewhat – a route not followed by Navdanya and TSRD which have deep roots in their respective localities and in grassroots activism. Greenpeace and WWF are transnational actors which certainly merge their wide-ranging influence with concrete projects in particular localities. Nevertheless, their interventions are not to be understood as 'quasi-organic' that would stem from the tradition of activism firmly rooted in the social origin of the organization. The latter rather applies to Navdanya whose activities unfolded from and are perpetually related back to the birth place of Vandana Shiva; or the TSRD whose scope is limited by the origins of its founders and a shared regional identity.

Here it could be argued that Vandana Shiva, for example, is a highly globalized and far-traveled campaigner for her cause. That being said, the character of her work is still quite different from that of other large NGOs. Navdanya does not rely on a group of professional and mobile, but ultimately replaceable campaigners that would secure the global influence of the organization. Vandana Shiva rather operates a 'one-woman-show' conveying a quite defensive political message: She urges transnational concerns to not intervene in 'traditional' approaches of Indian farmers. Instead of contributing to a national energy transition or trying to influence international law, she contests the dislocation brought about by a liberalized economy.

Within this dispersed landscape of actors, the economization of climate protection has to be emphasized. The latter is evident from, above others, the growth rates in the market segment of regenerative energy. Due to the potential capacity of the market and the state to create a transition of the energy system one could ask: Is an Indian climate movement really necessary to reduce the CO_2 emissions in India? Can climate activism, to take another perspective, concentrate then on the complicated, yet hardly profitable field of adaptation? We note two considerations opposing such a division of labor between businesses, the state, and social movements. On the one hand it remains to be seen how effective relevant business strategies and trade options really are. How fast and thorough will photovoltaic capacities be introduced and how important will energy efficiency be when the post-Kyoto protocol eventually fails and the lucrative Clean Development Measures are terminated? On the other hand, certain forms of carbon dioxide avoidance actually influence the living standard of poor populations positively. The campaign by Greenpeace in Bihar and partially Navdanya's organic agriculture highlights that. If anything, these dynamics entail negotiations of 'climate-friendly' lifestyles and low-emission energy generations that are somewhat incompatible with the speed and returns expected by the Indian boom economy.

To summarize, instead of an Indian climate movement the present is characterized by a multitude of civil society groups and campaigns. While they engage in related problems and conflicts surrounding climate change, the multitude is not bracketed by a unifying antagonism. An ecologically sound transformation of Indian society can hardly be envisioned without pressure from civil society. The complex terrain of vulnerabilities with its diverse, yet comparable hazardous potentials and realities is most likely the common ground from which cooperation and a shared identity of an Indian climate activism could emerge. As noted at the outset, South Asia is among the global regions most affected by climate change. Yet the state only partially registers this threat and is far from adequately engaging in it. The slow and hardly visible character of climate change certainly contributes to its fading behind the lure of rapid growth. From the marginal zones of media attention, however, a concerted outcry by the vulnerable may soon appear. Integrating different guises of vulnerability and being aware of the shared condition to be unable to adapt, such an outcry would enable offensive demands for improvement and effective safety measures. With the unfolding of such a movement, the existing, sporadic networks of NGOs could enter into a broad Indian climate movement. They would follow particular interests to a lesser extent than is the case now. In the dynamics of such a collective uprising, the concrete actions would be determined by the needs of the vulnerable.

Notes

1 For remembrance: The emerging and developing countries have signed the Kyoto protocol, but they are not committed to any reductions of their carbon dioxide emissions.

2 The boom of renewables is thus not free from political settings. Although there is no central system of promotions like in Germany, India has several state programs which support the engagement of Indian enterprises in this sector. Further information about this issue is available in a study of the German Federal Ministry for Economy and Technology. See www.icon-institute.de/films/Finanzierungsstudie 20EE Indien ICON 2012.pdf (accessed 10/15/2012).

3 See www.gtai.de/GTAI/Navigation/DE/Trade/maerkte,did=558762.html (accessed 11/18/2012).

4 See www.klimaretter.info/energie/nachricht/11918-solarkraft-indien-knackt-1-gigawatt (accessed 11/17/2012).

5 Greenpeace accompanies international climate policy and tries to tighten India's position in the negotiations. Foremost the collaboration of the so-called BASIC states (Brazil, South Africa, India, and China) is observed carefully. The following demands were directed to the summit of BASIC in Delhi in April 2010: Most vulnerable states shall be supported through the transfer of technology and resources; the 30 billion dollars which were promised in Copenhagen shall be paid immediately; the so called 'gigatonne gap' shall be closed soon to adjust the CO_2 reductions of BASIC in compliance with scientific measures; furthermore the four states shall confirm a post-Kyoto agreement (see www.greenpeace.org/india/en/news/greenpeace-calls-for-basic-lea/, accessed 11/18/2012).

6 At the same time Greenpeace points towards the urgency and directness of dangerous climate change. According to its own appraisal climate change is already killing 150,000 people a year (Greenpeace 2009a: 7).

7 Cit.: www.greenpeace.org/india/en/news/energy-r-evolution-what-it-w/ (accessed 12/20/2012).

8 Further information can be found at www.greenpeace.org/india/en/Multimedia/Photos/Energy-Revolution-Bihar-The-Renewable-Energy-Way/ (accessed 12/14/2012).

9 Vandana Shiva and the leader of the climate campaign wrote:

> Climate Change is not a uni-dimensional linear phenomenon. It is complex with many factors interacting to create an unstable climate. Changes in patterns of snowfall and rainfall and changes in temperature are altering the very conditions, which have maintained the fragile and young Himalaya as a stable and sustainable ecosystem on which 40% of humanity depends, largely through the gift of water.
>
> *(Shiva and Bhatt 2009: xi)*

10 The illustrated book *Schicksale des Klimawandels* tries to document the evidence of ongoing changes in the Himalaya (Braschler and Fischer 2011).

11 Astonishingly Navdanya also adopts among these citations a heavily criticized prognoses of the IPCC. According to that the water supply from bedrocks will collapse in the next 20–30 years.

12 Climate change is described by Navdanya as a factor of the extraordinary. It's a hidden source of power which is recreating rural livelihoods, an overthrow, that is suspending well-established modes of agricultural production which became dysfunctional regardless of several attempts of adaptation. The example of apple production in the Himalaya region is therefore very significant. The shift of the snow

line made it impossible in the former Himalayan production sides. For many but not for all peasants such a change in environmental conditions is a complete disaster. Some of the affected apple producers searched proactively for new cultivation methods. They had success with planting other fruits, while apple production shifted to higher layers.

13 The proper name of the document is *People's Participatory Climate Adaptation and Mitigation Plan*. Its political relevance for municipalities and local administration hasn't been investigated.

14 According to Navdanya such an adaptation serves at the same time the mitigation of climate change. Organic farming releases less carbon dioxide than conventional methods because the soil gets less plowed and stays richer in microbiological activity.

15 Amita Baviskar has convincingly shown that the success of particular environmental movements in India is highly dependent on their ability to relate to translocal narratives and elites (Baviskar 2001, 2005). A considerable part of the influence gained by the TSRD rests, therefore, on a muting of the considerable differences that marked Tagore's and Gandhi's approaches to development and freedom (see, e.g., Dasgupta 2004).

16 See www.wwfindia.org/about_wwf/reducing_footprint/climate_change_and_energy/solution/adaptation_and_impacts/climate_witness (accessed 11/19/2012).

17 Voice, agency, and witness are central notions of postcolonial theory. Their intricacies are beyond the scope of this paper. That much has to suffice: As it is often the case also in postcolonial writing, the instrumentalization of voices and subjects is not avoided by the initiative. Neither are, therefore, the problematic politics of representation.

18 For centuries nothing less than dramatic environmental changes has characterized the everyday life in the Sundarbans (Chakrabarti 2009; Sarkar 2010). It is, therefore, a problematic reduction to frame contemporary experiences of these changes solely in the idiom of climate change.

19 Beyond the increase in disastrous events, three processes are particularly relevant for the Sundarbans: sea level rise; changes in rainfall patterns and temperature; and, finally, changes in marine ecosystems. While sea level rise will result in land subsidence that takes on disastrous proportions (Harms 2012), meteorologists actually point to possible improvements of the conditions of paddy cultivation in Eastern India through changed rain patterns and increasing heat. The consequences for marine ecosystems are still highly uncertain.

References

Agarwal, A. and Narain, S. (1991) *Global Warming in an Unequal World*, New Delhi: Centre for Science and Environment.

Alley, K. (2003) *On the Banks of the Gangā: When Wastewater Meets a Sacred River*, Ann Arbor: University of Michigan Press.

Apffel-Marglin, F. and Parajuli, P. (2000) Sacred Grove and Ecology: Ritual and Science, in Key, C. and Evelyn Tucker, M. (eds.) *Hinduism and Ecology: The Intersection of Earth, Sky, and Water*, Distributed by Harvard University Press for the Center for the Study of World Religions, Harvard Divinity School.

Baviskar, A. (2001) Written on the Body, Written on the Land: Violence and Environmental Struggles in Central India, in Peluso, N. and Watts, M. (eds.) *Violent Environments*, Ithaca and London: Cornell University Press.

Baviskar, A. (2005) Red in Tooth and Claw? Looking for Class in Struggles over Nature, in Ray, R. and Katzenstein, M. F. (eds.) *Social Movements In India: Poverty, Power, and Politics*, Oxford: Rowman & Littlefield.

Betz, J. (2012) Indiens nationale und internationale Klimapolitik, *GIGA Focus Asien, 5/2012*, Hamburg: GIGA.

Braschler, M. and Fischer, M. (2011) *Schicksale des Klimawandels*, Berlin: Hatje Cantz.

Chakrabarti, R. (2009) Local People and the Global Tiger: An Environmental History of the Sundarbans, *Global Environment*, 3: 72–95.

Chakrabarty, D. (2009) The Climate of History: Four Theses, *Critical Inquiry*, 35(2): 197–222.

Das, V. (2006) *Life and Words: Violence and the Descent into the Ordinary*, first edition, New Delhi: Oxford University Press.

Dasgupta, U. (2004) *Rabindranath Tagore: A Biography*, New Delhi: Oxford University Press.

Dash, S. K. (2007) *Climate Change: An Indian Perspective*, New Delhi: Cambridge University Press.

Escobar, A. (1995) *Encountering Development: The Making and Unmaking of the Third World*, Princeton: Princeton University Press.

Gardiner, S. (2011) *A Perfect Moral Storm: The Ethical Tragedy of Climate Change*, Oxford: Oxford University Press.
Government of India (2008) *National Action Plan on Climate Change*, New Delhi: Prime Minister's Council on Climate Change.
Greenough, P. (1998) Hunter's Drowned Land: An Environmental Fantasy of the Victorian Sundarbans, in Groove, R. (ed.) *Nature and the Orient: Environmental History of South and Southeast Asia*, New Delhi: Oxford University Press.
Greenough, P. (2003) Pathogens, Pugmarks, and Political 'Emergency': The 1970's South Asian Debate on Nature, in Greenough, P. and Tsing, A. (eds.) *Nature in the Global South: Environmental Projects in South and Southeast Asia*, Durham: Duke University Press.
Greenpeace (2007) *Hiding Behind the Poor: A Report by Greenpeace on Climate Injustice*, Bangalore: Greenpeace India.
Greenpeace (2009a) *Energy Revolution. A Sustainable India Energy Outlook*, Bangalore: Greenpeace India.
Greenpeace (2009b) Energy [R]evolution. What it will take to (em)power Kalavatis across India. Accessed 10/15/2012 from www.greenpeace.org/india/en/news/energy-r-evolution-what-it-w/.
Guha, R. (2007) *India After Gandhi: The History of the World's Largest Democracy*, Gurgaon: Picador.
Guha, R. and Martinez-Alier, J. (1997) *Varieties of Environmentalism: Essays North and South*, London: Earthscan.
Harms, A. (2012) Squatters on a Shrinking Coast: Environmental Hazards, Memory and Social Resilience in the Ganges Delta, in Luig, U. (ed.) *Negotiating Disasters: Politics, Representation, Meanings*, Frankfurt am Main: Peter Lang.
IEA (International Energy Agency) (2012) *Understanding Energy Challenges in India: Policies, Players and Issues*, Paris: IEA.
Indian Network for Climate Change Assessment (2010) *Climate Change and India – A 4×4 Assessment: A Sectoral and Regional Analysis for 2030s*, New Delhi: Ministry of Environment & Forests.
Jalais, A. (2008) Unmasking the Cosmopolitan Tiger, *Nature and Culture*, 3(1): 25–40.
Jalais, A. (2010) *Forest of Tigers: People, Politics and Environment in the Sundarbans*, New Delhi: Routledge.
Kanjilal, T. (2000) *Who Killed the Sundarbans?* Calcutta: Tagore Society for Rural Development.
Karanth, U. (2007) Sacred Groves for the Twenty-First Century, in Rangarajan, M. (ed.) *Environmental Issues in India: A Reader*, Noida: Pearson.
Lele, S. (2012) Climate Change and the Indian Environmental Movement, in Dubash, N. (ed.) *Handbook of Climate Change and India: Development, Politics and Governance*, London and New York: Earthscan.
Naess, A. (1995) Self-Realization: An Ecological Apporach to Being in the World, in Drengson, A. and Inoue, Y. (eds.) *The Deep Ecology Movement: An Introductory Anthology*, Berkeley: North Atlantic Books.
Ramesh, J. (2012) Foreword by Jairam Ramesh, in Dubash, N. (ed.) *Handbook of Climate Change and India: Development, Politics and Governance*, London; New York: Earthscan.
Rawat, R. S. (2010) Participatory Research on Impacts of Climate Change on People's Livelihood in Garwhal Himalaya, in Shiva, V. (ed.) *Climate Change at the Third Pole*, New Delhi: Navdanja.
Sarkar, S. C. (2010) *The Sundarbans: Folk Deities, Monsters and Mortals*, New Delhi: Social Science Press.
Shiva, V. (1993) *Monocultures of the Mind: Perspectives on Biodiversity and Biotechnology*, London: Zed Books.
Shiva, V. and Bhatt, V. (2009) Himalayan Ecosystems and Climate Change: An Overview, in Shiva, V. and Bhatt, V. (eds.) *Climate Change at the Third Pole*, New Delhi: Navdanya.
Srinivas, M. N. (2004) The Indian Village: Myth and Reality, in Madan, Vandana (ed.) *The Village in India*. Oxford: Oxford University Press.
Toman, M., Chakravorty, U., and Gupta, S. (2003) *India and Global Climate Change*, London: Oxford University Press.
Wagner, C. (2009) Indien: ein schwieriger Partner in der internationalen Klimapolitik, in Dröge, Susanne (ed.) *Die internationale Klimapolitik. Prioritäten wichtiger Verhandlungsmächte*, Berlin: Stiftung Wissenschaft und Politik.
WWF India (2010) Voices of Change: Stories from Ladakh and Sundarbans. Accessed 10/15/2012 from http://awsassets.wwfindia.org/downloads/voices_of_change.pdf.

13

CHINA'S EMERGING CLIMATE CHANGE MOVEMENT – FINDING A PLACE TO STAND

Patrick Schroeder

1 China and the climate change challenge

The task of engaging and cooperating with large developing countries to support their national and local climate change mitigation and adaptation actions is becoming increasingly important. China in particular – with about 9.05 billion tons of CO_2 emissions, or 28 percent of the world's total, from fossil fuels in 2011 (Global Carbon Project 2012) the world's largest emitter of greenhouse gases – has been gaining more and more attention by media, research, and the global public. The rapid increase of CO_2 emissions in China is often used as an argument in industrialized countries that there is no point in taking action, as the emissions of big emitters cancel out reductions in Western countries. This argument, however, does not hold, because China is taking action.

The country's current 12th Five-Year Plan (2011–2015) includes many elements for a low-carbon development pathway and progressive thinking on setting up emissions trading schemes similar to that in Europe (e.g., Han et al. 2012). Furthermore, the rapid development of China's renewable energy sector and investment in clean energy technologies – China became the global leader in renewable energy in terms of annual investment in 2010 – and will likely remain the global leader in the coming decades (REN21 2013), are also key developments for climate protection.

In addition to top-down government action which pushes for renewable energy, energy efficiency, and greening economic growth, the successful transition to a low-carbon economy and society will, in addition to technology innovation, also require much more participation from the public. Raising public awareness and countering the strong influence of emission-intensive industries on climate policy-making will be crucial. In this regard, China is not fundamentally different from Europe, the US, or any other country where a vast majority of the public is not overly concerned about climate change, and emission-intensive industries are lobbying strongly against climate policies.

The development of China's climate change movement has not received much attention from the academic research community. Although China does not (yet) have a strong and mature civil society and environmental movement comparable to that of other countries pushing for a transition to a low-carbon economy and society from the bottom up, Chinese civil society and CSOs do deserve more attention and credit than they currently receive. Civil engagement in climate protection is a very dynamic field and has undergone new developments in recent years which will be discussed in more detail in the remainder of this chapter. However, whether these activities will develop into a full-fledged climate change movement remains to be seen.

2 Who's who in China's climate change movement?

China's emerging climate change initiatives by civil society have their roots in the environmental green community which emerged in the mid-1990s (Yang 2005). The number of civil society organizations (CSOs) working on environmental issues has been increasing steadily in recent years. However, the overall share of environmental CSOs is still very low (see Table 13.1).

Despite the increasing number of environmental CSOs and increasing numbers of public protests, China's environmental movement is still in early stages of formation (see Stanley and Yang 2006). The number of Chinese CSOs working specifically on climate change related issues is even smaller. According to estimates of experts working in the area the number is around 50. While the CSO initiatives do not constitute a movement, they would be at the core of any future climate movement. Many of these are based in Beijing, close to the central government departments and think tanks. Being closer to the political center as they are, it is mainly the CSOs based in Beijing which are engaged in national climate change and energy policy lobbying, while the environmental CSOs from other cities and regions are less involved on a national level, but work on local solutions for low-carbon development. CSOs working on climate change exclusively apply "soft" approaches and do not use confrontational elements associated with climate change movements in other parts of the world. China's emerging civil climate change movement therefore lacks the element of "contentious collective action" (Tarrow 1994: 3).

The number of committed individuals shaping the climate movement is still quite low. These individuals are mostly the younger generation of environmentalists, many of which have studied abroad and are familiar with the climate change movements in the US and Europe. Furthermore, a number of low-carbon entrepreneurs can also be counted towards the drivers or key persons of China's climate change movement. A prominent example would be Huang Ming, CEO of the Himin Solar Group in Dezhou city.

Furthermore, through the Internet and particularly the Chinese *weibo*, a microblog hybrid version of Twitter and Facebook, which is one of the most popular sites in China, used by well over 30 percent of Chinese Internet users, online debates about climate change issues are emerging. The Internet shapes public opinion, particularly those of the younger generation. Furthermore, through media and environmental journalist blogs, the issue of climate change is increasingly made known among the general public. The bilingual website *Chinadialogue*[2] has also become an

Table 13.1 Registered environmental CSOs in China, 2007–2010

	Social organizations	*Nonprofit enterprises*	*Foundations*	*Total number*	*Share of total number of CSOs*
2007	5,330	345	34	5,709	1.48%
2008	6,716	908	28	7,652	1.85%
2009	6,702	1,049	35	7,806	1.81%
2010	6,961	1,070	47	8,078	1.81%

Source: China Ministry of Civil Affairs (2012).[1]

important web portal for international communication for Chinese CSOs to post news, articles, opinion pieces about climate change and related issues in China and globally.

In addition to CSOs one could also count carbon market professionals into the climate change movement. Through the dominant role and proactive engagement of China in the Clean Development Mechanism from 2006 to 2012, China supplied 60 percent of all 1,200 million certified emission reductions (CERs) issued by the end of 2012 (Point Carbon 2013). Many professionals have built up knowledge on climate change and technical understanding of market-based mechanisms on how to implement mitigation actions.

The motivations why people are getting engaged to do something about climate change include solidarity with China's rural population which would be strongly impacted by future unchecked climate change (CCICED 2009). Engagement on the international level is motivated by influencing China's 'going out strategy' and contributing to China becoming a responsible player in the multilateral system.

3 Actors, organizations, and networks

Examples of CSO initiatives to raise awareness about climate change and promoting low-carbon development is the Green Commuting Fund, initiated by the China Association for NGO Cooperation (CANGO) and the Environmental Defense Fund (EDF). The fund carried out activities to raise awareness and increase public acceptance towards the restrictions on car use in more than 20 cities and cooperated with the organizers of the Beijing Olympics in 2008, the Shanghai Expo in 2010, and the Guangzhou Asian Games in 2011. Through this campaign, which involved a number of innovative approaches to draw the interest of the public and enterprises, such as low-carbon commuting metro cards and online carbon calculators, the concept of 'green commuting' (绿色出行 *lv se chu xing*) has become widely known and is now even being promoted across China by municipal governments.

Other types of projects include the establishment of low-carbon rural eco-villages in Sichuan Province by Global Village of Beijing, forest carbon projects by Shanshui Conservation Centre, and supporting small and medium sized companies to measure energy consumption and calculate greenhouse gas emission data by the Institute for Environment and Development (IED).

To move beyond individual projects with localized impact to more coordinated CSO climate action, the China Civil Climate Action Network (CCAN) was established in 2007, with the network secretariat based at CANGO. The establishment of the CCAN network is a continuation and formalization of this process of cooperation between environmental CSOs. Prior to setting up the network in 2007, environmental CSOs were not very interested in the topic of climate change, as it seemed too remote from the local environmental problems CSOs had been working on. Furthermore, the issue of climate change was perceived as a donor-driven issue, with only limited funding being available, potentially setting CSOs in a competitive relationship with each other, rather than promoting collaboration. Therefore, the initial goals of the network were defined as providing capacity-building opportunities for domestic Chinese environmental groups on climate change, introducing Chinese CSOs to the international climate change negotiations and commencing work on national climate and energy policies – all areas in which China's CSOs had not been very actively engaged in until the network had been established.

4 Key events

It is probably fair to say that the crucial event for the Chinese climate CSOs was the UNFCCC Tianjin conference in October 2010, where more than 60 Chinese CSOs joined forces and organized a multitude of activities under the slogan of 'Green China, Race to the Future' (绿色中国, 竞跑未来 *Lv se zhong guo, jing pao wei lai*) (Dong 2010). It is quite telling that this level of activity was only possible in the context of an international UN conference and through support of the Global Campaign for Climate Action (GCCA). Although Chinese media were present and participated in CSO activities, the CSOs' efforts remained more or less unnoticed by the Chinese general public in Tianjin, due largely to government restrictions which limited both the scope and visibility of CSO activities.

More recently, the UNFCCC negotiations in Durban in 2011 and Doha in 2012 were a key event for the CCAN network and other CSOs – several side events jointly organized with the Chinese delegation showed that CSOs have been acknowledged by the Chinese government as important stakeholders in climate change.

5 Strategies and activities

The groups within the Chinese CSOs who are working on climate change have initiated a new long-term climate campaign 'C+ Action – Beyond Government Commitment, Beyond Climate, Beyond China.' C+ Action is an initiative by 40 CSOs based on the idea of mobilizing stakeholders to take actions beyond government commitments. The project description states that "C+ encourages progressive climate actions from all sectors and individuals. Under C+ framework, companies, schools, urban communities and households, villages as well as individuals could join our initiative" (CCAN 2011a). The campaign has the goal of engaging Chinese stakeholders to take more action than that required by government, especially in regards to energy efficiency and emissions reduction. As such the C+ Initiative would have a key role in the creation of a civil climate change movement. The issue of MRV ('measurable, reportable, and verifiable') will be a central element of this campaign, through which CSOs will not only encourage stakeholders to take action, but also offer cooperation on the issue of transparency of information. Chinese CSOs will develop independent tools and methodologies for the measurement, reporting, and verification of emissions reduction by organizations, companies, and communities, acting as 'third party verifiers.'

Other CSOs are exploring options for on-the-ground low-carbon projects and developing campaign strategies to engage the general public and public policymakers. These will include issues on promoting sustainable consumption and production, low-carbon commuting, or the relationship between the growing municipal solid waste problem and climate change, particularly focusing on waste incineration plants. For these issues, CSOs are building on existing knowledge and expertise while adding the additional dimension of climate change thereto. Here, initiatives overlap with those of other CSO networks, especially those projects relating to municipal waste management issues. Waste incineration in particular is a cross-cutting issue that is also relevant to climate change as most waste-to-energy plants in China currently have very poor environmental performance standards, construction processes are not transparent, and their emissions contribute significantly to local air pollution (Balkan 2012).

For the reaction of a real climate change movement, engaging the public is crucial. The importance of 'public participation' (公共参与 *gong gong can yu*) is frequently stated in official government documents in the context of climate change mitigation, for instance, China's National Climate Change Program from 2007 (NDRC 2007). The government tries to encourage public participation for energy savings and emissions reductions (节能减排 *jie neng jian pai*), with, so far, only limited success, as the concept generally remains a rather empty slogan for most citizens. As China's National Climate Change Program calls for "giving full play to the initiatives of social communities and non-governmental organizations," real and independent bottom-up 'public activism' (公共行动 *gong gong xing dong*) could certainly add much to current government efforts in mobilizing the public to become active climate protectors, rather than reluctantly following yet another restriction, such as limits to private passenger car usage in cities like Beijing and Shanghai.

However, to bring debates about climate change to the attention of the majority of people, much more work is still needed. One major challenge for CSOs shaping public opinion is the growing influence of the idea that climate change and low-carbon development are, in fact, a conspiracy (e.g., Gou 2010). Adherents claim that climate change is a plot by Western powers to restrict China's development and limit China's growing influence in world affairs. This kind of thinking has found its way to the mainstream, especially since the Copenhagen conference in 2009 after which China received an unjustified bashing by the Western media (e.g., Lynas 2009). It adds another layer of complexity to the already difficult problem of climate skepticism which is leaking into China from the US.

Engaging in policy-making for climate change is an important objective of CSOs. The Climate Change Department of the National Development and Reform Commission is, at the time of writing, working on establishing a National Climate Change Law, and is seeking public comments and recommendations. Chinese CSOs prepared input and recommendations into the policy-making process for this important legislation (CCAN 2011b). The engagement and provision of inputs to the process is ongoing.

Most recent developments in China's CSO climate change movement also see much closer cooperation between local domestic CSOs of the CCAN network and the international groups based in China – particularly WWF, Greenpeace, and Oxfam. This has not always been the case. While in Europe and the US, these international organizations were incremental in setting up climate action networks, in China, they have not yet taken this leadership role. One key reason has been that international CSOs have been simply too focused on addressing their own causes independently, whereby another reason has been domestic CSOs' desire for self-emancipation, thereby rejecting being led by international CSOs. Furthermore, in the Chinese socio-political context, it is politically difficult for international CSOs to take the lead in CSO networks.

US CSOs such as the Environmental Defense Fund, Natural Resources Defense Council, and The Nature Conservancy are also beginning to contribute their resources and expertise to support the Chinese CSO climate change networks and improve collaboration for a common goal. European environmental CSOs are currently not present in Beijing through representative offices, which makes closer cooperation more challenging. Cooperation with international CSOs in China on issues like media work, joint-messaging, or climate policy, is a double-edged sword. While it does provide opportunities for capacity development and learning from colleagues working in international CSOs, and increases influence and supports capacity development of local CSOs, there is also the danger that the international or US-based CSOs may take the lead while local CSOs simply 'tag along.'

6 Joining global networks – navigating between North and South

Internationally, Chinese CSOs are now also actively engaging with other actors, especially with CSOs from other countries through the various global CSO networks. As there is not *one* unified global CSO climate change movement, Chinese CSOs are still in the process of positioning themselves in the complex landscape of global civil society. The global CSO climate change movement is fragmented into different networks with different approaches and strategies, which in many cases are united only by the common goal of preventing dangerous climate change.

It is often remarked, in regards to the negotiations, that the devil is in the technical detail on how to design effective mechanisms to achieve the required emissions reductions. Not only are the technical issues important, but also considerations of equity and politics between the global North and South. When it comes to participation in international climate change CSO networks, Chinese CSOs need to navigate between sometimes competing global networks and campaigns, such as Climate Action Network International (CAN-I), Climate Justice Now!, Third World Network, 350.org, or Global Campaign for Climate Action (GCCA). Association with each network, in most cases, means taking sides in political questions such as historical responsibility, equity, burden sharing between North and South, or the choice for or against market-based mechanisms, such as emissions trading systems, or advocating complete changes of global socio-economic systems. Although Chinese CSOs as representing a developing country feel close to the CSO networks of the global South, one issue that divides them is the antiglobalization and anti-free trade ideology that underpins the thinking of many Southern CSOs such as Focus on the Global South or Third World Network.

Initially, one of the motivations for founding CCAN was the establishment of a contact point for the Climate Action Network International (CAN-I) in China. While CCAN, in many cases, already acts as the focal point for CAN, formal integration through the establishment of a 'China Node' within CAN has not yet occurred. Reasons for this include a number of factors, including the still-limited capacity of CCAN CSOs, internal disagreements over which organization should host the node, and the concern of being too closely associated with an international CSO network and positions. CAN also closely observes the government's performance during the negotiations and expresses criticism in the form of the 'Fossil of the Day' award by publicly exposing bad performance of countries during the negotiations.

Bilateral international exchange and dialogue is another important element where Chinese CSOs have ramped up their efforts and continue to engage actively in civil society 'climate diplomacy.' For example, in 2010, the German Asia Foundation organized a climate change study-tour and conference for Chinese CSOs (Ang et al. 2010). Regular dialogues and exchanges between European and Chinese colleagues have been established in the context of the climate change negotiations. These dialogues try to bring Chinese CSOs together with European organizations experienced in working with Chinese civil society groups and European CSOs and think tanks active in the climate change negotiations. Since 2012 European and Chinese CSOs have engaged in staff exchange activities on topics such as low-carbon development, climate finance, climate education, and carbon markets (CCAN 2013).

Topics of interest and discussion are positive examples from Europe, such as Germany's nuclear exit-strategy and 'Energiewende' or the EU's energy roadmap, China's energy policies and recent developments, such as strategies of China's current 12th Five-Year Plan for the period 2011–2015. Through these exchanges, Chinese CSOs support the dissemination of good examples for low-carbon development which are happening in China. As much as Chinese CSOs learn about developments in Europe, so do CSOs from Europe learn about China – in many cases they still lack a basic understanding of China's climate change or energy-related policies.

But unsolved problems and deep-lying issues are also being discussed in these CSO exchanges. Issues like 'carbon leakage' and alternative consumption-based CO_2 accounting approaches, which would account for embodied emissions of international trade, have been topics of discussion between CSOs in both Europe and China. Because about 22.5 percent of the emissions produced in China in 2004 were exported, on net, to consumers elsewhere (Davis and Caldeira 2010), solutions for how to share responsibility for emissions among producers in China and consumers elsewhere in an equitable way (Schroeder 2011) are an important but not easily solvable challenge for the future. Very relevant to this issue are questions about the need to realize sustainable lifestyles and well-being in the context of 'harmonious society' (和谐社会 *he xie she hui*) and how this stands in contrast to China's current development model focusing on resource-intensive rapid growth. Furthermore, future debates will circle around the question of compatibility between China's ecological modernization (Mol 2006) and the concept of Ecological Civilization (生态文明 *sheng tai wen ming*) restated by Hu Jintao at the 18th Party Congress of the Communist Party of China in November 2012 as a new development paradigm (Xinhua 2012). These issues are closely related to the European debates about de-growth strategies, all of which will very likely receive much more attention in the coming years by both European and Chinese researchers and environmentalists.

7 EU–China CSO cooperation – putting ideas into practice

More concrete cooperation between European and Chinese on climate change has been somewhat difficult to initiate. European environmental CSOs have, so far, not been very proactive in seeking engagement beyond dialogue. Cooperation on low-carbon city developments is another real opportunity for European–Chinese CSO cooperation. As many municipal governments are beginning to use existing structures and channels of sister-city partnerships to have dialogues on climate change and low-carbon development, CSOs can begin to explore options for city level cooperation. Noteworthy examples are the cooperation partnerships between the cities of Bonn and Chengdu (Germanwatch and E3G 2013) and between Wuxi and Duesseldorf (LCFC 2013), where local CSOs and think tanks are developing concrete models for collaboration mechanisms between Chinese and European CSOs, research institutes, and city municipalities for low-carbon city development. The partnerships also focus on bottom-up low-carbon communities and neighborhood initiatives, which are mostly neglected in urban planning processes.

The East Asia Civil Society Environment Network,[3] which comprises CSOs from China, South Korea, and Japan, has since 2010 held climate change as an important topic for dialogue and exchange. For instance, in July 2010, an 'East Asia Workshop on Low-carbon Communities' was held in Beijing, and in October 2010, at an CSO conference in Kwangju, South Korea, CSO representatives from the three countries issued a trilingual joint statement on climate change. In September 2011, a trilateral CSO dialogue forum titled 'Low-carbon East Asia – the road to its achievement' ([低炭素東アジア]実現への道 *Tei tanso higashi ajia jitsugen e no michi*) (FOE Japan 2011) was held in Tokyo on the topic of energy security, nuclear safety, and renewable energy. China's rapid development of renewable energies was of particular interest for the Japanese and Korean CSOs and the public. In June 2013, Chinese CSOs hosted the next Low-carbon East Asia Forum meeting in Hangzhou city at the Hangzhou Low-carbon Technology Museum, the first of its kind in China, which opened in 2012.

Chinese CSOs are also engaged in dialogues with other colleagues from the BASIC countries (Brazil, South Africa, India, and China), a coalition that had initially been formed by the four governments and has growing influence on the negotiation context. This move by the governments was subsequently followed by CSOs to keep civil society up to date about BASIC country positions.

8 Opening windows of opportunity – widening political space

Chinese environmental CSOs continue to grow despite the environmental, social, and political challenges they face – climate change being the biggest. These challenges also create numerous opportunities. However, Chinese climate CSOs now find themselves in a situation where they need to tackle too many issues at the same time, which creates the danger of overstretching and losing focus.

On the positive side, contrary to other trends in China, where civil rights activists have become more restricted in recent years and months, the political space for CSO activities in the field of environmental protection and climate change has been widening, particularly since the UNFCCC Tianjin conference, where Chinese government officials (likely for the first time) noticed domestic Chinese CSO activism and the work CSOs are doing on low-carbon development. As a result, prior to the COP 16 Cancun conference a meeting between Xie Zhenhua, Vice-Minister of the NDRC, and environmental CSOs (including, among others, grassroots organizations and international environmental groups based in Beijing) took place for a dialogue and exchange of ideas. Minister Xie further encouraged CSOs to continue their active engagement, both nationally and on the international stage. This encouragement has to be seen in the context of the contribution Chinese CSOs have made towards messaging of China's low-carbon actions and achievements. In Tianjin, the government experienced that Chinese CSOs have high credibility with the international community and media, and CSOs can therefore support the government in bridging its credibility gap which lies at the heart of the MRV discussions. As a result of improved relationships with NDRC, in the latter half of 2011, Chinese CSOs have provided input in the form of recommendations to China's climate change legislation – a move which has been welcomed by the climate change department at the NDRC. In 2012 at the climate conference in Doha, CCAN and NDRC hosted a joint side event.

9 Remaining challenges

Politically, there are signals that new space has opened for CSOs, particularly since the political dialogue with the NDRC. But are CSOs in a position to use the new political space? In terms of capacity and professionalism, Chinese CSOs need to acquire the necessary technical skills and knowledge to be able to contribute to future expert consultations and climate policy discussions in order to exert influence on climate policies. Being taken seriously by experts and policy-makers will only be possible if CSOs further specialize in climate change-related areas.

Shaping public opinion remains yet another area of concern. In many cases, CSOs are only discussing climate change issues among themselves, without reaching out to the public. Creating a climate change movement would particularly require this type of outreach. For cooperation and dialogue with CSOs of other countries on very specific political or technical topics, more capacity-building is needed. For example, a number of European CSOs working on issues of emissions trading and energy policy would like to cooperate more closely. However, due to the lack of expertise and resources, many requests for partnerships and collaboration have so far remained unanswered.

Understandably, many Chinese CSOs, like CSOs in other countries, do not see the benefit of putting large amounts of resources into following the international climate negotiation process and prefer to continue focusing on concrete local action. Overcoming the disconnection of concrete on-the-ground projects for low-carbon development and the international dimensions of climate change will be one of the new challenges for Chinese CSOs. Engaging in discussions on issues of fairness and equity to eventually achieve a common understanding is particularly crucial. Reaching a global deal on climate change will require agreement on how to share out remaining 'carbon space' and CSO dialogue has an important role to play in this process (see for instance Athanasiou 2013). One of the future goals of CCAN – becoming an important civil society think tank on climate change issues – will require not only implementing local low-carbon projects and engaging sporadically in international dialogues, but providing concrete inputs to climate change policies on both national level and the international negotiations. More importantly, in the future it will require dealing with potentially sensitive or contentious political topics such as China's growing per capita emissions, China's future contribution to climate change, or green trade barriers in a professional and sensitive manner.

Diversification of funding sources is another major challenge. So far, funding for CSO climate activity comes mainly from foreign foundations or networks. CCAN activities, for instance, are funded by various international foundations and the 'Green China, Race to the Future' campaign in 2010 received financial support from the Global Campaign for Climate Action (GCCA). Funding for international exchanges of CCAN is mainly provided by the German Federal Ministry of the Environment (BMU). More recently, the Chinese foundation Society for Ecology and Entrepreneurs (SEE) has provided some support to CSO work on climate change and for further capacity-building of committed individuals. US foundations like the Energy Foundation are also an emerging financial supporter of environmental groups working on energy and climate. However, international foundations increasingly tend towards allocating their funds for cooperation with government bodies and think tanks rather than CSOs. While continued international support is crucial, the dependency of Chinese CSOs on foreign donors is not a very sustainable situation. While the increasing closeness to NDRC holds the danger of being instrumentalized by the Chinese government, it also offers opportunity to access government funds for action on low-carbon development, thereby reducing the dependency on international funding.

10 Future perspectives and potentials

The funding situation shows how difficult it is for Chinese CSOs to find a place to stand. Becoming a strong, independent Chinese CSO network would also mean having mature and independent position on climate change. Independent would mean independent from the government position, but also independent from international financial supporters and expectations of where Chinese CSOs should stand or what they should think or do. Commenting on the position paper on climate change from 2010, which was signed by more than 60 Chinese CSOs and was published during Tianjin and the COP 16, a number of foreign commentators remarked that the position paper simply echoes the official position of the Chinese government. While it is true that many elements are identical to the government position, the view that CSOs are simply a mouthpiece of the government is very limited and does not fully consider the complexities and evolving relations between CSO, the Chinese government, and international actors. When it comes to public international positions, Chinese CSOs are indeed in a somewhat difficult position: While on the international stage they do support the position of their government, domestically they try to push for more ambitious action of industry and government. Due to this situation, the CSOs have developed a somewhat knotty relationship with the delegation: On the one hand, they are standing up for their country's position in CSO dialogues, while on the other hand they are worried that their increasing engagement within global civil society could result in repercussions back home. This is a reason why, so far, Chinese CSOs are not yet exactly clear on how to articulate what they want in the context of the international negotiations and have so far restrained from to trying to proactively influence negotiations.

Chinese CSOs have also not yet made up their minds about important issues, such as whether to support CSOs' positions of 1.5 degree or 2 degree targets. This is also a difficult issue as CSOs in China, as in other countries, look to scientists for guidance – and there are few Chinese climate change scientists who support the 1.5 degree target.

11 Conclusions

There is no doubt that public awareness, to some degree also public concern, and civil engagement about the issue of climate change is increasing. The different stakeholders are increasingly engaged in networking both domestically and internationally. CSOs through their multitude of concrete solution-oriented climate change projects receive increasing recognition of the Chinese authorities as important players. It is, however, at the time of writing not foreseeable whether the current initiatives by CSOs, entrepreneurs, academics, and online communities will shape into a climate change movement that goes beyond soft advisory approaches and would be characterized in traditional terms of contentious political action including mass mobilizations, protests against construction of coal-fired power stations, public pressure lobbying.

For a strong Chinese climate change movement to influence the international and national climate change debates, it would be necessary for those organizations and individuals which can provide political leadership and those which shoulder the political responsibility for the climate movement to work closely together. Although many respectable efforts have been made, China has not (yet) achieved the shift towards a low-carbon development path. A more mature civil society with a climate change movement that has more freedom to act and freely discuss contentious issues, would very likely support effective implementation of China's national climate chance mitigation and adaptation actions.

Then again, the recent positive developments in China beg the question whether a climate change movement does necessarily require a contentious political element. Given the different political, cultural, and social background, a domestic climate change movement with Chinese characteristics might well become as effective as their international counterparts. This will to a large degree depend on how successful the Chinese government will be in the implementation of its targets and low-carbon development policies.

Notes

1 Official government statistics accessed through personal contacts of the CSO Management Department at the Ministry of Civil Affairs.
2 See www.chinadialogue.net/ (accessed 05/27/2013).
3 See www.enviroasia.info (accessed 05/29/2013).

References

Ang, C., Heidel, K., and Wong, S. (eds.) (2010) *I Could Feel Climate Change. Climate Change and China: Civil Society Perspectives*. Werkstatt Ökonomie and EU-China Civil Society Forum. Accessed 05/26/2013 fromwww.eu-china.net/german/Materialien/Ang-Chin-By-Heidel-Klaus-Wong-Staphany-Eds._2010_I-could-feel-climate-change.html.

Athanasiou, T. (2013) The Climate Talks: Could an Equity Tipping Point Be on the Horizon? Accessed 05/26/2013 from www.ecoequity.org/2013/05/the-climate-talks-could-an-equity-tipping-point-be-on-the-horizon/.

Balkan, E. (2012) The Dirty Truth About China's Incinerators, *Guardian*, July 4. Accessed 05/26/2013 from www.guardian.co.uk/environment/2012/jul/04/dirty-truth-chinas-incinerators.

CCAN (China Civil Climate Action Network) (2011a) C+ Initiative – Beyond Government Commitment, Beyond Climate Change, Beyond China. Accessed 10/10/2011 from www.c-can.cn/en/node/674.

CCAN (China Civil Climate Action Network) (2011b) Zhong guo min jian zhu zhi dui yu zhong guo ying dui qi hou bian hua li fade jian yi (中国民间组织关于中国应对气候变化立法的建议), Beijing. Accessed 05/26/2013 from www.c-can.cn/zh-hans/node/677.

CCAN (China Civil Climate Action Network) (2013) China-EU NGO Exchange Fellowship 2013, Beijing. Accessed 05/26/2013 from www.c-can.cn/en/node/730.

CCICED (China Council for International Cooperation on Environment and Development) (2009) Rural Development and its Energy, Environment and Climate Change Adaptation, Policy Research Report 2009. Accessed 05/26/2013 from www.cciced.net/encciced/policyresearch/report/201205/P020120529358134723461.pdf.

Davis, S. and Caldeira, K. (2010) Consumption-based Accounting of CO_2 Emissions, in *PNAS*, doi: 10.1073/pnas.0906974107 PNAS. Accessed 03/15/2012 from www.pnas.org/content/early/2010/02/23/0906974107.abstract.

Dong, W. (2010) CSOs Urge Global Effort in Low-carbon Race. Accessed 04/15/2013 from www.chinadaily.com.cn/china/2010–10/06/content_11380530.htm.

FOE Japan (2011) Tei tanso higashi ajia jitsugen e no michi. Friends of the Earth Japan. Accessed 05/26/2013 from www.foejapan.org/energy/news/evt_110902.html.

Germanwatch and E3G (2013) Sino-European Partnership on Low Carbon and Sustainable Urban Development. Accessed 05/26/2013 from http://low-carbon-partnerships.org/en/home.

Global Carbon Project (2012) Carbon Budget and Trends 2012. Accessed 04/15/2013 from www.globalcarbonproject.org/carbonbudget.

Gou, H. (2010) *Di tan yin mou* (低碳阴谋), Shanxi: Economic Press.

Han, G., Olsson, M., Hallding, K., and Lunsford, D. (2012) *China's Carbon Emission Trading: An Overview of Current Development*, Stockholm Environment Institute & FORES, Stockholm: FORES.

LCFC (2013) Low Carbon Future Cities – A Sino-German Cooperation on Integrated Climate and Resource Proof Urban Development, Wuppertal Institute for Climate, Environment and Energy. Accessed 05/26/2013 from www.lowcarbonfuture.net/about/project/.

Lynas, M. (2009) How Do I Know China Wrecked the Copenhagen Deal? I Was in the Room, *Guardian*, December 22. Accessed 05/26/2013 from www.guardian.co.uk/environment/2009/dec/22/copenhagen-climate-change-mark-lynas.

Mol, A. (2006) Environment and Modernity in Transitional China: Frontiers of Ecological Modernization, *Development and Change*, 37(1): 29–56.

NDRC (2007) China's National Climate Change Program, National Development and Reform Commission. Accessed 05/26/2013 from www.ccchina.gov.cn/WebSite/CCChina/UpFile/File188.pdf.

Point Carbon (2013) Low CER Prices Leave Chinese CDM Contracts in Disarray. Accessed 05/26/2013 from www.pointcarbon.com/aboutus/pressroom/pressreleases/1.2199929.

REN21 (2013) *Renewables Global Futures Report*, Paris: REN21.

Schroeder, P. (2011) Sustainable Consumption and Production in Global Value Chains, in Harris, P. (ed.) *China's Responsibility for Climate Change: Ethics, Fairness and Environmental Policy*, Bristol: The Policy Press.

Stanley, P. and Yang, D. (2006) An Emerging Environmental Movement in China?, *The China Quarterly*, 186 (June 2006): 333–356.

Tarrow, S. (1994) *Power in Movement*, Cambridge: Cambridge University Press.

Xinhua (2012) CPC Advocates Building 'Beautiful' China, *Xinhua News*, August 11. Accessed 05/26/2013 from http://news.xinhuanet.com/english/special/18cpcnc/2012–11/08/c_131959465.htm.

Yang, G. (2005) Environmental CSOs and Institutional Dynamics in China, *The China Quarterly*, 181(2): 46–66.

14

CLIMATE JUSTICE IN, BY, AND FOR AFRICA

Patrick Bond

1 Introduction

The African component of the climate justice movement is critical, given the continent's extreme vulnerability to climate change, and the 'climate debt' it is owed, having not (South African and Nigerian-based corporations aside) contributed to the world's greatest man-made crisis. The broader Climate Justice movement emerged in part from Africa during the early 2000s, fusing a variety of progressive political-economic and political-ecological currents. Africa soon became the source of some outstanding examples of social mobilization against climate change, its sources, and its impacts. One network, the Pan African Climate Justice Alliance, built a coalition of 500 groups between 2008 and 2013.[1]

Africa has, for centuries, been exploited and looted, and hence an extra amount of intellectual, political, and solidaristic care is required when considering climate justice here (Bond 2006). Exemplifying carelessness is Larry Summers, who in December 1991 as World Bank chief economist, wrote (or at least signed a memo written by Lant Pritchett) in defense of trading pollution, "the economic logic behind dumping a load of toxic waste on the lowest-wage country is impeccable and we should face up to that … African countries are vastly *under*-polluted [original emphasis]."[2] The quote assists us in assessing debates about carbon trading, as noted below.

This chapter begins by reviewing damage anticipated in Africa as a result of climate change, and then turns to the way Africans have been let down by elites – in the Northern capitals and financial markets, as well as in African capitals. Even some well-meaning advocates of climate justice are promoting strategies such as carbon trading that are just as environmentally racist as Summers' ideas, when applied to Africa. As a result, a genuine Climate Justice movement is developing anticapitalist values in several sites, including Durban in 2004 when the Durban Group for Climate Justice was founded.

The stakes could not be higher. Aside from drowning small island states and drying out the glacier, snow, and water supplies of the Andes, Himalayas, and other mountains, it is widely understood that the most harmful climate change process is 'cooking the continent' of Africa, to cite the book by Nigerian activist Nnimmo Bassey (2011). According to UN IPCC director R.K. Pachauri: "[C]rop net revenues could fall by as much as 90 percent by 2100" (Pachauri 2008). Climate damage to Africa will include rapid desertification, more floods and droughts, worse water shortages, increased starvation, climate refugees jamming shanty-packed megalopolises, and the spread of malarial and other diseases. The danger is imminent, for eight of the 20 countries that the Centre for Global Development expects to be most adversely affected by extreme weather events by 2015 are African: Djibouti, Kenya, Somalia, Mozambique, Ethiopia, Madagascar, Zambia, and Zimbabwe. In the Horn of Africa, those affected are anticipated to include 14 percent of Djiboutis, 8 percent of Kenyans, 5 percent of Ethiopians, and 4 percent of Somalis (Wheeler 2011: 15). In 2006, Christian Aid estimated that 182 million Africans were at risk of premature death due to climate change this century (Christian Aid 2008). According to African Union (AU) official Abebe Hailegabriel, "Trillions of dollars might not be enough in compensation. Thus, there must be an assessment of the impact before the figure" (Redi 2009).

In global climate negotiations, however, the most important African negotiator – and largest CO_2 emitter (responsible for more than 40 percent of the continent's CO_2) – is South Africa (Bond et al. 2009). Aside from ostensibly preventing climate change that could have an especially devastating impact at home, Pretoria's climate negotiators have two conflicting agendas. The first is increasing Northern payments to Africa, a longstanding objective of the New Partnership for Africa's Development, when during the early 2000s Thabo Mbeki unsuccessfully requested 64 billion dollars per annum in aid and investment concessions. South African negotiators have also led the G77, and are on record from August 2009 demanding that

> at least 1 per cent of global GDP should be set aside by rich nations in order to help developing countries conduct research, improve flood control, protect their coastlines, create seed banks and take other steps to cope with the severe storms and droughts linked to climate change. The money also could help poor countries obtain technology to reduce their carbon emissions.
>
> *(Sapa-AP 2009)*

The second objective, in contrast, is increasing South Africa's own rates of CO_2 outputs around 2030–2035, when the Long-Term Mitigation Scenario – South Africa's official (albeit nonbinding) climate strategy – would come into effect. Only then are absolute emissions declines offered as a scenario. In the meantime, Pretoria earmarked more than 100 billion dollars for emissions-intensive coal-fired and nuclear-fired electricity generation plants, which would exacerbate Africa's climate crisis, requiring more resources from the North for adaptation.

The way that both mitigation and adaptation narratives are unfolding poses a great threat to the victims, since the valuation of life and planetary ecology through market mechanisms will punish those without market access many times over, especially in Africa where Resource Curse mechanisms leave those with political power closely overlapped with those whose extractive economic interests are opposed to environmental justice. That means that even when court-based justice is sought, in the form of an ecological debt payment, the danger emerges that political elites and other rentiers will capture the funding. The alternative, as argued in the conclusion, is that the funds should go directly to victims of ecological damage, in the form of a Basic Income Grant piloted in Namibia. One reason to reroute climate finance away from corrupt state managers and their allies in the aid industry, is the utterly unreliable way in which Africa's elites have addressed climate.

2 Africans collaborating with injustice

The AU's erratic stance on climate was evident in the period June–November 2009, in which a militant position was adopted against the causal drivers of climate change, and large climate debt payments were demanded. But in December 2009, the Copenhagen Accord dashed hopes of a global deal to cut greenhouse gas emissions and, ironically, the relatives of two signatories, Barack Obama and Jacob Zuma – the Luo and Zulu in rural Kenya and South Africa's KwaZulu-Natal, respectively – will be among its first victims. Ethiopian dictator Meles Zenawi, the AU's primary climate representative prior to his death in 2012, proclaimed in September 2009: "If need be we are prepared to walk out of any negotiations that threatens to be another rape of our continent" (in the same way Africans left the World Trade Organization to collapse in Seattle in 1999 and Cancun in 2003) (Ashine 2009). But Zenawi did not walk out. He walked off his plane in Paris on the way to Copenhagen, into the arms of Nicolas Sarkozy, who persuaded Zenawi to relent (Vidal 2009). The side deal, according to Mithika Mwenda of the PanAfrican Climate Justice Alliance (Pacja), had the effect of "undermining the bold positions of our negotiators and ministers represented here, and threatening the very future of Africa" (Reddy 2009).

Not only did Zuma and Zenawi surrender on emissions cuts, but also on demanding full payment of the North's climate debt to the South. "Meles wants to sell out the lives and hopes of Africans for a pittance," said Mwenda. "Every other African country has committed to policy based on the science." On December 8, Pacja protesters chanted in the main Bella Centre auditorium: "Two degrees is suicide! One Africa! One degree!" (Friends of the Earth International 2009). Zenawi was soon unveiled by WikiLeaks as a convert to the Copenhagen Accord, as a result of pressure applied by the US State Department on February 2, 2010, when Zenawi asked for more North–South resources in return, while also defending his tyrannical rule to the State Department's Johnnie Carson.[3]

In one Copenhagen session, the lead G77 negotiator, Lumumba Di-Aping, "sat silently, tears rolling down his face," according to a South African report. Di-Aping said simply: "We have been asked to sign a suicide pact," explaining that in his home region, it was "better to stand and cry than to walk away."

Speaking in measured tones, Di-Aping first attacked the 2°C warming maximum that most rich countries currently consider acceptable. Referring continuously to science, in particular parts of the latest IPCC report (which he referenced by page and section) he said that 2°C globally meant 3.5°C for much of Africa. He called global warming of 2°C "certain death for Africa," a type of "climate fascism" imposed on Africa by high carbon emitters. He said Africa was being asked to sign on to an agreement that would allow this warming in exchange for ten billion dollars, and that Africa was also being asked to "celebrate" this deal.

He then went on to forthrightly address the weakness of many African negotiating delegations, noting that many were unprepared and that some members were either lazy or had been "bought off" by the industrialized nations. He singled out South Africa, saying that some members of that delegation had actively sought to disrupt the unity of the bloc. He said that civil society needed to hold its negotiators to account, but warned of a long and difficult struggle for a fair climate deal (words to the effect of "you have no idea of the powers that are arrayed against you," spoken in the tone of someone who has spent years interacting with these powers). He said that people all over the world had to be made aware of what a bad climate deal means for Africa ("I am absolutely convinced that what Western governments are doing is *not* acceptable to Western civil society"). He explained that, by wanting to subvert the established post-Kyoto process, the industrialized nations were effectively wanting to ignore historical emissions, and by locking in deals that would allow each citizen of those countries to carry on emitting a far greater amount of carbon per year than each citizen in poor countries, would prevent many African countries from lifting their people out of poverty. This was nothing less than a colonization of the sky, he said. "$10 billion is not enough to buy us coffins" (Welz 2009).

Di-Aping asked, poignantly: "What is Obama going to tell his daughters? That their relatives' lives are not worth anything?" Agreeing with leading US climate scientist James Hansen that the Copenhagen deal on offer was "worse than no deal," Di-Aping concluded: "I would rather die with my dignity than sign a deal that will channel my people into a furnace" (Gilbertson and Santos 2009). In the final plenary session, Di-Aping called the Copenhagen Accord "an incineration pact in order to maintain the economic dependence of a few countries. It's a solution based on values that funnelled six million people in Europe into furnaces." He was strongly condemned by Europeans including UK Labour Party politician Ed Miliband for losing his diplomatic cool (Vidal and Watts 2009).

3 Climate justice

The need to lose any semblance of cool when it comes to climate was expressed not only by Di-Aping and Pacja at Copenhagen, but by the Climate Justice movement since the early 2000s. The concept of climate justice addresses the intersections of 'green' ecological concerns and 'red' socio-economic ethical considerations, ranging from public policy deliberations to political practices. Climate justice follows directly from the much older environmental justice movement that especially emphasized the racial and class injustices of pollution in the United States (Bullard 2009; Schlosberg 1999).

The first known conference based on the term 'climate justice' was a 2000 event in The Hague sponsored by the New York group CorpWatch (Karliner 2000). Four years later, the Durban Group for Climate Justice was launched, and for many years remained an important strategic listserve for those opposed to carbon trading and other 'false solutions' to the climate crisis.[4] The sometimes inchoate advocacy movement known as Climate Justice Now! (CJN!) began in 2007, and played a role in grassroots environmental advocacy as well as global-scale United Nations climate summits.[5] The highest-profile of these, with 100,000 protesters demanding a strong agreement from negotiators, was in Copenhagen. In contesting mainstream environmentalists, Danes and other Europeans formed a Climate Justice Alliance (CJA) whose 'Reclaiming Power' protest was severely repressed by Danish police protecting the United Nations Framework Convention on Climate Change Conference of the Parties (COP) 15.[6]

At that time, Anne Petermann (2009) of the Global Justice Ecology Project defined the concept as follows:

> Climate Justice is the recognition that the historical responsibility for the vast majority of greenhouse gas emissions lies with the industrialized countries of the Global North. It is the understanding that peasants, indigenous peoples, fisher-folk, women and local communities have been disproportionately affected by climate change, also by the fossil fuel industry and by false solutions to climate change, including tree plantations, genetically modified organisms like crops, large scale hydro projects and agro-fuels. These are also the people least responsible for climate change. Climate Justice recognizes that instead of market based solutions, the sustainable practices of these peoples and communities should be seen as offering the real solutions to climate change. Climate Justice is the fundamental knowledge that climate change cannot be addressed through corporations and the market as these are the entities that caused the problem in the first place.

Shortly after the Copenhagen summit's well-recognized failure, the Bolivian government led by Evo Morales and his then UN Ambassador Pablo Solon hosted a 2010 conference in Cochabamba, attended by 35,000 activists, including 10,000 from outside the country.[7] This was important partly because of attempts to more deeply incorporate within mainstream climate politics a commitment to carbon markets and offset payments, especially through the Reducing Emissions from Deforestation and Forest Degradation (REDD) project. The Cochabamba conference (see chapter by Kruse in this book) adopted several demands that were anathema to mainstream climate politics. The Bolivian government struggled to put these (and a few others) into official UN texts:

- 50 percent reduction of greenhouse gas emissions by 2017;
- stabilizing temperature rises to 1°C and 300 parts per million;
- acknowledging the climate debt owed by developed countries;
- full respect for human rights and the inherent rights of indigenous people;
- universal declaration of rights of Mother Earth to ensure harmony with nature;
- establishment of an International Court of Climate Justice;
- rejection of carbon markets and commodification of nature and forests through the REDD program;
- promotion of measures that change the consumption patterns of developed countries;
- end of intellectual property rights for technologies useful for mitigating climate change; and
- payment of 6 percent of developed countries' GDP to addressing climate change.[8]

As Edgardo Lander (2010) explained in his review of the Cochabamba conference:

> [s]truggles for environmental or climate justice have managed to bring together most of the most important issues/struggles of the last decades (justice/equality, war/militarization, free trade, food sovereignty, agribusiness, peasants' rights, struggles against patriarchy, defense of indigenous peoples' rights, migration, the critique of the dominant Eurocentric/colonial patterns of knowledge, as well as struggles for democracy [etc.]. All these issues were debated in Cochabamba and, to some degree, present in the Cochabamba Peoples' Agreement.

A regroupment of the Climate Justice Now! network began in 2013, after former negotiator Salon – subsequently director of Focus on the Global South – helped gather activists at the Tunis World Social Forum. The divisions between CJN! insiders and outsiders within the COPs became less important, as the underlying drive to promote local activism was reasserted, and as the differences between CJN! and other climate activists also became clearer.

Climate justice politics typically lines up against the for-profit economy as the underlying cause of the climate crisis, resulting in an uncompromising opposition to market-based strategies. There is also an ambitious emphasis on emissions cuts (far greater than those proposed by UN negotiators), and a marked disdain for the inadequacy of official global-scale and most national climate mitigation efforts. In a rejection of technocratic and tinkering responses to climate change, there is a hostility to technological fixes and geo-engineering, and instead a search for prefigurative post-carbon lifestyles and production systems. Climate Justice activists seek explicit alliances with activists specializing in food sovereignty and land access (La Via Campesina is typically central within Climate Justice mobilizations), decommodified water, renewable energy, economic justice, and other overlapping struggles – given how many issues have climate implications. This is a feature of a broader orientation to the politics of the Global South (not just North–South power adjustments), and an openness to fusing traditional Left and radical environmental politics with new 'Rights of Nature' strategies, ecofeminist and ecosocialist philosophies, and horizontalist political strategies and tactics, especially in concrete sites of struggle.

The climate justice advocacy challenge to orthodox market-oriented environmentalism and failed insider deal-making had surfaced as a formal movement beginning in 2007, at the UN COP 13 climate summit in Bali, Indonesia. By then it was obvious that the era of extreme global state failure and market failure – i.e., because the 'externality' of pollution remained unaccounted for within capitalist production, trade, consumption, and disposal – would continue unabated. Within months, these failures were amplified by a world capitalist crisis that had broken out in East Asia and soon threatened the world economy. The basic dilemma was the inability of global leaders to solve major environmental, geopolitical, social, and economic problems; none of significance was properly addressed in world summits after 1987, the year that the Montreal Protocol on the ozone hole banned chlorofluorocarbons.

The inadequacy of global climate negotiations, and the turn by the United Nations towards 'Type Two Partnerships' involving corporations, together generated enormous frustration in civil society. Indeed, by the time of the 2002 World Summit on Sustainable Development in Johannesburg, many activists had come to the conclusion that the UN was part of the problem, not the solution. This frustration was dramatized by a march of 30,000 against that UN summit in Johannesburg, from a poor neighborhood in Alexandra across to Sandton (the wealthiest suburb in Africa) where the convention was held (Death 2010). This was at a time when South Africa had become the world's most unequal major country; Johannesburg had become a major site of conflict over water privatization; and carbon trading experimentation had begun in nearby Durban (Bond 2002, 2012; McDonald 2002).

In short, it was in part because of experiences in Africa that climate justice arrived on the international scene as a coherent political approach in the wake of the failure of a more collaborative strategy between major environmental NGOs and the global managerial class. The first effort to generate a global climate advocacy movement in civil society was the Climate Action Network (CAN), which still has a small presence in Africa. But from 1997 in Kyoto, CAN adopted as its core strategy an emphasis on regular UN interstate negotiations aiming at minor, incremental emissions reductions augmented by carbon trading and related offsets. The cul-de-sac of CAN's commitment to carbon trading was confirmed when Friends of the Earth International broke away in 2010, with strong inputs by its African affiliates. But even before this, at the time of the December 2009 COP 15, CAN's critics from the climate justice movement made the case for an alternative strategy with such force that they gained half the space reserved for nongovernmental delegations in Copenhagen's Bella Centre.

Core concerns of climate justice activists expressed at Copenhagen and since have included the decommissioning of the carbon markets so favored by elites, massive investments in renewable energy, a thoroughly reformed agricultural system, public transport and other transformative infrastructure, production and disposal technologies. Strategically, however, it soon became evident that the next stage of the climate justice struggle was necessarily to retreat from a naively overambitious reform agenda at the global scale, given the adverse power balance. Instead of politely asking UNFCCC delegates to save the planet, the challenge for climate justice activists from Copenhagen in 2009 into the future would be to whistle-blow and prevent further deterioration in global environmental governance at the annual COPs, but in the meantime to intensify both local and national activism.

4 The African carbon trading wedge

It is, however, worth acknowledging that the semantics of climate justice are hotly debated, and that this has become evident in interpreting Africa's experience and strategies. The founder of the Mary Robinson Climate Justice Foundation, for example, favors keeping the carbon trading option on the strategic table. In spite of all the evidence to the contrary, Robinson argued in a 2011 London School of Economics lecture that carbon trading is "finally starting to reap dividends for Africa and least developed countries" and that "the experience gained through the design and implementation of successful regional cap-and-trade programs is hugely valuable if shared with developing country regional groups" (Robinson 2011). She provided no justification for these claims, and while the European Union's regional cap-and-trade scheme was collapsing in value by 90 percent, several efforts made by this author to discern what evidence lies behind her optimism came to naught.

Others with backgrounds in social justice struggles but compromised political agendas had the same agenda, including Valli Moosa, South Africa's former environment minister (1999–2004). He promoted carbon trading as minister at the critical 2002 World Summit on Sustainable Development in Johannesburg, and in the latter half of the 2000s, Moosa went on to preside over the International Union for the Conservation of Nature (IUCN), headed the South African branch of the World Wildlife Fund, and chaired the boards of both the continent's largest energy company and CO_2 emitter, Eskom, and from 2013, the world's largest platinum company, AngloPlats. In the capacity as Eskom chair in 2007–2008, he was implicated, as a member of the African National Congress (ANC) financing committee, in unethically channeling tens of millions of Rands in earnings to the ruling party by signing Eskom purchase orders for Medupi's new boilers in a way that directly benefited the ANC, which in turn was financed by the controversial World Bank loan. The SA government's Public Protector acknowledged that his role was "improper" (Mabuza 2009).

Moosa's successor as South Africa's minister of environment and tourism, Marthinus van Schalkwyk, a youth spy for the white apartheid regime during the 1980s, took control of the National Party in the late 1990s and then dissolved it into the ANC in exchange for the ministerial position. In 2009, he was demoted to tourism minister. An enthusiastic proponent of the Kyoto Protocol's Clean Development Mechanism (CDM) carbon trading strategy, Van Schalkwyk argued in 2006 that

> The 17 CDM projects in the pipeline in Sub-Sahara Africa account for only 1.7 per cent of the total of 990 projects worldwide. To build faith in the carbon market and to ensure that everyone shares in its benefits, we must address the obstacles that African countries face.[9]

At the International Emissions Trading Association Forum in Washington a year later, he insisted, "an all-encompassing global carbon market regime which includes all developed countries is the first and ultimate aim."[10] Van Schalkwyk was nominated by South Africa to replace Yvo de Boer as UN climate negotiations director in early 2010, but his candidacy failed at the last moment, as Costa Rican carbon trader Christiana Figueres got the position.

In the highest-profile African case of NGO support for carbon markets, Wangari Maathai, the former Kenyan deputy environment minister and Nobel Peace Prize laureate, such conflicts were not a factor. Prior to her death in 2011, Maathai also promoted carbon trading through her own Greenbelt Movement in the expectation that CDMs and emerging proposals for REDD would reward tree-planting in both her indigenous strategy as well as mono-cultural

Table 14.1 Africa's low relative share of CDM projects, 2011

Region	*Number of CDM projects*	*Share of CDM projects*
Latin America	1,094	14.2%
Asia & Pacific	6,212	80.8%
Europe and Central Asia	84	1.1%
Africa	**217**	**2.8%**
Middle East	85	1.1%
Total	7,692	100%

Source: Author's creation according to http://cdmpipeline.org/cdm-projects-region.htm (accessed 05/15/2013).

timber plantations. She was also the leading proponent of the document 'Africa speaks up on Climate Change,' which fed into the 'African Climate Appeal,' a statement which insists upon more CDM finance with fewer strings attached, especially for afforestation:

> African governments should ensure that there is equity in geographical distribution of CDM projects and that this is entrenched in the international policy process. They should negotiate for the requirement of upfront funding of CDM projects to be waived for many African countries who cannot afford it. The appeal calls upon African countries to embark on the development of CDM capacities and projects including capacity building and development of centers of incubation for CDM projects. African governments should explore possibilities of accessing grants to provide upfront funding for CDM projects and also project development and financing through bilateral arrangements.
>
> *(Maathea 2009: 4)*

Maathai criticized three existing funds – the Special Climate Change Fund, the Least Developed Countries Fund, and the Bali Adaptation Fund – because these funds have not been able to address concerns of African countries on adaptation, namely:

> [a]ccess, adequacy and equitable geographical distribution. The funds are largely inadequate and inappropriately structured; currently relying on a 2 percent levy on CDM projects. Access to the funds has been made difficult, among others, by bureaucratic bottlenecks of the Global Environmental Fund and the World Bank.
>
> *(Maathea 2009: 4)*

Maathai's appeals for a more generous and efficient system for Africa were never properly satisfied, for the bank continued to play the most critical role in carbon market stimulation notwithstanding the market's ongoing crisis. A proliferation of new Bank funds did not change the basic calculus: CDMs 'can't deliver the money' to Africa (see Table 14.1).

5 African resistance to carbon trading

Frustration with CDMs in Africa reached a critical mass as early as 2004 when the Durban Group for Climate Justice gathered for an historic meeting. A global civil society network, the Durban Group,[11] was formed to oppose carbon trading's 'privatization of the air.' From the vantage point

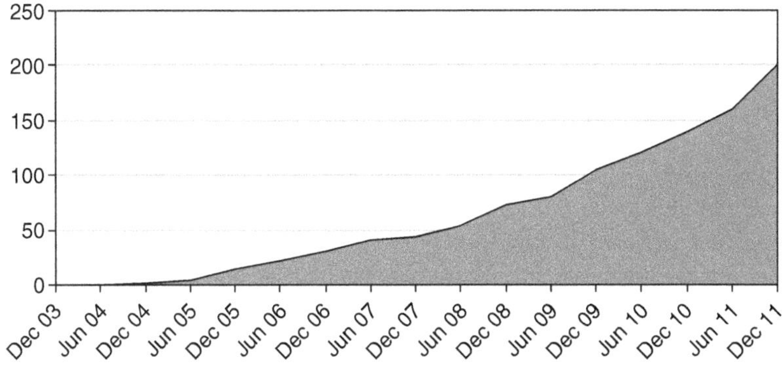

Figure 14.1 Number of CDM projects in Africa, 2003–2011 (*source:* Author's creation according to http://cdmpipeline.org [accessed 05/15/2013]).

of an austere Catholic mission on Durban's highest central hill, the Glenmore Pastoral Centre, a score of the world's critical thinkers and activists for environmental justice convened at the Swedish Dag Hammarskjold Foundation and deliberated over the neoliberal climate fix for several days. Participants worried that the main test case, the EU's Emissions Trading Scheme, not only failed to reduce net greenhouse gases there, but suffered extreme volatility, an inadequate price, the potential for fraud and corruption, and the likelihood of the market crowding out other, more appropriate strategies for addressing the climate crisis (Bond 2012).

The Durban Group critique can be summed up in eight points (Bond 2012):

- the idea of inventing a property right to pollute is effectively the 'privatization of the air,' a moral problem given vast, growing differentials in wealth inequalities;
- greenhouse gases are complex and their rising production creates a non-linear impact which cannot be reduced to a commodity exchange relationship (a tonne of CO_2 produced in one place is accommodated by reducing a tonne in another, as is the premise of the emissions trade);
- the corporations most guilty of pollution and the World Bank – which is most responsible for fossil fuel financing – are the driving forces behind the market, and can be expected to engage in systemic corruption to attract money into the market even if this prevents genuine emissions reductions;
- many of the offsetting projects – such as mono-cultural timber plantations, forest 'protection' and landfill methane-electricity projects – have devastating impacts on local communities and ecologies, and have been hotly contested in part because the carbon sequestered is far more temporary (since trees die) than the carbon emitted;
- the price of carbon determined in these markets is haywire, having crashed by half in a short period in April 2006 and by two-thirds in 2008, by another 50 percent during 2011, thus making a mockery of the idea that there will be an effective market mechanism to make renewable energy a cost-effective investment;
- there is serious potential for carbon markets to become an out-of-control, multi-trillion dollar speculative bubble, similar to exotic financial instruments associated with Enron's 2002 collapse (indeed, many former Enron employees populate the carbon markets);
- as a 'false solution' to climate change, carbon trading encourages merely small, incremental shifts, and thus distracts us from a wide range of radical changes we need to make in materials extraction, production, distribution, consumption, and disposal; and

- the idea of market solutions to market failure ('externalities') is an ideology that rarely makes sense, and especially not following the world's worst-ever financial market failure, and especially not when the very idea of derivatives – a financial asset whose underlying value is several degrees removed and also subject to extreme variability – was thrown into question.[12]

Durban remains an important guinea pig, not only for hosting the COP 17, but for initiating SA's lead CDM pilot, the Bisasar Road landfill. There, methane from rotting rubbish is converted to electricity and fed back into the municipal grid. The CDM was set up illegally because it fails the crucial test of its validity for raising international funding, 'additionality.' It was always assumed that the 15 million dollar estimated cost of the project would not be justified by the small amount of electricity fed into Durban's municipal supply, and hence that the funds would have to come from external sources. But as noted by journalists who visited Bisasar during the COP 17, at least one relevant Durban official now concedes that the Bisasar Road methane-electricity project would have gone ahead without the external credits.

After helping set it up, the World Bank refused in August 2005 to take part in marketing or purchasing Bisasar Road emissions credits. The reason was growing awareness of Durban's notorious environmental racism, via activism and an environmental impact assessment challenge. In March 2005, just as the Kyoto Protocol came into force, a *Washington Post* front-page story revealed how community organizer Sajida Khan suffered cancer from Bisasar Road's toxic legacy (Vedantam 2005). Back in 1980, the landfill – Africa's largest – was plopped in the middle of Durban's Clare Estate suburb, across the road from Khan's house, thanks to apartheid insensitivity. Instead of honoring African National Congress politicians' promises to close the dump in 1994, the municipality kept it open when 15 million dollars in emissions financing was dangled. After Khan died in mid-2007 after her second bout with cancer – which she believed was landfill-induced – Clare Estate civic pressure to close Bisasar subsided (Bond 2010).

Similar controversy surrounds REDD, which, in theory, sells investors forest protection. But notwithstanding disagreements in civil society, it was increasingly seen by African advocacy networks such as Durban's TimberWatch as a boon to voracious commercial forestry and a danger to indigenous peoples, given that proper safeguards were not adopted. In 2013 at the World Social Forum in Tunis, such pressure led to a 'No REDD in Africa' network accusing proponents of contributing to a potential "genocide."[13]

While some African civil society groups were supportive of REDD, other offsets, and carbon trading opportunities, in late 2011, Pacja in Nairobi and the Institute for Security Studies (ISS) in Cape Town issued an extensive report on African CDMs. One of the contributors, Yacob Mulugetta of Surrey University, expressed the problem succinctly:

> The argument that carbon trading offers real benefits to the poor in Africa is simply not credible. What is puzzling is the persistence of the proponents of carbon markets, who continue to cling onto these ideas in the face of mounting evidence that carbon trading does not deliver results commensurate to the effort invested in it.... Fundamental inequality is behind the climate problem, and the search for solutions must involve industrialized societies making fundamental structural changes to their lifestyles, energy practices, and their production and consumption systems.
>
> *(Mulugetta 2011: 29)*

The Pacja/ISS report was scathing of using subsidies meant for supporting Africans, to instead prop up the world carbon market, especially via the African Development Bank's (AfDB) African Carbon Facility:

> It promises to buy post-2012 credits in order to 'maintain private sector confidence' in the ailing carbon market, as well as providing debt financing for the development of new projects. The AfDB is also offering an African Carbon Support Programme, which was launched in November 2010. This is supported by the Fund for African Private Sector Assistance, a joint initiative of Japan, Austria and the AfDB to promote private sector development. The aim is to support potential project developers throughout the whole CDM process, from formulating the original project idea through to advice on credit sales. The running of the programme has been outsourced to Carbon Limits, the Norwegian consultancy which developed the Pan Ocean Gas Utilization Project.
>
> *(Reddy 2011)*

In addition, Pacja/ISS expressed concern about the African Carbon Asset Development Facility (a joint venture of the United Nations Environment Programme, Standard Bank of South Africa, and the German Federal Environment Ministry), which provides technical support and small grants for project developers to establish 15 CDM projects. As Pacja/ISS argue, this strategy "diverts scarce public resources away from directly addressing climate change" (cited in Allen et al. 2012: 96).

In sum, the emissions markets were the wrong idea (a neoliberal strategy) in the wrong place (financial markets) at the wrong time (the 2000s era of repeated bubbles and bursts).

6 African campaigns against climate change

In addition to opposing cap-and-trade schemes, there are numerous other concrete African climate justice campaigns. Some defend against land grabs that promote energy-negative and disruptive land grabs. Others contest mega-dams (such as the Gibe in Ethiopia, the Mpande Nkuwa in Mozambique, the Bujagali in Uganda, and the Inga in the Democratic Republic of the Congo) in part because methane emissions from rotting vegetation is so climate destructive. Others address land and agricultural adaptation from the standpoint of rural peasant women producers. Others attack the underlying causes of Africa's emissions in minerals-extraction sectors which consume the most coal-fired and oil-related electricity.

Some campaigns aim to 'leave the oil in the soil, coal in the hole, tarsand in the land and shale gas under the grass,' exemplified in South Africa's Karoo region where Shell fracking has been fought, and in Nigeria where Niger Delta activists have intimidated oil companies through both nonviolent and armed struggle. In the former category, Environmental Rights Action in Port Harcourt insists on an end to extraction and exploration on grounds of the climate threat, and the descendants of Ken Saro-Wiwa and eight of his Ogoni movement comrades won a precedent-setting out-of-court-settlement with Shell in June 2009 that may scare off other oil firms.

In the latter category, even after an amnesty in 2009 had a divide-and-conquer effect, the Movement for the Emancipation of the Niger Delta (MEND) continued to kidnap foreign oil workers, demanding they vacate the Delta for good. After a combined struggle of this type, Shell was evicted from Ogoniland in June 2008, 13 years after the company arranged for Saro-Wiwa's execution, and a year later, Saro-Wiwa's family (and those of eight others executed at the same time) won 15.5 million dollars from Shell in an Alien Tort Claims Act case settled out of court, a large amount of which was recommitted to movement building. An estimated 1.5 million tonnes of oil have spilled since Delta drilling began in the late 1950s, the equivalent of an Exxon Valdez spill each year, costing more than 5 billion dollars in annual environmental damage. In his closing statement at the trial prior to his execution, Saro-Wiwa demanded that Shell vacate the area: "The military dictatorship holds down oil-producing areas such as Ogoni by military decrees and the threat or actual use of physical violence so that Shell can wage its ecological war without hindrance" (cited in Wiwa v. Shell 2009).

At the time, Shell executives met with the Nigerian High Commission in London, stating that if the "Ogoni virus spreads to other areas in the Delta it would be the end of the oil business" (cited in Wiwa v. Shell 2009). In court, the plaintiffs had alleged that Shell hired Nigerian police for internal security; that Shell purchased vehicles and arms for the military; that Shell requested military support to build a pipeline through Ogoniland; that Shell assisted and financed the Nigerian military to repress the resistance of the Ogoni people; and that the firm participated in the arrest of Saro-Wiwa and others on fake murder charges and bribed witnesses to produce false testimony (Bond and Sharife 2009). The exposure by WikiLeaks of Shell's ongoing influence in Lagos politics suggests that the Ogoni virus must spread far faster if Nigeria is to be free, and if both the Niger Delta's ecology and the climate are to be spared (Bassey 2010).

Indeed, most remarkably, rather than letting such destruction rest at the scale of the local, Environmental Rights Action led the climate justice movement in Nigeria, West Africa, and globally at Copenhagen to a much deeper critique of ecological responsibility. In opposition to the shifting, stalling, and stealing that characterizes economic and environmental commodification in their own region, and in relation to world financial and oil markets, Environmental Rights Action and its visionary leader Nnimmo Bassey – who was also Friends of the Earth International chairperson for several years – jumped scale to demand that the oil be left in the soil and under the Gulf of Guinea water, given the threat to the planet.

MEND is a variegated force, including some who are apparently opportunistically self-seeking in mere financial terms, with merely criminal backgrounds. But whatever their motives, they are effective, for in 1997, Nigeria Business estimated that 20 percent of the Niger Delta's oil's extraction was disrupted the year before, just as MEND started operations (Umezuruike 2007). According to a US-based magazine, *TerroristPlanet.com*, by February 2010, even after an amnesty was implemented by the central government, MEND was a formidable insurgency:

- Since it first emerged in early 2006, MEND, which says it is fighting for a larger share of Southern Nigeria's oil revenue to go to local people, has cut Nigeria's oil production by more than one-quarter.
- Swarm-based maneuvers: Guerrillas are using speed boats in the Niger Delta's swamps to quickly attack targets in succession. Multiple, highly maneuverable units have kept the government and Shell's defensive systems off-balance defending the sprawling network. Radically improved firepower and combat training: allowing guerrillas to overpower a combination of Shell's Western-trained private military guards and elite Nigerian units in several engagements.[14]
- Effective use of system disruption: Targets have been systematically and accurately selected to completely shut down production and delay and/or halt repairs, and the guerrillas are making effective use of Shell's hostages to coerce both the government and the company.[15]

In another innovative example of resistance to climate change, the Cape Town-based Alternative Information and Development Centre hosts a 'Million Climate Jobs' campaign involving leading progressive labor officials from the metalworkers, transport workers, and municipal workers unions, fusing with community activists and environmentalists. As their website explained:

> Climate jobs are decent, primarily publicly driven jobs that directly reduce the causes and impacts of climate change. Climate jobs involve:
>
> - the building of renewable energy power stations such as wind, solar, wave, and tidal;
> - the building of public transport networks that would reduce the need for private cars and fossil fuel intensive trucks;
> - energy efficiency in construction and housing, including retrofitting;
> - transforming industrial agriculture;
> - reforming production and consumption; and
> - addressing the energy intensification of the economy.
>
> Significant jobs would be created in the related areas of research, education, and training, to ensure the country has the skills to undertake the transition to a low carbon, labor absorbing and socially developed sustainable future economy. By linking climate change to the transition to a sustainable, low-carbon economy that provides meaningful mitigation opportunities, the Campaign will stimulate the interest of many sectors of South African society who do not realise the immediate relevance of climate change.[16]

Another case of civil society resistance to climate change involves a growing critique of the 'Green Economy,' in which the Gabarone Declaration of May 2012 and Rio+20 of June 2012 featured prominently, as some African governments sought to commodify nature in order to save it. In contrast, Pacja signed a declaration against the Green Economy as "the privatization of nature" at the 2013 World Social Forum. Pacja and indigenous people's allies committed to "recognize our responsibilities to protect Mother Earth and work towards a new paradigm that rejects the property right regimes with a new set of laws that recognizes the Rights of Nature – the Rights of Mother Earth."[17] To that end, advocacy for the Global North's repayment of ecological debt is one potential disincentive to multinational corporations which wreck the environment. According to Joan Martinez-Alier,

> the notion of an Ecological Debt is not particularly radical. Think of the environmental liabilities incurred by firms (under the United States Superfund legislation), or of the engineering field called 'restoration ecology,' or the proposals by the Swedish government in the early 1990s to calculate the country's environmental debt.
>
> *(Martinez-Alier 2002: 261)*

But the sums involved are potentially vast. As Martinez-Alier puts it,

> [t]ropical rainforests used for wood exports have an extraordinary past we will never know and ongoing biodiversity whose destruction we cannot begin to value.... [However,] although it is not possible to make an exact accounting, it is necessary to establish the principal categories and certain orders of magnitude in order to stimulate discussion. If we take the present human-made emissions of carbon, [this represents] a total annual subsidy of $75 billion is forthcoming from South to North.
>
> *(Martinez-Alier 2002: 261)*

Leading eco-feminist Vandana Shiva (Shiva 2005) and former South Centre director Yash Tandon[18] estimate that wild seed varieties alone account for 66 billion dollars in annual biopiracy benefits to the US. There are dozens of examples of biopiracy in Africa, according to a 2005 study commissioned by the Edmonds Institute and African Centre for Biosafety (McGown 2006).[19]

A partial ecological debt accounting was published by environmental scientists in early 2008, and counted 1.8 trillion dollars in concrete damages over several decades (Srinivasan et al. 2008). Co-author Richard Norgaard, ecological economist at the University of California, Berkeley, generated a crucial finding: "At least to some extent, the rich nations have developed at the expense of the poor, and, in effect, there is a debt to the poor."[20] The study included factors such as greenhouse gas emissions, ozone layer depletion, agriculture, deforestation, overfishing, and the conversion of mangrove swamps into shrimp farms, but did not (so far) succeed in calculating other damages, for example, excessive freshwater withdrawals, destruction of coral reefs, biodiversity loss, invasive species, and war.

7 Conclusion

Yet ecological debt advocates must be very careful to not adopt the Green Economy's approach to valuing nature so as to generate new markets. The idea of ecological debt – especially related to climate damage – entails collective reparations and an accountability system that generates structural change, as opposed to merely a 'fine' for wrongdoing. In other words, the 'fine' (polluter pays) must be followed by a *ban* on further pollution, for there to be restorative justice. At all costs, the approach termed 'Payment for Ecosystem Services' should be avoided, so as to halt the growth of markets in nature, offsets, and carbon trading.

How, then, to pay the climate debt? It became clear to many civil society groups in recent decades that postcolonial African governments were too easily corrupted, just as were UN agencies and aid (and even international NGO) bureaucracies. The solution to the payment distribution problem appeared in 2009: the idea of simply passing along a monthly grant – universal in amount and access, with no means-testing or other qualifications – to each African citizen via an individual Basic Income Program payment. According to Germany's *Der Spiegel*, the village of Otjivero, Namibia, was an exceptionally successful pilot for this form of redistribution:

> It sounds like a communist utopia, but a basic income program pioneered by German aid workers has helped alleviate poverty in a Namibian village. Crime is down and children can finally attend school. Only the local white farmers are unhappy.... "This country is a time bomb," says Dirk Haarmann, reaching for his black laptop. "There is no time to lose." Haarmann and his wife Claudia, both of them economists and theologians from western Germany, were the ones who calculated the basic income for Namibia. "The basic income scheme," says Haarmann, "doesn't work like charity, but like a constitutional right." Under the plan, every citizen, rich or poor, would be entitled to it starting at birth.
>
> *(Krahe 2009)*

There was an enormous impact from BIG (Basic Income Guarantee) on social indicators, including food, education, and health security, and domestic violence, begging, and sex work (and with it, transfer of HIV and STDs) declined considerably (cited in Sharife and Bond 2013). The first priority would be to supply a Basic Income Program to Africans who live in areas most adversely affected by droughts, floods, or other extreme weather events. Logistically, the use of Post Office Savings Banks or rapidly introduced Automated Teller Machines would be sensible, although currency distortions, security, and other such challenges would differ from place to place. The Namibian case has much to recommend it, in part because it is among the driest sites in Africa.

Such a strategy would be just an emergency salve on a burning problem: how to ensure that the greenhouse gas 'polluters pay' in a manner that first, compensates their climate change victims; that second, permits transformation of African energy, transport, extraction, production, distribution, consumption, and disposal systems; and that third, in the process assures the 'right to development' for Africa in a future world economy constrained by emissions caps. Extremely radical changes will be required in all these activities in order not only to ensure the safety of the species and planet, but also that Africans are at the front of the queue for long-overdue ecological and economic compensation, given the North's direct role in Africa's environmental damage.

The contemporary argument for climate debt to be paid is simply the first step in a long process, akin to decolonization, in which the master – the polluting global North (South Africa included) – must know that the tools with which his house were built, including the profit motive and markets, are not and cannot be effective in remodeling for a new society. And he must also know that not only is it time to halt the reliance on fossil fuels, but having 'broken' the climate, it is his responsibility to foot the clean-up bill. And that, finally, must be done without creating new market strategies to privatize nature and contribute to world financial chaos and corruption; it must be done to achieve justice, via direct payments to the victims of the crisis the master has created. These are some of the lessons of climate justice, done in, by, and for Africa.

Notes

1 See www.diakonia.se/sa/node.asp?node=4675 (accessed 06/01/2013).
2 Reprinted at www.whirledbank.org (accessed 06/01/2013).
3 See *Guardian*, 'US Embassy Cables. US Urges Ethiopia to Back Copenhagen Climate Accord,' December 3, 2010.
4 See www.durbanclimatejustice.org/ (accessed 06/01/2013).
5 See www.climate-justice-now.org (accessed 06/01/2013), see also Kaara (2010).
6 See www.climate-justice-action.org (accessed 06/01/2013).
7 See http://pwccc.wordpress.com (accessed 06/01/2013).
8 See http://cochabamba2010.typepad.com/blog/2010/08/the-proposals-of-peoples-agreement-in-the-texts-for-united-nations-negotiation-on-climate-change.html (accessed 06/01/2013).
9 See South African Minister outlines priorities for Nairobi Climate Meetings, accessed 06/01/2013 from www.stakeholderforum.org/index.php?id=11sept.
10 Marthinus van Schalkwyk: 'Speech to the International Emissions Trading Association,' accessed 06/01/2013 from www.polity.org.za/.../sa-van-schalkwyk-international-emissions-trading-association-forum-26092007–2007–09–26.
11 See www.durbanclimatejustice.org/ (accessed 06/01/2013).
12 The analyses of emission market contradictions authored by Larry Lohmann are probably the best (see Lohmann 2006, 2009a, 2009b, 2009c, 2009d, 2010).
13 See http://climatespace2013.wordpress.com/2013/04/08/launch-of-no-redd-in-africa-network-redd-could-cause-genocide/ (accessed 06/01/2013).
14 One of Shell's private military operators was captured as a hostage.
15 See www.terroristplanet.com/2010/02/terrorist-group-profile-the-movement-for-the-emancipation-of-the-niger-delta/ (accessed 06/01/2013).
16 See One Million Climate Jobs, 'Why One Million Climate Jobs?' 2011, accessed 06/01/2013 from www.climatejobs.org.za/component/content/article/37-climate-jobs/54-why-one-million-climate-jobs.html.
17 See www.pacja.org/index.php?option=com_content&view=article&id=302%3Aworld-urged-to-reject-false-solutions-for-green-economy&Itemid=130 (accessed 06/01/2013).
18 See www.globalpolicy.org/socecon/develop/devthry/well-being/2000/tandon.htm (accessed 06/01/2013).

19 They include three dozen cases of African resources – worth many billions of dollars – captured by firms for resale without adequate 'Access and Benefit Sharing' agreements between producers and the people who first used the natural products, such as a diabetes drug produced by a Kenyan microbe, the South African and Namibian indigenous appetite suppressant Hoodia, and groundnuts from Malawi, Senegal, Mozambique, Sudan, and Nigeria.

20 See *Guardian*, 'Rich Countries Owe Poor a Huge Environmental Debt,' January 21, 2008, www.guardian.co.uk/science/2008/jan/21/environmental.debt1 (accessed 06/01/2013).

References

Allen, F., Amisi, B., Bond, P., et al. (2012) *The CDM in Africa Cannot Deliver the Money. Why the Carbon Trading Gamble and Clean Development Mechanism Won't Save the Planet from Climate Change and How African Civil Society Is Resisting*. A report by the University of KwaZulu-Natal Centre for Civil Society (SA) and Dartmouth College Climate Justice Research Project (USA). Accessed 09/12/2013 from http://cdmscannotdeliver.files.wordpress.com/2012/04/ccs-dartmouth-ejolt-cdms-cannot-deliver-the-money-web.pdf.

Ashine, A. (2009) Africa Threatens Withdrawal from Climate Talks, *The Nation*, Nairobi, September 3.

Bassey, N. (2010) So Shell Is Everywhere, *Next*, December 16. Accessed 06/01/2013 from http://234next.com/csp/cms/sites/Next/Home/5654242–46/oil_politics_so_shell_is_everywhere.csp.

Bassey, N. (2011) *To Cook a Continent*, Oxford: Pambazuka Press.

Bond, P. (2002) *Unsustainable South Africa: Environment, Development and Social Protest*, London: Merlin Press.

Bond, P. (2006) *Looting Africa: The Economics of Exploitation*, London: Zed Books.

Bond, P. (2010) True Cost of Durban's Waste Strategy, *The Mercury*, February 2.

Bond, P. (2012) *Politics of Climate Justice: Paralysis Above, Movement Below*, Pietermaritzburg: University of KwaZulu-Natal Press.

Bond, P. and Sharife, K. (2009) Shell Oil: Guilty in World Court of Public Opinion, *Economic & Political Weekly*, July 18.

Bond, P., Dada, R., and Erion, G. (2009) *Climate Change, Carbon Trading and Civil Society*, Pietermaritzburg: University of KwaZulu-Natal Press.

Bullard, N. (2009) The State of Climate Politics, speech delivered at a Rosa Luxemburg Foundation conference in Johannesburg, South Africa, November 19.

Christian Aid (2008) Developing Countries Demand Compensation for Climate Change. Accessed 06/01/2013 from www.christianaid.org.uk/pressoffice/pressreleases/august2008/climate_change_talks_accra.aspx.

Death, C. (2010) *Governing Sustainable Development: Protests, Partnerships and Power at the World Summit*, Abingdon: Routledge.

Friends of the Earth International (2009) Report from Copenhagen: Africa Demands that Wealth Nations Step it up, December 8. Accessed 06/01/2013 from www.foe.org/international-work?page=2.

Gilbertson T. and Santos, R. (2009) Confronting the Climate Circus, *Red Pepper*, December 12. Accessed 06/01/2013 from www.redpepper.org.uk/Confronting-the-climate-circus/.

Kaara, W. (2010) Reclaiming People's Power in Copenhagen 2009: A Victory for Ecosocialist Ecofeminism, *Capitalism Nature Socialism*, 21(2): 107–111.

Karliner, J. (2000) Climate Justice Summit Provides Alternative Vision, *CorpWatch*. Accessed 04/12/2013 from www.corpwatch.org/article.php?id=977.

Krahe, D. (2009) A New Approach to Aid: How a Basic Income Program Saved a Namibian Village, *Der Spiegel International*, August 10.

Lander, E. (2010) Reflections on the Cochabamba Climate Summit. Accessed 06/01/2013 from www.tni.org/article/reflections-cochabamba-climate-summit.

Lohmann, L. (2006) Carbon Trading: A Critical Conversation on Climate Change, Privatisation and Power, *Development Dialogue*, 48, Uppsala: Dag Hammarskjold Foundation.

Lohmann, L. (2009a) Climate as Investment. Accessed 06/01/2013 from www.thecornerhouse.org.uk/pdf/document/Climate percent20as percent20Investment.pdLohmann.

Lohmann, L. (2009b) Neoliberalism and the Calculable World: The Rise of Carbon Trading, in Birch, K., Mykhnenko, V., and Trebeck, K. (eds.) *The Rise and Fall of Neoliberalism: The Collapse of an Economic Order?*, London: Zed Books.

Lohmann, L. (2009c) Regulatory Challenges for Financial and Carbon Markets, *Carbon & Climate Law Review*, 3(2): 161–171.
Lohmann, L. (2009d) Toward a Different Debate in Environmental Accounting: The Cases of Carbon and Cost-benefit, *Accounting, Organisations and Society*, 34(3–4): 499–534.
Lohmann, L. (2010) Uncertainty Markets and Carbon Markets: Variations on Polanyian Themes. Accessed 06/01/2013 from www.thecornerhouse.org.uk/pdf/document/NPE2high.pdf.
Maathea, W. (2009) African Climate Change Appeal. Nairobi. Accessed 06/01/2013 from www.ecobuild-africa.com.
Mabuza, E. (2009) Valli Moosa-Eskom 'Conflict of Interest,' *Business Day*, February 19.
Martinez-Alier, J. (2002) *The Environmentalism of the Poor*, Cheltenham: Edward Elgar.
McDonald, D. A. (ed.) (2002) *Environmental Justice in South Africa*, Athens, OH: Ohio University Press.
McGown, J. (2006) *Out of Africa: Mysteries of Access and Benefit Sharing*, Edmonds, Washington: Edmonds Institute in cooperation with the African Centre for Biosafety. Accessed 06/01/2013 from www.ecobuild-africa.com.
Mulugetta, V. (2011) Climate Change and Carbon Trading in Africa, in Reddy, T. (ed.) *Carbon Trading in Africa: A Critical Review*, Cape Town, Institute for Security Studies.
Pachauri, R. K. (2008) Summary of testimony provided to the House Select Committee on Energy Independence and Global Warming, US Congress, Washington, DC. Accessed 03/15/2013 from globalwarming.house.gov/tools/assets/files/0342.pdf.
Petermann, A. (2009) What Is Climate Justice?, Vermont, Global Justice Ecology Project. Accessed 04/12/2013 from http://globaljusticeecology.org/climate_justice.php.
Redi, O. (2009) Africa Builds United Position for Copenhagen, *InterPress Service*, August 25. Accessed 06/01/2013 from www.ipsnews.net/2009/08/environment-africa-builds-united-position-for-copenhagen/.
Reddy, T. (2009) From African Walk Out to Sell Out, *Climate Chronicle*, December 18.
Reddy, T. (ed.) (2011) *Carbon Trading in Africa: A Critical Review*, Cape Town: Institute for Security Studies.
Robinson, M. (2011) Protecting the Most Vulnerable, speech at the London School of Economics Centre for the Study of Human Rights, London, March 10. Accessed 04/12/2013 from www.mrfcj.org/news_centre/2011/mary_robinson_lecture_lse.html.
Sapa-AP (2009) SA on Climate Change: 'No Money, No Deal,' *Mail&Guardian*, August 5. Accessed 06/01/2013 from http://mg.co.za/article/2009–08–05-sa-on-climate-change-no-money-deal.
Schlosberg, D. (1999) *Environmental Justice and the New Pluralism*, Oxford: Oxford University Press.
Sharife, K. and Bond, P. (2013) Payment for Ecosystem Services versus Ecological Reparations: The 'Green Economy,' Litigation and a Redistributive Eco-debt Grant, *South African Journal on Human Rights*.
Shiva, V. (2005) Beyond the WTO Ministerial in Hong Kong, ZNet Commentary, December 26.
Srinivasan, U., Carey, S., Hallstein, E., Higgins, P., Ker, A., Koteen, L., Smith, A., Watson, R., Harte, J., and Norgaard, R. (2008) The Debt of Nations and the Distribution of Ecological Impacts from Human Activities, *Proceedings of the National Academy of Sciences of the United States of America*, 105(5). Accessed 06/01/2013 from www.pnas.org/content/105/5/1768.
Umezuruike, E. (2007) The Niger Delta Challenge: A Global Perspective, *Nigeria Business*, January 23. Accessed 06/01/2013 from www.thenigeriabusiness.com/column6.htmld.
Vedantam, S. (2005) Kyoto Credits System Aids the Rich, Some Say, *The Washington Post*, March 12.
Vidal, J. (2009) Copenhagen: Head of African Bloc Calls on Poorer Nations to Compromise over Climate Funding, *Guardian*, December 16.
Vidal, J. and Watts, J. (2009) Copenhagen Closes with Weak Deal that the Poor Threaten to Reject, *Guardian*, December 19.
Welz, A. (2009) Emotional Scenes at Copenhagen: Lumumba Di-Aping @ Africa Civil Society Meeting – 8 Dec 2009, blog post, December 8. Accessed 06/08/2013 from http://adamwelz.wordpress.com/2009/12/08/emotional-scenes-at-copenhagen-lumumba-di-aping-africa-civil-society-meeting-8-dec-2009.
Wheeler, D. (2011) *Quantifying Vulnerability to Climate Change: Implications for Adaptation Assistance*, Center for Global Development Working Paper 240, Washington, DC. Accessed 06/01/2013 from www.cgdev.org/content/publications/detail/1424759.
Wiwa v. Shell (2009) The Case Against Shell. Accessed 06/01/2013 from http://wiwavshell.org/about/about-wiwa-v-shell/.

15
ACTIVIST PROFILE – BILL McKIBBEN

Hans Baer

Bill McKibben is a renowned American environmentalist and the Shumann Distinguished Scholar at Middlebury College in Vermont. As an undergraduate student at Harvard University, he served as the president of the *Harvard Crimson*, the campus student newsletter. McKibben's first book *The End of Nature* which initially was serialized in *The New Yorker* propelled him to fame as an important environmental writer, in large part because it was regarded to be the first book about climate change targeted at a general audience (McKibben 1989). McKibben has since then authored numerous other books. In terms of climate change, other important books that he has written are *Deep Economy* (McKibben 2007a), *Fight Global Warming Now* (McKibben 2007b), *Eaarth* (McKibben 2010), and *The Global Warming Reader* (McKibben 2011). In *Deep Economy*, McKibben (2007a) argues that humanity must move beyond the growth paradigm as the paramount economic objective and calls for localization of energy, food, and even cultural production. In *Eaarth*, he urges people to create societies and communities that concentrate on the essentials, such as small agriculture, rather than on many of the luxuries associated with modernity which has been highly dependent on relatively inexpensive fossil fuels. McKibben's writing places him squarely within the green social democratic wing of the climate movement rather than the much smaller democratic eco-socialist wing (Baer 2012: 250–292). He has advocated carbon trading and until 2012 was open to nuclear power as a purported climate change mitigation strategy.

McKibben is a scholar-activist in the sense that he 'walks the talk' in terms of the environmental issues, particularly climate change, about which he is deeply passionate. In late summer 2006 he assisted in a five-day march across Vermont to demand action on climate change. In January 2007 McKibben (2007b: xi) and others launched a website, StepItUp2007.org in which they "asked people across the country [the United States] to demand that Congress cut carbon emissions 80 percent by 2050." This effort resulted in the organization of some 1,400 climate rallies in all 50 US states. On March 2, 2009 McKibben allowed himself to be arrested in a march on a coal-fired power plant in Washington D.C.

McKibben and others formed 350.org as an effort to create a global climate movement and to return the planet back to an atmospheric carbon dioxide concentration of 350 ppm in contrast to its present 390 ppm. In 2010 he and 350.org created the 10/10/10 Global Work Party which has convened thousands in events in 188 countries. In 2011 and 2012 McKibben led a massive environmental campaign against the proposed Keystone XL pipeline project. He attended

Rio+20 – the UN Conference on Sustainable Development – in June 2012 where he co-signed a letter critical of the conference organizers titled 'The Rio+ we don't want' (Schuetze 2012).

I have had the opportunity of hearing McKibben speak on two occasions and even chatting with him before his two addresses in Melbourne. The first time was on May 11, 2009 when he presented a public lecture titled 'Because the World Needs to Know – Building an International Climate Movement' at the University of Melbourne. The second time was on May 26, 2010 when he spoke to some 125–150 people at Environment Victoria in its campaign to close down Hazelwood power plant, one of the dirtiest coal-fired power plants in Australia. In my brief interaction with him on both occasions, I found McKibben to be an unassuming and personable individual as he was in his presentations. In the case of the second presentation, he poked fun at himself by noting "I am sorry to have to come and go. My carbon footprint rivals that of Hazelwood." McKibben probably is the leading climate activist in the world, one who recognizes that the developed world needs to pare back in terms of its consumption but has been reluctant to call for the transcendence of global capitalism in order to achieve a socially just and environmentally sustainable world.

References

Baer, H. A. (2012) *Global Capitalism and Climate Change: The Need for an Alternative World System*, Lanham, MD: AltaMira Press.

McKibben, B. (1989) *The End of Nature*, New York: Random House.

McKibben, B. (2007a) *Deep Economy: The Wealth of Communities and the Durable Future*, New York: Times Book.

McKibben, B. (2007b) *Fight Global Warming Now: The Handbook for Taking Action in Your Community*, New York: Holt Paperback.

McKibben, B. (2010) *Eaarth: Making a Life on a Tough New Planet*, Melbourne: Black, Inc.

McKibben, B. (2011) *The Global Warming Reader*, New York: Penguin Books.

Schuetze, C. (2012) Sustainability Q&A: Bill McKibben. Accessed 05/27/2013 from http://rendezvous.blogs.nytimes.com/2012/07/18/sustainability-q-a-bill-mckibben.

16
ACTIVIST PROFILE – NAOMI KLEIN

Matthias Dietz

Naomi Klein is one of the most popular activists in the climate movement. She is more a voice and intellectual leading figure of the movement than an organizer or practical leader. Her influence and popularity primarily stem from her texts, speeches, and interviews about climate change and the climate movement. One of her most known texts is 'Capitalism vs. the Climate' published in the US magazine *The Nation* in November 2011 (Klein 2011).

Her intellectual role is caused by her profession and past. Naomi Klein is a left-wing journalist and author from Canada who was already popular before her engagement in the climate movement. In the year 2000 she had her breakthrough with the book *No Logo: Taking Aim at the Brand Bullies* (Klein 1999) which has sold more than one million copies and has been translated into 25 languages. The book criticizes the power of transnational corporations, their negative influence on production conditions in poor countries, and the consumption behavior in rich countries. It became the "bible" (*New York Times*) of the global justice movement, because it described the negative consequences of neoliberalism in a clear and powerful way. In the following years Klein was intensively engaged in the global justice movement, but also worked on topics such as the war in Iraq and antiracism.

In 2008 research on negotiations about reparations of the Global North for the Global South due to slavery brought her into contact with the concepts of climate debt and climate justice (Stephenson 2012). Her interest in the topic was raised such that she visited the Climate Conference in Copenhagen in 2009 where she was attracted by the climate movement. Since then Naomi Klein has taken part in the activity of the movement in different ways: For example, she is member of the board of the network organization 350.org, reported from the Alternative Climate Summit in Cochabamba (Klein 2010), and participated in the campaign against the Keystone XL pipeline and the solidarity actions for Tim deChristopher.

At first glance, one could easily assign Naomi Klein to the radical part of the climate movement: First, she criticizes big environmental NGOs like the Environmental Defense Fund for their support of economic and consumption based solutions of the climate crises. She is an opponent of carbon trading and products, which are supposed to be green and eco-friendly, but nevertheless require resources and cause emissions (Revkin 2011). Second, she holds the opinion that an adequate reaction to the challenge of climate change requires big changes in individual behavior as well as in the economy and the role of the state (Revkin 2011). For her, reduction of consumption, 'aggressive regulation' of emissions, massive investment in public transport as well as new economic guidelines are necessary for a successful reaction to climate change. Third, she is particularly critical of the fossil fuel industry and also criticizes capitalism itself. In an interview with Bill Moyers she recently said: "climate change is, I would argue, the greatest single free-market failure. This is what happens when you don't regulate corporations and you allow them to treat the atmosphere as an open sewer" (Sweet 2012). In her opinion not only market failures were responsible for the climate crises, but also the growth paradigm of capitalism (Revkin 2011). In 'Capitalism vs. the Climate' she wrote: "Fact [is] that the airborne waste of industrial capitalism is causing the planet to warm" (Klein 2011).

Analyzing her opinions, Naomi Klein rather seems to support a system change solution of the climate crises instead of a solution within the current system. Nevertheless, she is not so radical to demand a revolution (Common Dreams 2012) and does not share all positions of the radical climate groups such as Rising Tide. In contrast to their positions, Naomi Klein voices her support for the transfer of climate friendly technology to developing countries because of their right of development. She also supports market solutions if these, for example, "encourage renewable energy" (Sweet 2012). And for her it is a conservative argument that has the aim of threatening people if capitalism and climate protection are described as opponents which cannot exist at the same time.

On the whole, Naomi Klein seems to have a huge importance and even a bigger potential for the climate movement. She has a high popularity and positive image inside and outside of the movement which she uses to raise the awareness for climate change and the necessary steps to react to it. This awareness has – at least in the US – declined in the last years (Sweet 2012). Furthermore, she seems to have a kind of integrating function for the movement. Naomi Klein's positions are radical, but not so radical that moderate actors or the public automatically refuse them. Similar to Bill McKibben and his network 350.org, she is an orientation point and authority for both radical and moderate actors and therefore could have a mediating and uniting effect on the nature of the climate movement that seems to be disparate in some ways. Furthermore, she also enables bridges to other movements she knows such as the global justice movement, the antiwar movement, or the Occupy movement. Naomi Klein does not treat climate change as an isolated topic, but sees it in huge contexts with justice, economic, North–South, and ideological questions.

References

Common Dreams (2012) Naomi Klein: 'If You Take Climate Change Seriously, You Have to Throw Out the Free-Market Playbook.' February 29. Accessed 03/15/2013 from www.commondreams.org/headline/2012/02/29-4.

Klein, N. (1999) *No Logo: Taking Aim at the Brand Bullies*, New York: Picador.

Klein, N. (2010) A New Climate Movement in Bolivia, *The Nation*, May 10. Accessed 03/14/2013 from www.thenation.com/article/new-climate-movement-bolivia#.

Klein, N. (2011) Capitalism vs. the Climate, *The Nation*, November 28. Accessed 03/15/2013 from www.thenation.com/article/164497/capitalism-vs-climate#.

Revkin, A. (2011) Naomi Klein in the *New York Times*: Inconvenient Climate Conclusions, *The Nation*, December 8. Accessed 03/14/2013 from www.thenation.com/blog/165058/naomi-klein-new-york-times-inconvenient-climate-conclusions.

Stephenson, W. (2012) Interview: Naomi Klein on Motherhood, Climate Justice, and the Failures of the Environmental Movement. Accessed 03/15/2013 from http://blog.thephoenix.com/BLOGS/phlog/archive/2012/12/14/interview-naomi-klein-on-motherhood-climate-justice-the-failures-of-the-environmental-movement.aspx.

Sweet, D. (2012) Naomi Klein on Capitalism and Climate Change. Accessed 03/20/2013 from http://occupyamerica.crooksandliars.com/diane-sweet/naomi-klein-capitalism-and-climate-cha.

17
ACTIVIST PROFILE – JAMES HANSEN

Hans Baer

Like the social sciences, Steven Vanderheiden (2008) observes, climatology has the potential to serve as a form of social critique instead of acting largely as a descriptive endeavor. Indeed, various climate scientists, such as James Hansen, the Director of NASA's Goddard Institute of Space Studies, based in New York, and David Karoly at the University of Melbourne, have become vocal climate activists. The summer of 1988 witnessed a series of record-breaking heat-waves and droughts in many US regions. In response, congressional hearings on planetary issues were held in Washington, DC, which propelled Hansen into the national limelight when he testified, along with other scientists, that a century-long trend of global warming had resumed after a period of leveling off during the 1950s, 1960s, and early 1970s (Stevens 1999: 131). At the hearings, Hansen stated "with 99 percent confidence" that the planet was undergoing a significant and long-term warming trend and suggested that the greenhouse effect was the culprit (quoted in Weart 2003: 155). These developments prompted the environmental movement, which until then had only expressed passing concern about climate change, to take a closer notice and ultimately to engage in greater action. More recently, Hansen (2007a) has predicted the possibility of disintegration of West Antarctica and/or Greenland ice sheets in a "rapid, non-linear fashion," resulting in a "rise of seal level of more than 5 meters (14.4 ft) by 2095." He also argued in 2007 that increases above 1.7°C would be "highly disruptive," therefore humanity needs to reduce atmospheric CO_2 below 390–350 ppm to avoid irrevocable damage to human societies and the planet.

The George W. Bush administration placed pressure on the Environmental Protection Agency in Summer 2003 to alter its 600-page report by editing out all references to the potential impacts of climate change on the US. It unsuccessfully ordered James Hansen not to speak in public forums about the grave implications of climate change. In open defiance of the effort to gag him, he started a public lecture a week before the 2004 US presidential election with the following words: "I have been told by a high government official that I should not talk about dangerous anthropogenic interference with the climate, because we do not know how much humans are changing the Earth's climate or how much change is dangerous. Actually we know quite a bit" (quoted in Pearce 2006: 82).

In terms of the political economy of research funding, Hansen (2007b: 31) has gone so far as to suggest that those climate scientists who tend to downplay the dangers of climate change are more likely to obtain funding. He along with others have proposed that a "carbon tax ... be implemented at the wellhead, mine gate or port of energy" given the flaws in other market mechanisms, particularly emissions trading schemes and carbon offsetting (Frank 2009: 36). Hansen (2009) also advocates the phasing out of coal, reforestation, and rapidly developing alternative energy, including fourth-generation nuclear power that would theoretically rely upon fast breeder reactors, as climate change mitigation strategies.

Over the past several years, Hansen, who was born on March 29, 1941, has proven that it is never too late to become a social activist. In January 2009 in an interview, he stated,

> We cannot now afford to put off change any longer. We have to get on a new path within this new administration [the Obama administration]. We have only four years left for Obama to set an example to the rest of the world. America must take the lead.
>
> *(quoted in McKie 2009)*

Hansen in particular has challenged the coal industry in calling for the phasing out of coal power completely by 2030. He has been arrested three times for his involvement in protest rallies targeting the fossil fuel industry, the first time on June 23, 2009, along with 30 other protestors, for allegedly obstructing police and impeding traffic in a rally against mining in Raleigh County, West Virginia, in which protesters attempted to enter the Massey Energy Company facility; the second time in September 2010 in front of the White House in a protest calling for a ban on open-pit coal mining; and the third time on August 29, 2011, along with 142 other activists, at a rally in front of the White House that called upon President Barack Obama to reject the Keystone XL pipeline extension slated to transport synthetic crude oil from the Athabasca Oil Sands in Canada to various US sites. While many climate activists have expressed concern about Hansen's somewhat uncritical stance on nuclear power, Hansen is a unique politically engaged scientist who is highly credible in both his efforts to unravel the dangers of anthropogenic climate change and to combat climate change denialism, who continues to research and struggle at a point in life when most people have retired into obscurity.

References

Frank, C. (2009) The Bankruptcy of Capitalist Solutions to the Climate Crisis, *Capitalism Nature Socialism*, 20: 32–43.

Hansen, J. (2007a) Huge Sea Level Rises Are Coming – Unless We Act Now, *New Scientist.com* News Service, July 25.

Hansen, J. (2007b) Climate Catastrophe, *New Scientist*, 298: 30–34.

Hansen, J. (2009) *Storms of My Grandchildren: The Truths about the Coming Climate Catastrophe and Our Last Chance to Save Humanity*, New York: Bloomsbury Press.

McKie, R. (2009) We Have Only Four Years Left to Act on Climate Change – America Has to Lead, *Guardian*, January 18. Accessed 05/27/2013 from www.guardian.co.uk/environment/2009/jan/18/obama-climate-change.

Pearce, F. (2006) *The Last Generation: How Nature Will Take Her Revenge for Climate Change*, London: Transworld Publishers.

Stevens, W. K. (1999) *The Change in the Weather: People, Weather, and the Science of Climate*, New York: Delta Trade Paperbacks.

Vanderheiden, S. (2008) Introduction, in Vanderheiden, S. (ed.) *Political Theory and Climate Change*, Cambridge, MA: MIT Press.

Weart, S. R. (2003) *The Discovery of Global Warming*, Cambridge, MA: Harvard University Press.

18
ACTIVIST PROFILE – AL GORE

Hans Baer

Al Gore's film and accompanying book *An Inconvenient Truth* (Gore 2006), perhaps along with the *Stern Report* authored by Nicholas Stern (2007), a former World Bank economist, propelled climate change into public consciousness around the world. *An Inconvenient Truth* earned $49 million at the box office and became the seventh-highest-grossing documentary in US film history. The film derived from a slide show on climate change that Gore had been showing around the world. For his highly acclaimed film, Gore and the Intergovernmental Panel on Climate Change were jointly awarded the Nobel Peace Prize in 2007. While some climate scientists may quibble about certain details in Gore's presentations, the vast majority along with many government leaders express appreciation to him in his efforts to bring his message to the attention of millions of people. German Environment Minister Sigmar Gabriel distributed 6,000 DVDs of *An Inconvenient Truth* to German schools and the film has been shown in schools in other countries. Climate skeptics, most of whom are politically ultraconservative, have repeatedly sought to undermine his efforts. In the UK, climate skeptics challenged the government's distribution of the film and managed to convince a judge that it purportedly contained nine scientific doubts which had to be pointed out before showing the film in schools.

Unfortunately, while he served as Vice-President in the Clinton administration, in large part, the emissions trading paradigm that underlies the Kyoto Protocol was "successfully pressed on the UN by the US government, advised by US economists, US NGOs and US business, in the 1990s" (Lohmann 2006: 48). Gore argued that the US Congress would not ratify the Kyoto Protocol if a carbon tax rather than an emissions trading scheme were adopted. Ironically, while the Clinton administration signed off on the Protocol, the measure was defeated 95–0 in the US Senate. While to his credit, Gore popularized the findings of climate science, his solutions are framed very much within the parameters of green capitalism and ecological modernization by advocating carbon trading, green consumerism, tree plantations, and techno-fixes as sufficient climate change mitigation strategies. In *Earth in the Balance*, the first edition of which appeared in 1992, Gore (2007: 307–337) proposes implementation of a Global Marshall Plan, which would entail the following elements: (1) stabilization of the world population; (2) the development and sharing of 'appropriate technologies'; and (3) the development of a 'new global economics.' Gore (2007: 346) argues that the definition of GNP should be changed to include environmental costs and benefits, or treated as what mainstream environmental economists term an *externality*. In his most recent book, *Our Choice: A Plan to Solve the Climate* (2009), Gore lays out his views on ecological modernization, which includes an overall endorsement of energy efficiency, retrofitted buildings, hybrid cars, great reliance on public transport, and renewable sources of energy (solar, wind, and geothermal) as climate change mitigation strategies. He feels ambivalent about carbon capture and storage and nuclear power as viable climate change mitigation strategies. While Gore (2009: 123) expresses concern about producing ethanol for motor vehicles from corn and even sugar, he argues that

> the second generation technology for producing ethanol – when it becomes commercially available – has a significant advantage over the first generation technology; instead of using food crops, it will make liquid fuels from perennial grasses, fast-growing trees, and waste steam with a high cellulose content.

In the wake of his film, Gore created The Climate Project[1] which eventually trained some 3,500 activists to show it around the world. TCP has branches in the US, Canada, Australia, the UK, India, Spain, Indonesia, China, and Latin America. Inconvenient Youth[2] formed in 2010 as the brainchild of Mary Doerr, a 16-year-old TCP trainee, who convinced Gore that teens assist in leading efforts to solve the global climate crisis. Gore is also the founder and current chair of the Alliance for Climate Protection (est. 2006). In June 2011 the Climate Project and the Alliance for Climate Protection merged and became the Climate Reality Project.[3] The Project sponsored an expedition in February 2012 to Antarctica with civic and business leaders, climate activists, and concerned citizens from numerous countries in order to observe first-hand the effects of climate change. Included in the trip were Virgin Airlines magnate Richard Branson and Christian Figueres, the Secretary of the United Nations Framework Convention on Climate Change (FCCC).

While Gore has been primarily the target of criticism from climate skeptics, progressive environmentalists have critiqued him on various points, not only the limitations of his green capitalistic solutions but also the tremendous amount of energy and greenhouse emissions that he generates at his Tennessee farm and in his travels by private jet around the world and his failure to promote vegetarianism. To his credit, however, Gore has advocated civil disobedience in the construction of coal-fired power plants. Furthermore, the former Vice-President in the Clinton administration for eight years recovered from his narrow defeat in the 2000 Presidential election to George W. Bush during which he was often described as 'wooden' and in which he even lost his home state of Tennessee of which he had served as Senator for many years by becoming several years later a media star as a result of his generally highly acclaimed film *An Inconvenient Truth* and thereby achieving greater global recognition as a climate activist than he ever did as a politician. In essence, Gore has proved to be a rather unique and contradictory personality who blends moderate stances on climate change mitigation and whose own lifestyle contributes to greenhouse gas emissions while occasionally calling for civil disobedience in challenging the fossil fuel industry.

Notes

1 See presenters.climaterealityproject.org (accessed 10/15/2012).
2 See www.inconvenientyouth.org (accessed 05/30/2013).
3 See www.climaterealityproject.org (accessed 05/30/2013).

References

Gore, A. (2006) *An Inconvenient Truth: The Planetary Emergency of Global Warming and What We Can Do About It*, London: Rodale.
Gore, A. (2007) *Earth in the Balance: Forging a New Common Purpose*, second edition, London: Earthscan.
Gore, A. (2009) *Our Choice: A Plan to Solve the Climate Crisis*, London: Bloomsbury.
Stern, N. (2007) *The Economics of Climate Change: The Stern Review*, Cambridge, UK: Cambridge University Press.

19
ACTIVIST PROFILE – TIM DECHRISTOPHER

Matthias Dietz

Tim DeChristopher is a student and climate activist from the US. He became popular after he had participated in an auction for gas and oil drilling rights in Utah in December 2008. He bought 13 parcels of land for 1.7 million dollars, although he had no money to pay for them. He also participated in other biddings on the day and thus raised their price. The aim of DeChristopher's special form of protest was to act against the oil and gas industry and to prevent local drilling. Because of the manipulation of an auction and false testimony, in July 2011 he was sentenced to a fine of 10,000 dollars and two years in prison.

The case of DeChristopher led to national and international coverage. Popular climate activists such as Bill McKibben, Naomi Klein, and James Hansen published statements of solidarity. Other activists collected money to support DeChristopher in his trial. Furthermore, several protests were organized by different climate groups. One of them was the NGO Peaceful Uprising which was cofounded by DeChristopher at the beginning of 2009. For example, on June 23, 2009, there was a 'Day of Solidarity Action for Tim DeChristopher.' Among other events, a demonstration in front of the court building in Salt Lake City as well as the 'Chicago Solidarity Rally for Tim DeChristopher' took place. Media companies from different countries reported about the trial. The *New York Times* (Kaufman 2009), the British *Guardian* (van der Zee 2011), the *Australian Financial Review* (Dobner 2011), and the German *Spiegel* (Neitsch 2009) published articles on DeChristopher.

The auction, which was interrupted by DeChristopher, had a certain political significance. As well as other auctions it was enabled by one of the last orders of the Bush administration. Shortly after inauguration, the Obama administration revoked more than half of the treaties based on these auctions. It argued that in many cases the sold parcels lay next to environmental protection areas, which is prohibited by law (Neitsch 2009). Nevertheless, this had no influence on the trial against DeChristopher.

Because of his special form of protest and the resulting trial, DeChristopher became one of the most famous climate activists in the US. The climate movement made him into a symbol and leading figure for civil disobedience and the fight for climate justice. One activist said about him: "The story of Tim DeChristopher is not just an important one to those fighting climate change and seeking climate justice, he is a symbol of democracy. An advocate for free speech and justice, he deserves solidarity" (Oden 2011). DeChristopher became an important point of orientation and led to a new dynamic in the climate movement because he caused discussions, cooperations, and activity. Due to DeChristopher, the members of the climate movement developed a closer relationship to each other. DeChristopher and his organization Peaceful Uprising used the trial and the resulting media coverage for publishing several statements and calls to action. DeChristopher regards his participation in the auction as a legitimate form of protest and denies having done something unrightful or wrong (Neitsch 2009). From his point of view, radical protest and civil disobedience are necessary because demonstrations or lobbying cannot stop the harmful exploitation and burning of fossil fuels. Furthermore, he criticizes companies for making profits from the destruction of the climate.

These statements mirror a general tendency in the climate movement. After the UN conference of Copenhagen the strategic approach of the movement was shifting from moderate to radical forms of protest (see for example RTNA 2010). For DeChristopher, this also leads to "confrontation and will involve sacrifice" (Goodell 2011). Personal commitment and the readiness to accept sacrifices would be necessary to implement the new strategy and to stop climate change. From this point of view, DeChristopher has become a kind of leading figure or vanguard activist for a new or at least rising approach of climate action.

There is another general pattern which can be observed in the case of DeChristopher. In the past there were also other trials against climate activists who blocked power stations or airports. While the affected companies demanded strong punishment and high payments to compensate for their losses and to prevent further protest, the activists tried to prove the necessity of their actions (van der Zee 2011). They had not intended to harm something or someone, but aimed at preventing existential damage to the environment and the climate. One example is the 'Kingsnorth Six' from Great Britain who climbed on the chimney of a coal power station and disrupted the power production, but were found not guilty in court. In the case of DeChristopher, the judge did not accept this necessity argument. The verdict against DeChristopher seems to be one of the harshest against a climate activist ever, although the judge did not use his full scope.

On the whole, DeChristopher is an example for a radical climate activist whose protest not only had a direct effect – it delayed oil and gas drillings – but also caused huge indirect effects. The trial against him led to public attention for his aims, mobilized the climate movement, and intensified the relations among its members. The protest of a single person created a huge and lasting effect.

References

Dobner, J. (2011) Green Hero Goes to Jail, *The Australian Financial Review*, July 28.

Goodell, J. (2011) Meet America's Most Creative Climate Criminal, *Rolling Stone*, July 7. Accessed 12/12/2012 from www.rollingstone.com/politics/blogs/national-affairs/meet-america-s-most-creative-climate-criminal-20110707.

Kaufman, L. (2009) Drilling Leases Scrapped in Utah, *New York Times*, February 4.

Neitsch, Peter (2009) US-Student gegen Konzerne: der völlig mittellose Ölbaron, *Unispiegel*, April 6.

Oden, B. (2011) Columbus Solidarity with Tim DeChristopher. Accessed 12/15/2012 from www.wearepowershift.org/blogs/columbus-solidarity-tim-dechristopher.

RTNA (Rising Tide North America) (2010) The Climate Movement Is Dead. Long Live the Climate Movement. Accessed 01/31/2013 from http://risingtidenorthamerica.org/wp-content/uploads/2010/03/climatemovement_longlive.pdf.

van der Zee, B. (2011) Tim DeChristopher on Trial for Sabotaging Oil and Gas Land Auction, *Guardian*, March 1.

20
ACTIVIST PROFILE – ANNA ROSE

Hans Baer

Lawyers Anna Rose and Amanda McKenzie created the Australian Youth Climate Coalition (AYCC) in November 2006. Members are easily spotted at climate movement events in their sky-blue and white AYCC 'Climate Action Now' t-shirts. AYCC claims to have over 70,000 members and to function as a climate action coalition for about 27 other youth organizations, including the Australian Student Environment Network, Engineers Without Borders, the Youth Food Movement, Law Students for a Just Community, the National Indigenous Youth Movement of Australia, the National Union of Students, the Sai Youth (Multi-Faith Youth Network), the Student Organised Network for Architecture, and Youth Christian Students. The AYCC wants the Australian government to: (1) "reduce Australian emissions at home"; (2) "invest in sustainability education"; and (3) "do our fair share internationally [in terms of climate change mitigation]."[1]

In contrast to numerous more radical grassroots climate activists, Anna Rose's moderate stance on climate change mitigation strategies appears to have propelled her into the position of being the most visible climate activist in Australia today. She represented AYCC at the '2020 Summit' convened by Australian Labor Party Prime Minister Kevin Rudd in April 2008 purportedly designed to open "a window on our democracy to let some fresh air in." Indeed, Rose noticed the disproportionate presence of coal industry representatives (*Green Left Weekly*, April 30, 2008, p. 10). She recalled that

> I found myself in the climate stream with the representatives of coal-mining companies including Xstrata and Shell, yet ... no-one from Friends of the Earth, the Australian Conservation Foundation, Greenpeace, Climate Action or any of the State Conservation Councils.
>
> *(quoted in Rintoul 2008: 3)*

In 2009 Rose was a joint recipient of the 2009 Environment Minister's Young Environmentalist of the Year Award.

Rose achieved national coverage in Australia in her role as a co-star in a documentary titled 'I Can Change Your Mind on Climate Change' that aired on Australian Broadcasting Corporation television in 2012. The other co-star in the film is Nick Minchin, a Liberal Party powerbroker, former federal senator, and a staunch climate skeptic. Rose and Minchin traveled extensively to numerous locations around the world and conversed with various prominent climate scientists and climate skeptics in a trip that undoubtedly generated thousands of 'carbon miles.' Rose's travels with Minchin are chronicled in a book titled *Madlands* (Rose 2012). The places and people that Rose and Minchin visited included: Moree (a country town in New South Wales); Perth, where they visited a couple who are climate skeptics; Mauna Loa in Hawaii, where they visited Matthew England, an Australian climate scientist on a four-month research assignment; Berkeley, where they visited Professor Richard Muller (a renowned climate skeptic); the University of California – San Diego where they visited Naomi Oreskes, a geologist, historian of science, and co-author of the highly acclaimed *Merchants of Doubt: How a Handful of Scientists Obscured the Truth on Issues from Tobacco Smoke to Global Warming*; MIT, where they visited Professor Richard Lindzen, a meteorologist and climate skeptic; Washington, DC, where they visited Rear Admiral David Titley, the chief US Navy oceanographer, and Marc Morano, a Republican media personality and blogger; New York City, where they visited Anthony Leiserowitz, Director of the Yale Project on Climate Change Communication, and Bjorn Lomborg (2007, 2010), the provocative skeptical environmentalist; London, where they visited Zac Goldsmith, the former editor of *The Ecologist* magazine and a current Conservative member of the British Parliament; Geneva, where they visited experimental particle physicist Jasper Kirkby who works at the headquarters of CERN (European Organisation for Nuclear Research) and Irene Khan, a trustee of the Climate Justice Foundation; Norwich, England, where they visited Mike Hulme (2009), the founder and former director of the Tyndall Centre for Climate Change Research and the author of *Why We Disagree About Climate Change*; and the Great Barrier Reef, where they visited Ove Hoegh-Guldberg, the director of the Global Change Institute at the University of Queensland. Rose undoubtedly had the best of intentions in engaging in an entertaining debate with a prominent Australian climate skeptic in what proved to be a dizzying romp on three continents which resulted in a film documentary and a book published by a prestigious university press. My worry is that she may have unwittingly played into the mass media's pattern of presenting the debate on climate change as a 50/50 proposition between two equally valid perspectives when in reality the vast majority of climate scientists argue that climate change is not only real but due to anthropogenic activities.

Note

1 See www.youthclimatecoalition.org and http//:www.aycc.org.au (accessed 05/30/2013).

References

Hulme, M. (2009) *Why We Disagree about Climate Change: Understanding Controversy, Inaction and Opportunity*, Cambridge: Cambridge University Press.

Lomborg, B. (2007) *The Skeptical Environmentalist: Measuring the Real State of the World*, Cambridge: Cambridge University Press.

Lomborg, B. (2010) *Cool It: The Skeptical Environmental's Guide to Global Warming*, Cambridge: Cambridge University Press.

Rintoul, I. (2008) 2020 – The Summit Vision in Hindsight, *Solidarity*, Issue 3 (May): 8.

Rose, A. (2012) *Madlands: A Journey to Change the Mind of a Climate Sceptic*, Melbourne: Melbourne University Press.

21
ACTIVIST PROFILES – DAVID SPRATT AND PHILIP SUTTON

Hans Baer

David Spratt and Philip Sutton (2008a, 2008b) propelled themselves into prominence in the Australian climate movement by releasing *Climate Code Red*, a monograph initially in loose spiral binding by Friends of the Earth Australia in 2008 with the subtitle 'The Case for a Sustainability Emergency' later republished in an expanded book by Scribe Publishing in Melbourne with the subtitle 'The Case for Emergency Action.' Spratt and Sutton are organic intellectuals, not university academics, and were major figures in the grassroots climate movement before *Climate Code Red* and have become even more significant since the publication of the book.

Spratt is a longstanding peace and environmental activist. In 2008, with FoE Climate Justice Campaign Director Damien Lawson, Spratt established the Climate Action Centre, located in the Victorian Trades Hall, to strengthen the movement by developing, supporting, and forging links between groups. The Centre provided informative materials, such as a '4 Degrees Hotter' primer in February 2011. It also served as an important physical space where local climate action groups and networks, including the Climate Emergency Network and Union Climate Action Network, could meet. Spratt paid the Centre's rent until it closed sometime in 2011. Along with Cam Walker of FoE, Spratt maintains the CarbonEquity website. He regularly releases reports and commentaries on climate science and climate politics, often through CarbonEquity, for example 'The Big Melt: Lesson from the Arctic Summer of 2007' (October 2007).

Sutton, who never completed the veterinary science degree he started at the University of Sydney in the 1970s, has been described as a 'green business leader,' having formed Environmental Management Systems to assist corporations in managing their environmental impacts. He presents himself as a 'sustainability strategist.' By working unpaid for the Australian climate movement for many years, Sutton effectively gave most of his inheritances to the movement. He is Convenor of the Greening Strategic Institute, Convenor of the Climate Emergency Network, co-founder of Safe Climate Australia, and the past-President of the Sustainable Living Foundation and the Australia New Zealand Society for Ecological Economics. His work focuses on "swift achievement of an ecologically sustainable economy and strategies for rapid mobilization." Sutton has been a key player in the Transition Decade campaign (T10) which is committed within the 2010 decade to creating a "safe climate outcome" (Transition Decade, n.d.). The first two years of T10 focus on encouraging whole communities to say 'yes' to this objective, the subsequent eight years to obtain government commitment to the plan, and the last years to adoption of an "emergency transition mode." On February 14, 2010, T10 was launched publicly to over 700 people at Melbourne Town Hall and another 500 around Australia in smaller events.

In much the same way that 'code red' signals to hospital workers that a patient needs advanced life-support, Spratt and Sutton's *Climate Code Red* argues that climate change sends a strong warning that societies and their leaders must declare a 'climate emergency,' which transcends 'business and politics as usual' and indicates the will to prevent a global climate catastrophe. To create a safe climate, Spratt and Sutton contend that humanity needs to reduce the current global temperature to less than 5°C above the preindustrial level and greenhouse gases to less than 320 ppm CO_2. Part I reviews the scientific evidence indicating the seriousness of climate change. Part II discusses debates about emissions reduction targets and proposes reductions consistent with a safe-climate future, and Part III calls for a rapid transition to a sustainable economy. Spratt and Sutton are deeply critical of neoliberalism, under which the 'corporate agenda runs politics,' ensuring that capitalism is environmentally unsustainable. They envisage a green social democracy, a highly regulated capitalism, because governments, they maintain, have a crucial role to play in planning, coordinating, and implementing the transition to a safe climate.

Crucial to implementation of a climate emergency will be the energies and active support of a popular climate movement, involving community-action groups, unions, churches, neighborhoods, schools, local government, political parties, even corporate elites. *Climate Code Red* has prompted constructive debate within the Australian climate movement. Ben Courtice (2008) in his review in *Green Left Weekly* published by the Socialist Alliance argues that Spratt and Sutton overlooked important questions such as: "Can the planet be saved without first abolishing capitalism?" Is a "state of emergency compatible with democracy?" Because of the bold argument of *Climate Code Red*, it helped to galvanize the Australian climate movement. The book has not only provided a significant text for the movement but has enabled its co-authors, longstanding and relentless climate activists who have worked with a wide array of groups, to become even more useful leaders for the movement. They are both in high demand and speak frequently to local climate action meetings and large climate action conferences, as well as to socialist and trade union conferences concerned with climate change.

References

Courtice, B. (2008) Climate Change: The Case for Immediate Action, *Green Left Weekly*, August 23. Accessed 30/05/2013 from www.greenleft.org.au.

Spratt, D. and Sutton, P. (2008a) *Climate Code Red: The Case for a Sustainability Emergency*, Melbourne: Friends of the Earth Australia.

Spratt, D. and Sutton, P. (2008b) *Climate Code Red: The Case for Emergency Action*, Melbourne: Scribe.

Transition Decade (n.d.) The Transition Decade Strategic Framework, leaflet, 2pp.

22
ORGANIZATION PROFILE – CLIMATE ACTION NETWORK INTERNATIONAL

Heiko Garrelts

Climate Action Network International (CAN-I) is a transnational/global organization and network of nongovernmental organizations (NGOs). It is one of the oldest, biggest, and most popular actors of the climate movement worldwide. With its continuous activities, it has been of significant weight so far. Climate Action Network was founded in 1989 and started joint activities during the second world climate conference in 1990. It has been growing steadily and consists currently of more than 700 NGOs in 95 countries. Climate Action Network is organized into regional and national "nodes" (Duwe 2001: 179). Each node is responsible for its own governance and procedures, in conducting joint policy and advocacy work within its given country or region. The regional networks include Climate Action Network Europe, United States, and South Asia among others.[1] These networks consist of member organizations, among them different national groups of 350.org, different groups of the humanitarian organization CARE, environmental NGOs such as Friends of the Earth and the World Wildlife Fund as well as other groups engaged in issues of urban sustainability, environmental justice, and other issues. According to Duwe (2001: 179), the network nodes differ from each other insofar as CNE and US-Climate Action Network run steady offices with full-time staff in Brussels and Washington respectively. The capacity of US-Climate Action Network and CNE puts the main part of the international coordination into their hands in addition to facilitating the members of their regions. In contrast, most of the other coordinators are members of regional NGOs and have to fulfill their Climate Action Network job alongside their usual obligations with the consequence of less time for networking activities (Duwe 2001: 179).

According to its own statement, Climate Action Network and its members "place a high priority on both a healthy environment and development that 'meets the needs of the present without compromising the ability of future generations to meet their own needs' (Brundtland Commission). Climate Action Network's vision is to protect the atmosphere while allowing for sustainable and equitable development worldwide" (CAN 2013a).

An important milestone in the organization's history was the Kyoto Protocol negotiations. Here, the organization's position consisted of four main objectives (Betsill 2002: 53). First, the network argued that the Protocol should include commitments for industrialized countries to reduce their greenhouse gas (GHG) emissions amounting to 20 percent below 1990 levels by 2005. Second, it argued for strong review and compliance mechanisms to enhance the implementation of the commitments contained in the Protocol. Third, Climate Action Network opposed proposals to allow industrialized Parties to meet their commitments through emissions trading. Fourth, it also objected to proposals to allow Parties to get credit for emissions absorbed by sinks such as forests. Especially with regard to the issue of emission trading, Climate Action Network has revised its position. Since the Kyoto Protocol entered into force in 2005, CAN is among those civil society organizations which subsequently *adopted* the market based instruments[2] and focused their activities on further developing, designing, and implementing them. Recently, CAN has published a report on possible changes to the modalities and procedures for the Clean Development Mechanism (CDM) (CAN 2013b). The issue of sinks remains contested and the respective position depends on single CAN member organizations. Another milestone of CAN's member structure is the thirteenth Conference of the Parties (COP 13). At this occasion, many significant NGOs left CAN and founded Climate Justice Now! This is a coalition with a significantly stronger system-critical programmatic. To a relevant degree, this split the concerned community based groups in the South (see chapters by Kössler and Dietz in this book).

Despite this, CAN is still active at the local, national, and international level (Lipschutz and McKendry 2011: 375). At local level, member groups engage in projects and educational efforts with an intention to engage various publics adopting behaviors and practices that will reduce GHG emissions. At the national level, CAN member organizations mainly provide education, conduct research, and analyze data in the effort to influence governments to adopt climate friendly policies. Besides this approach, some target corporations and corporate associations and try to alter business practices and attitudes. With regard to the international level, it is CAN's aim "to put forward a strong and consistent message from the environmental community and to create a single dominant environmental voice in the climate debates" (Lipschutz and McKendry 2011: 375). For this reason, CAN is most active at meetings of the United Nations Framework Convention on Climate Change. CAN activities include the lobbying of delegates, criticizing their actions as well as making recommendations where necessary. One highlight of the conference activities is CAN's daily newsletter *ECO*, presenting the views of (parts of) the environmental NGO community on a variety of issues. *ECO* makes public what is happening in the negotiations. In addition, *ECO* functions as a political forum for promoting CAN's own positions. Shaming, an important instrument for nongovernmental organizations in general (Keck and Shikking 1998), is demonstrated by the 'Fossil of the Day' and the 'Fossil of the Year.' Since 1999, Climate Action Network's member organizations have regularly judged 'Fossil of the Day' awards. The awards are given to countries that have performed badly in the climate change negotiations. These awards are presented daily during climate talks. This judgment is complemented by the 'Fossil of the Year' award. For instance in 2012, the Colossal Fossil went to Canada and New Zealand. After a five-year "reign" as the Colossal Fossil, it "seems Canada is refusing to bow out gracefully into the irrelevance that comes with being an historic climate laggard. They, instead, stood strong for inaction throughout the UN climate talks challenged only by New Zealand" (CAN 2013c).

With regard to the impacts of CAN's activities, a broader perspective seems appropriate. There is little evidence of direct influence, in the form of the uptake of particular proposals for the Kyoto Protocol (Betsill 2002; Bulkeley and Newell 2010). In contrast, CAN has successfully worked to bridge tensions between European and American environmental organizations and between groups from the North and the South present at the international negotiations. This "ability to bridge differences between the quite different environmental organisations" has according to Lipschutz and McKendry (2011: 375), arguably been CAN's most relevant achievement. Another crucial role is the provision of the concerned public with consistent and timely updates on the course of climate negotiations. Therefore, CAN efforts have helped "to make the negotiating process more transparent and to force a certain amount of accountability on states and their representatives through shaming" (Lipschutz and McKendry 2011: 376). Despite these activities, years of international climate negotiations have had little impact in GHG emissions. Consequently, environmental organizations and networks such as CAN have "begun to shift their efforts increasingly towards local projects designed to influence, if not transform norms, beliefs, behaviours and practices" (Lipschutz and McKendry 2011: 376) in the context of global warming. As a result, the significant rise in public awareness and in the burgeoning environmental consciousness is another crucial impact of CAN's activities (although, of course, these phenomena cannot fully be attributed to CAN) (Lipschutz and McKendry 2011: 379).

Notes

1 The complete list contains CAN Australia (CANA), CAN Canada (RAC-Canada), CAN China, CAN East Africa, CAN Eastern Europe, Caucasus and Central Asia, CAN Europe, CAN France (RAC-France), CAN Japan (Kiko Network), CAN Latin America (CAN-LA), CAN North Africa (RAC-Maghreb), CAN Pacific, CAN South Asia (CANSA), CAN Southern Africa (SACAN), CAN South Africa, CAN South East Asia (CAN-SEA), CAN United States, CAN West Africa.

2 According to Obertür and Ott (2000: 114) CAN had been divided before: while European NGOs voted for state-related steering, NGOs from the US argued in favor of market based instruments.

References

Betsill, M. (2002) Environmental NGOs Meet the Sovereign State: The Kyoto Protocol Negotiations on Global Climate Change, *Colorado Journal of International Environmental Law and Policy*, 13: 49–64.

Bulkeley, H. and Newell, P. (2010) *Governing Climate Change*, London: Routledge.

CAN (2013a) About CAN. Accessed 05/30/2013 from www.climatenetwork.org/about/about-can.

CAN (2013b) Views on Possible Changes to the Modalities and Procedures for the Clean Development Mechanism. Accessed 05/27/2013 from www.climatenetwork.org/sites/default/files/can_submission_cdmrevision_march2013_final.pdf.

CAN (2013c) Canada and New Zealand Tie for the Infamous Colossal Fossil 2012 Award. Accessed 05/27/2013 from www.climatenetwork.org/fossil-of-the-day/canada-and-new-zealand-tie-infamous-colossal-fossil-2012-award.

Duwe, M. (2001) The Climate Action Network: A Glance behind the Curtain of a Transnational NGO Network, *Review of European Community and International Environmental Law*, 10(2): 177–189.

Keck, M. E. and Shikking, K. (1998) *Activists Beyond Borders: Advocacy in International Politics*, Ithaca, NY: Cornell University Press.

Lipschutz, R. and McKendry, C. (2011) Social Movements and Global Civil Society, in Dryzek, J., Norgaard, R., and Schlosberg, D. (eds.) *Oxford Handbook of Climate Change and Society*, Oxford: Oxford University Press.

Obertür, S. and Ott, H. E. (2000) *Das Kyoto-Protokoll. Internationale Klimapolitik für das 21. Jahrhundert*, Opladen: Leske & Budrich.

23
ORGANIZATION PROFILE – CLIMATE JUSTICE NOW!

Matthias Dietz

Climate Justice Now! (CJN!) is one of the biggest network organizations in the climate movement. It was founded in 2007 by 30 organizations during the UN Climate Change Conference in Bali. Since then the network has grown rapidly and today unites several hundred member groups. These come from different regions of the world with an emphasis on the Global South. Popular international members are 350.org, Friends of the Earth, and Attac. Groups from the South include FASE from Brazil, La Confederación Campesina del Perú, the Centre for Environmental Justice from Sri Lanka, and the Third World Network from Ghana (CJN! 2013a).

CJN! has a strong presence during the yearly UN Climate Change Conferences where its representatives take part in the official negotiations, but are also involved in protests and alternative events. For example, 300 accredited representatives of CJN! joined the negotiation process in the conference center in Copenhagen in 2009. Nevertheless, members of the network also participated in a protest action called 'Reclaim Power.' A march of activists tried to enter the Conference Center in order to hold an alternative summit there (Rizvi 2009). Apart from the UN climate conferences, CJN! was also present at the Alternative Climate Summit in Cochabamba in 2010, the World Social Forum in Bélém in 2009, and organized the 'World Bank Out of Climate Finance Campaign' in 2010 (CJN! 2010). Furthermore, the network sometimes also joins the meetings of the UN for the preparation of the Climate Change Conferences (CJN! 2013b).

What makes CJN! special is not only the majority of its members coming from the Global South, but also the critical programmatic and perspective of the network. This can be observed from the foundation of the network. It came into existence by splitting off from the largest climate network CAN during the Climate Change Conference in Bali in 2007. The climate groups which founded CJN! justified their secession with the dominant position of big Western NGOs in CAN. They claim that these Western NGOs blindly support the Kyoto Protocol and that they do not really care about Southern interests and positions (Passadakis 2009). As a consequence, CJN! demands a new orientation of the climate negotiations as well as of civil society. Countries and people from the Global South should be supported in their struggle against global warming because they are especially vulnerable to climate change, but have not caused it. "Huge financial transfers" (CJN! 2013c) and respect for the right of self-determination and food sovereignty (see organization profile of La Via Campesina in this book) would be necessary. Furthermore, the industrial countries should lower their emissions by reduced consumption and a cessation of the extraction of fossil fuels. CJN! demands "social, ecological and gender justice" (CJN! 2013c). By holding these positions, the network clearly takes a justice-oriented perspective and refers to the concept of climate justice. The close relation of CJN! to the concept of climate justice is also expressed in the name of the network. It is probably based on the document 'Climate Justice Now! The Durban Declaration on Carbon Trading' which was developed and signed by a number of climate groups in 2004.

Several other positions of CJN! result from this South and justice centered perspective. The network is an opponent of economic instruments for climate protection such as carbon trading and carbon offsets. One of its arguments is that carbon markets are not working because there are too many and free circulated certificates. As a consequence, Western companies could make profits by carbon trading without reducing emissions. Furthermore, projects for climate protection in the South would strengthen the dependence from Northern countries and often ignore the rights of local people. Due to the integration of these economic instruments in the Kyoto Protocol, CJN! rejects the protocol as well as the enlargement of its period of validity (Casey 2012).

The network expresses its critical positions in statements, declarations, press releases, and articles which are published on its website. A popular document derives from the foundation period of the network and is entitled 'The Climate Justice Now! Principles' (CJN! 2013c). Apart from the participation in the UN negotiation process and the mobilization of summit oriented protest, publications are a third important way of acting for CJN.

The organizational structure of CJN! is quite diffuse. It describes itself as a decentralized network. It is run by volunteers. It doesn't have professional staff, nor an official headquarter or even an office. The coordination of the network seems to be based intensively on the Internet. The members of CJN! mostly communicate by using websites, blogs, and email lists. Nevertheless, there are also some rare hints at meetings of the network during the UN Conferences (CJN! 2009). Only sometimes are names of individuals such as Assad Rehman or Nicola Bullard mentioned in the context of the network. The website of CJN! does not contain any contact names or coordination persons. As a consequence, some kind of anonymity as well as a certain looseness of the relations among the members seem to be characteristic of CJN!.

Finally, CJN! could be categorized as a huge and consistently active climate network holding positions which are critical and oriented towards the global South. It is one of the big names in the climate movement. Nevertheless, its influence and importance are difficult to judge. CJN! seems to have a loose structure and is intensively focused on the UN negotiation process. This limits its influence and the credibility of its programmatic. Rejection of economic instruments and critique of the UN negotiation process are not really convincing if one takes part in these negotiations instead of building up alternatives or acting on different levels. This is especially true against the background of a declining importance of the UN process for the climate movement. Furthermore, diplomatic advocacy work and revolutionary rhetoric do not fit together. Considering all these points, the description of CJN! clearly reveals some ambivalence and weakness of the network. Nevertheless, its impressive member list and radical rhetoric have already had an empowering and motivating effect on the climate movement, although its acting is less influential and spectacular.

References

Casey, M. (2012) At Climate Talks, UN Warns of Thawing Permafrost Increasing Global Warming; Potential Hazards of Thaw Have Until Now Not Been Factored into Climate Models: Report, *The Associated Press*, November 27.

CJN! (Climate Justice Now!) (2009) Report of the CJN! Strategy Meeting 4 & 5 October 2009, Bangkok Thailand. Accessed 09/07/2012 from www.climate-justice-now.org/report-of-the-cjn-strategy-meeting-4–5-october-2009-bangkok-thail.

CJN! (Climate Justice Now!) (2010) World Bank Out of Climate Campaign. Accessed 03/20/2013 from www.climate-justice-now.org/category/climate-justice-movement/worldbank-out-of-climate-campaign/.

CJN! (Climate Justice Now!) (2013a) CJN! Network Members. Accessed 03/15/2013 from www.climate-justice-now.org/cjn-network-members-november-2010/.

CJN! (Climate Justice Now!) (2013b) History. A Short History of Climate Justice Now!. Accessed 03/15/2013 from www.climate-justice-now.org/about-cjn/history/.

CJN! (Climate Justice Now!) (2013c) Principles. Accessed 03/15/2013 from www.climate-justice-now.org/em-cjn/mission/.

Passadakis, A. (2009) Auf dem Weg nach Kopenhagen, *taz*, November 10.

Rizvi, H. (2009) Civil Society Delegates Excluded from Climate Summit, Gassed, Beaten, *Environmental News Service*, December 16.

24
ORGANIZATION PROFILE – LA VIA CAMPESINA

Matthias Dietz

La Via Campesina is an international network supporting the rights of small farmers. During the years it has become more and more focused on climate politics. The Spanish name of the network means 'the way of the farmers.' It was founded in 1996 and currently has 148 member organizations from almost 70 countries. By describing itself as peasant movement the network claims to be a social movement (Desmarais 2007). This category seems to have some justification because La Via Campesina shows movement characteristics such as high growth dynamic and the frequent use of protest as a political instrument. Furthermore, the member organizations have developed a common identity. Groups such as Confederación Campesina Peru, Landless Peoples Movement South Africa, or Farmers from the Austrian Alps share the idea of food sovereignty. They demand that small farms should be sustained and that the local population should be able to decide what and which technique it wants to plant with. Despite their different living conditions and countries of origin, the members of La Via Campesina have developed a common understanding of rural farming and rural life. They act together to realize this idea and support each other in conflicts regardless of national borders or huge geographical distances.

The engagement of La Via Campesina in climate politics can be explained in huge parts by its concept of food sovereignty. First, the members of the network oppose huge plantations for generating agro fuels in order to save carbon emissions. For them, these plantations destroy small farming structures and local food production. Second, they expect climate change to raise the vulnerability of small farmers: "We are now at the worst moment for agriculture and small farmers and for nature. The impacts of climate change are steadily worsening, leading to harvest failures, destruction of habitats and homes, hunger and famine and loss of lives" (La Via Campesina 2011). Nevertheless, the network does not just see climate change as a threat for small farmers, but underlines the adaptation experience and low emissions of traditional local agriculture. It advertises traditional agriculture as an efficient instrument in the fight against global warming (Hamer 2009).

As a consequence of the importance that La Via Campesina attaches to climate change, the network has been present at each UN Climate Change Conference over the last years and is part of the international climate movement. It fights for climate justice and against so-called false solutions such as agrofuels and carbon trading. During the UN conference in Durban in December 2011, La Via Campesina organized a worldwide day of action, with the title '1,000 Durbans for Climate Justice.' The network also published a declaration, which heavily criticized the negotiation process: "Here in Durban ... developed countries are working hard to escape their historical responsibility and not to pay their climate debt" (La Via Campesina 2011). The attraction that the concept of climate justice holds for La Via Campesina can be explained by the vulnerability of its members (especially in the global south) to climate change as well as by its wish for local self-determination. La Via Campesina also took part in the Alternative Climate Summit in Cochabamba, Bolivia in 2010. Up to now it has shared the radical demands of the conference such as high compensation payments of the industrialized countries or a global warming threshold of one degree. Due to its membership in different networks of climate groups and its involvement in several Days of Actions, the engagement of La Via Campesina is not limited to the UN conferences but spreads over the whole year.

La Via Campesina is of high importance for the climate movement. With almost 150 member organizations, representing some 100,000 individuals, it is a big player with a broad and well-connected basis. As a consequence, the organizational structure of the network is decentralized. It is separated into different world regions. In addition, the headquarters of the network periodically moves from country to country. Because of the huge number of members from South America and Asia, La Via Campesina is the strongest voice of the Global South in the international climate movement. The network is very active in mobilizing climate protest also because it relates climate change to other topics of its agenda such as agriculture, women, and human rights (La Via Campesina 2012).

All in all, La Via Campesina is a huge actor with a high credibility. This credibility stems from its broad basis and a programmatic which fits to the members of the network. These characteristics and its rising interest for climate change make La Via Campesina an important and in future possibly leading actor in the climate movement.

References

Desmarais, A. (2007) *La Via Campesina: Globalization and the Power of Peasants*, London: Pluto Press.

Durban Declaration (2004) Climate Justice Now! The Durban Declaration on Carbon Trading. Accessed 12/06/2012 from www.durbanclimatejustice.org/durban-declaration/english.html.

Hamer, E. (2009) Copenhagen: Peasant Farmers Can Save the Planet, *The Ecologist*, December 15.

La Via Campesina (2011) La Via Campesina Declaration in Durban. Accessed 01/15/2013 from http://viacampesina.org/en/index.php/actions-and-events-mainmenu-26/-climate-change-and-agrofuels-mainmenu-75/1147-la-via-campesina-declaration-in-durban.

La Via Campesina (2012) Main Issues. Accessed 12/06/2012 from http://viacampesina.org/en/index.php/main-issues-mainmenu-27.

25
ORGANIZATION PROFILE – FRIENDS OF THE EARTH INTERNATIONAL

Johannes Kruse

Friends of the Earth International (FoEI) is a transnational environmental NGO describing itself as "the world's largest grassroots environmental network" (FoEI 2011a). The organization was founded by David Brower in the United States in 1969. Before that, Brower, an environmentalist, had long played a prominent role in leading the Sierra Club. After leaving the Sierra Club due to conflicts about campaign goals and tactics, Brower's aim was to establish an organization that would be truly international, decentralized, and more actively engaged with the media (Wapner 1996: 121–122). According to Brower, Friends of the Earth (FoE) should "have as its specific purpose the task of waging political battles to protect the environment" (quoted in Burke 1982: 105). In 1969, FoE established an office in San Francisco and later also opened offices in Paris, London, and Sweden. In 1971, FoE groups from these four countries founded the transnational umbrella organization FoEI. Today, FoEI is one of the largest transnational environmental NGOs in the world with member organizations in over 70 countries, which – taken together – have more than two million individual members. Its annual budget of a bit more than two million euro is funded by two main sources: (i) donations and core funding provided by foundations, aid organizations, and governments; (ii) fees paid by its member NGOs (FoEI 2012a). Compared with other environmental NGOs and even some of its own members, FoEI's budget is only modest. The structure of FoEI is that of an international umbrella organization or federation: All decisions about its international work are made at biennial assemblies in which every member NGO has equal speaking and voting rights. In contrast to other transnational NGOs, voting rights are not coupled to membership numbers or financial contributions. The participants of these assemblies also elect the chair and the board of FoEI (FoEI 2012b). Its structure makes FoEI rather a network of NGOs than one hierarchical organization. Furthermore, this structure secures the independence of its national member NGOs and arguably grants Southern NGOs more influence than in other transnational environmental NGOs (Doherty 2006).

In the late 1980s, FoEI began working on climate change and was one of the first environmental NGOs identifying it as an important issue of environmental politics. Already in 1987, FoE in England, Wales, and Northern Ireland planned a campaign to raise awareness on the issue. However, it took a back seat to campaigns on acid rain and on the ozone hole (Wynne et al. 2001). In the following year, FoE published two reports on the impacts of rising greenhouse gas concentrations in the atmosphere and provided policy recommendations (Jackson and

Roberts 1989; Karas and Kelly 1989). Despite this early activism on the national level, it was only in 1997 that climate change also became the target of an international campaign of FoEI (Walk and Brunnengräber 2000: 113). Many member organizations from the global South were initially hesitant to work on the topic, because they feared that campaigns on current ecological injustices would become less important (Doherty 2006: 872–873). Today, however, climate change forms a core pillar of FoEI's work and is also addressed by campaigns on other issues such as biofuels, energy, forests, biodiversity, food sovereignty, water questions, and resistance against the further exploitation of oil and gas.

In analyzing FoEI's work on climate change, one can identify four principles on which its activities are based: (1) International campaigns regularly link international goals with concerns on the local and national level. In this way, FoEI has combined its principled rejection of the role of the World Bank in climate finance with its resistance against the Medupi coal-fired power plant; a plant that is being built in South Africa and has been financed with loans from the World Bank (FoEI 2011b). Linking national and international concerns has three advantages for FoEI's work: It allows the organization to enlist its local and national groups for international campaigns, to adjust its campaigns to its grassroots members' demands, and to ensure that campaigns aim for change that benefits them. (2) Furthermore, FoEI actively challenges what it identifies as 'neoliberal policies' and the influence of big corporations. Therefore, FoEI criticizes the role of international financial institutions in climate politics and rejects market-based mechanism. The organization has, for instance, criticized the influence of big corporations on the negotiations on climate and sustainable development and has demanded to "reclaim the UN from corporate capture." In a report, FoEI (2011a) complained about "corporate pressure exerted over national and international climate and environmental policy" and made fighting this 'corporate capture' a priority of its future work. (3) As part of its international work, FoEI has for a long time been lobbying international institutions and is accredited as an observer at a number of institutions including the United Nations Framework Convention on Climate Change (UNFCCC). In this way, FoEI delegations have participated in the international climate negotiations since the first conference of parties in 1995. In the last years, however, FoEI has increasingly set different priorities and has decided that its "participation in international fora will … become secondary to our resistance work" (FoEI 2008). Despite these new priorities, FoE has been present at all climate negotiations since then. (4) FoEI aims at addressing social and political inequality and has adopted environmental justice as a guiding principle for its work (Doherty 2006). At the core of this understanding is the idea that "environmental questions are also always distributive questions" (Mittler 2002: 189). As a result, FoEI (2007) was one of the first signatories of the declaration 'What's Missing from the Climate Talks? Justice!' and has participated in organizing the Climate Justice Now! network (CJN!). One of the central demands of FoEI is that developed countries should support developing countries with far-reaching financial and technological transfers (FoEI 2012c). In 2010, FoEI withdrew from the Climate Action Network, the largest network of NGOs working on climate change, and has since then become a full member of CJN! (Reitan 2011). With its focus on the victims of climate change, its critique of neoliberalism and market-based mechanisms, the demand for energy sovereignty, and the resistance against fossil fuel energy, FoEI is one of the most prominent voices on climate change within civil society.

References

Burke, T. (1982) Friends of the Earth and Conservation of Resources, in Willetts, P. (ed.) *Pressure Groups in the Global System: The Transnational Relations of Issue-Orientated Non-Governmental Organisations*, London: Frances Pinter.

Doherty, B. (2006) Friends of the Earth International: Negotiating a Transnational Identity, *Environmental Politics*, 15(5): 860–880.

FoEI (Friends of the Earth International) (2007) What's Missing from the Climate Talks? Justice! Press Release. Accessed 10/18/2012 from www.foei.org/en/media/archive/2007/whats-missing-from-the-climate-talks-justice/.

FoEI (Friends of the Earth International) (2008) Our Strategic Plan, *Annual Report 2007*. Accessed 09/04/2011 from www.foei.org/en/resources/publications/annual-report/2007/our-strategic-plan/our-strategic-plan.

FoEI (Friends of the Earth International) (2011a) Who We Are. Accessed 05/22/2011 from www.foei.org/en/who-we-are.

FoEI (Friends of the Earth International) (2011b) *World Bank: Catalysing Catastrophic Climate Change – The World Bank's Role in Dirty Energy Investment and Carbon Markets*, Amsterdam: Friends of the Earth International.

FoEI (Friends of the Earth International) (2012a) *Financial Report 2011*, Amsterdam. Accessed 10/19/2012 from www.foei.org/en/resources/publications/annual-report/annual-report-2011/annual-report-2011/at_download/file.

FoEI (Friends of the Earth International) (2012b) Our Structure. Accessed 10/19/2012 from www.foei.org/en/who-we-are/about/structure.

FoEI (Friends of the Earth International) (2012c) UN Climate Talks: Urgent Progress Still Not in Sight. Accessed 12/12/2012 from www.foei.org/en/media/archive/2012/un-climate-talks-urgent-progress-still-not-in-sight.

Jackson, T. and Roberts, S. (1989) *Getting out of the Greenhouse: An Agenda for UK Action on Energy Policy*, London: Friends of the Earth.

Karas, J. H. W. and Kelly, P. M. (1989) *The Heat Trap: The Threat Caused by Rising Levels of Greenhouse Gases*, London: Friends of the Earth.

Mittler, D. (2002) Globalisierung und Nachhaltigkeit: Friends of the Earth International als umweltpolitischer Teil der globalisierungskritischen Bewegung, in Walk, H. and Boehme, N. (eds.) *Globaler Widerstand: Internationale Netzwerke auf der Suche nach Alternativen im globalen Kapitalismus*, Münster: Westfälisches Dampfboot.

Reitan, R. (2011) Coordinated Power in Contemporary Leftist Activism, in Olesen, T. (ed.) *Power and Transnational Activism*, Abingdon: Routledge.

Walk, H. and Brunnengräber, A. (2000) *Die Globalisierungswächter: NGOs und ihre transnationalen Netze im Konfliktfeld Klima*, Münster: Westfälisches Dampfboot.

Wapner, P. K. (1996) *Environmental Activism and World Civic Politics*, Albany: State University of New York Press.

Wynne, B., Simmons, P., Waterton, C., Hughes, P., and Shackley, S. (2001) Institutional Cultures and the Management of Global Environmental Risks in the United Kingdom, in Clark, W. C., Dickson, N., Jäger, J., and van Eijndhoven, J. (eds.) *Learning to Manage Global Environmental Risks, Volume 1: A Comparative History of Social Responses to Climate Change, Ozone Depletion, and Acid Rain*, Cambridge, MA: MIT Press.

26
ORGANIZATION PROFILE – CLIMATE ALLIANCE

Heiko Garrelts

The Climate Alliance of European Cities with Indigenous Rainforest Peoples was founded in 1990 (before climate change started to be a relevant policy issue). Basically, it is a European network of cities, municipalities, districts, and indigenous peoples of the Amazonas Region. The latter are represented by the organization Coordinadora de las Organizaciones Indígenas de la Cuenca Amazónica (COICA). In addition to COICA, more than 1,600 member cities, municipalities, and districts, further NGOs and organizations from 20 European countries were organized in the Climate Alliance in January 2013. Compared to other European transnational municipal climate protection networks (hereafter called 'TMNs'), in particular the Cities for Climate Protection (CCP)[1] and the Energie-Cités,[2] the Climate Alliance is the largest. Most of the member cities are from continental European countries, especially from Germany, Austria, and the Netherlands.[3] The members of Climate Alliance engage in activities referred to commonly as the global climate protection and set the following three goals: reducing greenhouse gas emissions, supporting indigenous rainforest peoples, and conserving the tropical rainforests and their biological diversity (Climate Alliance 2013a).

Just like other TMNs (see Kern and Bulkeley 2009: 317), the Climate Alliance is structured in a way that reflects the process of combining centralization and decentralization which is characteristic of Europeanization. According to Mayer-Ries (2000: 143), this also applies to the Latin America side: the establishment of new alliances between local indigenous actors and other levels where actors emphasize the ideas of autonomy of the indigenous towards the national state – apart from democratization and decentralization processes. The Climate Alliance is managed by an international board[4] and coordinated by the European Secretariat, complemented by the Climate Alliance Office in Brussels and national and regional coordination offices (Climate Alliance 2013b). Further political levels and sectors are relevant thereby: Climate Alliance speaks of connecting the different local, national, European, and international political levels in the field of climate protection, biodiversity, and indigenous peoples. Central topics of the association, key activities, and positions are discussed at the international annual conference and general assembly. The network is financed mainly by membership fees that have to be paid by every member municipality and district. The amount is calculated as follows: the fee is €0.0066 per capita or a minimum overall contribution of €200 or a maximum overall contribution of €15,000. Municipalities and districts of some countries in Central and Eastern Europe pay 50 percent of the regular fee. The partner organization of the Amazonas Region is non-contributory. However, 3 percent of all membership fees provided for supporting environment protection is put at COICA's disposal (Feldt 2010: 201). Further sources of finance are project funds provided by the European Union or national authorities. Since the network was founded, the program of Climate Alliance has been deliberately based on content and tone of international agreements (Mayer-Ries 2000: 146). This is also highlighted by the goals set in 2006 aimed at reducing greenhouse gas emissions. Accordingly, the members of Climate Alliance commit themselves to reducing their greenhouse gas emissions continuously, with the aim being to cut CO_2 emissions by 10 percent every five years. Halving their per capita emissions (baseline year 1990) by 2030 at the latest is the agreed milestone (Climate Alliance 2013c). Furthermore, "In the long-term, Climate Alliance members aim at a sustainable level of 2.5 tons CO_2 equivalent emissions per capita and year by energy saving, energy efficiency and the use of renewable energy sources" (Climate Alliance 2013a). As "reaching these goals, however, cannot be achieved by measures taken by municipalities alone ... concerted efforts by all decision-making levels (EU, national states, regional/province governments, municipalities)" (Climate Alliance 2013a) are required in the sense of a prerequisite.

Climate Alliance has developed a number of tools and methods to support local authorities in their climate protection efforts including the so-called "Climate Compass," the "ECORegion," and the "Climate Cities Benchmark" (Climate Alliance 2013d). The Climate Compass is to provide a methodological basis for the development of local climate protection action plans. The internet-based tool ECORegion is to support municipalities in terms of calculating their carbon dioxide emissions and monitoring the achievement of the carbon dioxide reduction goals set within the frame of (Climate Alliance) members' self-commitment. Ultimately, the Climate Cities Benchmark is to support municipalities regarding the systematic development, implementation, and modification of climate protection activities (Climate Alliance 2013d).

Other activities[5] include political campaigns against malpractices such as threat to indigenous territories posed by oil and mining concessions. Furthermore, the network is active in the field of public relations in various forms with a thematic focus on the promotion of renewable energy in the Amazonas Region (Feldt 2010: 200). After all, small projects of indigenous people are financed by the Climate Alliance (Feldt 2010: 200). Special characteristics of Climate Alliance are the "regime and process linkages" at and between all levels (Mayer-Ries 2000: 149). For example, activities of the municipal network are associated with European public procurement and in particular with the purchase of tropical wood. Currently, Climate Alliance runs an (EU-funded) project which is supposed to contribute to the achievement of the Millennium Development Goals (MDGs)[6] (Climate Alliance 2013f: 16). In Germany, Climate Alliance was the founding member of the Forest Stewardship Council (FSC). Climate Alliance advised municipalities with respect to dealing with different certificates of the timber market (Feldt 2010: 201). Climate Alliance, together with other organizations, promoted the prohibition of trade in illegally logged timber in addition to links of fair trade. According to Feldt, Climate Alliance activities contributed to the raising awareness of relevant issues in the European public and policy, so that they were supported – for example the ILO Convention 169, by the relevance of indigenous people for the preservation of biological diversity and forests as well as campaigns for supporting land, territorial, and human rights (Feldt 2010: 203). With its various activities and members, Climate Alliance has a considerable potential to relate and also link local, global as well as environmental developmental and climate policy aspects (Mayer-Ries 2000: 157).

In climate policy since the Climate Conference in Copenhagen, increasing importance is attached to transnational networks at subnational level, among them municipal networks (Bulkeley and Newell 2010; WBGU 2010). Networks such as the described Climate Alliance or ICLEI's Cities for Climate Protection[7] originated in the early 1990s as both sustainability and climate change began to become relevant issues for local authorities.[8] According to Bulkeley (2011: 474), concrete success depends on various factors such as resource availability, leadership, access to relevant sectors, and incorporation in multilevel systems among others.[9] Another trend worth mentioning when it comes to the role of the local level is the rapid development of the Transition Town movement. Inspired by the practices of permaculture, alternative energy generation, organic food production, and the use of open-space technology to facilitate participation, the movement defines itself as a response to two critical twenty-first century concerns: the approach of peak-oil production, and the ensuing socio-environmental problems associated with climate change (Mason and Whitehead 2012).

Notes

1 See also www.iclei.org/index.php?id=800 (accessed 02/10/2013).
2 See also www.energy-cities.eu/ (accessed 02/10/2013).
3 According to Kern and Bulkeley (2009: 317) the expansion of Climate Alliance towards for example France or Great Britain is considerably restricted as the official language mainly used in the network is German.
4 The board is elected by the general assembly for a two-year term. It consists of at least four and at most thirteen persons. The indigenous partners are represented with one or two persons in the board (Climate Alliance 2013e).
5 For an overview on all current activities see the 2012 Activity Report (Climate Alliance 2013f).
6 See www.cities-for-mdgs.eu/ (accessed 05/28/2013).
7 See www.iclei.org/ (accessed 05/28/2013).
8 Apart from environmental problem solving, broader trends such as globalization have "both provided the means and the rationale for the involvement of a range of actors in seeking to shape the governance of a range of local to global problems" (Bulkeley and Newell 2010: 55–56). Another broader trend is the discourse on reducing the role of the state, increasing the influence of the private sector, and promoting various forms of partnerships (Bulkeley and Newell 2010: 55–56).
9 In the opinion of Bulkeley, climate policy has presently often little importance at local and regional level compared to economic and social policy issues. Hence, this level does not significantly differ from national and international levels (Bulkeley 2011: 474).

References

Bulkeley, H. (2011) Cities and Subnational Governments, in Dryzek, J., Norgaard, R., and Schlosberg, D. (eds.) *Oxford Handbook of Climate Change and Society*, Oxford: Oxford University Press.

Bulkeley, H. and Newell, P. (2010) *Governing Climate Change*, London and New York: Routledge.

Climate Alliance (2013a) Our Objectives. Accessed 05/27/2013 from www.klimabuendnis.org/our-objectives0.html.

Climate Alliance (2013b) The Association. Accessed 05/27/2013 from www.klimabuendnis.org/association0.html.

Climate Alliance (2013c) Tangible Targets. Accessed 05/27/2013 from www.klimabuendnis.org/home.html.

Climate Alliance (2013d) Tools and Methods. Accessed 05/27/2013 from www.klimabuendnis.org/543.html.

Climate Alliance (2013e) Executive Board Methods. Accessed 05/27/2013 from www.klimabuendnis.org/executive-board-members0.html.

Climate Alliance (2013f) 2012 Activity Report 2013 Planning. Accessed 05/27/2013 from http://climatealliance.org/fileadmin/inhalte/dokumente/2012/CA_report2011_planning2012_web_.pdf.

Feldt, H. (2010) Zusammenarbeit zwischen den indigenen Organisationen der Regenwälder und europäischen Kommunen. Eine Bestandsaufnahme des Erreichten im Klima-Bündnis, in Feldt, H. and Müller-Plantenberg, C. (eds.) *Gesellschaftliche Bündnisse zur Rückgewinnung des Naturbezugs. 20 Jahre Klimabündnis*, Kassel: Universität Kassel.

Kern, K. and Bulkeley, H. (2009) Cities, Europeanisation and Multi-level Governance: Governing Climate Change Through Transnational Municipal Networks, *Journal of Common Market Studies*, 47(2): 309–332.

Mason, K. and Whitehead, M. (2012) Transition Urbanism and the Contested Politics of Ethical Place Making, *Antipode*, 44(2): 494–516.

Mayer-Ries, J. (2000) Globales Arrangement für eine lokale Politik nachhaltiger Entwicklung – das 'Klima-Bündnis,' in Prittwitz, V. von (ed.) *Institutionelle Arrangements in der Umweltpolitik. Zukunftsfähigkeit durch innovative Verfahrenskombination?*, Opladen: Leske und Budrich.

WBGU (German Advisory Council on Global Change) (2010) Climate Policy Post-Copenhagen: A Three-Level Strategy for Success, Policy Paper 6, WBGU: Berlin.

27 ORGANIZATION PROFILE – 350.ORG

Brian Tokar

From modest beginnings in Vermont's Lake Champlain Valley, the group 350.org has emerged as the nexus of an activist network that is international in scope, and speaks to the potential for well-coordinated, worldwide actions to raise awareness about the effects of global climate chaos.

The origins of the 350 campaign are traced to a call by Vermont-based author Bill McKibben and his Middlebury College colleague John Elder for a five-day walk in September of 2006 to highlight the unfolding climate crisis and urge local politicians to act.[1] Some participants walked nearly 50 miles, starting at a cabin in McKibben's home town that once belonged to the poet Robert Frost. On the final day, approximately 1,000 people walked into the center of Burlington, Vermont, where several candidates for statewide office signed a pledge to support an 80 percent reduction in greenhouse gases by the year 2050.

The walk's unanticipated public appeal inspired a call for climate walks across the United States the following spring. The Step It Up campaign, as it was called, organized at least 1,400 events in all 50 US states (Fernandez 2007). Later that year, Step It Up organizers collaborated with a Washington, DC-based climate advocacy alliance known as 1Sky to organize two more nationwide events focusing on demands to curtail climate pollution, invest in renewable energy, and create green jobs across the country.[2]

During the summer and fall of 2008, McKibben and several of his former Middlebury students announced the creation of 350.org, inspired by climate scientist James Hansen's paper suggesting that 350 parts per million (ppm) is the maximum atmospheric CO_2 concentration compatible with a relatively stable climate (Hansen et al. 2008). 350 was launched with a series of full-page ads in prominent newspapers declaring that 350 is "the most important number in the world," together with a campaign to urge the winner of the 2008 US presidential race – whether Obama or McCain – to attend the upcoming UN climate conference in Poznan, Poland.[3]

In 2009, 350 organized its first international day of climate action, making highly strategic use of the Internet and social media. On October 24, prior to the UN climate conference in Copenhagen, there were reportedly over 5,200 events in 181 countries, hundreds of which were photographed and highlighted on the 350.org website.[4] A year later, 350's 'Global Work Party' featured some 7,000 projects supporting efforts around the world to create locally rooted alternatives to the continued use of fossil fuels. By this time, 350 had established a strong reputation for colorful, photogenic demonstrations in a wide variety of cultural contexts, united around the imperative of reducing atmospheric carbon.

With this increased visibility came new challenges for the 350 team. Some long-time activists were skeptical of 350's politically cautious posture and apparent unwillingness to confront the political institutions and corporations responsible for neglecting the climate issue. A few writers speculated that powerful donors were exercising too much influence on 350 and other high-profile climate advocacy groups, while others saw the group's caution more as a reflection of the founders' personal inclinations.[5] McKibben's moderate stances on controversial issues such as carbon trading and nuclear power (which he renounced in 2012) also raised concerns among other climate activists.

In 2011, after merging with 1Sky and calling for another worldwide day of actions symbolizing the need to limit atmospheric CO_2 to 350 ppm, the organization took a significant turn toward more challenging forms of political engagement. In July of 2011, McKibben joined with other prominent writers and activists to call for two weeks of civil disobedience in front of the White House to protest the proposed Keystone XL pipeline, which was slated to transport tar sands oil from Alberta, Canada to US refineries along the Gulf of Mexico.[6] Over 1,200 people were arrested in Washington by the end of that summer and, even though the arrest scenario was staged in collaboration with the DC police, the campaign greatly increased the profile of the tar sands issue in the US.

In November 2011, some 12,000 people circled the White House to express opposition to the tar sands pipeline, this time with more explicit organizing support from both 350 and the Indigenous Environmental Network, which brought representatives of Cree communities directly threatened by the exploitation of the tar sands. Two months later, President Obama denied a permit for the pipeline to cross into the US from Canada and 350 was declaring victory (350.org 2012). While the Keystone XL pipeline continued to be at the center of a protracted political battle with an uncertain conclusion, and even though Obama pushed to speed up the construction of one domestic portion of the pipeline, these events appeared to embolden 350 toward greater political challenges.

By the middle of 2012, 350.org boasted a staff of a dozen international field organizers, eight campaign coordinators in several regions of the United States, and another 16 individuals performing various organizational functions.[7] Its annual budget had grown to over $2 million, with support from almost 30 foundations.[8] It was challenging US tax subsidies for fossil fuel corporations, campaigning against potential tar sands pipelines in New England and in Texas, and engaged in a wide variety of international initiatives to illustrate the 350 ppm goal for CO_2 in the earth's atmosphere. In the late fall of 2012, McKibben and a team of experienced organizers toured more than 20 US cities to challenge university investments in fossil fuel corporations. With its impressive worldwide activist network, well-coordinated global actions, and its increasing involvement in determined direct actions against climate-destructive policies and new energy megaprojects, 350.org has raised the bar for climate activism in the US and served as a central catalytic force in the global climate movement.

Notes

1 McKibben's book, *The End of Nature* (McKibben 1989) was the first popular book on global warming in the US.

2 1Sky's website lists 643 allied organizations, including 80 business supporters. See www.1sky.org/learn/allies (accessed 11/15/2012).

3 Bill McKibben email, October 7, 2008; see www.350.org/en/node/434 (accessed 11/15/2012).

Advertisements announcing the focus on 350 parts per million first appeared in the *New York Times*, *International Herald Tribune*, and the *Financial Times* in June of 2008.

4 See www.350.org/en/story (accessed 11/15/2012).

5 The sharpest published critiques of 350 and the tar sands campaign have come from Canadian activists, particularly Cory Morningstar (Morningstar 2011) and Macdonald Stainsby (Stainsby 2011).

6 Bill McKibben email, July 3, 2011. Co-signers of the tar sands call to action included climate scientist James Hansen, writers Wendell Berry and Naomi Klein, and Indigenous Environmental Network Executive Director Tom Goldtooth.

7 See www.350.org/en/our-team (accessed 01/23/2013).

8 See www.350.org/2011-annual-report-financial-information (accessed 01/23/2013). The supporters mostly appear to be mostly family foundations with low public profiles, except for the Rockefeller Brothers Fund, Rockefeller Family Fund, the Overbrook Foundation, and the Clif Bar Family Foundation.

References

350.org (2012) Victory: Take the Keystone Pledge. Accessed 12/16/2012 from http://act.350.org/sign/keystone-pledge.

Fernandez, M. (2007) For the Environment, Rallies Great and Small (and Unusual Attire), *New York Times*, April 15, 2008.

Hansen, J., Sato, M. Kharecha, P., Beerling, D., Masson-Delmotte, V., Pagani, M., Raymo, M., Royer, D., and Zachos, J. (2008) Target Atmospheric CO_2: Where Should Humanity Aim?, *Open Atmospheric Science Journal*, 2: 217–231.

McKibben, B. (1989) *The End of Nature*, New York: Random House.

Morningstar, C. (2011) Rockefellers' 1Sky Unveils the New 350.org. Accessed 12/15/2012 from http://climatesoscanada.org/blog/2011/04/18/rockefellers-1sky-unveils-the-new-350-org-more-more-delusion.

Stainsby, M. (2011) A Tar Sands Partnership Agreement in the Making? *Canadian Business*, August 1, 2011. Accessed on 08/30/2013 from http://canadiandimension.com/articles/4070/.

28 ORGANIZATION PROFILE – RISING TIDE

Brian Tokar

Rising Tide describes itself as "a grassroots international network of groups and individuals who take direct action to confront the root causes of climate change and to promote local, community-based solutions to the climate crisis" (Rising Tide n.d.). Originally formed to organize protests around the November 2000 conference of the UN Framework Convention on Climate Change (UNFCCC) in the Hague, Rising Tide currently lists six regional affiliates – North America (US and Canada), UK, Finland, Mexico, Ecuador, and Australia – as well as organizing collectives in several US states and regions.

Made up mostly of youthful activists with roots in decentralist and anti-authoritarian political traditions, Rising Tide has supported numerous direct action campaigns against the fossil fuel industry as well as a variety of corporate-driven "false solutions" to the climate crisis. On several occasions, it has openly challenged more centrist and politically accommodating environmental NGOs. Rising Tide has organized and trained participants for many high-profile direct actions, especially in the US and UK, and is also noted for its efforts in publishing and popular education.

A Rising Tide statement of principles from 2007 that was supported by all six affiliated groups, proclaims that

> the roots of the climate crisis lie in the current global economic system and its endless pursuit of economic growth at all costs ... as a matter of survival, we must confront capitalism and decrease our dependence on the industries and institutions that are destroying the planet.
>
> *(Rising Tide n.d.)*

The statement calls for a socially just transition to a low-carbon economy and to "dismantle the systems of oppression that permeate our culture and ourselves." Rising Tide advocates for strong links with environmental justice groups worldwide, supporting repayment of the Global North's social and ecological debts to the South and an end to restrictive controls on migration.

Rising Tide's North American network (RTNA) traces its origins to Mountain Justice Summer, a nonviolent direct action campaign initiated in 2006 against the practice of "mountaintop removal" coal mining in the Southern Appalachian Mountains (Shapiro 2010). Rising

Tide's North America's core principles include a commitment "to Earth-centered, community-based solutions to the climate crisis that foster local autonomy and self-sufficiency."[1] They focus on opposition to market-oriented climate measures, such as tradable emissions permits and offsets, along with technological "false solutions" to the climate crisis such as nuclear power and mega-hydro dams, encouraging autonomous, nonviolent direct actions to challenge these developments. RTNA has perhaps most influenced climate activists in the US through its publications, especially the popular critical pamphlets, 'The Copenhagen Climate Talks: Deal or No Deal' (2009), 'The Climate Movement Is Dead: Long Live the Climate Movement' (2010), and 'Hoodwinked in the Hothouse: False Solutions to Climate Change' (2010).[2] During 2011, it produced a regular newsletter titled 'Burning Issues,' and has since relied mainly on electronic updates to report on actions and convey its critical outlook on climate issues.

The direct actions carried out and supported by Rising Tide's UK and US affiliates speak to the broad scope of energy and climate issues that have sparked public controversy in recent years. In the UK, recent actions have focused on major oil companies such as BP and Shell, the resurgent nuclear industry, and the country's continuing dependence on coal.[3] Rising Tide has organized blockades of drilling rigs, coal mines, refineries, and petrol stations, occupied land adjacent to proposed new nuclear power plants, and disrupted several corporate conferences. It was instrumental in the UK Climate Camp movement, which organized high profile actions between 2006 and 2010 at power plant sites, Heathrow Airport, London's financial district, and the Edinburgh headquarters of the Royal Bank of Scotland.[4] An affiliated group known as the 'Greenwash Guerrillas' has used street theater and satirical performance art to challenge corporations that attempt to 'greenwash' their image through arts sponsorships and other public relations efforts.[5] Rising Tide Australia shut down the world's largest coal exporting port for a day in 2010, blockaded corporate offices and the federal parliament, and also advanced litigation that forced coal mining companies to disclose their carbon emissions (Evans 2010).

In the US, Rising Tide continues to be involved in action campaigns against the coal industry, especially the practice of mountaintop removal, whereby more than 500 mountaintops have been detonated to expose coal seams beneath the topsoil and gravel; the residue is often simply dumped in nearby streams. In close collaboration with residents of coal-dependent communities, Rising Tide has joined with Earth First and several newer formations to mount ongoing campaigns of public education, popular mobilization, and direct action (RTNA 2010). In Montana and other Western states, it has organized to block the transportation of heavy equipment to be used in the tar sands fields of Alberta. In late 2012, the RTNA website listed contact information for 54 local affiliates and allies across the US and Canada, of which 10–12 formally identify as Rising Tide groups.[6]

To highlight the proliferation of corporate "false solutions" to climate change, Rising Tide North America has also organized against industrial-scale biomass and wind projects that are seen as exceptionally destructive to local ecosystems, and on one occasion occupied the Washington, DC headquarters of Environmental Defense, which has long been a leading advocate for carbon trading as an alternative to regulation of greenhouse gases (see RTNA 2008). Rising Tide groups have organized several climate action camps in the US, focused on popular education and activist training, and were central to the effort to create a national Mobilization for Climate Justice in the lead-up to the 2009 UN climate conference in Copenhagen (see chapter by Tokar in this book). As new campaigns emerge to challenge expanded fossil fuel extraction in the US and beyond, Rising Tide's network of decentralized, all-volunteer collectives serves as an effective catalyst for direct action to protect the climate and for a systemic critique of the root causes of the climate crisis.

Notes

1 See www.risingtidenorthamerica.org/features/principles/ (accessed 12/15/2012).

2 See www.risingtidenorthamerica.org/category/rising-tide-climate-publications-and-climate-multimedia/page/2/ (accessed 12/15/2012).

3 See http://risingtide.org.uk/ (accessed 12/15/2012).

4 See www.risingtidenorthamerica.org/about-rising-tide-north-america/our-history/ (accessed 06/18/2012); see also Shifter (2011).

5 See http://greenwashguerrillas.wordpress.com (accessed 12/15/2012).

6 See Scott Parkin, personal communication, September 13, 2012, and www.risingtidenorthamerica.org/features/north-america-contacts/ (accessed 12/15/2012). The latter estimate includes some groups that were expected to announce more formal ties with RTNA in the coming months.

References

Evans, G. (2010) A Rising Tide: Linking Local and Global Climate Justice, *Australian Journal of Political Economy*, 66(11): 199–221.

Rising Tide (no date) The Rising Tide Coalition for Climate Justice Political Statement. Accessed 12/15/2012 from http://risingtide.org.uk/about/political.

RTNA (Rising Tide North America) (2008) Climate Activists Invade DC Offices of Environmental Defense. Accessed 12/15/2012 from www.culturechange.org/cms/content/view/255/68/.

RTNA (Rising Tide North America) (2010) Appalachia Rises Against Mountaintop Removal, *Burning Issues*, No. 2, Winter 2010/2011.

Shapiro, T. (2010) *Mountain Justice: Homegrown Resistance to Mountaintop Removal, for the Future of Us All*, Oakland: AK Press.

Shifter, A. (2011) Remember, Remember: Climate Camp, *Shift Magazine*, 12. Accessed 01/23/2013 from http://shiftmag.co.uk/?p=461.

29
ORGANIZATION PROFILE – ENERGY ACTION COALITION

Brian Tokar

The Washington, DC-based Energy Action Coalition (EAC) is a national alliance of some 50 youth-led environmental and social justice groups, focused on climate awareness and advocacy for renewable energy. It is best known for the series of 'Power Shift' climate conferences that have brought thousands of students and supporters to Washington to hear well-known speakers, attend workshops, lobby members of Congress, and participate in high-profile public actions.

EAC's first initiative was the Campus Climate Change, launched in 2005 and focused on increasing the use of renewable energy on college campuses across the US.[1] In November of 2007, it organized the first large Power Shift conference, which brought 6,000 young people to Washington for an event that highlighted the potential for a clean energy economy and green jobs and aimed to bring these issues into the 2008 presidential campaign. In early 2009, twice as many people participated in the second Power Shift event, and at least 3,000 of those converged on the streets surrounding a coal-fired power plant on the outskirts of the city at the end of the conference.

Following a call to action from authors Bill McKibben and Wendell Berry, the Power Shift participants had pledged to blockade the entrances to the power plant, risking arrest to urge Congress to shut it down.[2] Three days prior to the planned action, the Democratic Congressional leadership pledged to convert the plant to cleaner-burning natural gas by the end of the year. With the success of the action a somewhat foregone conclusion, people still assembled outside the plant for nearly four hours and no one was arrested (Environmental News Service 2009). For numerous participants it was their first experience of a large, potentially risky demonstration, and many were emboldened to return to their campuses and develop action campaigns of their own. Some participants also urged that future actions should seek more support from community-based groups in the places most affected by destructive climate and energy policies (Russell 2009).

In April 2011, the third Power Shift conference attracted some 10,000 participants, and focused on demands for Congress and President Obama to resist Republican initiatives to weaken the federal Environmental Protection Agency. The Agency's top administrator, Lisa Jackson, was a featured speaker, along with Al Gore and the prominent green jobs advocate, Van Jones. The speaker who generated the most enthusiasm, however, was activist Tim DeChristopher, who had offered numerous false bids at a controversial 2008 auction of oil and gas leases on public lands in Utah. DeChristopher was facing an extended prison sentence for disrupting and eventually nullifying the sale, and he offered an impassioned plea for students to put their careers and convenience aside to fight injustice and stop corporate crime (see DeChristopher 2011; Johnson 2011).

The Coalition has also been actively involved in the campaign to stop the Keystone XL pipeline and in youth-led voter registration drives. It has also pledged support for environmental justice, endorsing the principles of the people-of-color-led environmental justice and climate justice movements.[3] The EAC is also engaged in active partnerships with several larger organizations, including Greenpeace, the Rainforest Action Network, Indigenous Environmental Network, the Sierra Club's student and youth coalitions, and several state-level Public Interest Research Groups.[4] At the UN's Durban climate conference in 2011, and again at the 'Rio+20' environmental summit in 2012, it joined with young activists from around the world and supported a student activist from Vermont who disrupted a speech by Todd Stern, the US government's leading envoy to the UN climate negotiations.[5]

Describing itself as 'The Hub of the Youth Climate Movement,' EAC proclaims that

> Working with hundreds of campus and youth groups, dozens of youth networks, and hundreds of thousands of young people, Energy Action Coalition and its partners have united a burgeoning movement behind winning local victories and coordinating on state, regional, and national levels in the United States and Canada.
>
> (*EAC n.d.*)

In both the domestic and international arenas, it has provided an important entry point for new activists, as well as an effective public platform for the voices of a younger generation that may have the greatest stake in organizing to confront the global climate crisis.

Notes

1 See www.energyactioncoalition.org/about/our-work (accessed 12/15/2012).
2 This protest event was called 'Capitol Climate Action – March 2, 2009'; see www.greenpeace.org/usa/en/news-and-blogs/campaign-blog/capitol-climate-action-march-2–2009/blog/25472 (accessed 12/15/2012).
3 See www.energyactioncoalition.org/about (accessed 01/23/2013).
4 See www.energyactioncoalition.org/partners (accessed 01/23/2013).
5 See www.wearepowershift.org/blogs/why-i-stood-and-spoke-out (accessed 01/23/2012).

References

DeChristopher, T. (2011) The Climate Movement Needs to Step It Up, *Yes!Magazine*, April 18. Accessed 12/15/2012 from www.yesmagazine.org/people-power/tim-dechristopher-the-climate-movement-needs-to-step-it-up.

EAC (Energy Action Coalition) (no date) About Us. Accessed 01/23/2012 from www.energyactioncoalition.org/about.

Environment News Service (2009) Anti-Coal Protesters Block Gates to Capitol Power Plant, Press Release, March 2. Accessed 01/23/2013 from www.ens-newswire.com/ens/mar2009/2009–03–02–03.html.

Johnson, B. (2011) Power Shift 2011: Tim DeChristopher on the Chamber's Corporate Crimes. Accessed 01/23/2013 from http://thinkprogress.org/climate/2011/04/18/174996/dechristopherchamber-crimes.

Russell, J. K. (2009) In Context: Capitol Climate Action Victory. Accessed 01/23/2013 from http://its-gettinghotinhere.org/2009/03/03/in-context-capitol-climate-action-victory.

30 ORGANIZATION PROFILE – PLANE STUPID

Heiko Garrelts

Plane Stupid is a British network of grassroots groups that take nonviolent direct action against aviation expansion. The name of the organization is a wordplay combining 'plane' and 'plain stupid.' Its logo shows a fist which brings down a plane.[1] The group's work is based on volunteer activities and is financed through donations. According to its own statements, Plane Stupid is organized horizontally without formal organization structures and without official media representatives (Plane Stupid 2013a). In this perspective, the organization is a wide network of self-organized regional groups acting for example in London, Glasgow, and Manchester while communicating mainly by use of social media such as Twitter in addition to operating its own blog.[2] In Ireland there is a sister organization named Plane Mad. Recently, a German sibling movement has been established with the assistance of British activists. This group is for instance co-operating with environmental protection organizations in Munich, such as Robin Wood.[3] In France, Plane Stupid opposes the planned airport project Notre-Dames-des Landes near Nantes. Clearly, Plane Stupid allows like-minded groups explicitly to take actions in their name as long as those groups are accountable and act nonviolent and in line with the key requirements of the organization.

Starting point for action is the following: UK aviation's CO_2 emissions have grown substantially in past decades, doubling from 16.9 Mt CO_2 in 1990 to 34.7 Mt CO_2 in 2009. UK aviation CO_2 emissions are forecast to grow steadily over the next 20 years, growing from 33.4 Mt CO_2 in 2010 to 47.6 Mt CO_2 by 2030, within the range 43.2–53.4 Mt CO_2 (Department for Transport 2011: 60). Despite the ambitious policy goals to reduce the carbon dioxide emissions in Great Britain by 80 percent by 2050, political decision-makers have recently authorized airport expansions by building additional runways several times. Accordingly, the central demands of Plane Stupid aim to stop the airport expansion, to prohibit all flights with a flight distance below 500 km and to ban advertisements for the aviation industry. Moreover, the organization requires that the Great Britain aviation industry, where several low-cost airlines are located, is no longer exempted from value added taxes. In general, the *political change* towards sustainable transport, mobility, and jobs is considered to be necessary (Plane Stupid 2013a; Honigstein 2008).

Since its foundation, the following larger actions have been taken by Plane Stupid: In September 2005 a group of activists disturbed an international aviation conference in London,[4] in the following years the organization occupied either areas of several British airports or their surrounding areas such as the Midlands in 2006, Manchester in 2007, Stansted in 2008, and Aberdeen in 2009. Moreover, access to the headquarters of the low-cost airline EasyJet and of the airport owner as well as holder of BAA[5] were blocked. In addition, some flights with private planes were forced to be cancelled. Equally, there were also opposing activities in the House of Commons and in the Scottish Parliament. Plane Stupid supported the 'Camp for Climate Action' in Heathrow and cooperated with the residents whose houses were threatened by demolition as a consequence of the airport expansion. When occupying Stansted airport – the protest action was taken at the same time as the 14th Conference of the Parties within the United Nations Framework Convention on Climate Change held in Poznan (Poland) – the activists had banners with slogans such as 'Climate Emergency' and 'please DO something'. Such actions usually ended with activists temporarily arrested (Plane Stupid 2013b).

Apart from the direct actions against concrete building projects (by way in sharp contrast to motives according to the motto: 'Not-in-my-backyard'; see Plane Stupid Germany 2013) the group organized a cinema campaign in 2009. For this purpose, a video clip called 'Polar Bear,' lasting around one minute, was produced:[6] Along the skyscrapers in an office district, dozens of polar bears are tumbling out of the skies. Shortly after the start, the camera angle changes and the bears are approaching dangerously close to the viewer. In the background you can hear the sharp blasts of an obviously low-flying jet plane. The commercial ends with the statement: "An average European flight produces over 400 kg of greenhouse gases from every passenger ... that's the weight of an adult polar bear." While the organization Campaign for Better Transport, also from Great Britain,[7] comes up with alternative ways of transport such as buses and trains, and aims at a change in awareness by providing (enlightening) information, Plane Stupid focused on shock effects by using the 'Splatter' genre.[8] The polar bear, icon for the vulnerability of the Earth in times of climate change (Manzo 2009), serves drastically as the warning voice. According to Tollmann, the video clip includes the following message:

> [T]his time the polar bear does not stay on ice floes without touching the human living environment. On the contrary, in an almost surrealist style, the video clip brings two issues together [meaning the wildlife and the human life-world], which at first glance have nothing in common. Secondly, ... the polar bears 'physically' hit the cold hard surface of the capitalistic architecture. Natural and built environment come together in a radical way.... And the statement of the individual 'offender' or victim, is a much more direct criticism of capitalistic structures.
>
> *(Tollmann 2012, my translation)*

The impacts of these pictures cannot be clearly determined. Obviously, the campaign is a subject of particular *attention* – although it receives little *support* – which is fairly often the case in the context of environmental protests (see Korte 2011: 190). In the YouTube Forum on the video clip 'Polar Bear,' mainly mocking, critical, and sarcastic comments prevail. At least at this level, there is hardly any discussion on the essential argument of the video. Media science studies regarding the response to the occupation of Stansted (Gavin 2010) conclude that occupying the runway was a success as long as media attention and headlines are the decisive criterion. But, the reporting was mainly concentrating on *episodic* aspects instead of *thematic* ones (see Korte 2011: 462). For example, the news on this action referred to the chaos caused at the airport or was framed as an issue of flight safety: An article in the *Mirror* (December 9, 2008, cited in Gavin 2010: 463) carried the title: 'Could They Stop Air Terrorists? Flight Security Fears as Protesters Close Stansted.' The two founders Joss Garman and Richard George (students at that time) had also learnt that there was no lack in public response to the group's activities. They were involved in the actions of Plane Stupid and were fairly often arrested by the police. Afterwards, they became, so to speak, "pop stars of the climate movement" in Britain (Honigstein 2008). For instance, Garman was on the BBC's *Newsnight* and had to change his mobile number afterwards due to countless requests.

In general, Plane Stupid is among the actors which require setting a fundamental political change in times of the climate crisis. In doing so, the focus is on the specific problem area of air traffic. There is some similarity between the actions taken by Plane Stupid and those taken by the groups that oppose the "symbols of unsustainability" which forms part of civil disobedience in a spectacular and nonviolent way (North 2011: 1582). These symbols include road building projects, nuclear facilities, or cultivation of genetically modified plants (see Gavin 2010; Korte 2011). The main focus is on direct action which is taken by somewhat small organized groups of experienced activists who work more or less in secret. Consequently, there is a distinct difference between the actions of Plane Stupid with their exclusive character and forms of actions which aim to involve numerous activists such as climate camps or demonstrations (North 2011: 1591–1592). Finally, there are strong indications that a European network involving critics of air traffic is being developed.

Notes

1 The logo visualizes the organisation's slogan: 'Bringing the aviation industry back down to earth!'
2 See www.planestupid.com/blog (accessed 01/28/2013).
3 The starting point, among others, was the Munich referendum held on June 17, 2012 to prohibit building a third runway at Erdinger Moos airport.
4 Protest actions were taken several times, sometimes in a humorous way. At a dinner event for aviation managers in 2007 Plane Stupid activists in fine suits mingled with the society and successfully sabotaged the event. Even the police could not differentiate between activists and managers (Honigstein 2008).
5 With the 'British Authority Act' law the British Airports Authority was established in 1966, which took responsibility for four state-owned airports (a.o. Heathrow). In the following years more facilities – for example in Edinburgh – were established. As a result of the 'Airports Acts' the company was privatized and listed on the stock market in 1986. Today BAA Limited also runs several airports in other countries as for example in the US.
6 See www.youtube.com/watch?v=fxis7Y1ikIQ (accessed 01/28/2013).
7 See www.bettertransport.org.uk/ (accessed 01/28/2013).
8 In the video clip shock effects were produced several times, as it does not end with polar bears *falling* from the sky – which is not least certainly an intended reminiscence about September 11, 2001 where people jumped from the World Trade Center. Furthermore, the clip shows polar bear bodies which *hit* objects or the ground, which breaks a taboo (Tollmann 2012).

References

Department for Transport (2011) UK Aviation Forecast. Accessed 05/30/2013 from www.icao.int/environmental-protection/Documents/ActionPlan/UK_Forecasts_en.pdf.

Gavin, N. (2010) Pressure Group Direct Action on Climate Change: The Role of the Media and the Web in Britain – A Case Study, *The British Journal of Politics and International Relations*, 12(3): 459–475.

Honigstein, R. (2008) Turbulenzen beim Check-in. Plane Stupid, Anti-Flug-Initiative, London. Mit fantasievollen Aktionen kämpft die britische Gruppe 'Plane Stupid' gegen die Klimakiller über den Wolken, *greenpeace magazin*, January. Accessed 01/28/2013 from www.greenpeace-magazin.de/magazin/archiv/1-08/09-plane-stupid/.

Korte, M. (2011) Der Köder muss dem Fisch schmecken und nicht dem Angler, in Brunnengräber, A. (ed.) *Zivilisierung des Klimaregimes. NGOs und soziale Bewegungen in der nationalen, europäischen und internationalen Klimapolitik*, Wiesbaden: VS Verlag.

Manzo, K. (2009) Imaging Vulnerability: The Iconography of Climate Change, *Area*, 42(1): 96–107.

North, P. (2011) The Politics of Climate Activism in the UK: A Social Movement Analysis, *Environment and Planning*, 43(7): 1581–1598.

Plane Stupid (2013a) About Us. Accessed 01/28/2013 from www.planestupid.com/aboutus.
Plane Stupid (2013b) Actions. Accessed 03/20/2013 from www.planestupid.com/actions.
Plane Stupid Germany (2013) Über uns. Accessed 01/28/2013 from www.planestupid-germany.de/de/ueber-uns.html.
Tollmann, V. (2012) Ein Bild von einem Bär, *Freitag*. Accessed 01/28/2013 from www.freitag.de/autoren/der-freitag/ein-bild-von-einem-bar.

31
ORGANIZATION PROFILE – GERMANWATCH

Heiko Garrelts

Germanwatch is a nonprofit and nongovernmental organization (NGO) based in Bonn, Germany. It was founded in 1991. Germanwatch's motto is 'Observing, Analysing, Acting – for Global Equity and the Preservation of Livelihoods.' This is its mission statement:

> The situation of marginalized people in the South forms the starting point for our engagement for sustainable development. The political and globalized market structures of the North, as well as their resource-intensive mode of production, which is now being increasingly imitated, are influencing human lives worldwide. We advocate for a political, economic and social framework which can ensure a future for the people of the South, who are being pushed to the margins of society through unbridled globalisation and whose very existence is threatened by the loss of their ecological and economic foundations of their livelihoods.
>
> (*Germanwatch n.d.*)

The issues Germanwatch focuses on are climate change, world food, land use and trade, corporate accountability, financing climate as well as development and education for sustainable development. The organization is funded by membership fees, donations, grants from the Foundation for Sustainability (Stiftung Zukunftsfähigkeit) and by several public and private donors. As of August 2012, the organization had 575 members (Germanwatch 2012). Germanwatch is part of Climate Action Network Europe (CAN-E). Activities are lobbying, campaigning, public relations, and education.

Annually at the UN Climate Change Conferences, Germanwatch together with CAN-Europe presents the Climate Change Performance Index (CCPI).[1] The latest CCPI (Germanwatch 2013a) evaluates the climate protection performance of 58 countries, responsible for over 90 percent of global energy-related CO_2 emissions. The climate change performance is measured through 15 different indicators that are combined into one single composite indicator. They are classified into four categories: emissions, efficiency, renewable energies, and climate policy. The policy components of the CCPI are based on expert assessments[2] solicited by Germanwatch. The 2013 edition of the CCPI for the first time includes data on emissions from deforestation which are based on the FAO Global Forest Resource Assessment 2010 (Germanwatch 2013b: 5–6). To list just a few findings of the latest survey, no country made it into the first three spots on the list "due to a lack of ambition to reach the goal of keeping global warming below 2 degree Celsius" (Germanwatch 2013c). With Denmark, Sweden, and Portugal on top, Germanwatch rejects drawing a 'black and white picture': While Denmark and Sweden show better policy ratings and relatively low emissions, Portugal's high rank is mainly derived from lower emissions due to the economic crisis, as is the case in several countries. The EU as a whole presents a mixed finding and the "positions of EU countries as frontrunners are at stake" (Germanwatch 2013c). The two biggest emitters of carbon dioxide worldwide, the US and China, are still ranking comparably low. The United States climbed up in the 2013 Index, "partly due to decreased emissions through the economic crisis and the massive exploration of shale gas" (Germanwatch 2013c). China's emission level has risen. However, China's massive investments in renewable energies (see chapter by Schroeder in this book) are expected to show an impact shortly. Consequently, its emissions trend could slow down in the near future and lead to better results (Germanwatch 2013c). The inclusion of emissions from deforestation has shifted the position of those countries with a relevant share of these emissions such as Brazil and Indonesia (Germanwatch 2013a: 10).

In addition to the CCPI, Germanwatch publishes the Global Climate Risk Index. This index analyzes to what extent countries have been affected by the impacts of weather-related extreme events (storms, floods, heatwaves, etc.). The most recent available data from 2011 as well as for the period 1992–2011 were taken into account (for latest findings, see Harmeling and Eckstein 2013). With respect to issues of adaptation to climate change, the role of Germanwatch as host for the Adaptation Fund NGO network[3] has to be mentioned. This network aims at supporting NGOs in a number of developing countries by increasing their capacity to handle all relevant issues around the implementation of the fund-related projects throughout the implementation period (Adaptation Fund NGO Network 2012).

Among the Germanwatch campaigns worth being mentioned here is the short film 'The Bill,' which was produced with financial support from the German Ministry for Economic Cooperation and Development: Three friends meet in a bar and start talking about their lives. Big car utilization and holidays spent in Namibia are just two examples of a range of 'climate sins.' With the appearance of the barmaid a kind of bill that was not expected is presented.[4]

In sum, Germanwatch focuses on dialogues with policy makers and actors in the market sphere as well as on information and mobilization of the civil society. With its professional publications, the organization is often conceived as a think tank (see Bernauer and Böhmelt 2013), which distinguishes it from other organizations focusing for example on protest. Especially with the Climate Change Performance Index (CCPI), Germanwatch contributes to efforts by NGOs to identify 'leaders' and 'laggards' among nation states and their governments. Thus, transparency is supposed to be enhanced which might then enhance compliance with international norms. The involvement of Germanwatch in other policy networks (e.g., related to adaptation) aims at capacity building especially among actors from the Global South.

Notes

1 Both industrial countries and countries in transition (which are Annex I parties to the Framework Convention on Climate Change adopted in Rio 1992 and as such accept a special responsibility) as well as all countries that emit more than 1 percent of global CO_2 emissions are included in the Index (Germanwatch 2013a: 4).

2 More than 230 experts from NGOs contributed to the CCPI 2013, through the evaluation of those policies (Germanwatch 2013a: 18).

3 This network refers to the Adaptation Fund (AF) as established under the Kyoto Protocol to finance adaptation actions in developing countries.

4 For the short film (with English subtitles), see http://germanwatch.org/klima/film09e.htm (accessed 05/30/2013).

References

Adaptation Fund NGO Network (2012) Making the Adaptation Fund Work for the Most Vulnerable. Independent Insights from Vulnerable Developing Countries. Accessed 05/30/2013 from http://germanwatch.org/en/download/7568.pdf.

Bernauer, T. and Böhmelt, T. (2013) National Climate Policies in International Comparison: The Climate Change Cooperation Index, *Environmental Science & Policy*, 25: 196–206.

Germanwatch (n.d.) Mission Statement. Accessed 05/30/2013 from http://germanwatch.org/en/mission-statement.

Germanwatch (2012) Rechenschaftsbericht des Vorstandes 2011–2012, vorgelegt zur Mitgliederversammlung am 29. September 2012. Accessed 05/30/2013 from http://germanwatch.org/de/download/6843.pdf.

Germanwatch (2013a) *The Climate Change Performance Index Results 2013. Background and Methodology*. Accessed 05/30/2013 from http://germanwatch.org/en/download/7133.pdf.

Germanwatch (2013b) *The Climate Change Performance Index Results 2013*. Accessed 05/30/2013 from http://germanwatch.org/de/download/7158.pdf.

Germanwatch (2013c) High Emissions – Low Ambitions. Germanwatch's Climate Change Performance Index published at Doha's climate talks, Press release, December 3. Accessed 05/30/2013 from http://germanwatch.org/en/5876.

Harmeling, H. and Eckstein, D. (2013) *Global Climate Risk Index. Who Suffers Most From Extreme Weather Events? Weather Related Loss Events in 2011 and 1992 to 2011*. Accessed 05/30/2013 from http://germanwatch.org/en/download/7170.pdf.

PART III

Arenas, activities, and development of the climate movement

32

BETWEEN PRAGMATISM AND RADICALIZATION

NGOs and social movements in international climate politics[1]

Achim Brunnengräber

1 Introduction

The attributions were impressive: 'Our Hopes in Civil Society,' a 'Third Force between Market and Government,' or the 'Power of the Courageous' – that is how magazines titled their articles and issues written about NGOs (non-governmental organizations) in the mid-1990s. In the meantime it has become quieter around the NGOs. The supposed NGO-ization of world politics (FES 1996) did not last long and attention on NGOs falls off rapidly if they do not manage to create spectacular actions, loud protests, or humorous campaigns. Apart from that there are other quite interesting changes going on in the field of civil society. This is what the Occupy movement stands for and which began with the occupation of the Zuccotti parks in New York on September 17, 2011, and transformed into the role model for worldwide protest actions against the power of banks and the unequal distribution of wealth. The Occupy movement is still carefully interpreted as a starting point of a new cycle of movement activity (Brunnengräber 2012b). Roland Roth calls it a "return of utopia, situations and radical culture critical motives," similar to the motives of the 68's protests (Roth 2012: 22).

Evidently climate politics is also subject to societal change. There the protests against the official climate politics were radicalized even before the 'Arab Spring' and the Occupy movement. The agenda of the climate discourse was in terms of content clearly expanded to questions about global justice, global community goods, that are worth being protected, and democratic participation. Thus the situation in the field of civil society engaging in climate politics has become less clear as regards topic choosing and positioning. In this text I am going to deal with the causes and backgrounds of these developments and the new political geopolitics in the Global Climate Governance[2] in more detail. The aim is to show that the relations between civil society and state-owned and international institutions have changed and thus the state and civil society are not divided spheres, but can analytically be defined more precisely only by their general and specific interactions.

The following questions should be answered in this text: Which changes in the terrain of civil society can be defined more precisely between the NGO-ization of the 1990s and the dynamics of movement development within the last half decade? Are processes of social differentiation ascertainable within civil society? Are new lines of conflict and forms of protest developing so that one has good reason to speak of a new cycle of protest activity within the Climate Governance? What are the reasons for that? Are the world political crises and new contradictions restructuring the field of conflict between cooperation and resistant politics as well as between national and international politics? And which ambivalences can be named in the field of civil society? After a short overview I will answer these questions by making five observations and finally present a conclusion.

2 A short prehistory

International politics has been accompanied by NGOs and their transnational networks ever since the UN Conference for Environment and Development 1992 in Rio de Janeiro (UNCED). NGOs were also involved in the development of the Framework Convention on Climate Change (FCCC) that was signed during the UNCED conference 1992, and that came into force in 1994 after the national state-run process of ratification. Governments, diplomats, and international institutions since then are no longer the only actors on stage of global politics (Altvater and Brunnengräber 2002; Leggewie 2003). The 'revolution' in communication via the use of the Internet, Twitter messages, or Facebook contributed considerably to the fact that the state-run representation is completed with a civil societal authority of control. NGOs in climate politics as in other international politics nowadays are part of the Facebook generation. Whether Climate Action Network International, World Wide Fund for Nature, or Germanwatch, they use Facebook and the Internet in order to raise the awareness for the globally deplorable state of affairs. They scandalize, blame and shame, present their own studies, and develop their own approaches of action.

NGOs are seen as 'advocates' for those who do not have any voice although they are already affected by climate change since they lack resources and political participation (*advocacy*). Their campaigns against coal-fired power stations, the capture and storage of carbon below the ground (CCS), and the monopoly of the big energy group as well as their campaigns for the energy change and the remunicipalization of the energy supply are well known. They pursue the goal of changing the behavior of governmental and private-sector actors as well as of society as a whole (*campaigning*). In addition, their members have a considerable know-how that in contrast to politicians is not limited to only one legislative period. For that reason alone NGOs can give advice on climate political instruments or climate diplomacy – almost as climate political memory (*expertise*). This expert's report can be passed on in various ways: by giving advice to politicians and institutions, through the media, by publishing own brochures, or by *lobbying*. The commitment of NGOs shows undoubtedly that international climate politics is inconceivable without actors of civil society.

However, nowadays the affirmative acting of NGOs, which seems to follow institutionalized rules in order to fight against climate change, is being criticized. Networks of movements have been built up that keep a distance from political institutions and try to extend their agenda. The Bali Principles of Climate Justice and the development of the network Climate Justice Now! during the climate negotiations of 2007 in Bali, changed the civil societal field of climate politics considerably. The 'absolute power' of the relatively moderate Climate Action Network (CAN) disappeared. The radical wing of internationally active NGOs "was strong enough to provoke open conflict" with CAN (see Bullard and Müller [2012: 56] as well as the chapter by Bedall and Görg in this book). The climate camps, which took place in Great Britain in 2006 for the first time, belonged also to the new dynamics of the movements (Frenzel 2011). In the camps flexible and mobile groups of action and single active participants articulate their positions and live together with only a low degree of institutionalization. Apart from these actors and structures that are oriented towards the grass roots, in these new networks of movements also numerous NGOs are involved who are accredited to official UN negotiation processes. Politics, which deals with discursive struggles over problems and possible solutions, is becoming more confusing; and the arguments within civil society lead to more conflicts.

3 First observation: the return of the state

In the global arena the international institutions continue to be a guideline for mobilizing actors of civil society. Regarding this they almost seem to have a constituent character: In the Occupy movement the institutions of the financial market are denied the power to regulate. In the Global Climate Governance it is the yearly climate conferences that are important binding agents for the networking processes and alliances. However, the focus on the official negotiations is not undisputed. In civil society it is debated whether the orientation to the official agenda is still strategically and politically tenable or not. The tight political corset that the NGOs were squeezed into during the climate conference in Doha at the end of 2012 has reopened such debates.[3] In the end such processes show the magnetic force of attraction that international conferences have and refer to the challenges that actors of civil society are facing. Regarding the dealings with and acting within the institutions they have to decide between the poles of critical cooperation and radical resistance. In Doha the audible critique of state-run restrictions was rather restrained.

How can this be explained? Apart from the appearance of critical attitudes towards the neoliberal approaches of international politics and the resulting refusal of official conferences, climate political debates are increasingly held (again) on other levels of acting than on the international one. International climate politics is no longer considered as a showpiece project consisting of a Global Governance architecture that makes it possible to interpret uniformly and solve consensually the problems of supposedly global dimensions. Such a normative hope existed in the 1990s. But nowadays the performance of international institutions and their cooperative approaches for solving the existing world problems have become more doubted than ever (Altvater and Brunnengräber 2011). Not least of all was the 'Rio+20' conference in June 2012, which was adequate proof of the fact that the Governance imperative of international cooperation was removed from states acting singularly. These findings led to the consideration that despite globalization one could refer once more to the state. Furthermore, in the public opinion the question of who is responsible for solving problems and for acting in the public interest is regularly answered by referring to the state. In climate politics also the state is again normatively overvalued. The German Advisory Council on Global Change (WBGU) recommends that

> in order to revitalize the multilateral climate process, policy-makers and civil society in Europe have to take on a self-confident leading role in global alliances with selected 'climate pioneer' countries and that more intensive support should be provided for civil society initiatives.
>
> *(WBGU 2010: 3)*

The same actors who have failed in the group of actors under the name of the UN now should revive the climate politics in new constellations. Numerous actors from the spectrum of movements also address their claims to the state. However, a crucial difference between NGOs focused on conferences and social movements is that for the latter the affirmative reference to international institutions is not or no longer of importance. Rather, the new generation standing behind the movements articulates expectations to the state to become engaged (see also Long et al. 2010: 238; Reitan 2011: 65–66).

The contextualization of climate camps, which are site-specific and thus exact for regional and national realities, is also in this sense an expression for renationalization of protest events, although different functions and tasks are attributed to the state (see below). Furthermore, the remunicipalization of water and energy supply, single aspects of the energy change, social fights against the production of agro fuel, or the project for decentralized production of renewable energy sources are again clearly bound to national politics, more than was the case in the last two centuries of globalization, from 1992 to 2012.

National or rather local forms of mobilization as they appear now along with the new protest generation were treated as social peripheral phenomena for a long time. In the 1990s and also still in the early 2000s the established NGOs, which focused affirmatively on world politics and on global problems, played a vital role in media and in discursive framing of climate change. Especially in their classical functions as critical observers of negotiations and as creators of colorful pictures with their campaigns, NGOs as intimate experts on diplomacy of conferences were and still are demanded from the media. However, nowadays NGOs increasingly make specific problems and possible acting within national borders subject areas for discussion, which are relevant for the climate and have been neglected or even completely left out in UN negotiations so far. Thus what is taking place on the terrain of civil society is also observable in world politics: the return of the state (Jessop 2010).

4 Second observation: the consensus is cracking

The appearance of transnational NGOs marks the replacement of so-called 'new social movements' that were very successful in the peace movement or the women's movement in the 1970s and 1980s. The actors of the movement disappeared from the political stage, and institutionalized or professionalized themselves. As a consequence, the participation in institutions is hardly questioned any more. Both the weak mobilization of social movements and the limited national parties or other big organizations, as well as the democratic vacuum in the international system, apparently made another type of actor necessary that during the process of globalization combines different approaches and arenas of acting, i.e., the new type of actor could transnationalize and develop new ways on how to articulate interests.[4]

In this period of institutionalization especially the NGOs from the field of civil society participated in climate governance: Climate NGOs and their networks, the Climate Action Network (CAN) in particular, which was founded in 1989, already contributed intensively and continually to the processes of development and consolidation of international climate politics. Climate NGOs since then – in other issue areas as was earlier the case – are highly involved in different international institutions and rules. They accompany critically the Clean Development Mechanism, are in parts carrying it out themselves, or gave advice on the development of emission trading. They profit from the participation in international climate conferences since they can generate from promotional funds and donations as well as increasing attention from the public and media. In general, states approve of the participation of NGOs since the activities of civil society are situated within the institutional, topical, and ideological frame of international politics.

Regarding this subject the majority of organizations of the 'conference NGO' type behave cooperatively and critically. They point out failures in the round of negotiations, discuss contradictory interests of actors, or make suggestions for alternative solutions. Nevertheless, they always express certain dissatisfaction. For NGOs, there also seems to be no alternative to the effort of the international climate governance. The NGOs are in a certain way caught in the protocol of official processes which set structural limits to their presented critique. This can be explained by looking at the Kyoto Protocol of the United Nations Framework Convention on Climate Change. The NGOs' initial disapproval of market orientated mechanisms disappeared after the passing of the protocol and gave way to committed discussions about their concrete arrangement. The change of positions from fundamental critique to the participation in the development of technical details about the implementation is symptomatic of how NGOs examine the "internal scope of systems" (Walk and Brunnengräber 2000: 138). The sword of Damocles meaning the threat to be excluded from the negotiations hangs over the NGOs. The NGOs' critical position is accompanied by a general approval of international climate politics, although satisfying results are not produced. Their own interests reach so far that the NGOs plead for continuing the negotiations, although there is nothing left to be negotiated. Thus NGOs have a stabilizing effect and are significant "resources of legitimation" (Brunnengräber et al. 2001) on a level of symbolic politics that is determined by governments, but also by lobby associations of the private sector, in particular by transnational groups.

The NGOs' active participation on the state-run development of international politics is limited in their exertion of influence in a double sense. On the one hand, it comes up against limiting factors that are set up by powerful politics of self-interest (in order to overcome financial and state crisis) and by ecological topics that are neglected. On the other hand, the limit is set by inadequate instruments for climate protection. As regards the first conflict line, the opportunities to participate are not legally protected, but dependent on the governments' goodwill. Therefore, it is possible that after years of participatory generosity the international institutions will suddenly restrict the political participation if for example national interests and conflicts arise that are not meant to become published. This happened during the climate conference 2009 in Copenhagen when NGOs were denied access to the conference site (Rest 2011: 91). Therefore, it does not (always) seem to be necessary that NGOs legitimate their influence on international negotiations by participating and approving. In times of crises this becomes especially clear. Depending on how much the interests between the official negotiating parties are contradicting and on how much room for maneuver is left, NGOs have to attract attention for their topics, provided that this still seems to make sense at all.

Another limitation lies in the fact that NGOs are increasingly acting on contradictory areas because climate conferences and their mechanism only have partial success and their effects remain insufficient. Whether NGOs should constructively accompany politics of international institutions or not becomes obvious when reductions of greenhouse gas, which are binding under international law, are not successful or when the Clean Development Mechanism causes completely new social problems (Brouns and Witt 2008). The consensus of many years' standing among NGOs on the fact that international climate politics leave no alternatives was questioned against the background of this development. Among NGOs conservative forces want to prevent a reorientation from institutionalized approaches and programs of international climate politics to new political fields and approaches. Other NGOs are supporting exactly this strategy.

5 Third observation: radicalization in the field of civil society

The fact that the (conflict) situation in the field of civil society changes, can also be related to the changed politico-economic geography of the world (Brunnengräber 2012a). First the mobilizing force of the UNCED conference 1992 and the following conferences in the 1990s cannot be overestimated. "The good spirits" that accompanied the conference in Rio after the end of the East–West conflict, the peace dividend, and the optimistic mood concerning the solving of global problems set free resources and hope (Klein et al. 2005). However, during the 'Rio+10' conference in Johannesburg disillusionment had already come up that turned into disappointment and at the 'Rio+20' into frustration. The reason for that was that the world conferences about gender justice, human rights, climate protection, or protection of biodiversity did not live up to the expectations. On the contrary: In many issue areas of global politics the conditions deteriorated instead of improving (IPCC 2007; CBD 2010).

As a form of protest against the experience that international politics does not achieve more than the minimal consensus and operates very slowly, institution-critical and transnational connected movements came into being, or rather were reported on in the media (Juris 2008; Bedall 2011). The world financial and world trade politics, that governments and international institutions follow, was already criticized long before the appearance of the Occupy movement: Protests and demonstrations next to the place of negotiations were organized in order to question the legitimacy of certain agreements. In 1999, this happened in Seattle when the World Trade Organization stopped its negotiations without success, also in Prague where the annual conference of the International Monetary Fund and the World Bank was brought to an early end and in Goteborg where the European Head of Government was heavily criticized.

Since the G8 Summit in Heiligendamm and the following climate summit in Copenhagen, it has become obvious that also within climate politics oppositional forces were able to intensify their mobilization. Mass actions of civil disobedience or the mobilization of autonomous groups reveal a certain radicalization. On the other hand, negative attitudes can also be found towards the official process of international climate negotiations. These negative attitudes are not necessarily meant to be towards the state itself (see above). Instead it should be interpreted as an 'antagonistic' attitude towards dominant articulation of international climate political institutions and their interests which are rigid to instrumental, technological, and institutional modernizations. The protest of several climate NGOs is particularly addressed to neoliberal free market lines of international politics and its growth paradigm. The depiction of Angela Merkel as climate chancellor intensified this protest. Parts of civil society tried to delegitimize the politics in power. The climate camps, that were organized in several countries (primarily in the Global North) and that were limited in time, laid the foundations for protests. Nineteen climate camps, which took place until 2010 (Frenzel 2011), were considered to enable political debates about fair approaches to climate change and at the same time the camps stood for the beginning of civil disobedience against central greenhouse gas polluters (Rest 2011: 93). Questions regarding social ecology, balance of power, emancipation, justice, or lifestyle were treated in the camps and intensively discussed. These are questions that were not discussed in the institutions that are aligned along free market lines, nowadays also environmental political institutions (Brunnengräber 2006).

Not only because of their content, but also because of their forms of protest, the climate camps and climate demonstrations mark a period of radical, polarizing arguments with the deficits of international politics. The camps organized by the Occupy movement are not a single phenomenon, but rather – although institutionally unbound – one out of many extra-parliamentary initiatives that increasingly articulate their protest and have in common their critique of current circumstances: Students, the homeless, the unemployed who receive benefits (in Germany), trade unionists, opponents of nuclear energy, antifascists, or peace campaigners belong to this group. They were supported by the call 'Be outraged' formulated by Stéphane Hessel (Hessel 2011). Nevertheless, it would be incorrect to regard civil societal commitment of protest camps simply as 'outrage' and consequently prepolitical. Networks of movements and their camps are rather an expression of new political forms: They represent anti-institutional forms of articulation and criticize the participation in existing institutions.

6 Fourth observation: the creation of hybrid structures

A clear dividing line between NGOs and social movements, as used in this text, can only be drawn for the purpose of analysis. Several networks of movements that appear in the field of international institutions are on closer examination hybrid structures:

> These [hybrid network structures] are inclusive collectivities that favour diversity, linking multiple issues through horizontal networks and adopting social technologies that facilitate greater autonomy and leaderless networks that provide political capacities for communication. What have emerged are polycentric networks combining direct mass activism in diversely targeted campaigns proliferating via the internet. These networks are antagonistic to state and corporate actors ...
>
> *(Reitan 2011: 58)*

Reitan points out that there are clear structural and personal overlaps between NGOs and social movements (2011: 58). Thus the members of the NGO networks Climate Justice Now! (CJN!) cannot consequently be divided from grassroots oriented networks (Bedall 2011: 79). In addition, the motives for mobilization, such as in the context of alternative conferences in Cochabamba, are overlapping (Rest 2011: 96). NGOs as part of hybrid networks make the ambivalences more obvious that exist between cooperation and resistance. On the one hand, hybrid networks of movement orient themselves more into the direction of official institutions as soon as NGOs accompany the UN negotiations as observer organizations. On the other hand, these NGOs operate in a different governance structure than one of the conference NGOs mentioned above: They develop other forms of protest and disapprove of state monopolies (Rest 2011). This attitude was and is still expressed by the call for civil disobedience of CJN! and grassroots activists or in 2009 by the 'People's Assemblies on Climate Justice' during the climate conference in Copenhagen, by the Global Work Party of 350.org, or by the yearly Global Day of Action organized by Global Climate Campaign (350.org 2009; CJA 2009).

Furthermore, by looking at the cooperation partners or addressees we can only roughly structure the field of civil society in climate politics: A group of activists outside of the hybrid networks believes that movements have to build up cooperation with local authorities, trade associations, or trade unions. By contrast, within the networks civil disobedience, direct pressure on governments and elites and emancipative political approaches are visible. This group rarely sees the point in cooperating with state actors or in addressing their claims to governments since in their opinion they are partly responsible for the climate crisis. Instead they see the central challenge in developing separate autonomous "globality" (Bullard and Müller 2012). At the same time the articulation of protests in form of mass actions, such as civil disobedience against coal-fired power stations, is more dynamic now and has become part of discursive struggles over problems and solutions within civil society. Those who search for constructive participation in the emission focused climate regime are opposed to those who demand that fossil energy sources, whose incineration is the main cause for climate change, should be left in the ground (North 2011: 1589). But the experience is that protests outside of institutions in many cases are determining the political practice of the activities of civil society only temporarily. In order to cause concrete societal changes interactions with established institutions seem to be necessary. *Inside* and *outside* acting or pragmatism and radicalization cannot be strictly separated in political practice. They are rather powerfully intertwined.

Regarding the hybrid structures and processes in climate politics many examples can be mentioned that underline that NGOs and social movements again and again have to take a stand in areas of tension to act in, with, or against state-owned institutions: Examples are 'From Trade to Climate Caravan,' 'People's Movements on Climate Change,' or the civil mass disobedience 'Reclaim Power! Pushing for Climate Justice.' The climate groups and their campaigns are always confronted with state-owned institutions (CJA 2009; Bedall 2011; Bedall et al. 2011: 43–44; Rest 2011; Reitan 2011: 67–69) and try to find common grounds in order to get stronger. "This pragmatic hanging together of the left reflects the growing recognition of the complexity of globalization and power, as well as the multiple activist strategies needed to engage with and ultimately transform both" (Reitan 2011: 52).

On the one hand, within the hybrid networks there is principally a heterogeneous spectrum of actors: NGOs such as 'Focus on the Global South' or 'Indigenous Environmental Network' work together with grassroots oriented groups such as the climate camps that are active outside of the official processes. *On the other hand*, apart from the hybrid networks there are also affirmatively acting NGOs and their networks that do not give up their classic status as conference NGOs. Regarding their ideological profile it is hard to separate hybrid networks from conference NGOs. Conference NGOs have a tendency to consider the efficiency revolution, the extension of renewable energy, and technological modernization as central challenges. They attribute the responsibility for the implementation of these aims to governments. By contrast, hybrid networks are more systematically critical: They claim analysis of politico-economic dynamics between governments and fossil energy economy or between resource consumption and growth. In this group opposing attitudes such as anticapitalistic or antistate positions are more likely to be articulated. Apart from interdependences of conference NGOs and hybrid networks, changes regarding the current contents and articulation of their protests are noticeable.

7 Fifth observation: chances for more democracy

During the UN climate negotiations NGOs and social movements bring themselves to mind as the voice of the sovereign, thus pointing to democratic vacuums of international politics, and claim more participation. They point out that the existing forms of institutionalized democracy and its organizational forms are limited. In forms of civil disobedience, protests against the existing energy politics or in climate camps alternatives are made alive, in contrast to the existing elite rule and the solely representative deputizing of society and citizens in international climate politics: Open and free processes of agreement, transparent decision making, democratic forms of organization, less patriarchal rule, and gender equality can be realized and experienced, with all the problems that the efforts of emancipative everyday life entail. It is generally known that the creative potential of these new alliances is huge.

In contrast, processes of transnationalization come at a democratic and political price. International institutions structure the acting and determine strict approaches. This was exemplified in this text with the UN climate conferences. In order to correspond to democratic ideal institutions they would have to open up even more. But actors of civil society are not only 'reparatory instances' and 'providers of democracy,' they are also political subjects that can resist (out of necessity) in times of crises. The conflictive political style of hybrid transnational networks in climate politics is exemplary for this. In such times international institutions turn back to negotiations behind closed doors. They exclude NGOs from the negotiations as happened in Copenhagen (see Rest 2011: 91) or determine civil societal approaches such as in Doha. International climate governance keeps becoming less transparent. But in the field of civil society there is no cooperation among equals. The West European and North American NGOs have better opportunities to react to processes of exclusion than NGOs from the Global South. Due to their low resources the latter not only have a difficult position during the international climate conference, but also often cannot participate.

Therefore, it is important to consider that NGOs, social movements, and their networks always bear *lock-in-effects* and *exclusion-mechanisms*. Decentralized protests and campaigns that are organized at the same time in various places, such as the *global day of action* of 350.org in 2009 (350.org 2009) or the *global day of direct action for climate justice* coordinated by Climate Justice Action (CJA 2010), respond to these problems. Such forms of action have the advantage that the hurdles for political participation are lowered. New digital opportunities of communication guarantee that global networks can be renationalized while protests are transnationalized as never before. As a consequence the participants will have the feeling that they are part of a worldwide movement. Structural deficits of democracy, especially regarding those of international institutions, are not necessarily removed by the described forms of action and organization.

8 A new cycle of movement activity?

Multiple phenomena of crises on the financial and trade market of the world have considerable consequences on civil society. NGOs, which became the focus of attention because of international conferences, will no longer be needed if states worked on their 'hard' economic crises and protected their interests. In such times of crises NGOs should give advice or ease humanitarian catastrophes. At the same time actors of civil society are an integral part of institutionalized governance structures. It is obvious to assume that the 'offensive of smile' organized by these institutions in the 1990s was intended to increase their own legitimacy. This form of politics, which began with the UN world conferences, is overlapping – as argued in this text – with a new cycle of movement activity that consists of hybrid network structures and conference NGOs.

The new protest generation changes the internationalized structures of civil society actors and their relationship with the state. Earlier the state officially approved the participation of civil society, but now the statements are more moderate. In Copenhagen the access to the conference site was denied to the activists of 'Reclaim Power!' by using clubs, truncheons, and pepper spray. Blocked access roads to nuclear waste disposal sites in Wendland (Germany) or city squares that were occupied by members of the Occupy movement were cleared in order to regain public order. In these observations changes are revealed on the terrain of civil society: Dissatisfaction with the international system and with the capitalistic economic structures establishes new global ties. The protest is also aimed at representative forms of international politics that is far away from national and in particular subnational concerns. The national reactions to the protests of civil society are diverging dramatically: They reach from toleration and support to repression and criminalization (della Porta and Reiter 2011).

In the new cycle of the protest activity the renationalization produces new political practices. However, NGOs and social movements have a lack of detailed reactions in forms of civil approaches to multiple crises in general and the climate crises in particular. A common program is also missing and this leads to the fact that debates on content cannot be limited in certain frames from the start. As a consequence the model of political camps could rapidly become obsolete and the transnationalization of forms of protest and organization could come to nothing. In any case it is not certain that the observed developments on the field of civil society are sustainable. Their political influence in climate politics keeps – apart from partial successes – being low (North 2011).

From the women's movement, labor movement, and the environmental movement we know that certain corrections of existing unequal structures and the abolishment of injustice can be started and promoted by social movements. In order to strengthen the power of civil society and to guarantee continuity – as diverse cycles of movement activity have shown – processes of professionalization and institutionalization are necessary that at the same time bear several ambivalences. In climate governance there is rather an interim period which is intertwined with the multiple crises. A stabilization and formalization of civil society caused by cooperation with official international institutions is just as observable as new dynamics of social movements that articulate their protest outside of institutions. On the whole there is an unclear situation between climate governance and climate conflicts.

Notes

1 My thanks to Philip Bedall who provided me with helpful comments on the text.
2 The term Global Climate Governance comprises three central dimensions of governance: (i) the diversity of actors in market, government, and civil society; (ii) the multileveled character of politics that reaches from regional and national to global level; and (iii) the interdependence of political fields. Especially the first two dimensions are of relevance for this text. Therefore Global Climate Governance differs from international climate politics since the latter is more focused on the acting of governments and the official climate negotiations. In order to get an overview of Multilevel Governance and its usage as analytical instrument see various texts in Brunnengräber et al. (2008) as well as in Brunnengräber and Walk (2007).
3 Even the demonstration in support of more climate protection was considerably organized and controlled by the state in Doha.
4 The concept of Global Climate Governance explicitly includes the 'NGO-ization' of international politics into its perspective.

References

350.org (2009) International Day of Climate Action (24 October). A Day to Remember. Accessed 09/13/2010 from www.350.org/october24.

Altvater, E. and Brunnengräber, A. (2002) NGOs im Spannungsfeld von Lobbyarbeit und öffentlichem Protest, *Aus Politik und Zeitgeschichte*, B 6–7: 6–14.

Altvater, E. and Brunnengräber, A. (eds.) (2011) *After Cancún: Climate Governance or Climate Conflicts*, Wiesbaden: VS Verlag.

Bedall, P. (2011) NGOs, soziale Bewegungen und Auseinandersetzungen um Hegemonie. Eine gesellschaftstheoretische Verortung in der Internationalen Politischen Ökonomie, in Brunnengräber, A. (ed.) *Zivilisierung des Klimaregimes. NGOs und soziale Bewegungen als Akteure der nationalen, europäischen und internationalen Politik*, Wiesbaden: VS Verlag.

Bedall, P., Bricke, M., and Tzschiesche, S. (2011) Klima- und Energiekämpfe, in Bois, M. and Hüttner, B. (eds.) *Beiträge zur Geschichte einer pluralen Linken*. Accessed 15/10/2012 from www.rosalux.de/fileadmin/rls_uploads/pdfs/RosaLux/Rosalux_3–2010_web.pdf.

Brouns, B. and Witt, U. (2008) Klimaschutz als Gelddruckmaschine, in Altvater, E. and Brunnengräber, A. (eds.) *Ablasshandel gegen Klimawandel? Marktbasierte Instrumente in der globalen Klimapolitik und ihre Alternativen*, Hamburg: VSA.

Brunnengräber, A. (2006) The Political Economy of the Kyoto Protocol, in Panitch, L. and Leys, C. (eds.) *Coming to Terms with Nature*, London, New York, and Halifax: Merlin Press.

Brunnengräber, A. (2012a) Die neue Klima-Geopolitik. Konflikte und Chancen durch Deutungsverschiebungen, *Forschungsjournal Neue Soziale Bewegungen*, 25(2): 21–29.

Brunnengräber, A. (2012b) Ein neuer Bewegungszyklus. Von der NGOisierung zur Occupy-Bewegung, *Forschungsjournal Neue Soziale Bewegungen*, 25(1): 42–50.

Brunnengräber, A. and Walk, H. (2007) *Multi-Level-Governance. Umwelt-, Klima- und Sozialpolitik in einer interdependenten Welt*, Schriften zur Governance-Forschung des Wissenschaftszentrums Berlin (WZB), Band 9, Baden-Baden: Nomos.

Brunnengräber, A., Klein, A., and Walk, H. (eds.) (2001) *NGOs als Legitimationsressource. Zivilgesellschaftliche Partizipationsformen im Globalisierungsprozess*, Opladen: Leske + Budrich.

Brunnengräber, A., Burchardt, H.-J., and Görg, C (eds.) (2008) *Mit mehr Ebenen zu mehr Gestaltung? Multi-Level-Governance in der transnationalen Sozial- und Umweltpolitik*, Schriften zur Governance-Forschung des Wissenschaftszentrums Berlin (WZB), Band 11, Baden-Baden: Nomos.

Bullard, N. and Müller, T. (2012) Beyond the 'Green Economy': System Change, Not Climate Change? *Development*, 55(1): 54–62.

CBD (Convention on Biological Diversity) (2010) *Global Biodiversity Outlook 3*, Montreal. Accessed 12/15/2012 from www.cbd.int/doc/publications/gbo/gbo3-final-en.pdf.

CJA (Climate Justice Action) (2009) Call to Action! Reclaim Power – Join the Yellow (Badge) Bloc. No Decisions About Us Without Us – Climate Justice Now! Press Release. Accessed 02/15/2012 from www.climate-justice-now.org/call-to-action-reclaim-power-join-the-yellow-badge-bloc/.

della Porta, D. and Reiter, H. (2011) State Power and the Control of Transnational Protests, in Olesen, T. (ed.) *Power and Transnational Activism*, London and New York: Routledge.

FES (Friedrich Ebert Stiftung) (ed.) (1996) *Globale Trends und internationale Zivilgesellschaft oder: Die NGO-isierung der (Welt-)Politik?*, Bonn: Friedrich-Ebert-Stiftung.

Frenzel, F. (2011) Entlegene Orte in der Mitte der Gesellschaft. Zur Geschichte der britischen Klimacamps, in Brunnengräber, A. (ed.) *Zivilisierung des Klimaregimes. NGOs und soziale Bewegungen als Akteure der nationalen, europäischen und internationalen Politik*, Wiesbaden: VS Verlag.

Hessel, S. (2011) *Time for Outrage!* London: Quartet Books.

IPCC (Intergovernmental Panel on Climate Change) (2007) *Climate Change 2007: Climate Change Impacts, Adaptation and Vulnerability*, WG II Contribution to the IPCC Forum Assessment Report, Summary for Policymakers.

Jessop, B. (2010) The 'Return' of the National State in the Current Crisis of the World Market, *Capital & Class*, 34(1): 38–43.

Juris, J. (2008) *Networking Futures. The Movements Against Corporate Globalization*, Durham: Durham: Duke University Press.

Klein, A., Walk, H., and Brunnengräber, A. (2005) Mobile Herausforderer und alternative Eliten. NGOs als Hoffnungsträger einer demokratischen Globalisierung? in Klein, A., Walk, H., and Brunnengräber, A. (ed.) *NGOs im Prozess der Globalisierung. Mächtige Zwerge – umstrittene Riesen*, Bonn: Bundeszentrale für politische Bildung.

Leggewie, C. (2003) *Die Globalisierung und ihre Gegner*, München: C.H. Beck.

Long, S., Roberts, E., and Dehm, J. (2010) Climate Justice Inside and Outside the UNFCCC: The Example of REDD, *Journal of Australian Political Ecnonomy*, 66(11): 222–246.

North, P. (2011) The Politics of Climate Activism in the UK: A Social Movement Analysis, *Environment and Planning*, 43(7): 1581–1598.

Reitan, R. (2011) Coordinated Power in Contemporary Leftist Activism, in Oelsen, T. (ed.) *Power and Transnational Activism*, London and New York: Routledge.

Rest, J. (2011) Von der NGOisierung zur bewegten Mobilisierung. Die Krise der Klimapolitik und die neue Dynamik im Feld der NGOs und sozialen Bewegungen, in Brunnengräber, A. (ed.) *Zivilisierung des Klimaregimes. NGOs und soziale Bewegungen als Akteure der nationalen, europäischen und internationalen Politik*, Wiesbaden: VS Verlag.

Roth, R. (2012) Vom Gelingen und Scheitern sozialer Bewegungen, *Forschungsjournal Neue Soziale Bewegungen*, 25(1): 21–31.

Walk, H. and Brunnengräber, A. (2000) *Die Globalisierungswächter. NGOs und ihre transnationalen Netze im Konfliktfeld Klima*, Münster: Westfälisches Dampfboot.

WBGU (German Advisory Council on Global Change) (2010) Climate Policy Post-Copenhagen. A Three-level Strategy for Success. Accessed 05/23/2013 from www.wbgu.de/fileadmin/templates/dateien/veroeffentlichungen/politikpapiere/pp. 2010-pp. 6/wbgu_pp. 2010_en.pdf.

33

REFRAMING CLIMATE CHANGE

The Cochabamba conference and global climate politics[1]

Johannes Kruse

1 Introduction

In April 2010, more than 35,000 people attended the 'World People's Conference on Climate Change and the Rights of Mother Earth' in Cochabamba, Bolivia.[2] Invited by Bolivian president Evo Morales, participants discussed climate change, international climate policy, agriculture, food sovereignty, and the relationship between nature and humankind. By focusing on climate change and by attracting representatives of indigenous groups, social movements, and NGOs, the conference provided an alternative forum to the intergovernmental climate negotiations within the United Nations Framework Convention on Climate Change (UNFCCC). At the end of the Cochabamba conference, the participants agreed on a final declaration outlining their vision on the issues discussed: The so-called People's Agreement expresses an alternative understanding of climate change and its socio-economic and political causes. Furthermore, the agreement also proposes alternative solutions to tackle climate change.

In this chapter, I draw on frame analysis as an analytical tool to explore how the Cochabamba conference has defined climate change as a political problem. Frame analysis is a powerful analytical tool because it points to the social construction of environmental problems, highlights the relevance of different types of frames, including diagnostic and prognostic ones, and emphasizes the power of causal stories for attributing blame and identifying adequate solutions. Empirically, I explore these issues on the basis of the People's Agreement, the final declaration of the conference aiming at influencing both activism and diplomacy. In doing so, I set out to address two main questions: First, what were the developments leading up to the Cochabamba conference that motivated actors to establish and engage in an alternative forum outside the UNFCCC process? Second, what specific frames are invoked in the agreement, in what way do they differ from other framing patterns, and how successful are they in having an impact on global climate politics? In order to answer the first question, I will contextualize the Cochabamba conference in terms of developments in international climate policy, the political transformation of Bolivia in the 2000s, and with regard to the adverse climate change impacts expected for the Andean region. To address the second question, I will explore how the agreement frames climate change as an issue, what it recognizes as causing climate change, which actors it identifies as relevant, and what policy proposals it puts forward. In assessing the success of the Cochabamba conference, I will pay particular attention to frame resonance and the role of the audience.

The rest of this chapter is structured as follows: First, I will introduce frame analysis as a theoretical approach and analytical toolkit. In this section, I will pay attention to the importance of causal stories for 'politicizing' an issue and highlight factors that determine the success of any framing efforts. Second, I will contextualize the Cochabamba conference with regard to developments in international climate policy, changes in domestic politics in Bolivia, and the adverse climate change impacts expected for the Andean region. Third, I will analyze the frames used by the People's Agreement to describe climate change as an issue, identify its causes, and devise policy solutions. Finally, I will review the impact Cochabamba had on the climate movement and on the outcomes of the UN climate negotiations.

2 Framing and the power of causal stories

In this chapter, I draw on frame analysis as an analytical tool to explore how the People's Agreement describes climate change as an issue, identifies relevant actors, and puts forwards policy proposals to address climate change and its impacts. The role of frames, discourses, and narratives for environmental politics and environmental movements has long been acknowledged (Dryzek 1997; Hajer 1995; for an overview see Mühlhäusler and Peace 2006). A core assumption shared by all these approaches is that social problems are not simply 'out there' waiting to be solved by politics. Rather, empirical phenomena such as climate change first need to be identified as relevant problems that require political intervention in order to be solved (Hülsse 2007). In particular, it is complex phenomena that accommodate different framings. According to Zald (1996: 262) frames are "specific metaphors, symbolic representations and cognitive cues used to render or cast behavior and events in an evaluative mode and to suggest alternative modes of action." They are employed by political actors to put events into a context, interpret social problems, and devise solutions. By framing issues, actors try to raise other actors' interest in their concerns and strive to generate support and legitimacy for their policy goals (Koopmans and Statham 1999). By representing an issue in different ways, it can become attractive for the respective audience and entice their support. Even though the strategic-instrumental nature of framing has often been emphasized in the literature, frames cannot be reduced to such a perspective as they always are the result of complex cognitive and discursive processes.

The literature on framing has distinguished between three different types of frames: diagnostic frames, prognostic frames, and motivational frames (Snow and Benford 1988; Benford and Snow 2000). Diagnostic frames identify a problem, highlight causal mechanisms, and attribute blame or responsibility. Prognostic frames identify one or several solutions to the problem and formulate strategies, tactics, and goals. Motivational frames provide a "call to arms" by wrapping both diagnostic elements and prognostic elements in such a way that others are encouraged to take action (Snow and Benford 1988). For the diagnostic function of frames, causal ideas play a pivotal role, because it is only by identifying causal mechanisms that problems can be addressed by the political process:

> Causal stories have both an empirical and a moral dimension. On the empirical level, they purport to demonstrate the mechanism by which one set of people brings about harms to another set. On the normative level, they blame one set of people for causing the suffering of others. On both levels, causal stories move situations intellectually from the realm of fate to the realm of human agency.
>
> *(Stone 1989: 283)*

Different framings of the same issue do not only attribute different meanings to an issue, but also go beyond that by creating cognitive and normative maps that enable certain options for actions and rule out others (Surel 2000). Similarly, frames also highlight the importance of different actors: in Stone's words they point out those that 'are blamed' and those that 'are harmed.' This distinction forms the key element of any causal story. For the latter, the Copenhagen School of Security Studies has coined the term 'reference object.' The term refers to those entities that are thought to be threatened and seen as worth protecting (Buzan et al. 1998: 36).

The political success of any framing effort depends in large parts on the interplay between the quality of the used frames, the credibility of the claim-maker, and the audience that the frame is addressed to. A frame's quality can be assessed in terms of its internal consistency and its empirical credibility. The credibility of any claim-maker putting forward a particular frame depends on how their status and knowledge about the subject are perceived by others (Benford and Snow 2000: 619–622). Lastly, the framing approach has often been criticized for not adequately addressing the role of the audience, despite its importance for any framing effort to succeed (see also Benford and Snow 2000: 630; Davis 2002). This is particularly relevant, since it is the audience that assesses both the credibility of the used frames as well as the credibility of the claim-makers. On an individual level, frames are successful, if potential supporters can relate to them in their personal experience, interests, and ideas (Snow et al. 1986).

For the environmental realm, Hajer (1995) has pointed out that the environmental discourse as a whole is "fragmented and contradictory" and as a result has emphasized the social construction of environmental problems. As Biermann (2011) pointed out, climate change is an issue marked both by analytical uncertainties (scientific understanding) and normative uncertainties (which ethical principles should be applied). Therefore, climate change is an issue for which actors are particularly likely to emphasize different causal relationships, name different reference objects, and propose diverging solutions. The existing literature has distinguished three or four different framing patterns, on which political actors rely in describing climate change as an issue. Foran and Widick (2011), for instance, have pointed to three different framing patterns, namely a 'climate justice' frame, an 'environmental reform' frame, and a 'carbon industrial complex' frame. In a similar way, Schlichting and Schmidt (2012) have identified four different frames used by actors in climate politics. According to them, actors frame climate change either in terms of 'scientific uncertainty,' 'global economics,' 'ecological modernization,' or 'climate justice.' Stevenson and Dryzek (2012) in turn have distinguished four major discourses: 'mainstream sustainability,' 'expansive sustainability,' 'limits,' and 'green radicalism.' Even though these reconstructions of different framing patterns or discourses might differ in details, they identify very similar underlying conflict lines and actor constellations. As I will show in the following, the Cochabamba conference is an instructive example of framing patterns commonly identified as 'climate justice' or 'green radicalism.' In contrast to studies that compare different framing patterns or meta-narratives on climate change, I will focus on reconstructing the specific diagnostic and prognostic frames that can be found in the People's Agreement.

3 The road to Cochabamba

In this section, I will first contextualize the Cochabamba conference in terms of developments in international climate policy, political developments in Bolivia, and with regard to the adverse climate change impacts expected for the Andean region. The call for a People's Conference on climate change was in large parts a reaction to the perceived failure of the negotiation within the UNFCCC process. In particular, the 2009 climate change negotiations in Copenhagen were criticized for two reasons: *First*, the Copenhagen summit had dramatically disappointed the high expectations held by many participants and observers. Instead of agreeing on a legally binding agreement for the time after the first commitment period of the Kyoto Protocol, as envisioned by the Bali Action Platform adopted in 2007, state parties made only little progress: They only agreed upon a weak agreement stipulating nonbinding emission reduction targets (Dimitrov 2010; Rajamani 2010b). It was these insufficient results that propelled Evo Morales (2010) to refer in the invitation to Cochabamba explicitly to the "failure of Copenhagen." *Second*, the summit had also disappointed expectations in the negotiation process itself: Large parts of civil society and even some state delegates felt they were being excluded from the negotiations. Delegates of the ALBA countries, for instance, complained that they were not sufficiently included in the second week of the negotiations. The Danish presidency had set up a friends of the chair group after the negotiations had reached an impasse. While all negotiating blocs were represented, the ALBA countries were not included, neither as part of the G77 group nor independently. The Danish presidency made this decision fearing that including the ALBA countries would make it more difficult or even impossible to reach a compromise (Rajamani 2010b: 825–826). Already during the negotiations, Bolivia and others protested against the "dictatorial way" in which the negotiations were held (Dimitrov 2010: 811). Similarly, the participation of civil society representatives was also criticized as restricted. Due to an increased number of participants, a lack of organizations, and repressive reactions to protest activities, large parts of civil society were barred from attending the summit in the second week (Fisher 2010). Comparing this with the Cochabamba conference, Pablo Solón, then Bolivia's permanent representative to the United Nations and chief climate negotiator, pointed out:

> We invited ... 192 governments. We didn't do what they did in Copenhagen, because in Copenhagen they did only a meeting with about twenty-five to twenty-nine presidents, and they didn't invite the rest. We invited everybody, because we want to dialogue with everybody.
>
> *(quoted according to Democracy Now! 2010)*

Bolivia's motivation for hosting the conference cannot, however, be fully understood by just looking at the international climate negotiations. Instead one also needs to consider the profound political changes in Bolivia in the 2000s and the adverse climate impacts expected for the larger region. The choice to host the meeting in Cochabamba is symbolic because the city has been at the center of the so-called 'water war.' In 1999 and 2000, a protest movement successfully fought plans to privatize Cochabamba's water supply, grant a license to a transnational consortium, and increase water prices (Assies 2003). Participating in these protests was also an important stepping stone for the party of Bolivia's president Evo Morales. It was here that the *Movimento al Socialismo* (MAS) first asserted its claim to power (Horst 2009). When the MAS was finally voted into government in 2005, it was for the first time in Bolivia's history that an indigenous president took office. This propelled a radical transformation of Bolivia (Klein 2011). In addition to these political developments, the adverse impacts of climate change also play a role for Bolivia's engagement. A warmer climate in the tropical Andes is expected to increase glacier melting and thereby have a negative impact on the availability of water resources in Bolivia, Peru, Columbia, and Ecuador (Bradley et al. 2006). Despite substantial improvements in its education and health sector, Bolivia is still one of the poorest countries in the region and is dependent on foreign aid (Klein 2011: chapter 10). From 2000 to 2010, for instance, about 10 percent of all foreign aid meant for Latin America was sent to Bolivia, making Bolivia the second largest recipient of foreign aid in the region (OECD 2012). In light of the adverse climate change impacts on Bolivia, its high vulnerability due to its poor socio-economic performance, and its recent political development, the roles of Bolivia as the host of the conference and the choice of Cochabamba as host city are highly symbolic.

4 Reframing climate change as a political issue

The People's Agreement can be seen as an attempt to reframe climate change as an issue. In this section, I will analyze the diagnostic frames used in the agreement. In particular, I will describe what causal story is told, how blame is attributed, and what kinds of reference objects are identified. In doing that, I will focus the discussion on three aspects of the agreement: (i) capitalism being identified as the root cause of climate change; (ii) climate justice considerations attributing blame and responsibility; and (iii) reference objects which include developing countries and their populations but also eco-centric concepts such as 'Mother Earth' and 'living well.'[3]

The People's Agreement frames climate change as an issue that is predominantly caused by *capitalism*, that violates fundamental justice principles, and that poses an existential threat to both humanity and nature. The agreement points out that capitalism has, especially since the industrial revolution, considered nature one-sidedly as a source of resources. Furthermore, it argues that capitalism has reduced men and women to production and consumption tools, sparked off imperialism, and brought about the military-industrial complex. The authors of the agreement criticize industrialized countries and business for limiting the discussion about climate change to the increase in global average temperatures, instead of clearly identifying capitalism as its root cause. Resulting from such a diagnostic framing is the idea that climate change can only be successfully solved once capitalism has been abandoned as an economic order: "Humanity confronts a great dilemma: to continue on the path of capitalism, depredation, and death, or to choose the path of harmony with nature and respect for life" ('People's Agreement' 2010). The agreement argues that a radical change in the existing economic and political order as well as in the consumption and production patterns is necessary in order to overcome climate change. In this way, the agreement follows earlier statements of civil society pointing at the relationship between capitalism and climate change. In 2009, at an alternative counter-summit in Copenhagen, for instance, parts of civil society were demanding: "system change – not climate change" (Klimaforum09 2009).

Closely linked with this critique of capitalism is an interpretation of climate change from the *perspective of climate justice*. The concept of climate debt has a central position in this discourse referring to the asymmetric relationship between those countries that have caused climate change and those countries and peoples that will suffer the most from it: "Developed countries, as the main cause of climate change, in assuming their historical responsibility, must recognize and honor their climate debt in all of its dimensions as the basis for a just, effective, and scientific solution to climate change" ('People's Agreement' 2010).

The concept of climate debt is linked to earlier discussions about 'ecological debt' that were initiated by NGOs and social movements from the Global South in the 1990s. According to Goeminne and Paredis (2010: 708) the concept refers to all those activities that are "causing ecological damage elsewhere" and are "using ecosystem goods and services at the expense of equitable rights of others." Both of these aspects are picked up by the People's Agreement. With regard to the former, it points out that climate change is mostly caused by the historic emissions of industrialized countries, whereas the impacts of climate change are felt disproportionately by developing countries, peasants, and indigenous people. In addition, the agreement criticizes that industrialized countries have already emitted more than their fair share of greenhouse gases and demands a "decolonization of the atmosphere" by reducing and absorbing industrialized countries' emissions ('People's Agreement' 2010). According to the agreement, this should enable developing countries to exercise their right to development without further exacerbating climate change.

The agreement identifies two sets of *reference objects* that are thought to be threatened by climate change and worth saving: (i) the agreement emphasizes the vulnerability of developing countries, peasants, and indigenous peoples; (ii) in addition, the agreement also highlights the threat climate change poses to eco-centric entities such as 'Mother Earth' and practices of 'living well.'

> We propose to the peoples of the world the recovery, revalorization, and strengthening of the knowledge, wisdom, and ancestral practices of Indigenous Peoples, which are affirmed in the thought and practices of 'Living Well,' recognizing Mother Earth as a living being with which we have an indivisible, interdependent, complementary and spiritual relationship.
>
> *('People's Agreement' 2010)*

The concepts of 'living well' and 'Mother Earth' have both played a prominent role for the constitutional reforms in Ecuador and Bolivia in 2008 and 2009. The two concepts are based on indigenous thought (Gudynas 2011). The concept of 'living well,' for instance, refers to the Spanish term *vivir bien* and is described as a pluralistic concept with indigenous roots.[4] The concept challenges Western ideas of development and emphasizes the existence of alternative models of development (Fatheuer 2011). It underlines the idea of quality of life, while at the same time arguing that this is only conceivable within a community. In this context, it is important to note that nature itself and 'Mother Earth' are often included in this conception of community (Gudynas 2011). The concept of 'Mother Earth,' a translation of *pachamama*, sees nature as a living being. Hence, the People's Agreement is emblematic of an eco-centric conception of the relationship between nature and humankind. Such a conception differs from other approaches by granting nature intrinsic rights. According to Eckersley (1992: 26), such an environmental ethic "recognizes the moral standing of the nonhuman world and seeks to ensure that it, too, may unfold in its many diverse ways."

On the whole, the People's Agreement interprets capitalism and its modes of production and consumption as the root cause of climate change. By taking a perspective of environmental justice and climate debt, the agreement attributes the main bulk of responsibility to industrialized countries, while highlighting that the adverse impacts of climate change and the burden of adaptation are mostly borne by developing countries, peasants, and indigenous peoples. It is this disparity which creates a special responsibility of industrialized countries to act on climate change.

5 Alternative solutions for international climate policy

Following from its specific understanding of climate change as a political and economic problem, the People's Agreement proposes a set of alternative solutions for international climate policy. In discussing these, I will focus on the agreement's proposals for (i) climate mitigation, (ii) climate adaptation, and (iii) institutional reforms of international climate policy.

Regarding *climate mitigation* the People's Agreement refers to the text of the Framework Convention on Climate Change, which states that the ultimate objective of the convention is to stabilize "greenhouse gas concentrations in the atmosphere at a level that would prevent dangerous anthropogenic interference with the climate system" (UNFCCC 1992: Art. 2). The framework convention, however, leaves open what exactly is understood as a 'dangerous anthropogenic interference with the climate system.' As a result, the question of how to define the term 'dangerous interference' (and based on that, how to define emission reductions and temperature targets) has become one of the central issues of global climate politics (Lenton 2011). The People's Agreement answers this question as follows:

> Our vision is based on the principle of historical common but differentiated responsibilities, to demand the developed countries to commit with quantifiable goals of emission reduction that will allow to return the concentrations of greenhouse gases to 300 ppm [concentration in parts per million], therefore the increase in the average world temperature to a maximum of one degree Celsius.
>
> *('People's Agreement' 2010)*

In order to achieve this global goal, the agreement calls on all industrialized countries to agree to a second commitment period of the Kyoto Protocol (2013–2017) and to reduce their emissions significantly, that is, at least by 50 percent from the base year 1990. In the long term, a global agreement should define an overall goal for emissions reductions and then specify individual emission reduction targets for all developed countries. Again referring to the Framework Convention on Climate Change, the agreement argues that these individual targets should be based on the principle of "common but differentiated responsibilities" (UNFCCC 1992: Art. 3 and 4). Whereas the agreement backs the Kyoto Protocol, it rejects all market based mechanisms, carbon trading, and climate compensation measures as 'false solutions' that will not lead to emission reductions and will not mitigate climate change. "The carbon market has become a lucrative business, commodifying our Mother Earth. It is therefore not an alternative for tackle climate change, as it loots and ravages the land, water, and even life itself" ('People's Agreement' 2010).

In a similar way, other instruments of international climate policy such as Reducing Emissions from Deforestation and Forest Degradation (REDD+) are also rejected by the agreement because they are said to further the commodification of nature and to violate the rights of indigenous peoples. Asides from its criticism of market mechanisms, the People's Agreement also rejects the broad consensus on a two-degree target: Reviewing the implications of a two-degree increase in average global temperatures, the agreements argues that the consequences for African countries and many small island states as well as for indigenous peoples are unacceptable. One suggested mitigation measure is to stop the exploitation of gas and oil resources in rainforests.[5]

Climate adaptation forms the second main pillar of international climate policy. While the People's Agreement also addresses climate adaptation, it emphasizes that adaptation measures cannot be a substitute for drastic emission reductions. A central demand with regard to adaptation is that developed countries should acknowledge their climate debts and provide 6 percent of their GNP for adaptation measures in developing countries. These funds should come from public sources and be provided in addition to existing funds for development cooperation. They should be administered by a new body under the authority of the UNFCCC's conference of parties with significant involvement of developing countries. Furthermore, the agreement demands that developed countries "should assume responsibility for climate migrants" by welcoming them in their territory and granting them fundamental rights (see also chapter by Herbeck in this volume). Adaptation measures should also be supported by a far-reaching technological transfer: "Developed countries have a responsibility to share their technology with developing countries, to build research centers in developing countries for the creation of technologies and innovations, and defend and promote their development and application for 'living well'" ('People's Agreement' 2010). While taking a positive stance on technology transfer, the agreement is more critical towards any large-scale interventions in the climate systems such as climate or geo-engineering. These are described as 'false solutions' that only serve to aggravate the crisis.

The question of *institutional reform of international climate policy* is the third major issue touched upon by the People's Agreement. One central demand is the establishment of an international court for climate and the environment, which should charge and sentence those states, companies, and individuals that have not fulfilled their international obligations. In addition, the agreement refers to an international referendum that should decide on broader issues such as the future of economic order, demilitarization, and climate protection. The results should then be referred to the United Nations. The UN negotiation process should be complemented by a 'global people's movement' coordinating and uniting the activism of social movements around the world. In addition to these institutional reforms, the People's Agreement also calls for the recognition of a number of rights including the rights of people to be protected from the impacts of climate change, the rights of climate refugees and migrants, the right to development held by developing countries, and the rights of indigenous peoples. The agreement, thereby, references both rights that are already established and internationally recognized (such as the rights of indigenous peoples), but also rights, whose nature, extent, and content are still contested (Rajamani 2010a: 403). In addition, the agreement also advocates a number of eco-centric rights. It proposes to adopt a 'Universal Declaration of the Rights of Mother Earth,' which would protect, for instance, its "right to live and to exist" and its "right to regenerate its bio-capacity and to continue its vital cycles and processes free of human alteration" ('People's Agreement' 2010). Taken together, these demands aim at reforming global climate politics in two major ways: By establishing courts, the agreement advocates a legalization of climate politics. Furthermore, the agreement suggests to complement the top-down process of international negotiations with a bottom-up approach led by a 'global people's movement' on climate change.

6 Impact and relevance of Cochabamba for global climate politics

For many activists Cochabamba has become a central reference point because they are attracted by its far-reaching demands and the clear denouncement of capitalism as the root cause of climate change. In this way, Nnimmo Bassey, the former chair of Friends of the Earth International, has described the conference enthusiastically as a "turning point in the march to transform our world from the path of conflict, competition, exploitation and domination to a path of solidarity and dignity. It held a ray of hope for Africa" (Bassey 2012: 111). In those parts of civil society that do not share the anticapitalist frames expressed in the agreement, the conference has, however, raised only limited interest (Werdermann 2010). Moreover, Bolivia's attempts to be the mouthpiece of the climate justice movement have also sparked criticism. Eduardo Gudynas, for instance, has used the term 'neo-extractivism' to describe how progressive governments of Latin America exploit natural resources to fund their social policies (Gudynas 2010; see also Kennemore and Weeks 2011).[6] Some observers have also criticized procedural aspects of the conference pointing out that the conference and its outcomes were in large part predetermined by Bolivia. As an example, critics have noted that Bolivia excluded groups critical of its own policies from the conference: The participation of local and national groups wanting to address the contradiction between Bolivia's international agenda and its exploitation of natural resources and large-scale development and infrastructure projects was rejected by the argument that the conference was supposed to address global not national concerns (Turner 2010: 63–64).

In how far has Cochabamba had an impact on international climate policy? The ALBA countries, in particular Bolivia, have undertaken sustained efforts to promote some of the key elements of the People's Agreement on the international level. They have, for example, referred to the agreement within the climate change negotiations and the summit on sustainable development. In this way, Bolivia has submitted text passages for consideration in the 2010 intersessional meeting of the UNFCCC in Bonn and for the 2012 Rio+20 conference on sustainable development.[7] Some of these proposals were included in the negotiation text in Bonn, but only in brackets, as suggestions to be discussed. During the next round of UN climate negotiations in Cancún these proposals were not picked up again (Russell et al. 2012: 24–30). This seems to be symptomatic for the fate of the agreement: Rather than having a substantial impact on the international negotiations, it is often merely acknowledged. In the run-up to the UN climate negotiations in Cancún, Bolivia, for instance, complained that its draft texts were not considered (ENB 2010). Furthermore, Bolivia seems increasingly isolated in its opposition: While being part of a larger coalition preventing the adoption of the Copenhagen Agreement in 2009, Bolivia was rather isolated in its opposition to the results of Cancún. As a consequence, Bolivia's protest was not considered by the Mexican chair (Dimitrov 2010; Rajamani 2011).

7 Conclusion

The Cochabamba conference aimed at establishing a critical climate movement and at influencing the UN climate negotiations. The People's Agreement forms the basis for both of these goals: By presenting a coherent interpretation of what causes climate change, who is threatened by its adverse impacts, and how to fix the problem, the agreement has offered an alternative framing of climate change as an issue requiring political intervention. In particular it identified capitalism and its productions and consumption patterns as the root cause of climate change. It also highlighted the threat climate change poses to developing countries, peasants, and indigenous peoples and proposed far-reaching emission reduction targets and substantial financial and technological transfers to enable developing countries to adapt to global warming.

How can we evaluate the success of this attempt to reframe climate change? To answer this question, it is necessary to distinguish between social movements, NGOs, and climate activists on the one hand and the UN climate negotiations on the other. For parts of the former, particularly those that share its anticapitalistic stance, the conference has become a central reference point. Cochabamba's impact on international climate policy, however, has been much more limited. Whereas Bolivia and some of the other ALBA countries have frequently put forward its conclusions in the UNFCCC process, the impact on the outcome of the climate negotiations has been rather limited. Taking a framing approach perspective, one can identify two main reasons for this: First, the diagnostic and prognostic frames put forward in the People's Agreement are lacking resonance within the larger policy community. This can be illustrated by the position on capitalism expressed in the agreement: The capitalism critique evident in the agreement is not shared by many actors engaged in climate policy. Rather, as evident by the increased interest in concepts such as green economy, many actors see the market not as part of the problem, but as a possible solution. Second, the lack in resonance is not only due to fundamentally different ideas of what causes climate change, but also due to a lack of empirical credibility. Here, two central demands articulated in the People's Agreement can serve as examples: the emissions reductions and its related temperature target and the demands for climate finance. Against the backdrop of past emissions and the lack of ambition, the goal of reducing atmospheric carbon dioxide concentrations to 300 ppm and limiting global warming to one degree Celsius must be considered unrealistic. While these targets might form an adequate normative threshold for defining dangerous climate change, reaching them seems impossible. This becomes more explicit once one considers that even the two-degree target, which for long has been an almost consensus, is nowadays sometimes called into question as infeasible (Geden 2012; see also Randalls 2010). In a similar way, the demand to provide 6 percent of developed countries' GNP for adaptation measures seems unrealistic, given that these countries have not been willing to fulfill less ambitious goals regarding development cooperation.[8] Furthermore, assessing how much funding is needed for climate adaptation has led to varying results: The World Bank, for instance, has estimated the necessary amount of adaptation funding at US$70–100 billion per year, an amount that is roughly equivalent to the funds provided by developed countries for development cooperation (World Bank 2010).[9]

Concluding, I argue that the Cochabamba conference has become a central reference point for parts of civil society, while it had only limited impact on the UN climate negotiations. From a framing perspective, I pointed to the importance of the audience for any framing effort to succeed. How Cochabamba's success should ultimately be measured remains, however, an open question. Some might argue that the yardstick for Cochabamba's success should not be its impact on the intergovernmental negotiations in the short term, but rather its contribution to a transnational climate movement holding the potential to transform global climate politics in the long run. Whether that will be successful remains to be seen.

Notes

1 I would like to thank the editors of this volume, Matthias Dietz and Heiko Garrelts, and Frank A. Stengel for their most helpful comments.
2 For readability, I will use the terms 'Cochabamba conference' or 'conference' and 'People's Agreement' or 'agreement' as short forms to refer to the conference and its final declaration respectively.
3 The following quotes are from the English version of the People's Agreement, retrieved April 28, 2013 from http://pwccc.wordpress.com/2010/04/24/peoples-agreement/.
4 The Spanish version of the People's Agreement uses the term 'vivir bien,' which is in use in Bolivia. In other countries, including Ecuador, the term 'buen vivir' is used to refer to the same concept (Fatheuer 2011).
5 This suggestion is similar to the Yasuní-ITT initiative, in which Ecuador proposed to leave some of its oil resources in the ground in exchange for donations by industrialized countries matching 50 percent of the lost profits (Martin 2011).
6 However, as described above, the People's Agreement suggests abandoning the exploitation of oil and gas resources.
7 Compare the submissions by Bolivia for the Rio+20 summit (Bolivia 2011a) and for the UNFCCC Ad hoc Working Group on Long-term Cooperative Action (Bolivia 2011b).
8 So far, only few developed countries have, for instance, fulfilled the commitment to provide 0.7 percent of their GDP for development cooperation (Clemens and Moss 2007; Messner 2011).
9 The World Bank report, however, is based on lower greenhouse gas emissions and does not account for socio-political adaptation measures (Pauw et al. 2011).

References

Assies, W. (2003) David versus Goliath in Cochabamba: Water Rights, Neoliberalism, and the Revival of Social Protest in Bolivia, *Latin American Perspectives*, 30(3): 14–36.

Bassey, N. (2012) *To Cook a Continent: Destructive Extraction and the Climate Crisis in Africa*, Cape Town: Pambazuka.

Benford, R. D. and Snow, D. A. (2000) Framing Processes and Social Movements: An Overview and Assessment, *Annual Review of Sociology*, 26(1): 611–639.

Biermann, F. (2011) New Actors and Mechanisms of Global Governance, in Dryzek, J. S., Norgaard, R., and Schlosberg, D. (eds.), *The Oxford Handbook of Climate Change and Society*, Oxford: Oxford University Press.

Bolivia (2011a) Proposal of the Plurinational State of Bolivia for the United Nations Conference on Sustainable Development (Rio+20). Accessed 05/02/2012 from www.uncsd2012.org/rio20/content/documents/454Submission%20Plurinational%20State%20of%20Bolivia%20Rio20.pdf.

Bolivia (2011b) Submission by the Plurinational State of Bolivia on 'Market-based and Non-market-based Mechanisms,' Ad hoc Working Group on Long-term Cooperative Action under the Convention. *UN Doc. FCCC/AWGLCA/2011/MISC.2*. Accessed 02/08/2012 from http://unfccc.int/resource/docs/2011/awglca14/eng/misc02.pdf.

Bradley, R. S., Vuille, M., Diaz, H. F., et al. (2006) Threats to Water Supplies in the Tropical Andes, *Science*, 312(5781): 1755–1756.
Buzan, B., Wæver, O., and Wilde, J. D. (1998) *Security: A New Framework for Analysis*, Boulder: Lynne Rienner Publishers.
Clemens, M. A. and Moss, T. J. (2007) The Ghost of 0.7 Per Cent: Origins and Relevance of the International Aid Target, *International Journal of Development Issues*, 6(1): 3–25.
Davis, J. E. (2002) Narrative and Social Movements: The Power of Stories, in Davis, J. E. (ed.), *Stories of Change: Narratives and Social Movements*, Albany: State University of New York Press.
Democracy Now! (2010) *Bolivian UN Ambassador Pablo Solon on the World Peoples' Summit on Climate Change and Rights of Mother Earth*. Democracy Now! Accessed 06/11/2012 from www.democracynow.org/2010/4/19/bolivian_un_ambassador_pablo_solon_on.
Dimitrov, R. S. (2010) Inside UN Climate Change Negotiations: The Copenhagen Conference, *Review of Policy Research*, 27(6): 795–821.
Dryzek, J. S. (1997) *The Politics of the Earth: Environmental Discourses*, Oxford and New York: Oxford University Press.
Eckersley, R. (1992) *Environmentalism and Political Theory: Toward an Ecocentric Approach*, Albany: State University of New York Press.
ENB (Earth Negotiations Bulletin) (2010) Summary of the Bonn Climate Change Talks: 31 May–11 June 2010, International Institute for Sustainable Development. Accessed 06/14/2012 from www.iisd.ca/download/pdf/enb12472e.pdf.
Fatheuer, T. (2011) Buen Vivir: A Brief Introduction to Latin America's New Concepts for the Good Life and the Rights of Nature, Publication Series on Ecology, Berlin, Heinrich-Böll-Stiftung. Accessed 06/29/2012 from www.boell.de/downloads/Buen_Vivir_engl.pdf.
Fisher, D. R. (2010) COP-15 in Copenhagen: How the Merging of Movements Left Civil Society Out in the Cold, *Global Environmental Politics*, 10(2): 11–17.
Foran, J. and Widick, R. (2011) Climate Justice Movements: A Work in Progress, Manuscript for a Talk at the BREN School of Environmental Science and Management, UC Santa Barbara, Santa Barbara, CA. Accessed 11/28/2011 from www.iicat.org/climate-justice-movements-a-work-in-progress/.
Geden, O. (2012) Die Modifikation des 2-Grad-Ziels: Klimapolitische Zielmarken im Spannungsfeld von wissenschaftlicher Beratung, politischen Präferenzen und ansteigenden Emissionen, SWP-Studie, Berlin, Stiftung Wissenschaft und Politik. Accessed 06/17/2012 from www.swp-berlin.org/fileadmin/contents/products/studien/2012_S12_gdn.pdf.
Goeminne, G. and Paredis, E. (2010) The Concept of Ecological Debt: Some Steps Towards an Enriched Sustainability Paradigm, *Environment, Development and Sustainability*, 12(5): 691–712.
Gudynas, E. (2010) The New Extractivism of the 21st Century: Ten Urgent Theses About Extractivism in Relation to Current South American Progressivism, Americas Program Report, Washington, D.C., Center for International Policy. Accessed 02/24/2012 from www.iadb.org/intal/intalcdi/PE/2010/04716.pdf.
Gudynas, E. (2011) Buen Vivir: Today's Tomorrow, *Development*, 54(4): 441–447.
Hajer, M. A. (1995) *The Politics of Environmental Discourse: Ecological Modernization and the Policy Process*, Oxford: Clarendon.
Horst, R. (2009) Zielkonflikte zwischen Demokratie und ökonomischer Stabilität in Bolivien, in Jäger, T. (ed.), *Bolivien: Staatszerfall als Kollateralschaden*, Wiesbaden: VS Verlag.
Hülsse, R. (2007) Creating Demand for Global Governance: The Making of a Global Money-laundering Problem, *Global Society*, 21(2): 155–178.
Kennemore, A. and Weeks, G. (2011) Twenty-first Century Socialism? The Elusive Search for a Post-Neoliberal Development Model in Bolivia and Ecuador, *Bulletin of Latin American Research*, 30(3): 267–281.
Klein, H. S. (2011) *A Concise History of Bolivia*, 2nd edn., New York: Cambridge University Press.
Klimaforum09 (2009) System Change – Not Climate Change. A People's Declaration from Klimaforum09, Copenhagen. Accessed 06/26/2012 from http://declaration.klimaforum.org/files/declaration_english_screen.pdf.
Koopmans, R. and Statham, P. (1999) Political Claims Analysis: Integrating Protest Event and Political Discourse Approaches, *Mobilization: An International Quarterly*, 4(2): 203–221.
Lenton, T. M. (2011) Beyond 2°C: Redefining Dangerous Climate Change for Physical Systems, *Wiley Interdisciplinary Reviews: Climate Change*, 2(3): 451–461.
Martin, P. L. (2011) Global Governance from the Amazon: Leaving Oil Underground in Yasuní National Park, Ecuador, *Global Environmental Politics*, 11(4), 22–42.

Messner, D. (2011) Entwicklungspolitik als globale Strukturpolitik, in Jäger, T., Höse, A., and Oppermann, K. (eds.), *Deutsche Außenpolitik*, Wiesbaden: VS Verlag.

Morales, E. (2010) Invitation to the People's World Conference on Climate Change and Mother Earth's Rights. Accessed 06/11/2012 from www.cochabamba2010.org/allTexts/InvitationEvoMOrales.pdf.

Mühlhäusler, P. and Peace, A. (2006) Environmental Discourses, *Annual Review of Anthropology*, 35(1): 457–479.

OECD (Organisation for Economic Co-operation and Development) (2012) Development Aid at a Glance, Statistics by Region: 3. Accessed 08/15/2012 from www.oecd.org/dac/aidstatistics/42139289.pdf.

Pauw, P., Kaplan, M., and Ifejika Speranza, C. (2011) *Durbans "afrikanischer" Klimagipfel – Anpassung in Afrika*, Die aktuelle Kolumne, Bonn: Deutsches Institut für Entwicklungspolitik.

Rajamani, L. (2010a) The Increasing Currency and Relevance of Rights-Based Perspectives in the International Negotiations on Climate Change, *Journal of Environmental Law*, 22(3): 391–429.

Rajamani, L. (2010b) The Making and Unmaking of the Copenhagen Accord, *International and Comparative Law Quarterly*, 59(3): 824–843.

Rajamani, L. (2011) The Cancun Climate Agreements: Reading the Text, Subtext and Tea Leaves, *International & Comparative Law Quarterly*, 60(2): 499–519.

Randalls, S. (2010) History of the 2°C Climate Target, *Wiley Interdisciplinary Reviews: Climate Change*, 1(4): 598–605.

Russell, B., Pusey, A., and Sealey-Huggins, L. (2012) Movements and Moments for Climate Justice: From Copenhagen to Cancun via Cochabamba, *ACME: An International E-Journal for Critical Geographies*, 11(1): 15–32.

Schlichting, I. and Schmidt, A. (2012) Strategische Deutungen des Klimawandels: Frames und ihre Sponsoren, *Forschungsjournal Neue Soziale Bewegungen*, 25(2): 29–41.

Snow, D. A. and Benford, R. D. (1988) Ideology, Frame Resonance, and Participant Mobilization, *International Social Movement Research*, 1(1): 197–217.

Snow, D. A., Rochford, E. B., Worden, S. K., et al. (1986) Frame Alignment Processes, Micromobilization, and Movement Participation, *American Sociological Review*, 51(4): 464–481.

Stevenson, H. and Dryzek, J. S. (2012) The Discursive Democratisation of Global Climate Governance, *Environmental Politics*, 21(2): 189–210.

Stone, D. A. (1989) Causal Stories and the Formation of Policy Agendas, *Political Science Quarterly*, 104(2): 281–300.

Surel, Y. (2000) The Role of Cognitive and Normative Frames in Policy-making, *Journal of European Public Policy*, 7(4): 495–512.

Turner, T. E. (2010) From Cochabamba, a New Internationale and Manifesto for Mother Earth, *Capitalism Nature Socialism*, 21(3): 56–74.

UNFCCC (United Nations Framework Convention on Climate Change) (1992) *FCCC/INFORMAL/84*. Accessed 05/19/2011 from http://unfccc.int/resource/docs/convkp/conveng.pdf.

Werdermann, F. (2010) Cochabamba statt Kopenhagen?, *Klimaretter – Das Magazin zur Klima- und Energiewende*. Accessed 07/06/2012 from www.klimaretter.info/hintergruende/72-protest/5638-cochabamba-statt-kopenhagen.

World Bank (2010) *Economics of Adaptation to Climate Change – Synthesis Report*. Washington, D.C.: The World Bank. Accessed 07/14/2012 from http://climatechange.worldbank.org/sites/default/files/documents/EACCSynthesisReport.pdf.

Zald, M. N. (1996) Culture, Ideology, and Strategic Framing, in McAdam, D., McCarthy, J. D., and Zald, M. N. (eds.), *Comparative Perspectives on Social Movements: Political Opportunities, Mobilizing Structures, and Cultural Framings*, Cambridge: Cambridge University Press.

34
DEBATES AND CONFLICTS IN THE CLIMATE MOVEMENT

Matthias Dietz

This chapter focuses on the inner life of the climate movement. Behind its occurrence and activity stand internal processes of negotiation, weighing, conflicts, and other forms of communication. These processes should be revealed and analyzed. The analysis of the debates of the climate movement aims at finding out which topics and questions the members of the movement are dealing with, how they treat each other, and what can be concluded from these observations about the characteristics and development of the movement.

The debates of the climate movement can be separated into two fields. One part of the debates deals with questions of everyday life in the movement, the other part with more abstract and complex topics such as the state and development of the climate movement.

1 Micro-topics of everyday life

Debates of the first type are primarily held in a national or continental context. The level of conflict in these debates is quite limited. In the last few years members of the US and Canadian climate movement often have communicated about the topics of tar sands, mountaintop-removal, and fracking. These are new forms of extracting fossil fuels which require much energy and harm the environment. The climate movement is intensively engaged against these new forms of extraction. In newsletters, on websites, and on blogs climate groups publish calls for demonstrations, blockades, or other protest events. Furthermore, they report about realized protests. These reports mostly explain the negative impacts of the tackled projects and judge the protests as successful. Often the contributions to the debates also contain critique of companies and local administrations which are involved in the extraction projects. For example, several climate groups reported about the 'End Mountaintop Removal Week in Washington' in June 2012 (see Climatesciencewatch 2012). Besides, there were countless publications on the Keystone XL pipeline project (TckTckTck 2011). The protests against this project were also observed and supported by European climate groups (see for example Klimaretter.info 2011). The climate movement celebrated the temporary stop of the project by President Obama in January 2012 as a historic success. The network organization 350.org expressed its delight in the following message: "We won. You won. Thank You. After relentless campaigning, the Keystone XL pipeline has been killed" (350.org 2012). But there were also skeptical voices demanding rationality in observing the future of the project. Bill McKibben for example said: "This is a victory for Americans who testified in record numbers ... [but] we're well aware that the fossil fuel lobby won't give up easily" (McKibben in Henn 2012).

For climate groups in Europe and Australia, the extraction and burning of coal is a central topic. In the contributions to the debate, the huge extent of coal usage in countries such as Germany, Britain, and Poland as well as the negative climate impact of coal-fired power stations is criticized. Furthermore, the new technology of Carbon Capture and Storage (CCS) is resolutely rejected by the climate groups. Activists from the traditional German coal regions Brandenburg and Rhineland judge coal as the "dirtiest form of all fossil fuels." The extraction of coal would result in "expulsion and destruction of forests, villages, people and animals" (ausgeco$_2$hlt 2012). In Germany several citizen initiatives were started in order to prevent local experiments using the CCS technology. In their publications, the initiatives argue that CCS inhibits a real transformation of energy production, results in big profits for energy companies, and is not a safe technology. Due to porosity in the ground, CCS could pollute drinking water resources and damage the climate (Bürgerinitiative CO$_2$ntra Endlager Neutrebbin 2012). In Great Britain climate groups discuss successes of anticoal protests in their country and in Australia. There were many reports, solidarity statements, and finally success stories about the disruption of the biggest coal plant of Great Britain in Kingsnorth by climate activists in 2008 and the resulting trial. Members of Greenpeace climbed on a chimney of the power station and thereby interrupted the energy and emission production of the plant. As a consequence, they were charged in court, but not found guilty (Greenpeace 2008; Stewart 2009).

A common characteristic of the debates of the climate movement on fossil fuels in the different countries is the extremely negative judgment of the oil and coal industry. In the eyes of the climate activists, this "industry wields massive political influence and holds the greatest responsibility for the hundreds of thousands of deaths, extinction of species, and billions of dollars of damage" (Murray 2012). The members of the climate movement have created special names for the oil and gas industry – 'Big Oil' and 'King Coal.' Many climate activists hold the opinion that the fossil fuel industry has a very powerful political lobby and controls parliaments and governments (McKibben in Henn 2012). In their statements many climate activists consider the coal and oil industry as an enemy of the climate movement.

In particular, in Great Britain the debates of the climate movement also often deal with the climate camps, a mixture of protest event, educational workshop, and ecological lifestyle that began in 2006. In 2010 political groups participating in the climate camps discussed tensions and conflicts which had occurred among them. They did this in a public and constructive way by producing the publication 'Criticism without Critique. A Climate Camp Reader' (Shift Magazine and dysophia 2010). The pragmatic part of the discussions dealt with arguments in favor and against the selling of merchandising products or the adequate way of treating visitors (Page 2009; Beaumont 2010). One year later a discussion about the dissolution of the camps followed. In a public letter the organizers of the camps announced the end of their activity and justified their decision by exhaustion, conflicts, and the migration of activists to the movement against the cuttings of public expenditures (Climate Camp UK 2011a). On the whole, climate activists in Britain create the impression of being very interested and engaged in debates.

Another important topic for the members of the climate movement in several countries is the activity of police and courts against climate activists. For example, the climate activists intensively discussed the operations of undercover police officers in the British climate movement. The most popular case is the one of Mark Kennedy who pretended to be a climate activist for several years, but in fact was a police spy. In their statements on this topic the climate groups describe themselves as shocked and betrayed. They demand transparent intelligence about the police operations against them as well as a stop to the undercover strategy (Climate Camp UK

2011b; Eilperin 2012).[1] Furthermore, the climate activists criticize the interventions of the police during protest events as too harsh. Even years after mass arrests at the UN Climate Change Conference in Copenhagen climate groups reported about the resulting trials. Money was collected to support the involved activists and there was huge pleasure when a court fined the Danish state to pay reparations for activists who were unlawfully arrested (Climate Collective 2012). In these debates the climate movement presents itself as a peaceful and innocent actor and accuses the police of criminalizing climate activists (Murray 2012). The police are described as "liars" (Climate Camp UK 2011b) and "pigs" (Indymedia UK 2012) and are consequently also treated as an enemy of the climate movement. The climate activists feel they are unjustifiedly treated as eco-terrorists and consciously try to create a different, more positive image of themselves in public.

Apart from the judicial consequences concerning the protest activity of the Kingsnorth Six and the police operations in Copenhagen, the case of Tim DeChristopher is very popular in the climate movement. Climate groups from different countries expressed solidarity and protest when DeChristopher, a student from the US, was accused of manipulating an auction and of making false statement. DeChristopher was sentenced to two years in prison because he participated in an auction for oil and gasdrilling rights without having the money to pay the bought parcels (see activist profile of DeChristopher in this book). In the debates on the trial against DeChristopher, members of the climate movement accused the judge several times of being biased (see for example Hall 2009; Jensen 2011). Nevertheless, critique and rejection of courts and judges are not as intensive as of the police. The climate movement seems to use the trials against activists in order to generate public attention for its aims. Consequently, the statements and reports of the climate groups are not just intended to inform and mobilize members of the movement, but are directed towards a broader public (Hickman 2010).

The critical comments of the climate activists about police and trials are quite typical for social movements. Movements try to use the public to present themselves as relevant and positive actors (Rucht 2011: 21). Often the police are a classical opponent of social movements, or at least are constructed as such (Leßmann-Faust 2008: 10). Therefore, the comments of the climate activists about the police should not be overestimated. They are probably neither an expression of militarism nor a rejection of the state.

2 Macro-topics

Apart from these topics regarding the everyday life of the climate movement, the debates of the members also deal with more abstract questions about the orientation, strategy, and development of the movement. Which solutions of the climate crisis should the movement support? How much radicalism is necessary to realize its aims? And with whom should the movement cooperate? The debates on these questions are more controversial than the above described debates about the everyday life of the movement. They reveal lines of conflict, tendencies in the development and characteristics of the climate movement.

2.1 Searching for an adequate level of acting

The debates on the right way of acting of the climate movement had already started some years ago. They were intensified by the failure of the UN Climate Conference in Copenhagen in 2009. In the run-up to Copenhagen just a small minority of climate groups was opposed to the dominant orientation on the climate summits and instead demanded radical actions against the fossil fuel industry on the local level. Most climate and environmental groups such as TckTckTck,

350.org, 1 Sky, the Energy Action Coalition, CAN, and Greenpeace organized protests and campaigns to support a successful result of the UN summit. The failure of the summit to develop a post-Kyoto treaty led to a deep crisis of these groups, because their aims and hopes were dashed. One activist succinctly said: "I think the climate movement is in a really important period of rebuilding and developing a new identity" (Goodell 2011). The opponents of the UN negotiation process had attacked the majority of the movement before, during, and after Copenhagen in a confrontative way. They criticized the way of acting of many climate groups as well as their relations to economic actors. Furthermore, they denied acting under the banner of a united movement (MCJ 2010; RTNA 2010a: 3, 4, 6, 8, 17). Nevertheless, this polarization was reduced during the crisis of the climate movement resulting from the failure of Copenhagen. Due to their disappointment about politics and the UN negotiation process, several climate groups changed their orientation and began to focus on the local level. They began to follow the approach of acting practiced by the radical minority of the movement. As a consequence, the attention for international climate politics was lowered and a certain radicalization of the climate movement took place.

This period of crisis and reorientation came along with intensive discussions which are documented in printed form. Rising Tide North America published a 24-page-long pamphlet with the title 'The Climate Movement Is Dead. Long Live the Climate Movement.' This publication presented the organization's local and radical approach of acting and appealed to the other climate groups to take it over: "We propose an asymmetrical assault on the fossil fuel industry. Stop focusing on ... giant UN meetings" (RTNA 2010a: 11). The pamphlet was broadly received in the climate movement. Also mainstream media occasionally reported about the publication (see for example van der Zee 2010). One reason for the numerous reactions to the paper could lie in its strong and radical messages.[2] As a consequence of the crisis of the climate movement, even Bill McKibben and his organization 350.org changed their strategic approach. Until Copenhagen they preferred moderate and friendly campaigns such as worldwide days of action or mass letter sendings. Disappointed about the lack of success of cooperative action, McKibben wrote: "Making nice doesn't work. It was worth a try, but it didn't work" (McKibben 2010) and "movements require enemies ... the fossil fuel industry ... is Public Enemy Number One" (McKibben 2012). McKibben started to demand and to support the use of civil disobedience in the campaigns against the Keystone XL pipeline (Jacobsen 2012). More intensive forms of protest, civil disobedience, and the readiness to accept disadvantages such as trials and prison are also postulated by DeChristopher and his organization Peaceful Uprising (Goodell 2011). In addition, an exchange of public letters between the moderate and the more radical wing of the climate movement underlines the assumption that the different concepts of acting in the climate movement are converging. In an open letter the US campaign 1 Sky demanded processes of reflection and cooperation in the climate movement: "We should seek consensus where it is available and stand strong and undivided together, even as we discuss and use our differences to our strategic advantage wherever possible in the debate" (1 Sky 2010a). The authors of the letter argued in favor of an intensification of the activity of the movement at the grassroots level, but also emphasized the importance of national and international politics. A coalition led by Rising Tide publicly answered this letter. Their answer was also formulated in a positive and cooperative way. Nevertheless, the radical climate groups clearly stated that they favored exclusively a local level of acting in the climate movement. They justified this position with their distrust of politicians who they believe to be controlled by the fossil fuel industry. In addition, they judge the activity of climate and environmental groups at the national and international level as ineffective and therefore reject it (RTNA et al. 2010).

2.2 Ways of solving the climate crisis

Currently, a certain consensus about the level and form of acting seems to exist in the climate movement. Most groups favor acting locally against the fossil fuel industry and no longer trust in the UN negotiations. Nevertheless, there are huge differences and discussions regarding the right concept of solving the climate crisis. A reduction of the use of oil, gas, and coal is not enough to stop climate change since at least the question results how to produce energy alternatively. As a consequence, there is a huge debate in the climate movement on ways and concepts of solving the climate crisis. Apart from energy production, technology, lifestyle, the economic system, and questions of distribution are relevant topics of this discussion.

For example, one part of the climate movement supports the idea of green capitalism. This concept assumes that the capitalistic economy due to its growth and innovation dynamic is able to develop climate friendly products and technologies.[3] In their opinion, a green economy could not only reduce energy consumption and emissions, but also create new jobs. Supporters of this concept are groups such as Stop Climate Chaos, 1 Sky, or the UK Youth Climate Coalition (1 Sky 2010b) as well as political parties and syndicates (see for example Frieder Otto Wolf in Turbulence 2009). Nevertheless, in publications such as '20 Theses against Green Capitalism' other members of the climate movement criticize this concept. They argue that new products and a growing economy always result in additional consumption of energy and resources. Furthermore, from the point of view of the critics of green capitalism, companies are dominated by profit orientation so that they often just pretend to act in an environmentally friendly way. Apart from statements and pamphlets, this debate is also characterized by direct discussions. For example, in 2009 the magazine *Turbulence* published a discussion between a supporter and an opponent of the idea of green capitalism. It was titled 'Green New Deal: Dead End or Pathway Beyond Capitalism?'

Similar polarized positions in the climate movement exist regarding economic instruments for climate protection. Most popular is the argument about carbon trading. Like other economic instruments, carbon trading is integrated in the Kyoto Protocol. Markets and mechanisms for carbon trading have already been created in different countries. One supporter of carbon trading is the huge climate network CAN (CAN 2009). For them, carbon trading is an opportunity to integrate emissions effectively into the calculations of companies. This would change the acting of the companies and finally result in a broad reduction of emissions. Supporters of carbon trading often add the argument that the carbon offset mechanism enables climate friendly projects in developing countries. Opponents of carbon trading in the climate movement express their critique in texts such as '350 Reasons that Carbon Trading Won't Work.' They argue that the costs of carbon trading have to be paid by the consumers and not by the companies. Furthermore, they criticize the spreading of certificates for free, hint at speculations with carbon certificates, and the missing effect of carbon trading (Carbon Trade Watch 2009; Pusey and Russell 2010). In addition, projects initiated or financed by the carbon offset mechanism would often result in local conflicts on land grabbing, participation, or other issues. The radical part of the climate movement explicitly judges carbon trading as counterproductive and failed: "Carbon trading ... actually makes the climate and environmental crisis worse" (Carbon Trade Watch 2012: 14). In contrast to the conflict on green capitalism, where supporters and opponents have similar strength, a majority of the climate movement rejects carbon trading. Besides, the conflict on carbon trading is older than the conflict on green capitalism. The discussions about carbon trading already started in the 1990s, whereas the idea of green capitalism became popular from 2006 on.

There is also no consensus in the climate movement on the question of which form of energy production could replace oil, gas, and coal. Most members of the movement reject the idea that nuclear energy is a clean and climate friendly technology. This opinion is held by energy companies and by several governments. For many climate groups, nuclear energy and its possible extension is too dangerous, too expensive, and would harm the climate. As a consequence, several climate groups have been engaged in the mobilization of antinuclear protests (gegenstromberlin 2012; RTUK 2012a). In their statements and publications they describe and explain the disadvantages of nuclear energy (see, for example, Greenpeace 2007; IEN 2009; Tokar 2010). But there are also single supporters of nuclear energy in the climate movement. Two of them are the British journalists George Monbiot and Mark Lynas (see chapter by Graham-Leigh in this book). For a long time also Bill McKibben supported the idea of nuclear energy as a climate friendly technology (see activist profile McKibben in this book). Due to these discussions the members of the climate movement observed with huge interest the decision of the German government to phase out nuclear energy in consequence of the nuclear accident in Japan in Spring 2011: "Especially the English climate-movement is discussing controversially the German exit of nuclear power" (Kern 2012). Furthermore, the majority of the climate activists reject energy production through biofuels. In particular indigenous climate groups criticize this kind of energy production in their statements. They claim that it destroys small farms and foodstuff (IEN 2009; La Via Campesina 2012). Finally, in the debates on alternative forms of energy production, also solar and wind projects in desert regions are criticized because they harmed local eco systems (RTNA 2010b: 9).

2.3 False solutions versus positive solutions

Many instruments for climate protection, which in parts are already implemented, are criticized by climate activists. Their critique focuses on instruments such as carbon trading, subsidies for green companies, agro fuels, nuclear energy, CCS, technology transfer, and sometimes also solar and wind energy. Many climate activists call these instruments 'false solutions.' This term is very popular and is defined in publications and long lists. For example, a 15-page-long 'Indigenous Peoples' Guide. False Solutions to Climate Change' exists as well as a 28-page-long report with the title 'Hoodwinked in the Hothouse. False Solutions to Climate Change.' False solutions are defined as "manag[ing] the climate crisis without compromising profits, the power structure or the economic system" (RTNA and CTW 2010).[4] In the opinion of many members of the climate movement false solutions are not sustainable. For them, they are based on the assumption that Western science and technology can solve all problems of humankind. The amount of false solutions is quite high because even climate projects in developing countries, solar and wind energy, and the internationally accepted Kyoto Protocol are categorized as such. False solutions are attacked in an open and aggressive way: Climate projects in Southern countries are judged as "new form of colonialism" (IEN 2009) and politicians as well as employers supporting them are claimed to be liars and criminals (RTNA and CTW 2010). It can be concluded that there is deep distrust and selectivity in the climate movement of official suggestions to solve the climate crisis. Furthermore, many climate groups are characterized by a deep skepticism about technology.

The debates in the climate movement about ways and methods to prevent false solutions require much more time than the formulation and discussion of alternative suggestions, i.e., positive solutions. For example, the publication 'Hoodwinked in the Hothouse' contains 19 pages of false solutions, but only four pages of positive solutions. In the 'Indigenous Peoples' Guide' just one page out of 15 deals with positive solutions. Furthermore, the presented positive solutions – for example climate justice and system change – are often not very detailed and sometimes again consist of negations.

A common characteristic of the alternatives demanded in the debates of the climate movement instead of false solutions is that they aim at a more fundamental and comprehensive solution of the climate crisis. Climate groups such as Carbon Trade Watch, Climate Justice Now!, Rising Tide, MCJ, and many indigenous actors demand a "socio-ecological transition with global equity, justice and genuine democracy" (RTNA and CTW 2010). The concepts of mainstream NGOs focusing on emission reductions and reforms are not sufficient for them, even if these concepts include protests against the fossil fuel industry. This can be exemplified by the reactions to a recent publication of Bill McKibben whose positions, as shown before, became more radical in the last years. In a long article in the *Rolling Stone* magazine, McKibben (2012) wrote about the state of climate science, climate politics, and the climate movement. Members of the climate movement with several critical answers reacted on the article of McKibben (see Brunner 2012; Petermann 2012; Smolker 2012). They criticized him for ignoring or at least underestimating capitalism, elites, and governments as reasons of climate change:

> if you focus solely on eliminating fossil fuels without changing the underlying system, then very bad things will take their place because it is the system itself that is unsustainable. It is a system designed to ... [produce] gargantuan profits for an elite few: the so-called '1%.' Whether it's driven by fossil fuels or biofuels or even massive solar and wind installations, the system will continue to devour ecosystems, displace forest-based communities ... crush labor unions and generally make life hell for the vast majority of the world's peoples.
>
> *(Petermann 2012)*

A real solution of the climate crisis would make it necessary "not [to] consume excessively or destroy things, to recognize our common interests and act like it" (Smolker 2012) as well as "[to] lift up the community solutions" (Brunner 2012). These descriptions of an alternative way to solve the climate crisis are not very concrete or detailed and again use negations and critique in particular when rejecting capitalism as "deep crisis" (IEN 2009).

The rhetoric used in this debate is quite aggressive because it speaks of "giant" profits, "excessive" consumption, and "dreadful" consequences caused by false solutions of climate change. Not only capitalism and the current political system are attacked, but also rather moderate actors of the climate movement such as McKibben. Brunner (2012) criticizes that the ideas of McKibben wouldn't be holistic enough and could in the end harm the climate movement. For Petermann (2012), the argumentation of McKibben is contradictive and unconvincing.

The concept of climate justice, which often is referred to as an alternative solution of the climate crisis, particularly focuses on the normative dimension of a possible system change. By payments and reduction of emissions, industrial countries should compensate their historical climate debt and support vulnerable Southern countries (Goodman 2009: 503). Furthermore, inhabitants should be able to decide independently on the use of the land and resources of their region (IEN 2009). The traditional way of treating nature applied by small farmers would be sustainable and climate friendly. Investors, huge companies, and climate projects are rejected as carbon-colonialism and heteronomy (Goodman 2009; IEN 2009).

Among the climate groups the concept of climate justice is quite popular. Nevertheless, their publications and debates reveal differences in the definition of the concept. While groups such as 350.org and CAN see climate justice as one point of their multifaceted agenda or just use it for the demand to strengthen the position of Southern countries in the climate negotiations, other groups focus solely on the concept of climate justice. 'Climate Justice Now!,' 'Mobilization for

Climate Justice,' or the 'Durban Group for Climate Justice' have integrated the term into the name of their organization. This is a signal of the high importance that they attach to the concept. Apart from these three organizations, in particular indigenous groups such as La Via Campesina or the Indigenous Environmental Network pay high attention to climate justice. They treat it as a detailed concept, containing different rights, principles, and demands. Nevertheless, these details are often quite far-reaching, difficult to realize, and sometimes also contradictory. For example, the '10 Principles of Climate Justice' formulated by the US organization MCJ demand a better control of carbon trading and higher sensibility for the rights of US workers. Besides, the US should take over the leading role in the international fight against climate change (MCJ 2007). In contrast, the Indigenous Environmental network in its 'Four Principles for Climate Justice' rejects carbon trading and demands a more important role of indigenous actors in the fight against global warming (IEN 2010). Consequently, the concept of climate justice is defined in different ways by the members of the climate movement. Due to these differences and problems of implementation, the movement does not seem to agree on a coherent concept to solve the climate crisis. Nevertheless, the different definitions of climate justice are at least similar and more concrete than demands for system change. Furthermore, no huge conflicts over the concept of climate justice can be observed among the climate activists.

3 Characteristics of the climate movement

Several characteristics of the climate movement are revealed by their criticism of false solutions and the formulation of alternative solution proposals. First, capitalism is the concept of an enemy for many climate activists because it is often criticized and attacked in their debates. Second, the fights about false and positive solutions seem to be complex and contain ideological elements. Bill McKibben uses many figures when he predicts the carbon emissions in Gigatons and writes about the amount of small carbon particles in the atmosphere (McKibben 2012). The European activist Tadzio Müller is said to hold a neo-Marxian perspective (DVPW 2009: 47). In his comments he locates the development of the climate movement in huge historical contexts: "What the class antagonism was 80 years ago, the biocrisis is today, itself a product of an equally indissoluble antagonism" (Müller and Passadakis 2009: 57). A similar perspective is taken by the activists from Rising Tide North America when they write in one of their publications: "The same power that manifests itself as resource extraction in the countryside, manifests itself as racism, classism, and human exploitation in the city" (RTNA 2010a: 14). The Durban Group for Climate Justice uses references to Marxism. The following words are taken from an essay written by one of their members: "The rise of carbon trading over the last decade is most compellingly understood through Marxian political economy" (Bond 2009). In most cases these considerations are very theoretical and not related to the everyday life of the climate movement. The comments are quite long and contain no or only few references to the practical activity of the climate movement.

The complexity of the debates shed light on another characteristic of the climate movement. The members of the movement show a high level of reflection and self-consciousness. The climate movement sees itself as a movement and discusses its orientation and future. For example, the public letters exchanged between the radical and the moderate wing of the movement deal with the development of the climate movement (see Section 2.1). Several times the formulation 'movement building' is used (see for example 1 Sky 2010a; RTNA et al. 2010). Bill McKibben and his organization 350.org regularly appeal for extending and strengthening the climate movement. Their motto is "We're building a global movement to solve the climate crisis." Finally, also Rising Tide often reflects on the state and future of the climate movement (see for example RTNA 2010a).

The reflection processes in the climate movement are in parts characterized by a sensitive, selective, and negative perspective. This perspective was already noted in the debates on false solutions. Several climate groups hold the opinion that the climate movement needs a clear and unambiguous direction of development in order to prevent harmful steps in the wrong direction. Compromises would be a bad idea: "'Anything is better than nothing' will always lead to nothing because it is a declaration of our intention to compromise" (RTNA et al. 2010). Even the extension of protest activity demanded by Bill McKibben would be a wrong idea if it was not considered that everything on earth is connected with each other: "focus[ing] solely on eliminating fossil fuels without changing the underlying system [will result in] very bad things" (Petermann 2012).

Instead of harmful steps and false concepts such as the cooperative approach of many climate groups (RTNA 2010a: 3) a coherent political economic project should be developed (Müller in Turbulence 2009). The next steps of the climate movement should be considered seriously. For example, it would be useful to build up strategic relations with certain other movements. Some years ago in the debates of the climate groups the antiwar movement was considered to be the right partner. It was argued that the military is engaged in energy related conflicts and causes many emissions itself. During the economic and financial crisis, members of the climate movement demanded strategic cooperation with the Occupy movement. They justified this demand with the hint at the shared characteristic of rejecting capitalism (RTNA 2011; La Via Campesina et al. 2012). Alliances with environmental movements seem to be less interesting for the climate movement. This impression is created for example by the following slogan of a German protest event: "It's not about polar bears – unite social and ecological and battles" (Blockupy Frankfurt 2012). One can conclude that there are groups that are consciously trying to direct the climate movement.

4 Overview and conflict lines

The analyzed debates of the climate movement revealed characteristics as well as structures of cooperation and conflict existing within the movement. After Copenhagen many climate groups became skeptical about the problem-solving ability of politics. A British NGO summarized this changing attitude in the following sentence: "It's no good waiting for governments to enforce radical change – they simply won't do it" (RTUK 2012b). This resulted in a certain radicalization and reorientation towards local protest activity against the fossil fuel industry and forms of civil disobedience. Besides, the climate movement felt criminalized by the police because of mass arrests (Parkin 2012) and the use of police spies against them. A majority of the climate activists regard the oil, gas, and coal industry as opponents of the movement. A minority also judges the police and capitalism as enemies. The climate movement has a high self-consciousness and intensively reflects about its development. The resulting internal debates often deal with theoretical prospects, are quite controversial, and linked with complex ideological questions. Left-wing, Marxist analysis sees the development of the movement in historical contexts and demand well-considered and strategic useful decisions on further steps. They aim at a fundamental system change to solve the climate crisis. Capitalistic and postcolonial structures should be overcome to enable climate justice and sustainable climate protection. The selective, sensitive, and negative perspective, that a part of the movement has, results in the rejection of many existing instruments for climate protection. Other groups of the climate movement judge instruments, such as carbon trading and renewable energies, more positively and show less theoretical or ideological concern. They intend to tackle climate change in a broad and pragmatical way. As a consequence of the crisis of the climate movement, they just want to intensify their work, to stress the local level, and to enhance cooperation within the movement, but they do not aim at changing international politics or changing the economic system.

It is possible to put these results of the analysis of the debates of the climate activists together and to define the following conflict lines existing in the climate movement:

- Ideas of green capitalism and a green new deal are rejected by a radical critique of capitalism.
- Several often technical instruments of climate protection stand against a smaller number of ideas demanding radical system change. The former are often called false solutions, the latter positive solutions.
- There is another conflict line between the approach of supporting a political solution of the climate crisis at the international level and the approach of acting locally and creating direct effects for climate protection. Currently the second position is getting more popular in the climate movement so that this conflict line has lost polarization.
- Finally, a pragmatical, cooperative, and consensus-oriented way of acting stands against an intransigent approach which insists on detailed and strategical considerations before acting.

The different positions in these conflicts are supported by groups with similar strength. The supporters of radical positions – for example as regards the critique of capitalism – are in a minority, because the biggest organizations of the climate movement such as 350.org and CAN hold moderate positions. Nevertheless, this minority loudly raises their voices. In contrast, there are several moderate climate groups, which do not or almost not participate in the internal debates.

The conflicts in the debates of the climate movement are on a quite high, but not antagonistic level. The radical and moderate groups communicate with each other, as was shown by their open letters (see Section 2.1; 1 Sky 2010a), the reactions on McKibben's (2012) article in *Rolling Stone* magazine, the Climate Camp Reader (Shift Magazine and dysophia 2010) or the discussions about green capitalism (see Turbulence 2009). But the way of communication sometimes is quite aggressive and not characterized by respect for each other. In particular the radical climate groups criticize their opponents in a hard and in parts even polemic way. For example, Rising Tide called the campaign TckTckTck "FckFckFck" (RTNA 2010a: 19) and MCJ (2010) intended to grant a 'Fossil Fool Award' to TckTckTck due to its relations to companies. Furthermore, the radical climate groups don't seem to accept the compromises offered by the moderate groups.

5 Forms and ways of communication in the climate movement

Finally, it's interesting to look at the platforms and channels which are used for the communication in the climate movement. Many declarations, appeals, or comments are published on the websites of the climate groups. Parts of the debates of the movement are also published in newspapers and magazines such as the *Guardian* and *Rolling Stone* magazine. In many cases these publications are written by people participating in the climate movement or at least sympathizing with it. Discussions often take place on the websites of these newspapers and magazines by the use of the comment function (see for example van der Zee 2010, 2012). Besides, there are special magazines and other publications focused on civil society that often publish articles about the climate movement. Examples are *Indymedia*, *Altermedia*, *Shift* (UK), *Red Pepper* (UK), *Schattenblick* (Germany), *Critical Currents* (Sweden), *Counterpunch* (US), or *Turbulence* (UK). These magazines and news services are often located in Great Britain – a country which seems to be particularly sensitive to questions of climate change. Exclusively focused on the climate movement are blogs such as the "climateactioncafe. A space for discussion and analysis within the global climate movement," which is no longer active, or the blog of the German NGO Klima!Bewegungsnetzwerk. Finally, there are attempts of the climate movement to publish their own newspapers. During the climate conference in Copenhagen, Carbon Trade Watch, Earthlife Africa, and the Institute for Security Studies produced the 'Climate Chronicles.' Furthermore, the newsletter 'Burning Issues,' edited by Rising Tide North America, has similarities to a newspaper as regards layout, structure, and reporting style of the articles (RTNA 2010–2011).

On the whole, the forms of communication in the climate movement are quite diverse. They are neither hidden nor all located on the Internet. Apart from print and online media, the climate movement communicates at meetings and conferences. The meetings are often held in rather small and informal contexts (see chapter by Kössler in this book; CJN! 2009; CJA 2011). The conferences of the climate movement are bigger, more public, and therefore also documented in a better way. For years so-called Powershift conferences have been organized in North America and Australia. Young activists in particular take part in these conferences. For instance, the Powershift conference in April 2011 in Washington was attended by several thousand people and dealt with the construction of a clean economy and alliances with the environmental justice movement (Powershift 2011). In February 2012, the German NGOs gegenstromberlin, Climate Justice Leipzig, and ausgeco$_2$hlt organized a conference on the topic of energy battles and social movements. The participants of this conference wanted to develop common action strategies and discussed a possible cooperation with the anti-nuclear movement (gegenstromberlin 2012). One of the biggest conferences of the climate movement so far was the Alternative Climate Summit in Cochabamba, Bolivia in 2010. In Cochabamba, members of the climate movement and representatives of Southern states developed common demands on climate politics which were formulated in the 'People's Agreement of Cochabamba' (see chapter by Kruse in this book).

The different forms and channels of communication in the climate movement underline the intensity of its discussions and debates. The climate movement is heavily engaged in internal communication as well as in public relations work. In huge parts the debates of the climate movement seem to be public and are not hidden or locked away. Consequently, the hurdles for participating in these debates, at least with minor texts and comments, are quite low.

The debates at the described conferences underline the importance of the topics strategy development and alliances for the climate movement. The movement intensively searches for contact with other actors. It shows interest in the antinuclear movement, the environmental justice movement, the Occupy movement, as well as for Southern states. The period of reorientation of the climate movement doesn't seem to be completed yet. One reason for this could be that several climate activists judge the acting of the climate movement as too weak and ineffective. As a consequence, they search for partners in order to strengthen the movement.

Furthermore, the conferences confirm the assumption that the climate movement has an academic and scientific character. At the Powershift conferences or the Alternative Climate Summit in Cochabamba, keynote speakers as well as delegates took part, and long, often formal declarations were drafted. Publications of the climate groups such as 'Space for Movements? Reflections from Bolivia on Climate Justice, Social Movements and the State' (Building Bridges Collective 2010), 'Contours of Climate Justice. Ideas for Shaping New Climate and Energy Politics' (Brand et al. 2009), or 'What Does Climate Justice Mean in Europe? A Discussion Paper' (CJA 2009) also create a scientific impression due to their length, structure, and language.

It can be concluded that several academics and scientists seem to participate in the climate movement. One reason for this could be the complex and often prognostic character that the topic of climate change has.

6 Conclusion

The analyzed debates have enabled several insights into the climate movement. The topics, forms, and ways of the debates make it possible to conclude about the state, development, and characteristics of the climate movement.

The climate movement is engaged in several debates. The level of communication is quite high in the movement. There are intensive discussions among activists of one country, but also across national borders. A very vivid debating culture exists in Great Britain. The British climate activists discuss a lot. Furthermore, British news services, magazines, and newspapers are particularly sensitive to the topic of climate change. A part of the debates of the climate movement deals with topics of everyday life, but more important are fundamental questions about the state and future of the movement. This reveals a high level of self-consciousness and reflection in the climate movement. These characteristics have developed against the background of fights over an adequate concept of acting, the crisis and reorientation of the movement in the aftermath of Copenhagen, and the high level of education of many members of the movement.

The debates of the climate movement are on many points quite critical, controversial, and complex. Despite a certain consensus about focusing on the local level and acting against the fossil fuel industry, many members of the climate movement have controversial discussions about technical and economic instruments for tackling climate change. These instruments are intensively checked on possible disadvantages and often rejected by radical climate groups. They call them false solutions. Several climate activists reject capitalism and don't trust in the ability of politics to solve problems. For example, there are many publications of the climate movement which judge carbon trading as not working and counterproductive. The critique on and selectivity against existing instruments for tackling climate change sometimes also results in the rejection of wind or solar energy projects in desert regions. Marxian and neocolonial arguments are used to justify these rejections. As a consequence, moderate climate groups, of which several exist and which support ideas such as green capitalism, are openly attacked by the radical and capitalism-critical members of the movement. Positive solutions are discussed far less often than false solutions. The radical part of the climate movement defines positive solutions by the concept of climate justice and the aim of system change. They demand a transformation of capitalism and the end of heteronomy.

The debates of the climate movement on the right strategies to fight climate change are characterized by tensions and conflicts. As a consequence, different and not yet solved lines of conflict within the movement can be identified. For example, there are the conflicts 'green capitalism versus radical critic of capitalism' as well as 'false solutions versus positive solutions.' The supporters of the different positions communicate with each other, but do this in different ways. While moderate climate groups at least in parts are consensus oriented and want to mediate the conflicts in the movement, the radical groups often act in a confrontative way. Furthermore, their anticapitalistic and partly ideological positions create the impression that several former members of the global justice movement are active in the climate movement and still hold their former convictions and aims.

The complex debates among the climate activists could result in blockades of the movement. Radical climate groups demand prevention of wrong steps in the development of the movement and just want to cooperate with other movements if this seems to be of explicit strategic use. Before acting, the climate movement should critically check its options and anticipate possible consequences. This attempt of consciously directing the movement could have negative impacts. Long considerations can weaken the ability of a movement to act. Endless discussions stand in contradiction to spontaneous and pragmatical actions. Besides, the public image of the climate movement could be damaged by the impression that discussions and the rejections of political suggestions are more important for its members than practical progress and acting. Apart from the effects of the sensitive and negative perspective of a part of the movement, the question is raised whether their intentions to direct and control social movements are realistic. The research on social movements is opposed to this assumption because it describes movements as dynamic and plural actors (della Porta and Diani 2006: 26).

Nevertheless, it is also possible to see the high level of reflection and discussion in the climate movement in a positive way. Perhaps the climate movement is currently just in a specific period of intensified discussion, which will end in the near future and lead to a stronger movement with a dominant focus on acting. Furthermore, the ability to reflect could help the movement to recognize its weaknesses and to overcome them. For example, the discussions resulting from the failure of Copenhagen led to a certain consensus about the importance of local action and therefore had a positive effect. In the future, the movement could deal with the following questions and profit from answering them: Is there a causal link between capitalism and climate change? If so, how intensive and relevant is this link? Do the former members of the global justice movement use the climate topic for their purposes? If so, does this cause conflicts and blockades of the climate movement? One result of the reflections about these questions could be the realization that climate justice (if separated from basic questions on the economic system) is probably a more attractive, consensual, and specific leading concept for the climate movement than old and well-known anticapitalistic messages.

Notes

1 As a consequence of the exposure of Kennedy, two trials against climate activists were stopped. Kennedy had taken part in the protests the activists were charged with and probably acted as an "agent provocateur" (Antebi 2011).

2 In reaction to the cited article of van der Zee more than 150 comments were written on the website of the *Guardian* (see www.guardian.co.uk/environment/blog/2010/feb/25/climate-change-movement, last accessed 12/15/2012).

3 The suggestion of a 'Green New Deal' is often related to the idea of green capitalism. A green economy should be built up by huge public investments (CCC 2011; O'Beirne 2011).

4 Rising Tide North America calls politicians "the other false solution to climate change" (RTNA 2010a: 6).

References

1 Sky (2010a) An Open Letter to All People and Organizations Working to Combat Climate Warming. Accessed 09/28/2012 from www.1sky.org/openletter.

1 Sky (2010b) Solutions: 5 Million Green Jobs. Accessed 09/26/2010 from www.1Sky.org/about/1 Sky-solutions/green-jobs.

350.org (2012) Victory: Take the Keystone Pledge. Accessed 09/07/2012 from http://act.350.org/sign/keystone-pledge/.

Antebi, T. (2011) Climate Activist Trial Collapses Over Use of Undercover Police Officer, *The Ecologist*, January 10.

ausgeco2hlt (2012) Herrschaftskritisches Barrio. Herrschaftsfrei und Spaß dabei! Aufruf zum herrschaftskritischen Barrio vom 31.7–8.8 im Hambacher Forst. Accessed 09/07/2012 from www.ausgeco2hlt.de/klimacamp/klimacamp-2012/aufrufe/herrschaftskritisches-barrio/.

Beaumont, P. (2010) Too Self-Regarding to Be Effective, *Shift* Magazine and *dysophia* (eds.) *Criticism without Critique. A Climate Camp Reader.* Accessed 06/04/2010 from http://dysophia.files.wordpress.com/2010/01/cca_reader.pdf.

Blockupy Frankfurt (2012) Es geht nicht um Eisbären – soziale und ökologische Kämpfe zusammenführen! Accessed 09/28/2012 from http://blockupy-frankfurt.org/de/blockade/aktionsthemen/oekologie.

Bond, P. (2009) Carbon Trading, Space, Time and Eco-Social Contestation. Accessed 09/28/2012 from www.durbanclimatejustice.org/articles/carbon-trading-space-time-and-eco-social-contestation.html.

Brand, U., Lander, E., Bullard, N., and Müller, T. (eds.) (2009) *Contours of Climate Justice: Ideas for Shaping New Climate and Energy Politics*, Uppsala: Dag Hammarskjöld Foundation. Accessed 09/03/2013 from www.tni.org/tnibook/contours-climate-justice.

Brunner, K. (2012) The Dangers of Divestment Campaigns. Accessed 09/28/2012 from http://climate-connections.org/2012/07/24/three-responses-to-bill-mckibbens-new-article-global-warmings-terrifying-new-math/.

Building Bridges Collective (2010) Space for Movements? Reflections from Bolivia on Climate Justice, Social Movements and the State. Accessed 09/03/2013 from http://spaceformovement.files.wordpress.com/2010/08/space_for_movement2.

Bürgerinitiative CO2ntra Endlager Neutrebbin (2012) Hintergrund. Accessed 09/07/2012 from www.co2bombe.de/joomla/index.php/hintergrund.

CAN (Climate Action Network) (2009) Views on Possible Improvements to Emissions Trading and the Project-Based Mechanisms. Accessed 09/07/2012 from http://unfccc.int/resource/docs/2009/smsn/ngo/119.pdf.

Carbon Trade Watch (2009) 350 Reasons that Carbon Trading Won't Work. Accessed 09/07/2012 from www.tni.org/article/350-reasons-carbon-trading-wont-work.

Carbon Trade Watch (2012) Green Is the Color of Money: The EU ETS Failure as a Model for the 'Green Economy.' Accessed 09/07/2012 from www.carbontradewatch.org/downloads/publications/EU-ETS_Report-web.pdf.

CCC (Campaign against Climate Change) (2011) One Million Climate Jobs. The People's Petition for a Million Climate Jobs. Accessed 09/28/2012 from www.climate-change-jobs.org/petition.

CJA (Climate Justice Action) (2009) Reclaim Power in Copenhagen! A Call to Action. Accessed 09/28/2012 from www.climate-justice-action.org/news/2009/09/10/a-call-to-action/.

CJA (Climate Justice Action) (2011) Documents. The Minutes form the Past CJA International Meetings. Accessed 09/07/2012 from www.climate-justice-action.org/resources/documents/.

CJN! (Climate Justice Now!) (2009) Report of the CJN! Strategy Meeting 4 & 5 October 2009, Bangkok, Thailand. Accessed 09/07/2012 from www.climate-justice-now.org/report-of-the-cjn-strategy-meeting-4–5-october-2009-bangkok-thail.

Climate Camp UK (2011a) Metamorphosis: A Statement from the Camp for Climate Action. Accessed 09/07/2012 from www.climatecamp.org.uk/2011-statement.

Climate Camp UK (2011b) Second Police Spy Revealed – Climate Camp Responds: We Demand Answers. Accessed 09/07/2012 from http://climatecamp.org.uk/press/2011/01/13/second-police-spy-revealed-climate-camp-responds-we-demand-answers.

Climate Collective (2012) Danish National Court Rules Mass Arrest to Be Illegal. Accessed 09/07/2012 from www.climatecollective.org/post/187.

Climatesciencewatch (2012) End Mountaintop Removal Week in Washington, June 12, 2012. Accessed 09/07/2012 from www.climatesciencewatch.org/2012/06/10/end-mountaintop-removal-week-in-washington/.

della Porta, D. and Diani, M. (2006) *Social Movements. An Introduction*, Malden, Mass.: Blackwell.

DVPW (Deutsche Vereinigung für Politische Wissenschaft) (2009) Bericht vom DVPW-Kongress 'Politik im Klimawandel,' Kiel, September 21–25, 2009, in DVPW (ed.) Rundbrief der deutschen Vereinigung für Politische Wissenschaft, 141. Accessed 09/07/2012 from www.dvpw.de/fileadmin/docs/Rundbrief141.pdf.

Eilperin, J. (2012) As Eco-terrorism Wanes, Governments Still Target Activist Groups Seen as Threat. Accessed 09/08/2012 from http://itsgettinghotinhere.org/2012/03/11/wapofbi-targets-peaceful-anti-fracking-rising-tide-activists/.

gegenstromberlin (2012) SAVE THE DATE: 'Energiekämpfe in Bewegung!?!,' 3–5. Februar 2012. Accessed 09/07/2012 from www.gegenstromberlin.net/2011/12/28/save-the-date-energiekampfe-in-bewegung-3–5-feb-2012/.

Goodell, J. (2011) Meet America's Most Creative Climate Criminal, *Rolling Stone*, July 7. Accessed

12/15/2012 from www.rollingstone.com/politics/blogs/national-affairs/meet-america-s-most-creative-climate-criminal-20110707.

Goodman, J. (2009) From Global Justice to Climate Justice? Justice Ecologism in an Era of Global Warming, *New Political Science*, 31(4): 499–514.

Greenpeace (2007) Nuclear Energy. No Solution to Climate Change. A Background Paper. Accessed 09/07/2012 from http://archive.greenpeace.org/comms/no.nukes/nenstcc.html.

Greenpeace (2008) Breaking News: Kingsnorth Six Found Not Guilty! Accessed 09/07/2012 from www.greenpeace.org.uk/blog/climate/kingsnorth-trial-breaking-news-verdict-20080910.

Hall, R. (2009) Does Judge Benson Ruling Effectively Deny DeChrstopher a Fair Trial?, *Salt Lake Tribune*, November 21.

Henn, J. (2012) Bill McKibben on Keyston XL Announcement. Accessed 09/07/2012 from www.350.org/en/node/28028.

Hickman, A. (2010) Climate Activism: Is The Trial More Important than the Protest?, *The Ecologist*, August 26.

IEN (Indigenous Environmental Network) (2009) Indigenous Peoples' Guide False Solutions to Climate Change. Accessed 09/07/2012 from www.ienearth.org/docs/Indigenous_Peoples_Guide-E.pdf.

IEN (Indigenous Environmental Network) (2010) Four Principles of Climate Justice. Accessed 09/28/2012 from www.ienearth.org/docs/IEN_4_Principles_of_Climate_Justice.pdf.

Indymedia UK (2012) 'Pigs' Will Pay for Copenhagen. Accessed 09/28/2012 from www.indymedia.org.uk/en/2012/02/491869.html.

Jacobsen, R. (2012) Boil Over, *Outside Magazine*, September. Accessed 09/28/2012 from www.outside-online.com/outdoor-adventure/natural-intelligence/Boilover.html?page=all.

Jensen, D. (2011) De Christopher Sentence Riles Protesters, Spurs Arrests, *The Salt Lake Tribune*, July 26.

Kern, F. (2012) Revival of Coal-power in Germany? Accessed 09/28/2012 from http://hitchwiki.org/fabzgy/2012/06/29/revival-of-coal-power-in-germany/.

Klimaretter.info (2011) Keystone XL vor dem Aus, 11 November. Accessed 09/07/2012 from www.klimaretter.info/protest/hintergrund/9854-keystone-xl-vor-dem-aus.

La Via Campesina (2012) 'Peasant Farming Can Cool Down the Earth': An Interview with Chavannes Jean-Baptiste, Executive Director of Mouvement Paysan de Papaye, Durban, South Africa, December 2011. Accessed 09/07/2012 from http://viacampesina.org/en/index.php?option=com_content&view=article&id=1198:peasant-farming-can-cool-down-the-earth-an-interview-with-chavannes-jean-baptiste-executive-director-of-mouvement-paysan-de-papaye-durban-south-africa-december-2011&catid=48:-climate-change-and-agrofuels&Itemid=75.

La Via Campesina et al. (2012) Open Letter from Global Social Movements in Support of Occupy Wall Street. Accessed 09/28/2012 from http://climate-connections.org/2012/09/13/open-letter-from-global-social-movements-in-support-of-occupy-wall-street/.

Leßmann-Faust, P. (2008) Einleitung, in Leßmann-Faust, P. (ed.) *Polizei und Politische Bildung*, Wiesbaden: VS Verlag.

McKibben, B. (2010) Climate Change: It's Time to Talk, and Act, Tough, *Los Angeles Times*, August 4.

McKibben, B. (2012) Global Warming's Terrifying New Math, *Rolling Stone*, August 2.

MCJ (Mobilization for Climate Justice) (2007) 10 Principles for Just Climate Change Politicies in the US. Accessed 09/28/2012 from www.ejnet.org/ej/climatejustice.pdf.

MCJ (Mobilization for Climate Justice) (2010) EYES WIDE SHUT – TckTckTck Exposé from Activist Insider. Accessed 06/06/2010 from www.actforclimatejustice.org/2010/01/eyeswide-shut-tcktcktck-expose-from-activist-insider/.

Müller, T. and Passadakis, A. (2008) 20 Theses against Green Capitalism. Accessed 09/25/2010 from http://climateactioncafe.wordpress.com/2008/12/05/20-theses-against-green-capitalism/.

Müller, T. and Passadakis, A. (2009) Green Capitalism and the Climate: It's Economic Growth, Stupid!, in Brand, U., Lander, E., Bullard, N., and Müller, T. (eds.) *Contours of Climate Justice. Ideas for Shaping New Climate and Energy Politics*, Uppsala: Dag Hammarskjöld Foundation. Accessed 09/07/2012 from www.scribd.com/doc/24224164/Contours-of-Climate-Justice-Ideas-for-Shaping-New-Climate-and-Energy-Policy.

Murray, S. (2012) It Is the Coal Barons, Not Activists, Who Threaten Society, *National Times*, January 10.

O'Beirne, M. (2011) Climate Change, Ecosocialism and the Green New Deal, *The Ecosocialist*. Accessed 09/28/2012 from www.theecosocialist.com/1/post/2011/9/climate-change-the-green-new-deal-humanity-cannot-wait-for-socialism-to-replace-capitalism.html.

Page, N. (2009) Is the Camp for Climate Action Challenging or Embodying Capitalist Consumerism? In Climate Action Cafe Blog. Accessed 06/04/2010 from http://climateactioncafe.wordpress.com/2009/07/15/is-the-camp-for-climateactionchallenging-or-embodying-capitalist-consumerism/.

Parkin, S. (2012) The Battle over Tar Sands. How Quiet Environmental Uprisings are Spreading Across the Country, *Counterpunch*, Weekend Edition, July 27–29. Accessed 09/28/2012 from www.counterpunch.org/2012/07/27/how-quiet-environmental-uprisings-are-spreading-across-the-country/.

Petermann, A. (2012) The System Will Not Be Reformed. Accessed 09/28/2012 from http://climate-connections.org/2012/07/24/three-responses-to-bill-mckibbens-new-article-global-warmings-terrifying-new-math/.

Powershift (2011) Media & Press. About Powershift 2011. Accessed 09/07/2012 from www.wearepowershift.org/media.

Pusey, A. and Russell, B. (2010) The Climate Crisis or the Crisis of Climate Politics? *Perspectives* 2010. Accessed 09/03/2013 from: http://notesfrombelow.wordpress.com/2010/03/24/the-climate-crisis-or-the-crisis-of-climate-politics/.

RTNA (Rising Tide North America) (2010a) The Climate Movement Is Dead. Long Live the Climate Movement. Accessed 03/11/2010 from www.risingtidenorthamerica.org/wordpress/wpcontent/uploads/2010/03/climatemovement_longlive.pdf.

RTNA (Rising Tide North America) (2010b) False Solutions: Mojave Desert Threatened by Massive Solar Developments, *Burning Issues*, 2.

RTNA (Rising Tide North America) (2010–2011) *Burning Issues*, Newsletter. Accessed 09/28/2012 from www.risingtidenorthamerica.org/category/rising-tide-printed-newsletter/.

RTNA (Rising Tide North America) (2011) Occupy the Climate. Accessed 09/28/2012 from www.risingtidenorthamerica.org/2011/11/occupy-the-climate/.

RTNA and CTW (Rising Tide North America and Carbon Trade Watch) (2010) Hoodwinked in the Hothouse. False Solutions to Climate Change. Accessed 09/28/2012 from www.risingtidenorthamerica.org/special/hoodwinkedv2_WEB.pdf.

RTNA (Rising Tide North America) et al. (2010) To the Board and Staff of 1 Sky. A Letter from the Grassroots. Accessed 09/28/2012 from http://itsgettinghotinhere.org/2010/10/25/open-letter-to-1-sky-from-grassroots-organizations/.

RTUK (Rising Tide United Kingdom) (2012a) Anti-nuclear Action in Finland: Olkiluoto Blockade & Camp 6–13 Aug 2012. Accessed 09/28/2012 from http://risingtide.org.uk/content/anti-nuclear-action-finland-olkiluoto-blockade-camp-6–13-aug-2012.

RTUK (Rising Tide United Kingdom) (2012b) Positive Solutions. Accessed 11/28/2012 from http://risingtide.org.uk/resources/positivesolutions.

Rucht, D. (2011) Zum Stand der Forschung zu sozialen Bewegungen, *Forschungsjournal Neue Soziale Bewegungen*, 24(3): 20–47.

Shift Magazine and dysophia (2010) Criticism without Critique. A Climate Camp Reader. Accessed 09/28/2012 from http://dysophia.files.wordpress.com/2010/01/cca_reader.pdf.

Smolker, R. (2012) Fighting Fossil Fuels Must Include Opposing False Solutions. Accessed 09/28/2012 from http://climate-connections.org/2012/07/24/three-responses-to-bill-mckibbens-new-article-global-warmings-terrifying-new-math/.

Stewart, B. (2009) Kingsnorth Climbdown Is the British Climate Movement's Biggest Victory, *Guardian*, October 8.

TckTckTck (2011) The Great Tar Sands Stand-Off. 20 September 2011. Accessed 09/07/2012 from http://tcktcktck.org/2011/09/great-tar-sands-stand-off-infographic/10520.

Tokar, B. (2010) Toward a Movement for Climate Change. Accessed 02/08/2010 from www.social-ecology.org/wp/wpcontent/uploads/2009/12/brian1.Toward_a_Movement.pdf.

Turbulence (2009) Green New Deal: Dead End or Pathway Beyond Capitalism. Discussion between Tadzio Müller and Frieder Otto Wolf. Accessed 09/28/2012 from http://turbulence.org.uk/turbulence-5/green-new-deal/.

van der Zee, B. (2010) Is the Climate Movement Splintering?, *Guardian Online*, February 25. Accessed 09/07/2012 from www.guardian.co.uk/environment/blog/2010/feb/25/climate-change-movement.

van der Zee, B. (2011) Climate Camp Disbanded, *Guardian*, March 2. Accessed 09/07/2012 from www.guardian.co.uk/environment/2011/mar/02/climate-camp-disbanded.

van der Zee, B. (2012) Is Climate Protest Making a Return to the UK?, *Guardian Online*. April 27. Accessed 09/07/2012 from www.guardian.co.uk/environment/2012/apr/27/climate-protest-return-uk.

35

SMALL ISLAND STATES AND THE NEW CLIMATE CHANGE MOVEMENT

The case of Kiribati

Silja Klepp

1 Introduction

There has been continued emphasis that aspects of climate justice need to play a bigger role in the global negotiations dealing with climate change in order to overcome the stagnancy in agreement talks (Parks and Roberts 2010). This applies to the international negotiations on climate change in the case of the UNFCCC as well as to the debates on environmental or climate migrants, in which there exist no political mandates or judicial instruments for their protection.

Even in the island state of Kiribati in the central Pacific, the issue of climate justice influences the negotiation processes around anthropogenic climate change, and its results. The group of 32 atoll islands, which makes up Kiribati, has approximately 100,000 inhabitants and a population density of 136/km^2. The islands extend over 2,500 km along the Equator, but have a very small surface area. As they lie very flat in the water and almost all inhabitants live directly at the coast, the island state and its people are especially affected by the current and predicted results of climate change, such as stronger tropical storms and the rising sea level (Mimura et al. 2007: 687). By 2100, the sea level could rise by 124 cm (Rahmstorf 2010) with some researchers even speaking of a rise of more than 200 cm (Grinsted et al. 2009). Some prognoses by climate researchers state that the islands' sweet-water budgets are at great risk of salt-water contamination, as well as declining rainfall and longer droughts (WBGU 2006). The question of climate justice and "multidimensional inequality," which Kristina Dietz discusses in relation to the effects of climate change and the relationship between the Global North and the Global South (Dietz 2009: 186), apply particularly to islands such as Kiribati. They have a limited responsibility for anthropogenic climate change, due to low per capita emissions, yet they are the most strongly affected by its results. Along with strong socio-economic problems and few opportunities for financing adaptation operations, Kiribati suffers from additional structural inequalities on various levels. These include the effects of these islands of colonial history. Kiribati belonged to the Gilbert and Ellice Islands colony, which gained its freedom in 1979 and became the free states of Kiribati and Tuvalu. Another example is the limited access to resources and social services, as well as limited participation and arrangement opportunities of international as well as regional political negotiations (see Dietz 2009: 186). Adaptation opportunities such as the change in agriculture, which the inhabitants of Kiribati have brought about for a long time, may not be enough in the future to secure their existence on these islands. The uninhabitable nature of this territory needs to be considered in the medium and long term (Barnett and Adger 2003: 326; Risse 2009). Aside from the political and legal questions, as well as conflicts around migration and resettlement projects for the inhabitants of Kiribati, questions around sovereignty and the future of the Exclusive Economic Zone (EEZ) of Kiribati remain open (Esteban and Yamamoto 2010). In relation to these perspectives, Kiribati is active in various official and unofficial global and regional arenas of negotiation around climate justice, curbing emissions, and adaptation to climate change.

Overall, the social effects of climate change, and the uncertainties that come with the prognoses made by climate researchers, are a great political challenge for Kiribati as well as all other island states. The political measures related to these prognoses reach far into the future, and must consider the interests of coming generations. The island states of the Pacific agree on the point that climate change is a serious threat for them, and that industrial nations need to reduce their emissions drastically (McNamara and Gibson 2009: 482). In order to reach this goal and exert pressure primarily on the countries causing these emissions, diverse alliances and organizations have been created in the Pacific as well as other areas. The appropriate politics in relation to future scenarios are still under dispute (McNamara and Gibson 2009: 482). While governments of certain island states view international migration as an unacceptable option for their citizens, Kiribati is already taking an active role in negotiations around what Kiribati itself calls programs for climate migration (Bedford and Bedford 2010; Klepp 2012).

Kiribati is one of the first countries in the world in which issues such as climate justice and the search for strategies for climate migrants culminate and become tangible. Here people of all classes are affected by these issues, which are of existential importance for the future of all citizens, even though the problem of poverty determines the daily lives of many *I-Kiribati* (citizens of Kiribati). The government of Kiribati has become an important actor in global climate policy as well as movements around climate because it is itself stricken by these issues. Kiribati has a special role given to it by its own government and others due to its own fate as well as the future of other countries affected by climate change, which can be viewed with ambivalence. As this article shows, through the fight against climate change, the island state gains new resources and allies on the one hand, but neglects problems on the other hand. The debates around climate change can also have far-reaching consequences such as a lack of other adaptation strategies than migration, which may have a negative impact for the preservation of the island state. From its on-site research, this article can contribute to current debates based on Kiribati's citizens' own assessments and observations.

After a short introduction on the subject of climate change discourse in the Pacific, struggles around climate justice as well as the alliances and strategies in which Kiribati and other island states are involved, will be empirically described. Even the rather locally embedded level will be integrated, as a change in norms and values, based on discourses on climate justice, is taking place here. The island of South Tarawa, which was the area of my field research in April and May 2011, is the central island of the republic of Kiribati. Debates around climate change have spread throughout Kiribati, due to the devastating predictions for the islands. In relation to this, comments such as "*Australia should pay for our sewage system, since they are responsible for climate change and should now pay compensations*" can be heard. These result from an interesting combination of the discourses around the effects of climate change, the legacy of postcolonialism, and globalization. They also show that the discussions around the social consequences of climate change cannot be conducted separately from the controversies around North–South justice, even with the successful continuation of UNFCCC negotiations in mind.

2 The Pacific islands and the climate change discourse

The predictions of the devastating consequences of climate change determine the media and academic discourse around climate change in the Pacific region. The issues of these often technically oriented debates are "geographic objects" (Barnett and Campbell 2010: 2), such as coastlines, as well as statistics about sea level rise and precipitation. Their own interpretations on climate change and possible strategies of the islands' inhabitants, which they have been traditionally and currently using, in the context of environmental change, are barely addressed (Barnett and Campbell 2010: 2). The island states are seen as marginalized, vulnerable, and in need of development, as well as small, poor, and not designed for economic growth (Kempf 2009). These constructions are an expression of constellations of knowledge and power, which were reconfigured in the context of climate change. Tools of othering, used as tools of power during the colonial period, have gained new meaning. Carol Farbotko explains that they change the political agency of these countries and create a new "eco-colonial" perspective on these islands (Farbotko 2010: 58). Above all, the atoll island states of Kiribati and Tuvalu are characterized as "*Titanic States*" (Barnett and Campbell 2010: 168). Headlines such as "*Sinking islands cling to Kyoto lifebuoy*," "*Rising sea level forcing evacuation of island country*," or "*... will Tuvalu disappear beneath the sea ...*" (quoted after McNamara and Gibson 2009: 479), dominate media discussions. According to Barnett and Campbell, the Pacific islands have become a warning symbol for the effects of climate change as well as a field experiment for Western researchers, journalists, and organizations of international cooperation, which create a distorted image of victimization for the Pacific area, as well as a scandalizing image of the effects of climate change (Barnett and Campbell 2010: 177). On the other hand, it must be stated that some governments of the island states, foremost Kiribati's President Anote Tong, play an active role in the debates around climate change, and partly contribute to the image of sinking islands, as will become apparent below. Even in the day-to-day discussions on the islands, the subject of climate change is continually discussed, even though the people often have other pressing worries in need of a short-term solution.

McNamara and Gibson state that most Pacific island states reject the role of victim or supplicant on the international stage, and demand a stronger reduction in greenhouse gases (McNamara and Gibson 2009). Carol Farbotko emphasizes the increased position of moral power of countries such as Tuvalu and Kiribati, which had an important role in the negotiations in Copenhagen in 2009. Various alliances have been created, in order to bind together the efforts of small island states, such as in UNFCCC negotiations. Here, especially the international organizations *Alliance of Small Island States* (AOSIS) and *Small Island Developing States* (SIDS) must be named, which I will describe in detail later on.

The government of the republic of Kiribati, the first in the world to do so, researches strategies for finding medium- to long-term migration possibilities for all of its citizens. Therefore, apart from reductions in greenhouse gases and comprehensive resources for adaptation services, entry into the job and education markets outside of Kiribati are a focus for the support center of the government of Kiribati. In addition, prospects are obtained for resettling entire villages across national borders (Bedford and Bedford 2010; Klepp 2012). Various political and judicial levels and state and nonstate actors, which are now described in more detail, are included in negotiation processes for new rights and resources for environmental migrants in the Pacific area as well as adaptation to climate change.

3 The case of Kiribati – a sinking island?

Like other island states in the South Pacific, Kiribati is dependent on payments from international donors and money transfers from international migrants outside of the country. Therefore, Kiribati's economy, the same as its neighbors, is often described as a MIRAB economy. MIRAB is an acronym for *Migration, Remittances, Aid, and Bureaucracy*, and points to a high dependence on international donors, international migrants, and an inflated public sector (Marsters et al. 2006). In addition, subsistence farming, especially the fisheries, are an important economic factor. World Bank estimates for the Tarawa Atoll of Kiribati alone project the costs of climate change at $8–16 million for the year 2050. This would account for 17–34 percent of Kiribati's gross domestic product (Barnett and Campbell 2010: 21).

3.1 Negotiations around climate migration

Since his election in 2003, the island president Anote Tong has been appealing to the international community in various global and regional forums, to reduce their greenhouse gas emissions and take on their responsibility for the fate of the island as "victim of climate change" and for the "climate refugees of Kiribati" (BBC 2009). At the World Environment Day of the United Nations in June of 2008 in Wellington, New Zealand, Tong stated:

> None of the atolls that make up our country are more than two meters above sea level. We will be submerged by the end of the century, as things stand right now.... However, what happens before is equally devastating. Entire villages are being relocated as the sea encroaches upon the land. People are being asked to move from places that have been home for them. It's a human tragedy.
>
> *(Times of India 2008)*

Tong strives for a gradual evacuation of the island. Today, part of the population already lives in New Zealand and other countries, and sends remittances to Kiribati. When, how, and where these evacuations will take place in the future, is a question of climate justice, according to Tong (BBC 2009).

International migration and the full evacuation of the island in the long run play an important role in the deliberations of the government of Kiribati (Kempf 2009: 191; Bedford and Bedford 2010). In relation to this, migration is propagated as an adaptation strategy. For the migration strategies of Kiribati, President Tong is formulating a long-term plan, "so that when people migrate, they will migrate on merit and with dignity" (Risse 2009: 281). Aside from migration projects of individual citizens, they call for help from neighboring countries such as Australia and New Zealand, to initiate resettlement projects. However, none of these countries have yet made a formal agreement with Kiribati. New Zealand has created a yearly quota for the Pacific islands of Kiribati, Tuvalu, and Tonga for labor migration under the name *Pacific Access Category*. In this way, 75 migrants can migrate from Kiribati to New Zealand every year (Immigration New Zealand 2005). In addition, migration programs for New Zealand have been created in the area of harvest work (*Recognised Seasonal Employment* – RSE) and for Australia for the education and employment of nurses (*Kiribati Australia Nursing Initiative* – KINA). However, neither project acknowledges any connection between climate change, environmental change, migration, and a responsibility of industrial countries (McNamara and Gibson 2009: 482).

For some time, negotiations have been made with Fiji, which focuses on the resettlement of entire villages. On the border of the round of UNFCCC negotiations in Copenhagen in 2009, the president of Fiji was positive toward this option (Bedford and Bedford 2010: 90). In 2012, negotiations started for the acquisition of 2,000 hectares of land on Vanua Levu, the second largest island of Fiji. Fiji is made up of a group of islands that are mostly made of volcanic rock and are above sea level, meaning that it is affected less by the effects of climate change than Kiribati. According to media coverage, around 500 inhabitants of Kiribati are to resettle in Vanua Levu, in order to grow fruits and vegetables and raise animals, which will be exported to Kiribati. How these plans will be carried out will be seen in the future (BBC 2012).

There are opposing views to the active negotiations Kiribati is actively participating in, from other Pacific countries, as explained by McNamara. Some Pacific island governments reject the idea of describing migration as an adaptation strategy (McNamara 2009: 482), but rather that it is a form of failed adaptation (Campbell 2010). In academic discourses, the link between migration and climate change is disputed and considered one of the greatest political challenges (see Black 2001; Jönsson 2010; Nicholson 2011). Historically seen, these relations are not doubted, but the current nation-state system makes traveling across international borders much more difficult than it was in the past. Aside from debates around the acknowledgment of climate migrants, the government of Kiribati is also involved in several other alliances and forums, which will now be described.

3.2 Kiribati's fight for climate justice – diverse strategies and alliances

Kiribati has organizations similar to the international organizations, which push for regional and international climate justice and are active in UNFCCC negotiations. First among these is the *Alliance of Small Island States* (AOSIS), which came out of the organization of *Small Island Developing States* (SIDS). AOSIS is an alliance of island and coastal states, which are nearly all vulnerable to climate change. AOSIS is the negotiating voice for the *Small Island Developing States*, at the United Nations. The group currently has 42 members from the Pacific Ocean, Africa, the Caribbean, the Indian Ocean, and the South China Sea. At the UNFCCC negotiations and the discussions for the Kyoto Protocol, the AOSIS representatives are among the tireless advocates calling for effective climate protection policies. From the SIDS the 'Malé Declaration on Global Warming and Sea Level Rise' is a pioneer, which was brought into being in 1989 and was recreated by SIDS in 2007 as the 'Malé Declaration on the Human Dimension of Global Climate Change.' The declarations of SIDS and AOSIS express that their member countries have made relatively little contributions to climate change, but are now strongly affected by its consequences. They demand to be provided with the necessary tools and resources to self-determinedly deal with the effects of climate change and make their own decisions. Even within regional alliances, such as the *Pacific Island Forum* (PIF), the question of climate justice is always addressed, as in the 'Niue-Declaration on climate change' which was created in the frame of the 39th PIF meeting in 2008.

In addition, several religious institutions and actors are active in the debate around climate justice in the Pacific area. Representatives of Christian churches, which are organized in the *Pacific Conference of Churches* (PCC), which has its headquarters in Suva, Fiji, published the 'Moana-Declaration' in 2009, which calls for climate justice and contains conventions for environmental migrants and solutions for their resettlement and migration projects. The call is aimed explicitly at the countries of the *Pacific Islands Forum*, whose members include Australia and New Zealand. Here it becomes apparent that in the area of soft-law instruments, defined by Francis Snyder as "*those rules of conduct which, in principle, have no legally binding force but which nevertheless may have practical effect*" (Zerilli cited by Snyder 2010: 7), a growing number of instruments related to the debate around climate change and justice exist.

A large amount of campaigns for climate justice, in which Kiribati is involved, is initiated by NGOs. This has doubtlessly promoted the debates around climate justice and the vulnerability of Pacific island states. However, the NGOs' tendency to describe the islanders as victims of climate change has been criticized (McNamara and Gibson 2009: 479). One such example is a report written by the *Friends of the Earth Australia* in which they state that "the future of the Pacific Island states seems bleak" (cited in McNamara and Gibson 2009: 479). This representation of island states, which are often described as having no chance of survival, can also be politically harmful. Investments which would otherwise have been made might be withheld. In addition, Barnett and Adger point out that a too one-sided focus on doomsday scenarios might get in the way of directing the discourse toward adaptation strategies (Barnett and Adger 2003: 330). These victim discourses may produce numerous side effects, which I will detail more below.

NGOs such as *Many Strong Voices* (MSV) play a very different role. This coalition between the *Small Island Developing States* and the coastal arctic inhabitants, which was founded with the goal to fight climate change and its side effects, focuses on empowerment. Various actors of civil society and government are brought together, and experiences around climate change adaptation and political fights around climate justice are exchanged. In Kiribati, debates around climate change and climate justice take place primarily in church groups of various denominations, as the inhabitants of the island are very religious and, through my observations, there seems not be much civil engagement outside of this.

The government of Kiribati is also represented in the large international movement for climate justice, which focuses on human rights and the cultural and environmental rights of indigenous peoples and minorities, and was formed since the beginning of the twenty-first century (Tokar 2010: 8). In the frame of the official UNFCCC climate negotiations, this leads for example to the implementation of alternative events, to draw attention to the limited participation of smaller countries and indigenous peoples, and the opportunities they have for gaining influence (Baer and Reuter 2011: 3). The activities of this movement, whose members and actions are illustrated in this book, are often initiated by NGOs. For example, the network, *Climate Justice Now!* was created in the frame of the UNFCCC negotiations in 2007 in Bali by more than 30 NGOs and grassroots organizations. Similar movements joined at the People's Summit Klimaforum09 in Copenhagen in 2009, including *Climate Justice Now!* In the case of the *People's Conference on Climate Change and the Rights of Mother Earth*, in April 2010 in Bolivia, the government of Bolivia initiated the congress for a large movement for climate justice (see Kruse in this book). In this case, the same as with the aforementioned organizations and forums, the creation of a counter discourse against the UNFCCC negotiations was attempted, which did not focus on a reduction in greenhouse gases and a solution for climate change through a 'greening' economy, but rather demands a radical system change. *The People's Agreement*, created at the congress, demands far-reaching rights of self-determination such as access to water, land rights, and food production "through forms of production that are in harmony with Mother Earth and appropriate to local cultural contexts."

A growing number of voices in the Pacific area, in the context of the debates around climate justice, are demanding a new transnational solidarity and unity, to reduce the effects of climate change. One example of the new movements taking up this argument is *Pacific Voyaging*. In this case, traditional cross-border sea voyages are used in the creation of large sailing groups, coming from various countries in the Pacific, to raise awareness about the effects of unsustainable industry, climate change, and other environmental problems (Farbotko 2012). One demand, linked to postcolonial arguments, sees a dual responsibility for the former colonizing countries. Since they were the cause of anthropogenic climate change as well as the inequality and injustice that began during the colonial period, the former colonial powers are obliged to take in the inhabitants of the threatened islands of the Pacific (see also Campbell 2010: 25). Political arguments, based on the colonial period and tied to global discourses, are used, which connect to legal arguments.

The German social anthropologist Wolfgang Kempf, following Lazrus (2009) underlines the importance of acknowledging and implementing the abilities, experiences, and networks of the people of Oceania in the fight for climate justice and the adaption to climate change (Kempf 2009: 195). Too often, top-down measures, which neglect cultural and regional aspects, have been implemented in the climate change-adaptation projects in the Pacific area, and have worsened the situation.

In the demand for climate justice, especially in relation to financial compensation, increased resources are being set aside for countries particularly threatened by climate change (Tanner and Allouche 2011: 4). Over the last decades the international negotiations have, in this case, gained ground in the stagnating rights debates. Since 2003, infrastructure and sensitization programs for adaptation to climate change in Kiribati have been carried out, through the *Kiribati Adaptation Program* (KAP), with financial assistance from the *Global Environment Facility* (GEF) and other organizations and donors. Next is the *National Adaptation Program of Action* (NAPA), which is the cornerstone of Kiribati adaptation policy on a national level. However, even with a community-based approach, these programs are also criticized in that they are of too little use to the affected communities and inhabitants (Storey and Hunter 2010: 176).

Against the background of climate change, possibilities of an ever more fragmented political and legal landscape are being used. Sachs emphasizes the importance of dealing with climate change as a decidedly rights based issue, meaning that the problems, brought forth in often technically framed questions from scientific experts, should be reframed as human rights issue (Sachs 2009). The main actors, above all the government of Kiribati as well as the advisors and transnational networks and NGOs, who act as *knowledge brokers* (Merry 2006: 40), base themselves on concepts such as human rights, which have a global reach and can be used in moral and legal arguments. These *knowledge brokers* (Merry 2006: 40), who explain climate change, its effects in the area, and are supposed to mediate between the different actors during the project work, have gained importance in Kiribati, as almost every project and even the government of Kiribati itself, employ such *consultants*. They often bring these so-called 'universals,' who build the core of the current humanitarian projects (Tsing 2005: 7), into the debates. They increasingly become deciding factors in the context of climate change debates and demands for 'climate justice.' Interviews in Kiribati, conducted in April 2011, show the present state of the discourse around 'climate change' in Kiribati. Kaia Miller, an employee of the environmental department states that "The head of the unit, the minister actually said that ... 'Australia should buy us this and that, because climate change is all their fault'."[1] Here, delegations of responsibility for climate change, and the financial and legal demands that go with them, become clear.

3.3 Climate justice on the ground

Even on the local level, where the public and private spheres in the small island state of Kiribati often overlap, the interpretations and perceptions of the climate change debates are marked by discussions around climate justice. Since the settlement of the island group, the inhabitants of Kiribati, the *I-Kiribati*, have been confronted with a very dynamic environment. Environmental changes and climate variability are closely linked to spiritual ideas (Di Piazza 2001: 35). A living mythology exists in Kiribati, which includes the flora, fauna, weather phenomena, and the environment. Local patterns are based on cultural symbols and categories, which allow people to understand the changes in their environment, and to assess and influence them. In the frame of extreme situations and climate change, the meaning of *traditional ecological knowledge* (TEK) is often accentuated (Lazrus 2009). Berkes describes TEK as traditional knowledge and behavior passed down through generations, based on the relationship between individuals and their material and living environments (Berkes 1993: 3).

Interpretation systems such as TEK and their inherent ideals and norms not only influence the awareness and understanding of environmental impressions, but also the *agency* of the affected people, in the context of extreme situations (Kuruppu 2009: 805). They are defeated by continuous change processes and mirror changing power constellations (Lazrus 2009). Before the colonial period, important decisions were made in village assemblies (*Maneabas*), which were structured through age groups (Di Piazza 2001). Today several decisions fall to the democratically elected elites and above all religious leaders (Kuruppu and Liverman 2011). While the Catholic Church describes climate change as anthropogenically caused and actively advocates for adaptation services, the Seventh Day Adventists, for example, surmise that climate change is a punishment for humanity's sins and must, therefore, be passively accepted.[2] On a local level aside from Christian confessions, the government of Kiribati, local and regional media, and NGOs, influence the discourse around climate change. Media and NGOs partly construct doomsday scenarios, similar to those posited by Seventh Day Adventists. However, they support these scientifically. Aspects of climate change and the responsibility of emitting countries play a big role in the diverse adaptation workshops offered by the church, as well as other organizations for developmental cooperation and NGOs. However, the blame placement remains relatively diffuse, and lacks concrete rights based argument.

In addition, according to Connell, the danger in Kiribati is that the government of Kiribati uses climate change as a sort of 'garbage can,' in order to divert from failed policies (Connell 2003: 103). Environmental dangers are increasingly described as exogenous factors, which are out of the political control of the Pacific island states (Storey and Hunter 2010: 172). Therefore, climate change is described as a cause for various local environmental changes, often due to a lack in resource management. Ecological degeneration no longer counts as inadequate resource management, but as rather outwardly caused by climate change (Connell 2003: 103). Connell (2003) and Storey and Hunter (2010) assume that through the partly dramatic climate prognoses, the effects of non-sustainable development maintain a small media presence and get pushed into the background. Furthermore, Kuruppu and Liverman (2011) accuse the government of Kiribati of abusing the adaptation discourse to reproduce power dynamics in relation to local actors.

Initial studies indicate that many actors feel increasingly helpless toward predicted climate change. In spite of everything, few *I-Kiribati* are ready to emigrate (Kuruppu and Liverman 2011: 665–667). What stands out in Kiribati is that the government mainly brings forth the negotiations around climate migration. As various interview partners in Kiribati explained, on the other hand in the above-mentioned forums, the fight for climate justice is supported and carried out by a large part of the population.

4 Conclusion

The government of Kiribati has decided to proactively address the social effects of climate change, which are discussed as devastating for the island state, and plan a future for its citizens outside of Kiribati. What is critical in the case of islands such as Kiribati, which have to invest in adaptation services, is to pursue various adaptation strategies. This means a combination of tactics and a wide range of adaptation services in the fields of agriculture, water management, and coastal protection, as well as the support of citizens' individual migration projects in addition to state planned resettlements of village communities (Bedford and Bedford 2010: 93). Not all of these measures should be carried out by Kiribati alone, but through the lens of climate justice, demanding the support from various regional and global institutions and actors.

Overall, it becomes apparent, on the one hand, that due to the climate change discourse, more resources are being set aside for the development of Kiribati and the adaptation to climate change in the frame of international cooperation. On the other hand, there are also tendencies, which point out changes in norms and awareness of the government and inhabitants, such as the '*garbage can*' phenomenon, which could be unfavorable for the development of Kiribati. The change in knowledge, values, and norms, which seems to go hand in hand with the climate change discourse, influences processes and practices within the country, but will still play an important role in transnational negotiation processes for new resources and rights, such as in the form of new expectations. These negotiation processes have multilayered effects on the sustainable development of Kiribati. The negotiated migration programs, as well as victim debates could produce side effects, which are difficult to control. For policymaking, the challenge comes in creating positive conditions for large changes for the future of Kiribati, which at the same time will not negatively affect the development of Kiribati.

The representation of various forums and strategies, which Kiribati uses, shows that the government, the elites, such as church leaders, and civil society in Kiribati, connect to the diversity of international climate change and official climate negotiations. One option for Kiribati's commitment, in accordance with the division of the movement, which is asserted in the analyses of the global movement for climate justice, cannot be observed. Kiribati receives support from Western groups, which seek to mitigate climate change through a modernization of the economy, as well as from various NGOs and grassroots groups from the Global South and North, which seek an increased system change, and joins their campaigns. The island state is also represented in international organizations and takes part in official UNFCCC negotiations.

The negotiation efforts, which focus on a life without outside influence for current and future generations of *I-Kiribati*, a reduction of emissions, and on a fair distribution of adaptation services, are still in the early stages. They are often influenced by the negative attitude of countries that could potentially donate or take in citizens, above all the richer OECD countries, in the area of New Zealand and Australia. However, the successes of Kiribati need to be recorded, and instead of mainly subscribing to the role of victim of climate change, the government succeeds in achieving results through various negotiations, which have a positive effect on the situation of *I-Kiribati*, such as remittances from migration programs and increased financial support for adaptation projects.

In the question of the importance of Kiribati in the climate movement, the role of the island state becomes clear through a view on the circulating images and discourses, described above. A change in policy, as well as other areas, needs to be altered with stories of sinking islands and partly exaggerated horror scenarios. With this, Kiribati and other atoll islands become argumentative resources, fitting perfectly into the globalized media world, but which goes along with a relative public callousness and an erosion of the symbolic 'sinking island states.' For Kiribati, the side effects of the discourse are wide reaching and cannot be overlooked.

Notes

1 Interview with Kaia Miller, employee of the environmental department of Kiribati, on April 14, 2011 in her office. The name has been changed on request.

2 Interview with Catholic nun in Bairiki, Kiribati on April 15, 2011.

References

Baer, H. A. and Reuter, T. A. (2011) The Global Movement for a Safe Climate and Environmental Sustainability, *The Australian Journal of Anthropology*, 22: 255–277.

Barnett, J. and Adger, W. N. (2003) Climate Dangers and Atoll Countries, *Climatic Change*, 61: 321–337.

Barnett, J. and Campbell, J. (2010) *Climate Change and Small Island States: Power, Knowledge and the South Pacific*, London: Earthscan.

BBC (2009) The President's Dilemma: Should Kiribati's President Anote Give in to Climate Change? International Fund for Agricultural Development (IFAD), first published in the program 'Life on The Edge,' September 2, 2009.

BBC (2012) Kiribati Mulls Fiji Land Purchase in Battle Against Sea. Accessed 11/19/2012 from www.bbc.co.uk/news/world-asia-172958623.

Bedford, R. and Bedford, C. (2010) International Migration and Climate Change: A Post-Copenhagen Perspective on Options for Kiribati and Tuvalu, in Burson, B. (ed.) *Climate Change and Migration – South Pacific Perspectives*, Wellington: Institute of Policy Studies.

Berkes, F. (1993) Traditional Ecological Knowledge in Perspective, in Inglis, J. T. (ed.) *Traditional Ecological Knowledge. Concepts and Cases*, Ottawa: International Development Research Center.

Black, R. (2001) Environmental Refugees – Myth or Reality? New Issues in Refugee Research, Working Paper 34, Geneva: United Nation High Commissioner on Refugees.

Campbell, J. (2010) Climate Change and Population Movement in Pacific Island Countries, in Burson, B. (ed.) *Climate Change and Migration – South Pacific Perspectives*, Wellington: Institute of Policy Studies.

Connell, J. (2003) Losing Ground? Tuvalu, the Greenhouse Effect and the Garbage Can, *Asia Pacific Viewpoint*, 44: 89–107.

Dietz, K. (2009) Prima Klima in den Nord-Süd-Beziehungen? Die Antinomien globaler Klimapolitik: Diskurse, Politiken und Prozesse, in Burchardt, H.-J. (ed.) *Nord-Süd-Beziehungen im Umbruch. Neue Perspektiven auf Staat und Demokratie in der Weltpolitik*, Frankfurt/Main: Campus.

Di Piazza, A. (2001) Terre d'abondance ou terre de misère. Représentation de la sécheresse à Nikunau (République de Kiribati, Pacifique central), *L'Homme*, 157: 35–57.

Esteban, M. and Yamamoto, L. (2010) Vanishing Island States and Sovereignty, *Ocean and Coastal Management*, 53: 1–9.

Farbotko, C. (2010) Wishful Sinking: Disappearing Islands, Climate Refugees and Cosmopolitan Experimentation, *Asia Pacific Viewpoint*, 51(1): 47–60.

Farbotko, C. (2012) Skilful Seafarers, Oceanic Drifters or Climate Refugees? Pacific People, News Value and the Climate Refugee Crisis, in Moore, K., Gros, B., and Threatgold, T. (eds.) *Migrations and the Media*, New York: Peter Lang Publishing.

Grinsted, A., Moore, J. C., and Jevrejeva, S. (2009) Reconstructing Sea Level from Paleo and Projected Temperatures 200 to 2100 AD, *Climate Dynamics*, 34(4): 461–472.

Immigration New Zealand (2005) Pacific Access Category. Accessed 02/20/2012 from www.immigration.govt.nz/migrant/stream/live/pacificaccess.

Jönsson, G. (2010) The Environmental Factor in Migration Dynamics – a Review of African Case Studies. Working Paper 21, Oxford: International Migration Institute, University of Oxford.

Kempf, W. (2009) A Sea of Environmental Refugees? Oceania in an Age of Climate Change, in Hermann, E., Klenke, K., and Dickhardt, M. (eds.) *Form, Macht, Differenz: Motive und Felder ethnologischen Forschens*, Göttingen: Universitätsverlag Göttingen.

Klepp, S. (2012) *Klimawandel und Mobilität. Rechtliche Diskurse und Lösungsansätze für Umweltmigranten im Pazifikraum*, artec, Forschungszentrum Nachhaltigkeit, Universität Bremen, Artec-Paper 180, July 2012.

Kuruppu, N. (2009) Adapting Water Resources to Climate Change in Kiribati: The Importance of Cultural Values and Meanings, *Environmental Science and Policy*, 12(7): 799–809.

Kuruppu, N. and Liverman, D. (2011) Mental Preparation for Climate Adaptation: The Role of Cognition and Culture in Enhancing Adaptive Capacity of Water Management in Kiribati, *Global Environmental Change*, 21(2): 657–669.

Lazrus, H. (2009) The Governance of Vulnerability: Climate Change and Agency in Tuvalu, South Pacific, in Crate, S. and Nuttal, M. (eds.) *Anthropology and Climate Change: From Encounters to Actions*, Left Coast Press, Walnut Creek, US.

Marsters, E., Lewis, N., and Friesen, W. (2006) Pacific Flows: The Fluidity of Remittances in the Cook Islands, *Asia Pacific Viewpoint*, 47(1): 31–44.

McNamara, K. E. (2009) Conceptualizing Discourses on Environmental Refugees at the United Nations, *Population & Environment*, 29(1): 12–24.

McNamara, K. E. and Gibson, C. (2009) 'We Do Not Want to Leave our Land': Pacific Ambassadors at the United Nations Resist the Category of 'Climate Refugees,' *Geoforum*, 40(3): 475–483.

Merry, S. E. (2006) Transnational Human Rights and Local Activism: Mapping the Middle, *American Anthropologist*, 108: 38–51.

Mimura, N., Nurse, L., McLean, R. F., Agard, J., Briguglio, L., Lefale, P., Payet, R., and Sem, G. (2007) Small Islands, in Parry, M. L., Canziani, O. F., Palutikof, J. P., van der Linden, P. J., and Hanson, C. E. (eds.) *Climate Change 2007: Impacts, Adaptation and Vulnerability. Contribution of Working Group II to the Fourth Assessment Report of the Intergovernmental Panel on Climate Change,* Cambridge: Cambridge University Press.

Nicholson, C. T. (2011) Is the 'Environmental Migration' Nexus an Analytically Meaningful Subject for Research?, COMCAD Working Paper. COMCAD – Center on Migration, Citizenship and Development, Bielefeld.

Parks, B. C. and Roberts, T. J. (2010) Climate Change, Social Theory and Justice, *Theory, Culture & Society*, 27(2–3): 134–166.

Rahmstorf, S. (2010) New View on Sea Level Rise, *Nature Reports Climate Change* 4 (April).

Risse, M. (2009) The Right to Relocation: Disappearing Island Nations and Common Ownership of the Earth, *Ethics & International Affairs*, 23(3): 281–299.

Sachs, W. (2009) Cliamte Change and Human Rights, in Brand, U., Lander, E., Bullard, N., and Müller, T. (eds.) *Contours of Climate Justice. Ideas for Shaping New Climate and Energy Politics*, Critical Currents 6, Occasional Papers Series, Uppsala: Dag Hammerskjöld Foundation.

Snyder, F. (2010) *The EU, the WTO and China: Legal Pluralism and International Trade Regulation*, Oxford and others: Hart.

Storey, D. and Hunter, S. (2010) Kiribati: An Environmental 'Perfect Storm,' *Australian Geographer*, 41(2): 167–181.

Tanner, T. and Allouche, J. (2011) Towards a New Political Economy of Climate Change and Development, *IDS Bulletin*, 42(3): 1–14.

Times of India (2008) Tiny Atoll in Pacific Cries Out for Help. Accessed 09/03/2010 from http://timesofindia.indi-atimes.com/World/USA/Tiny_atoll_in_Pacific_cries_out_-for_-help/articleshow/3104328.cms.

Tokar, B. (2010) *Toward Climate Justice. Perspectives on the Climate Crisis and Social Change*, Porsgrunn: AK Press.

Tsing, A. (2005) *Friction: An Ethnography of Global Connection*, Princeton: Princeton University Press.

WBGU (Wissenschaftlicher Beirat der Bundesregierung Globale Umweltveränderungen) (2006) Die Zukunft der Meere – zu warm, zu hoch zu sauer, Berlin.

Zerilli, Filippo M. (2010) The Rule of Soft Law: An Introduction, *Focaal*, 56: 3–18.

36

AMBIGUOUS INVOLVEMENT: CIVIL-SOCIETY ACTORS IN FOREST CARBON OFFSETS

The case of the Climate Community and Biodiversity Standards (CCB)

Jonas Hein and Heiko Garrelts

1 Introduction

The ratification of the Kyoto Protocol led to the breakthrough of market-based instruments to fight climate change (McAfee 1999; Newell and Paterson 2010). International nongovernmental organizations (NGOs) allied in the Climate Action Network (CAN) initially criticized the market-based mitigation instruments of the Kyoto Protocol. Later on, many members of CAN subsequently adopted these instruments (Holz 2010; Unmüßig 2011; see also Bedall and Görg in this book). Currently, NGOs are taking on advisory, monitoring, and control functions in the further development of offsetting markets, and are even involved in the development of green financial products and designing emission reduction certificates for offset markets (Rest 2011: 87). These activities are primarily taking place on voluntary markets and, to a lesser extent, within the Clean Development Mechanism (CDM). By drafting *standards* and *safeguards* for the voluntary market, NGOs are providing the opportunity to trade emission permits in the first place. These standards are a decisive factor in the development of voluntary carbon markets as they allow emissions reductions to be compared and quantified.

We argue that these new transnational and private governance processes lead to new actor constellations between NGOs and market players and between NGOs and local communities (the latter being increasingly and equally affected by climate change and climate policy). Among others, this raises questions concerning the in- and output legitimacy of standards and safeguards for forest carbon offsets. Input legitimacy (see Klinke 2006: 44–46) refers to issues of access, participation, and influence. How are political processes shaped and which actors have the opportunity to participate? Output legitimacy refers to problem solving capacities (see Klinke 2006: 116–124) in our case of market-based instruments. Are forest carbon offsets "neutralizing" emissions? Have additionally pursued co-benefits such as sustainable development and biodiversity goals been achieved? And, finally, which implications do these developments have for the climate movement in general?

This contribution focuses on forest carbon offsets as an example of market-based instruments in international climate policy. These markets allow companies or individuals to "neutralize" their greenhouse gas emissions by investing in projects that reduce such emissions (Neeff et al. 2009: 7). This article proceeds as follows. First we outline various aspects and dimensions of offsets and offset markets. Then we analyze the role of standards for the functioning of offset markets taking the Climate Community and Biodiversity Standards (CCB) as an example for private offset standards. Two specific cases in Indonesia illustrate the problems that arise in implementing such projects locally. Finally, a number of conclusions are drawn. This article primarily adopts a secondary analysis approach, although the findings on Indonesia were obtained using other qualitative and empirical social research methods.[1]

2 The offsetting rhetoric and reality

Offsets are part of Kyoto Protocol mechanisms for achieving greater flexibility. The Clean Development Mechanism (CDM) permits industrialized countries (Annex 1 countries) to invest in emission-reduction projects in developing countries, allowing these industrialized countries to meet their national emissions reduction targets (UNFCCC 2007: 28; Bumpus and Liverman 2008: 203). In addition to CDM and emissions trading between Annex B countries, a large number of other carbon markets with binding emissions targets (cap and trade markets) have emerged over the last few years. The European Union (EU), New Zealand, Australia, and a number of US states, including California, run carbon markets that allow the trade of emission permits from offset projects (Corbera and Brown 2010: 1739). Emerging economies such as Indonesia and Thailand are also planning to develop carbon markets.

At the end of the 1980s, long before the Kyoto Protocol was signed, a number of private companies and NGOs primarily from the United States, started to invest on a voluntary basis in offset projects (Neeff et al. 2009: 8), thereby creating a carbon market for offsets[2] (Bumpus and Liverman 2008: 203; Corbera and Brown 2010). Typical CDM offset projects store and break down nitrous oxides and methane from landfill sites, replace conventional fossil fuel power stations with renewable energy sources, and invest in reforestation (Bumpus and Liverman 2008: 129). Offset projects on the voluntary market also permit investments in Reducing Emissions from Deforestation and Forest Degradation (REDD) projects. They convert emission reductions achieved through a wide range of projects and in many different sectors into tradable emissions permits (Gilbertson and Reyes 2009: 13). Thus, emission reductions become commodities. This complex process involves different actors such as standard developers, auditors, and supervisory authorities on different scales (Bumpus and Liverman 2008: 205).

The starting point of any offset project is the projection of a business as usual (BAU) or baseline scenario. A BAU scenario is a projection of greenhouse gas emissions without intervention of the offset project for a given time period (Bumpus 2011a: 615).[3] The difference between the BAU and the greenhouse gas emissions following project implementation represents the emissions reduction achieved by the project. The amount of emission reductions achieved by the project determines the amount of tradable emission permits.

In order to ensure that the project leads to effective and measurable emissions reductions, a number of mandatory verification steps for CDM projects have to be performed. Auditors from third parties have to verify that the offset is leading to additional emission reductions which would not happen anyway. The Executive Board of the CDM approves the auditing process and registers the project. The Executive Board of the CDM is accountable to the Conference of the Parties, serving as the meeting of the Parties to the Kyoto Protocol (3/CMP.6; UNFCCC 2012b). Offset projects on the voluntary market have to follow a similar procedure although they are not proved by the Executive Board of the CDM. By their voluntary nature, they have no legal obligation to report to any regulatory authorities.

In addition to emissions reductions, the offset concept has been associated with development and environmental goals and with the key principle of "common but differentiated responsibilities" (Boyd et al. 2009; UNFCCC 1998). According to Article 12 of the Kyoto Protocol, CDM is not only designed to reduce emissions, but also to support sustainable development processes in developing countries, for example, through the transfer of low carbon technologies (UNFCCC 1998, 2012a). Whereas emission reduction criteria are well defined by the UNFCCC, no clear criteria to measure sustainable development are defined by the UNFCCC (Gonzalez and Schomerus 2010: 14). Even though the voluntary carbon market and its offset projects are not governed by the UNFCCC or by national regulations (McAfee 2012: 113) they follow the same 'win-win' discourse and are supposed to combine poverty reduction, the provision of sustainable development, and climate change mitigation (CCBA 2008a; Neeff et al. 2009: 4).

Critical voices, many among them rooted in civil society, deny the 'win-win' rhetoric. The World Wide Fund for Nature (WWF) supports offsets as a tool for promoting low carbon development in the South, but criticizes large-scale offset projects that store and break down nitrous oxides and methane from landfill sites, leading to cost-effective emissions reductions with barely any benefit to local populations (WWF 2007). The WWF highlights that emission permits produced through large-scale offsets can have negative implications for local development since they reduce incentives for investments in renewable energy production or in project types with larger benefits for local communities. The Carbon Trade Watch argues that offset projects are creating revenues for polluting industries in the South while letting companies in the North maintain emission intensive activities (Carbon Trade Watch 2009). Having said this, offset projects may hinder the transformation towards a sustainable and low-emission economy in North and South (WWF 2007; Carbon Trade Watch 2009; Gilbertson and Reyes 2009: 13). Offset projects allow emission intensive companies and economies to maintain high emission levels while claiming that they are contributing to climate change mitigation (Gilbertson and Reyes 2009: 13). Offset projects are not necessarily leading to emission reductions since the buyer of the emission permits does so in order to avoid own emission reductions (Gilbertson and Reyes 2009: 53).

Forest carbon offsets are specific cases since they do not avoid fossil emissions. Rather, they aim to sequester additional carbon dioxide and to avoid deforestation. However, according to Wunder (2008) and Grainger et al. (2009), forest carbon offsets frequently displace deforestation geographically instead of avoiding deforestation totally (Wunder 2008: 67; Grainger et al. 2009: 974). Therefore, they are not providing additional emission reductions. Indigenous and smallholder-farming associations such as La Via Campesina, and NGOs such as Friends of the Earth International warn that forest carbon offsets lead to additional exclusions of marginalized groups if land rights of smallholders and forest dwellers are neglected (McAfee 1999; Phelps et al. 2010).

Finally, the uneven geographic distribution of existing CDM projects has been criticized. Forty-six percent of all registered CDM projects are located in China (UNFCCC 2011a: 5) and 35 percent of all voluntary market projects in North America (Peters-Stanley et al. 2011: 22). By contrast, just 5 percent of the voluntary offset projects and 3.3 percent of CDM projects (Africa and Middle East) are located on the African continent (Boyd et al. 2009: 822; Peters-Stanley et al. 2011: 21).[4] This distribution does not reflect development priorities. Nevertheless, this uneven distribution is not contradictive to the goals of the Kyoto Protocol, which clearly call for emission reductions in the most cost-effective locations (UNFCCC 1998).

3 Development of offset governance by NGOs

Carbon trade and offset mechanisms are based on the assumption that markets automatically deliver the most efficient mitigation options (McAfee 2012: 113). For protecting common goods through market-based instruments, complex carbon standards as control mechanisms have been developed (Fletcher 2010: 173; Merger and Pistorius 2011: 5). These mechanisms have to ensure that emissions reductions achieved in the different offset projects can be compared, quantified, and, as a result, traded (Merger and Pistorius 2011: 9). These standards consist of rules and regulations and are the basic prerequisites for the proper functioning of carbon markets (Merger and Pistorius 2011: 7). Market trading in emissions permits is a particularly abstract form of exchange. The traded goods (emissions permits) must first be created through a process of certification and verification, which is carried out by a number of nonstate actors such as audit companies and NGO-company hybrids such as the Rainforest Alliance (and state actors in the case of the CDM) on the basis of commonly defined standards. Standards are designed to reduce the information deficit between buyers and sellers (Eden 2011: 172), and to provide buyers with a guarantee that the offset projects are effectively reducing emissions. CDM is equipped with a formal structure of standards, regulations, and criteria that are covered by international agreements and managed by the signatory countries of the Kyoto Protocol (Bumpus and Liverman 2008: 138). Carbon standards for offset projects on the voluntary market have been developed by private companies, academic institutions, and NGOs in order to create comparable and tradable emission permits. These voluntary market standards claim to measure emission reductions reliably. Moreover, they often include sustainability safeguards. In the case of the CDM, emission permits are created according to UNFCCC-backed methodologies, but the absence of sustainability safeguards for ensuring compliance with Article 12 of the Kyoto Protocol (Gonzalez and Schomerus 2010: 15) encouraged NGOs to develop additional safeguards for CDMs. The general absence of formal regulation on the voluntary markets and the absence of sustainability safeguards for CDM have led to a high demand for credible standards (Neeff et al. 2009: 18). In order to close this "governance gap" and to encourage a critical, yet supportive approach to CDM (Holz 2010: 128), the Gold Standard[5] was developed in 2003 under the aegis of the WWF, SouthSouthNorth (SSN), and HELIO International, with the involvement of a wide range of other environmental groups (Gonzalez and Schomerus 2010: 15; Holz 2010: 128). This standard certifies CDM projects and voluntary market projects, and is designed to improve the credibility of offset projects and to ensure that they have a positive impact on all aspects of sustainability. Two years later, the Climate, Community and Biodiversity Standards (CCB Standards) were developed by the Climate, Community and Biodiversity Alliance (CCBA), which comprises international NGOs, global companies, and academic institutions. The CCB Standards and the Gold Standard both aim to link climate change mitigation projects with development goals and efforts to reduce poverty.

The most common standards for forest carbon offset projects on the voluntary market are the Verified Carbon Standard (VCS) and the CCB Standards introduced above (and discussed in detail below). The VCS lacks criteria for measuring the potential social and environmental impact of offset projects, and simply measures greenhouse gas emissions. Nonetheless, compliance with local and national environmental legislation is required (Kollmuss et al. 2008: 60). VCS was established in 2005 by the NGOs; The Climate Group, the International Emissions Trading Association (IETA), and the World Business Council for Sustainable Development. VCS' status is that of a nonprofit NGO whose objective is to provide quality assurance on the voluntary carbon market.[6]

4 The Climate, Community and Biodiversity Standards (CCB Standards)

The CCB Standards is an integrated project design standard for land-based offset projects.[7] Members of the CCBA are the NGOs Care, Conservation International, The Nature Conservancy, Rainforest Alliance, and the Wildlife Conservation Society, a number of research institutions,[8] as well as companies such as the oil giant BP, chip producer Intel, the forestry company Weyerhaeuser (CCBA 2010b), and the carbon consultancy provider GFA Envest. CCBA seeks to ensure that offsets certified against the CCB Standards reliably reducing greenhouse gas emissions have a positive impact on the living conditions of local populations and help to protect endangered species (CCBA 2008b: 7).[9]

The CCB Standards provide criteria for assessing Reducing Emissions from Deforestation and Degradation (REDD) projects, reforestation activities (A/R), and agroforestry offsets (CCBA 2008b: 2). The Standards' aim is to minimize negative tradeoffs (CCBA 2008b: 4) and ensure that maximization of carbon storage does not come at the expense of the local population and biodiversity (see Putz and Redford 2009: 400; Long 2010; Sikor et al. 2010). According to CCBA, the criteria of the Standards can be used as a guideline for project developers and as quality seal for customers who are searching for offsets that are combining mitigation with biodiversity conservation and local development (CCBA 2008b: 7; Neeff et al. 2009: 26). Additionally, CCBA highlights three key benefits of CCB-certified projects: First, the additional benefits created for the local population (often termed 'co-benefits') secure its goodwill and help to prevent conflicts between project developers and local communities.[10] Second, a free, prior, and informed consent (FPIC) is mandatory. FPIC is a specific consultation process guaranteed to indigenous groups by Convention 169 of the International Labour Organization (ILO).[11] This could help to avoid controversy regarding displacement of indigenous groups and local communities in the course of project implementation (CCBA 2008a: 12; CCBA 2008b: 4). Third, CCB-certified projects achieve higher prices on the voluntary carbon market (CCBA 2008b: 7; Neeff et al. 2009: 26).

The Standards and their criteria have been developed by the member organizations of the CCBA. Any decisions related to the Standards are taken by the CCBA and its members. This applies to both the criteria for the Standards and the composition of the Standard Committee (CCBA 2008b: 6; Kollmuss et al. 2008: 76). The Standard Committee assists the CCBA Director in developing criteria for the Standards. The CCBA appoints committee members and defines the rules for decision-making by the committee and the rules governing the competencies of the committee (CCBA 2008b: 6). The Standard Committee, which was convened to develop the latest version of the Standards document, included representatives of five companies operating on the carbon market, two representatives of indigenous associations, and three representatives of research institutions, as well as the CCBA members (CCBA 2008a: 2; CCBA 2008b: 6). Additionally, CCBA publishes the Standards on the website for public comments (CCBA 2008b: 7). At the time of writing (April, 2013) CCBA has invited the public to comment on the third edition of the CCB Standards.

The certification process for the CCB Standards is carried out by an independent auditor. However, this auditor is commissioned and paid by the project developer (Kollmuss et al. 2008: 78).[12] Auditors have to be accredited by the CCBA. Eligible auditors are the ones who have been accredited by the CDM Executive Board, by the Certification Body of the Forest Stewardship Council (FSC), or by the International Organization for Standardization (ISO; CCBA 2013). The certification process, called validation, consists of several steps which include field visits of the independent auditor, publication and dissemination of the Project Design Document (PDD) for public comments (for 30 days), revision of the PDD, and the preparation of the audit report by the auditor. If the auditor approves the project, the results of the audit and the validation statement are published on the CCBA website (CCBA 2010a: 8). Five years after validation the project has to be verified. The verification process also includes field visits and the publication and dissemination of the implementation plan (CCBA 2010a: 11). Verification and validation can be done by the same auditor.

The criteria that must be fulfilled as part of CCB certification are listed in the Project Checklist of the Standards document. The second edition of the CCB Standards has been available since 2008. The list of criteria consists of a General Section (G), which includes criteria for evaluating the original conditions in the project area, the project design, opportunities for community participation, the legal status of the project, and the consultation process (CCBA 2008a: 21). The Climate Section (CL) lists indicators to prove the additionality of the project (CBBA 2008a), or whether the emissions have merely been displaced (leakage) instead of reduced. Indicators referring to local communities affected by the project and indicators concerning the protection of endangered species are summarized in the Community Section (CM) (CBBA 2008a) and the Biodiversity Section (B) (CCBA 2008a: 28) respectively. In every section the project has to show net benefits. The Standards also contain an optional Gold Level Section (GL) for projects that are "explicitly pro-poor" (CCBA 2008a: 44), that create benefits for climate change adaptation, or that promote the conservation of key species. We would like to draw attention to the indicators listed in Table 36.1. The indicators presented in Table 36.1 have been selected in line with the aim of this paper, namely to address the consideration of input and output legitimacy of forest carbon offsets.

Since June 2009, all projects have to be certified in accordance with the second edition of the CCB Standards. Projects that have been certified in accordance with the first edition of the

Table 36.1 List of CCB Standards indicators

Input dimension	*Indicators*	
	G3.9	"Project proponents must play an active role in distributing key project documents to affected communities and stakeholders . . ."
	G5.2	"Document that the project has approval from the appropriate authorities."
	G5.3	"Demonstrate . . . that the project will not encroach uninvited on private property, community property . . . and has obtained the free, prior and informed consent of those whose rights will be affected by the project."
	G5.4	"Demonstrate that the project does not require the involuntary relocation of people or of the activities important for the livelihoods . . . . The project proponents must demonstrate that the agreement was made with the free, prior and informed consent of those concerned and includes provisions for just and fair compensation" ("including lands that communities have traditionally owned, occupied or otherwise used or acquired").
Output dimension	CL1	"The project must generate net positive impacts on atmospheric concentrations of greenhouse gases (GHGs) over the project lifetime from land use changes within the project boundaries."
	GL2.2	"Demonstrate that at least 50% of households within the lowest category of wellbeing (e.g., poorest quartile) of the community are likely to benefit substantially from the project."

Source: CCBA (2008a): 16, 20, 22, 34.

CCB Standards will continue to be verified on the basis of the older and less specific criteria. The first edition does not directly refer to FPIC and therefore does not acknowledge the rights of indigenous groups and local communities as strongly as the second version (CCBA 2005; CCBA 2008a: 20). According to Indicator G5.4 in the second edition, the land owners concerned are also entitled to FPIC and compensation, whereas the first edition simply points out that the project should not violate existing land titles and that residents may only be relocated on a voluntary basis (CCBA 2005: 12). According to Merger et al. (2011), the CCB Standards provide comprehensive guidelines for developing forest carbon projects that are participatory and environmentally sustainable pro-poor. WWF also recommends the concept of net positive social and environmental impacts (see Walter and Kahlert 2010: 16), promoted by the CCB Standards. Furthermore, by referring to the right to free prior and informed consent, the Standards ensure that the local population is consulted and that it participates in decision-making processes. Indicator G5.3 also addresses traditional and customary land ownership in addition to formal land ownership, ensuring that the CCB Standards take full account of diffuse and ambiguous land ownership structures in the project regions.

The criteria of the Standards cover most of the possible controversies related to forest carbon projects. However, we argue that the governance or input-dimension of the Standards shows a number of programmatic weaknesses. For example, the decision-making bodies of the CCB Standards offer no open or institutionalized opportunities for local communities affected by forest carbon projects to participate in the decision-making process of the Standards. As mentioned above, all members of the Standards Committee of the second edition of the Standards were appointed by CCBA members. In addition, the Standards Committee included only two representatives of groups potentially adversely affected by forest carbon projects: Gaia Amazonas and Indigenous Peoples of Africa Coordinating Committee (CCBA 2008a: 2). Furthermore, Kollmuss et al. (2008) are skeptical as to whether the public commenting periods are sufficient for properly identifying the potential environmental and social impacts of the project (Kollmuss et al. 2008: 36). In addition, Kollmuss et al. highlight that mutual interdependence between project developers and auditors may create conflicts of interests.[13] Verification and validation procedures can be conducted by the same auditor and probably weaken the credibility of the certification process. This applies in particular for most of the very broad qualitative criteria of the CCB Standards, since they leave plenty of room for interpretation (Kollmuss et al. 2008). The lack of an executive board supervising the validation reports may also undermine the impact of forest carbon projects since weak verification and validation may lead to positive certification of projects not contributing to the goals outlined in the Standards document. What is challenging for complex standards or safeguards is that they cannot include every possible issue which may come up on the ground. Field realities are complex, as the project sites in Indonesia show below.

5 REDD and CCB projects in Indonesia

Recent reports from two project sites in Indonesia indicate that certification from the CCB Standards neither guarantees additional emission reduction (Gaveau et al. 2009), nor community involvement (McCulloch 2010: 17; The Environmental Justice and Governance Research Lab 2011: 2), nor the adequate compensations of groups affected by the project (Friends of the Earth International 2010: 15; Morgan 2011: 2). The criteria of the CCB Standards cover these controversial issues technically. Their occurrence indicates either that different stakeholders may interpret and assess criteria differently or they indicate weak validation.

We start by examining reports and press articles on the Ulu Masen project in Aceh. The project intends to protect 750,000 hectares of forest thereby avoiding three million tonnes of carbon dioxide emissions per year (The Provincial Government of Nanggroe Aceh Darussalam 2007). According to the project developers, 20 percent of the revenues from emissions trading should support the sustainable livelihoods of local communities (The Provincial Government of Nanggroe Aceh Darussalam 2007: 42). The Provincial Government of Nanggroe Aceh Darussalam, Flora and Fauna International, and the carbon consultant company Carbon Conservation are implementing the project. The investment bank Merrill Lynch supports the project and holds a purchase option for emissions permits with a value of $9 million (McGregor 2010: 25). Merrill Lynch was praised for "innovative investment" by the *Wall Street Journal* and others. However, despite successful validation, the project's impact on climate change mitigation is debatable. A study by Gaveau et al. (2009) indicates that the project, validated according to the criteria of the first edition of the standard document, is not making any significant contribution to reducing deforestation, as the project is already protected from extensive deforestation due to its highly inaccessible location (Gaveau et al. 2009: 9). Indicator CL1 (in both the first and second edition) requires that certified projects have a positive impact on climate change mitigation. The example shows that even the technical and emission related criteria of the standards leave plenty of room for interpretation. The amount of emissions avoided by forest carbon projects depends on the calculated business as usual scenario. Scenarios are building on assumptions to predict future developments. Thus the decision-making process that leads to these assumptions is crucial. In addition, other studies point out that the criteria of the CCB Standards referring to the rights of local communities have not been satisfied despite the fact that the project has been certified successfully (Smartwood 2008; McCulloch 2010: 17). A study by the Environmental Justice and Governance Research Lab in Banda Aceh[14] highlights the lack of FPIC, revealing that 67 out of 69 village representatives interviewed in the project area were not adequately informed about the REDD project and its socio-economic consequences, nor were they involved in the project design process as required by the standard (The Environmental Justice and Governance Research Lab 2011: 2). While FPIC is not explicitly mentioned in the first edition of the Standards, Indicator CM1 ("In addition, local communities and other stakeholders should be engaged early on so that the project design can be revised based on their input"; CCBA 2005: 21) clearly requires that the local population should be involved in the project design process. Friends of the Earth International (2010) criticizes that the project's steering committee includes only one representative from local communities and fears that weak representation could increase the risk of unfair distribution of future revenues from carbon trade (Friends of the Earth International 2010: 24). At the same time, rumors circulated that the project area was protected against illegal deforestation activities by armed forces, which were former members of the Free Aceh Movement (ABC Rural 2008). Today the Ulu Masem project can be considered as failed. The project's CCBA validation expired (CCBA 2013; Lang 2013) and the NGO Flora and Fauna International dissociated itself from the project (SSNC 2013: 19).

The second project from which we have analyzed reports and press articles is the Rimba Raya project. The Rimba Raya project was initially set up to protect 91,000 hectares of peat swamp forest. Later on, the Ministry of Forestry reduced the project size by 50 percent issuing the other 50 percent to an oil palm company. In its original size the project could create revenues of up to $750 million assuming a carbon price of $10 per emission permit (Reuters 2010). The project intends to generate additional income for 14 local communities through community agro forestry activities. Additional revenues from carbon trade should support the Orang-Utan Foundation International (Infinite Earth 2010). The project has been CCB validated and VCS certified, and was praised as the first Indonesian REDD+ project on the voluntary carbon market Dewan 2010. Nonetheless, the audit was conducted prior to the shrinking of the project area and has not been audited for its actual size.[15]

Furthermore, this explicitly profit-driven REDD project shows that the stringent CCB indicators can be interpreted in different ways (Friends of the Earth International 2010: 15; Morgan 2011: 2). For example, indicator 5.4 highlights the need to provide fair and just compensation to actors who had to give up their livelihood strategy because of the project. Friends of the Earth International (2010) highlights that, while the project does provide healthcare and educational institutions, the extent to which the local population can benefit from the sale of REDD certificates on the voluntary carbon market is still unclear (Friends of the Earth International 2010: 15). Brihannala Morgan (2011) argues that the Rimba Raya project just gives handouts and gifts to the local population rather than compensating them adequately for income losses caused by the project (Morgan 2011: 2). Additionally, NGOs such as Friends of the Earth are criticizing the involvement of Gasprom and Shell accusing Shell in particular of greenwashing (Friends of the Earth International 2010: 15). The NGO argues that Shell seeks to deflect attention from its devastating oil extraction practices in the Niger Delta which are destroying the fragile delta ecosystem and the livelihood of the local population by investing in rainforest conservation and climate change mitigation in Indonesia (Friends of the Earth International 2010: 15).

The project sites reveal that implementing the CCB Standards, which are far reaching in principle, can be challenging and do raise questions at the local level: First, which group is actually entitled to FPIC, and second, which representatives are actually authorized to negotiate with the organizations implementing the projects? Such issues are difficult to condense into sets of criteria as they vary greatly between project sites. Regardless of the dilemma of establishing adequate criteria, NGOs such as Friends of the Earth act as watchdogs to monitor these processes. They highlight critical issues such as benefit sharing, identify cases of greenwashing by market actors, and draw attention to land conflicts. With regard to land conflicts, REDD has opened new debates on land and forest tenure rights, on political and economic participation, and on the empowerment of indigenous communities. The Indonesian association AMAN[16] uses the slogan 'No Rights, No REDD' to highlight the need for land reform and benefits from the attention currently being given to Indonesian forest policy by politicians, the media, and international donors.[17]

6 Conclusion

In this paper, we have analyzed NGO activities in the realm of forest carbon offsets. In particular, we paid attention to the input legitimacy and output legitimacy of forest carbon standards. In the following section, we summarize our findings and stress their relevance for the climate movement.

On the one hand, the input legitimacy of CCB Standards is lagging behind other and older approaches. For example, the Forest Stewardship Council (FSC), founded by WWF and other NGOs in 1993, offers far-reaching participation opportunities. Within the FSC, all organizations and individuals who identify with the goals of FSC have the opportunity to apply for membership and are represented at the General Assembly (central decision-making body); where they even have voting rights. The participation opportunities in CCB Standards are limited. CCBA *appoints* the members of the standard committee. Furthermore, the public commentary periods within the certification procedure of the Standards might not be sufficient to ensure participation of all affected stakeholders (see Kollmuss et al. 2008). On the other hand, CCBA attempts to improve opportunities for participation significantly. This, for example, applies to the inclusion of FPIC into the CCB Standards. Participation and the empowerment of local actors are key issues in ongoing negotiation processes in the policy arena of climate change mitigation and REDD. Put more concretely: The inclusion of FPIC into carbon standards and statements on local community rights within the Cancun Safeguards (Decision 1/ CP.16) offer new room for maneuver for groups on the local level. This new room for maneuver permits process inclusion and participation in decision-making processes and thus augments the input legitimacy for offset projects. Moreover, in particular REDD has pushed land tenure and the acknowledgment of customary forest tenure on the international agenda. In addition, decisions by UNFCCC, in particular the Cancun safeguards and national REDD strategies that clearly refer to rights of local communities can be used as soft law[18] mechanisms to legitimize claims. The REDD debate provides an opportunity to highlight historic structural inequalities in access to forest resources and land. To a certain degree the political struggle for carbon rights from offset projects become proxy conflicts over social and economic inclusion.

Referring to the *output* legitimacy of carbon standards, our article shows considerable deficits. Even the main objective of forest carbon offsets, namely avoiding greenhouse gas emissions, cannot not be guaranteed by the leading standards. In many cases, REDD offsets simply displace deforestation rather than avoiding deforestation and greenhouse gas emissions. Our first described forest carbon offset (Ulu Masen), for instance, fails to provide cost-effective emission reductions for industrialized countries. In any case, NGOs such as WWF, Greenpeace, and in principle all members of the Climate Action Network (CAN) are aware of this general problem. They claim that developed countries with their still much higher per-capita emissions compared to developing countries have to increase their domestic emission reduction commitments first, instead of relying on offsets. In the realm of forest carbon offset, NGOs have initiated to link offsetting of transnational polluters with sustainable development at the very local level. Carbon standards developed by NGOs established social and environmental safeguards for offsets. In *the first place* NGOs ensure that the offsetting industry *complies* with social and environmental safeguards. By ensuring this, the established NGOs have to respond to criticism expressed by other civil society organizations, the latter acting as watchdogs. Exactly this 'mechanism' might *in sum* lead to strong and reliable carbon standards – the one group of NGOs designing standards and the other group either criticizing these standards or just acting as watchdogs.

Within the climate movement, the assessment of market-based instruments for mitigating climate change is one of the major 'battle lines.' This finding has been stated repeatedly by other contributions in this handbook. A large group of *critical* civil-society actors categorically rejects these instruments. This group sets the formation of carbon markets in the wider context of finance-driven neoliberalism. They claim that current climate policies are characterized by a 'fetishization' of the market (see Newell and Paterson 2010). Organizations such as Rising Tide, Friends of the Earth, Carbon Trade Watch, and the Durban Group for Climate Justice express their criticism by using terms such as 'sub-prime carbon,' 'carbon crunch,' and 'toxic carbon' (Newell and Paterson 2010). The other group of rather *established* international NGOs, such as WWF, will become key players in the further development of offsetting processes within transnational and network-based clusters (Rest 2011: 879). The new clusters of the carbon economy consist of state actors, the private sector (such as banks, retailers, distributors, project developers, accountants, certifiers, auditors, and consultants), and academic players.[19] Our project sites studies illustrate how large international conservation organizations such as Flora and Fauna International intentionally are cooperating with market actors. NGOs and partners within the offsetting industry seek to improve the credibility of offset programs. The offsetting industry and buyers of emission permits are keen to do so as well, as they wish to avoid stricter regulations (Newell and Paterson 2010: 88). Within this network of actors, NGOs' activities are diverse. Their activities are no longer limited to lobbying and advocacy. Rather, NGOs increasingly also develop green financial products. For achieving their objectives, NGOs provide new markets and profit options. As a consequence, the boundaries between market actors and civil society organizations are blurring and some of the environmental NGOs become a rather new kind of *hybrid organization*. Referring to broader debates on environmental governance, NGOs become the *driving force* for the commodification of nature and ecosystem services (Castree 2003).[20] Finally, the fact that the CCB Standards explicitly aim to prevent potential conflicts (and, as a result, prevent any resistance from the population) indicates that NGOs in cooperation with other private-sector actors even adopt a *governing role*.

It is difficult to draw a concluding assessment with respect to these processes in the area of offsetting. For many observers, the described NGO activities are processes of hegemonial entanglement (Methmann 2011). We are skeptical towards such a 'conviction.' *Empirically*, positive regulatory effects of NGO-backed carbon standards and critical groups acting as watchdogs might be underestimated. With respect to the question 'what *should* be done?' we believe that such a 'conviction' neglects the dilemma currently faced by many civil society organizations. The carbon economy is being expanded vehemently despite all the crises and fraud cases that have occurred. Civil society organizations are faced with the dilemma of ultimately being barely able to manage and control these processes, yet having good reason to engage in them. After all, where no-one has final responsibility and where the state does not ensure or barely ensures that activities serve the common good, then civil-society organizations are the only currently identifiable corrective.

Notes

1 The lead author is examining the impact of market-based instruments in Indonesia as part of his doctoral studies.

2 A total of 131 megatons of carbon were traded on the voluntary market in 2010 (Peters-Stanley et al. 2011: 9). This corresponds to approximately one-quarter of the emissions permits traded as part of CDM (UNFCCC 2011a: 4).

3 In the forest sector, this calculation is based on complex models that use historic deforestation scenarios to predict future developments (Dudley 2010: 54).

4 Of this 3.3 percent, most projects are not located in 'typical' developing countries, but in Israel (34 percent) and South Africa (27 percent) (Boyd et al. 2009: 22). To counteract this trend, the EU recently decided that only CDM projects from least developed countries (LDCs) will be granted access to the European carbon market from 2013 onwards (European Commission 2012).

5 The Gold Standard does not certify forest carbon projects and, unlike normal CDM, permits only renewable-energy and energy-efficiency projects (Holz 2010: 127).

6 Recently, VCS and CCBA announced that they would cooperate and provide a process facilitating certification against both standards (VCS 2012).

7 The term 'land-based offset project' refers to any offset project based on land use type, which includes REDD projects, CDM reforestation projects (A/R), and sustainable agroforestry systems, such as those used in cocoa and coffee production.

8 These are the International Center for Agroforestry Research (ICRAF), the *Centro Agronómico Tropical de Investigación y Enseñanza* (CATIE), and the Center for International Forestry Research (CIFOR).

9 Unlike the VCS Standard, the CCB Standards do not require emissions reductions to be quantified and, consequently, do not provide for the issuing of tradable emissions certificates. As such, the CCB Standards document recommends that the VCS Standard be used to quantify greenhouse gas reductions.

10 This is worded as follows in the original document:

> In this way, the Standards help investors to minimize risks by identifying high-quality projects that are unlikely to become implicated in controversy. Multiple-benefit projects also create valuable goodwill and other ancillary returns for investors. Social and environmental benefits and sustainability are also an important means to reduce risks to the permanence of the climate benefits.
>
> *(CCBA 2008a: 7)*

11 Article 7 of Convention No. 169 is particularly relevant in this regard:

> Indigenous and tribal peoples have the right to decide their own priorities for the process of development as it affects their lives, beliefs, institutions and spiritual well-being and the lands they occupy or otherwise use, and to exercise control, to the extent possible, over their own economic, social and cultural development.

12 There are several stages to the validation process (see CCBA 2010a: 6) and each certified product must be verified every five years. Certification (referred to as validation by CCBA) and verification can be performed by the same auditor (Kollmuss et al. 2008: 78; CCBA 2010a: 11).

13 Auditors are paid by the relevant project developers, both for CDM projects and for projects on the voluntary market. However, CCB has no executive committee that decides whether to approve the project once the audit has been completed.

14 The Environmental Justice and Governance Research Lab is a newly opened center in Banda Aceh, Indonesia.

15 The approval process of the Indonesian Ministry of Forestry is controversial and complex in this context (see Greenomics 2012).

16 AMAN stands for *Aliansi Masyarakat Adat Nusantara.*

17 Interview with an AMAN activist on July 27, 2012 in Jakarta, Indonesia. Similar processes are being seen in Peru, where the AIDESEP (*Asociación Interétnica de Desarrollo de la Selva Peruana*) association is working. AIDESEP is currently using the awareness to designate new indigenous territories and is attempting to develop its own REDD projects (COICA 2012; Zelli et al. 2013).

18 'Soft law' refers here to regulations that are based on international agreements, but for which the implementation mechanisms are very limited (Hein and Moon 2013: 9).

19 Newell and Paterson (2010: 87) describe CDM as an enormous public – private partnership.

20 However, NGOs such as Friends of the Earth fear that commodification of nature will be particularly detrimental to local actors, and they act as watchdogs locally, unafraid of conflict with major market actors.

References

ABC Rural (2008) Carbon Scheme Will Help Save Aceh Forest. Accessed 06/25/2012 from www.abc.net.au/rural/news/content/200804/s2214030.htm.

Boyd, E., Hultman, N., Roberts, J. T., Corbera, E., Cole, J., Bozmoski, A., Ebeling, J., Tippman, R., Mann, P., Brown, K., and Liverman, D. M. (2009) Reforming the CDM for Sustainable Development: Lessons Learned and Policy Futures, *Environmental Science & Policy*, 12(7): 820–831.

Bumpus, A. G. (2011a) The Matter of Carbon: Understanding the Materiality of tCO2e in Carbon Offsets, *Antipode*, 43(3): 612–638.

Bumpus, A. G. (2011b) Carbon Colonialism? Offsets, Greenhouse Gas Reductions, and Sustainable Development, in Peet, R., Robbins, P., and Watts, M. (eds.) *Global Political Ecology*, London and New York: Routledge.

Bumpus, A. G. and Liverman, D. M. (2008) Accumulation by Decarbonization and the Governance of Carbon Offsets, *Economic Geography*, 84(2): 127–155.

Carbon Trade Watch (2009) Carbon Offsets, Carbon Trade Watch Fact Sheet 2. Accessed 05/26/2013 from www.carbontradewatch.org/downloads/publications/factsheet02-offsets.pdf.

Castree, Noel (2003) Commodifying What Nature? *Progress in Human Geography*, 27(3): 273–297.

Climate Community and Biodiversity Alliance (CCBA) (2005) *Climate, Community and Biodiversity Project Design Standards* (First Edition), Washington DC: CCBA.

Climate Community and Biodiversity Alliance (CCBA) (2008a) *Climate, Community and Biodiversity Project Design Standards* (Second Edition), Arlington, U: CCBA.

Climate Community and Biodiversity Alliance (CCBA) (2008b) Terms of Reference, Procedures and Work Plan for Revision of the Climate, Community & Biodiversity Standards. Accessed 06/25/2012 from www.climate-standards.org/pdf/CCBS_Revision_TORS_15_Feb_updated_13_June_2008.pdf.

Climate Community and Biodiversity Alliance (CCBA) (2010a) Rules for the Use of the Climate, Community & Biodiversity Standards. Accessed 06/25/2012 from www.climate-standards.org/pdf/CCB_Standards_Rules_Version_June_21_2010.pdf.

Climate Community and Biodiversity Alliance (CCBA) (2010b) Who We Are. Accessed 06/25/2012 from www.climate-standards.org/who/index.html.

Climate Community and Biodiversity Alliance (CCBA) (2013) List of Auditors Approved to Validate and Verify Projects Against the Climate, Community & Biodiversity (CCB) Standards. Accessed 05/26/2013 from https://s3.amazonaws.com/CCBA/Approved_CCBS_Auditors.pdf.

COICA (2012) AIDESEP Press Conference, 12/04/12. Accessed 02/07/2013 from http://issuu.com/edsonrosalesfigueroa/docs/aidesep_press_conference_04.12.12_cop18?mode=window&backgroundColor=%23222222.

Corbera, E. and Brown, K. (2010) Offsetting Benefits? Analyzing Access to Forest Carbon, *Environment and Planning*, 42(7): 1739–1761.

Dewan, A. (2010) Getting REDD+ Right, in Forests news, A blog by the Center for International Forestry Research. Accessed 16/10/2013 from http://blog.cifor.org/809/getting-redd-right-2#.Ul6EJxD-M9aQ.

Dudley, R. G. (2010) A Little REDD Model to Quickly Compare Possible Baseline and Policy Scenarios for Reducing Emissions from Deforestation and Forest Degradation, *Mitigation and Adaptation Strategies for Global Change*, 15(1): 53–69.

Eden, S. (2011) The Politics of Certification: Consumer Knowledge, Power, and Global Governance in Ecolabeling, in Peet, R., Robbins, P., and Watts, M. (eds.) *Global Political Ecology*, London and New York: Routledge.

European Commission (2012) International Carbon Market – Questions & Answers on Use of International Credits in the Third Trading Phase of the EU ETS (January 2012). Accessed 02/07/13 from http://ec.europa.eu/clima/policies/ets/linking/faq_en.htm.

Fletcher, R. (2010) Neoliberal Environmentality: Towards a Poststructuralist Political Ecology of the Conservation Debate, *Conservation and Society*, 8(3): 171–181.

Friends of the Earth International (2010) REDD: The Realities in Black and White. Accessed 02/07/13 from www.foei.org/en/resources/publications/pdfs/2010/redd-the-realities-in-black-and-white.

Gaveau, D. L A., Wich, S., Epting, J., Juhn, D., Kanninen, M., and Leader-Williams, N. (2009) The Future of Forests and Orangutans (Pongo Abelii) in Sumatra: Predicting Impacts of Oil Palm Plantations, Road Construction, and Mechanisms for Reducing Carbon Emissions from Deforestation, *Environmental Research Letters*, 4: 1–12.

Gilbertson, T. and Reyes, O. (2009) Reforming the CDM for Sustainable Development: Lessons Learned and Policy Futures Carbon Trading – How It Works and Why It Fails, Critical Currents Occasional Paper Series 7 (November 2009), Uppsala: Dag Hammerskjöld Foundation.

Gonzalez, J. G. and Schomerus, T. (2010) *Der Gold Standard als Garant für die Nachhaltigkeit von CDM-Projekten in Entwicklungsländern*. Arbeitspapierreihe Wirtschaft& Recht 5, Lüneburg: Leuphana Universität Lüneburg.

Grainger, A., Boucheret, D. H., Frumhoff, P. C., Laurance, W. F., Lovejoy, T., McNeely, J., Niekisch, M., Raven, P., Sodhi, N. S., Venter, O., and Pimm, S. L. (2009) Biodiversity and REDD at Copenhagen, *Current Biology*, 19(21): R974-R976.

Greenomics (2012) Indonesia's First REDD+ Project – Not a Done Deal, Despite Promoter's Claims. Accessed 02/07/13 from www.greenomics.org/docs/Report_201212_Greenomics_RRC_REDD_Not_a_Done_Deal.pdf.

Hein, W. and Moon, S. (2013) *Informal Norms in Global Governance – Human Rights, Intellectual Property Rules and Access to Medicines*, Surrey, England: Ashgate.

Holz, C. (2010) Möglichkeiten und Grenzen der Partizipation – CDM-Kritik in den UN-Klimaverhandlungen, in Voss, M. (ed.) *Der Klimawandel. Sozialwissenschaftliche Perspektiven*, Wiesbaden: VS-Verlag für Sozialwissenschaften.

Indonesian REDD+ Taskforce (2012) *REDD+ National Strategy*, Jakarta, Indonesia.

Infinite Earth (2010) Beyond Carbon, Beyond Sustainability – Introducing the World's First Fully Certified REDD++ Project: Rimba Raya Biodiversity Reserve. Accessed 09/03/2013 from http://www.infinite-earth.com/files/brochure.pdf.

Klinke, A. (2006) *Demokratisches Regieren jenseits des Staates. Deliberative Politik im nordamerikanischen Große-Seen-Regime*, Opladen: Verlag Barbara Budrich.

Kollmuss, A., Zink, H., and Polycarp, C. (2008) Making Sense of the Voluntary Carbon Market: A Comparison of Carbon Offset Standards, WWF Germany.

Lipschutz, R. D. and McKendry, C. (2011) Social Movements and Global Civil Society, in Dryzek, J. S., Norgaard, R. B., and Schlosberg, D. (eds.) *Oxford Handbook of Climate Change and Society*, Oxford: Oxford University Press.

Lang, C. (2013) REDD Plus or REDD 'Light'? Biodiversity, Communities and Forest Carbon. Redd-Monitor.org, February 21, 2013. Accessed 08/31/2013 from http://www.redd-monitor.org/2013/02/21/redd-plus-or-redd-light-biodiversity-communities-and-forest-carbon-certification/.

Long, A. (2010) Global Climate Governance to Enhance Biodiversity & Well-Being: Integrating Non-State Networks and Public International Law in Tropical Forests. Accessed 06/25/2012 from http://works.bepress.com/andrew_long/1/.

McAfee, K. (1999) Selling Nature to Save it? Biodiversity and Green Developmentalism, *Environment and Planning*, 17: 133–154.

McAfee, K. (2012) The Contradictory Logic of Global Ecosystem Services Markets, *Development and Change*, 43(1): 105–131.

McCulloch, L. (2010) Ulu Masen REDD Demonstration Project – The Challenges of Tackling Market Policy and Governance Failures that Underlie Deforestation and Forest Degradation, Forest Conservation Team Occasional Papers, Kamiyamaguchi, Japan: Institute for Global Environmental Strategies.

McGregor, A. (2010) Green and REDD? Towards a Political Ecology of Deforestation in Aceh, Indonesia, *Human Geography*, 3(2): 21–34.

Merger, E. and Pistorius, T. (2011) Effectiveness and Legitimacy of Forest Carbon Standards in the OTC Voluntary Carbon Market, *Carbon Balance and Management*, 6(1): 1–12.

Merger, E., Dutschke, M., and Verchot, L. (2011) Options for REDD+ Voluntary Certification to Ensure Net GHG Benefits, Poverty Alleviation, Sustainable Management of Forests and Biodiversity Conservation, *Forests*, 2(2): 550–577.

Methmann, C. (2011) NGOs in der staatlichen Klimapolitik. Zwischen diskursiver Intervention und hegemonialer Verstrickung, in Brunnengräber, A. (ed.) *Zivilisierung des Klimaregimes. NGOs und soziale Bewegungen in der nationalen, europäischen und internationalen Klimapolitik*. Wiesbaden: VS-Verlag.

Morgan, B. (2011) Community Engagement – Don't Ignore REDD's Impact on Communities, *Inside Indonesia*, 105(Jul–Sep).

Neeff, T., Ashford, L., Calvert, J., Davey, C., Durbin, J., Ebeling, J., Herrera, T., Janson-Smith, T., Lazo, B., Mountain, R., O'Keeffe, S., Panfil, S., Thorburn, N., Tuite, C., Wheeland, M., and Young, S. (2009) *The Forest Carbon Offsetting Survey 2009*, Eco Securitas, Conservation International, The Climate, Community & Biodiversity Alliance, ClimateBiz.

Newell, P. and Paterson, M. (2010) The Politics of the Carbon Economy, in Boykoff, M. T. (ed.) *The Politics of Climate Change. A Survey*, London, New York: Routledge.

Peters-Stanley, M., Hamilton, K., Marcello, T., and Sjardin, M. (2011) *Back to the Future – State of the Voluntary Carbon Markets 2011 – A Report*, New York and Washington: Ecosystem Marketplace.

Phelps, J., Guerrero, M. C., Dalabajan, D. A. et al. (2010) What Makes a REDD Country? *Global Environmental Change*, 20(2): 322–332.

Putz, F. E. and Redford, K. (2009) Dangers of Carbon-based Conservation, *Global Environmental Change*, 19(4): 400–401.

Rest, J. (2011) Von der NGOisierung zur bewegten Mobilisierung, in Brunnengräber, A. (ed.) *Zivilisierung des Klimaregimes. NGOs und soziale Bewegungen in der nationalen, europäischen und internationalen Klimapolitik*, Wiesbaden: VS Verlag.

Reuters (2010) Indonesia Project Boosts Global Forest CO2 Market. Accessed 02/09/2013 from www.reuters.com/article/2010/08/24/forests-carbon-redd-idAFSGE67N0D920100824?sp=true.

Sikor, T., Stahl, J., Enters, T., Ribot, J. C., Singh, N., Sunderlin,W. D., and Wollenberg, L. (2010) REDD-plus, Forest People's Rights and Nested Climate Governance, *Global Environmental Change*, 20(3): 423–425.

Smartwood (2008) Validation Audit Report For: Provincial Government of Nanggroe Aceh Darussalam – Fauna & Flora International – Carbon Conservation in Ulu Masen Ecosystem (Aceh Province, Indonesia). Accessed 25/06/2012 from www.climate-standards.org/projects/files/Carbon_Conservation_FFI_Provincial_Govt_of_Nanggroe_Aceh_Darussalam_CCB_Validation_Audit_Final_Feb_08.pdf.

Swedish Society for Nature Conservation (SSNC) (2013) Report: REDD Plus or REDD 'Light'? – Biodiversity, Communities and Forest Carbon Certification, Stockholm: SSNC.

The Environmental Justice and Governance Reserach Lab (2011) Selling the Wind, *Inside Indonesia*, 105(Jul–Sep).

The Gold Standard (2011) A Best Practice Methodology and a High Quality Carbon Credit Label for both Kyoto and Voluntary Markets. Accessed 06/25/2012 from www.cdmgoldstandard.org/about-us/who-we-are.

The Provincial Government of Nanggroe Aceh Darussalam (Aceh) (2007) Reducing Carbon Emissions from Deforestation in the Ulu Masen Ecosystem, Aceh, Indonesia, A Triple Bebefit Project Design Note for CCBA Audit. Accessed 02/11/2013 from https://s3.amazonaws.com/CCBA/Projects/Reducing_Carbon_Emissions_from_Deforestation_in_the_Ulu_Masen_Ecosystem/Final_Ulu_Masen_CCBA_project_design_note_Dec29.pdf.

UNFCCC (1998) *The Kyoto Protocol*, Bonn: UNFCCC.

UNFCCC (2007) *Uniting on Climate – A Guide to the Climate Change Convention and the Kyoto Protocol*, Bonn: UNFCCC.

UNFCCC (2011a) *Benefits of the Clean Development Mechanism*, Bonn: UNFCCC.

UNFCCC (2011b) The Cancun Agreements, Decision 1/CP.16.

UNFCCC (2012a) About Web. Accessed 06/25/2012 from http://cdm.unfccc.int/about/index.html.

UNFCCC (2012b) CDM Project Cycle. Accessed 06/25/2012 from http://cdm.unfccc.int/Projects/diagram.html.

Unmüßig, B. (2011) NGOs in der Klimakrise. Fragmentierungsprozesse, Konfliktlinien und strategische Ansätze, in Brunnengräber, A. (ed.) *Zivilisierung des Klimaregimes. NGOs und soziale Bewegungen in der nationalen, europäischen und internationalen Klimapolitik*, Wiesbaden: VS-Verlag.

VCS (2012) VCS, CCBA Join Forces on Forestry Projects. Accessed 09/03/2013 from http://v-c-s.org/news-events/news/vcs-ccba-join-forces-forestry-projects.

Walter, M. and Kahlert, G. (2010) *Forest Carbon Standards – a WWF Assessment Guide*, Frankfurt/Main: WWF Germany.

Wunder, S. (2008) How Do We Deal with Leakage, in Angelsen, A. (ed.) *Moving Ahead with REDD: Issues, Options and Implications*, Bogor: CIFOR.

WWF (2007) Is the CDM Fulfilling its Environmental Objectives? An Evaluation of the CDM and Options for Improvement Prepared for WWF, Öko-Institut.

Zelli, F., Erler, D., Frank, S., Hein, J. I., Hotz, H., and Santacruz-Melgarejo, A. M. (2013) *Reducing Emissions from Deforestation and Degradation in Peru – A Challenge to Social Inclusion and Multi-level Governance*, Bonn: German Development Institute.

37
IN THE PERIPHERY OF THE CLIMATE MOVEMENT: HUMANITARIAN ORGANIZATIONS

Johannes Herbeck

1 Introduction

Their perception as "normative and epistemic authorities" (Ecker-Ehrhardt 2007: iv) has built up the reputation of humanitarian organizations as a global consciousness over the past decades. In this role, humanitarian organizations could represent a valuable resource for the climate movement. They could carry positions and political aims of the movement into international debates and decision-making processes and could add the necessary moral credibility and weight to substantially influence such processes. They could give a voice to actual and future losers of climate change, as well as of adaptation and mitigation strategies of the international climate governance.

The environmental question is not a new issue for humanitarian organizations. The engagement for people affected by sudden or slow-onset changes in the environment has always been among the main challenges for those organizations, besides the support of war-affected people. This is especially true, as conflicts develop significantly stronger impacts under conditions of environmental change (Strömberg 2007; Fink and Redaelli 2011; Rodella-Boitreaud and Wagnery 2011). Already today, humanitarian organizations perceive to face new challenges, especially regarding the scope and frequency of emergency situations triggered or amplified by environmental hazards. Over the past few years, between 15 and 42 million people have been displaced by such hazards annually, with often a high share of displacements due to weather-related events that are at least potentially influenced by global climate change (IDMC 2012: 5). In comparison with 42.5 million people that were displaced by violent conflicts and political prosecution in 2011 (see UNHCR 2012), those numbers show the comparably high scope of environmental hazards as triggers of emergency and displacement situations. The United Nations Office for the Coordination of Humanitarian Affairs (OCHA) underscores that climate change is not perceived as a "distant future threat," but as a very real phenomenon that already affects societies all over the globe (OCHA n.d.).

Nevertheless, according to many of the big humanitarian organizations, climate change will play out especially in future and will develop into one of the major drivers of emergency situations in coming decades. In a shared press release issued by all relevant, globally acting humanitarian organizations in preparation of the Conference of Parties of the UNFCCC in Copenhagen, the heads of those organizations stated: "The scale of the potential humanitarian challenge presented by climate change in the future is huge. This is a defining moment to ensure that the challenge is not insurmountable and human suffering is minimized" (IASC 2009: 1). This article describes the positions and roles of humanitarian organizations in the climate change discourse. First, the notions of humanitarianism and humanitarian organizations are discussed, as well as diverging definitions, political spectra within the humanitarian community, and differing opinions regarding right humanitarian practice. In a second step, the positioning of selected humanitarian organizations towards climate change is presented, also regarding internal debates that led to the actual position of those organizations. In a final step, the consequences of the climate change debate for the future of humanitarianism and potential cooperation with actors of the climate movement are evaluated.

2 Humanitarianism – contours of a discourse field

There are no consistent definitions of the notions of 'humanitarian action' or 'humanitarianism.' Rather, humanitarianism can be interpreted as a loose discursive framing of different functions, forms, and interventions of national or transnational, governmental or nongovernmental institutions and in which ethical norms and political aims are interwoven. At the same time, the notion describes a set of actors and a field of activity of aid workers that are employed in crisis situations:

> In addition to being an ideology, a movement, a profession and a compassionate endeavour to provide assistance and protection to populations at risk, humanitarianism is also a set of institutions, a business and an industry that employs hundreds of thousands of individuals, in which actors compete for market share.
>
> *(Donini 2008: 30)*

Focusing stronger on the activities and aims of humanitarian organizations, humanitarian aid is often distinguished from the long-term efforts of the development community; humanitarian aid is then characterized as activity whose geographical and temporal scope is closely bound to a hazard event or violent conflict that has initiated the intervention. The Global Humanitarian Assistance (GHA) defines humanitarian aid as "aid and action designed to save lives, alleviate suffering and maintain and protect human dignity during and in the aftermath of emergencies" (GHA 2012: 87). The short-term character as a central characteristic of humanitarian aid is constrained in the same source, when the GHA states that humanitarian aid is

> intended to be 'short-term' in nature and provide for activities in the 'immediate aftermath' of a disaster. In practice it is often difficult to say where 'during and in the immediate aftermath of emergencies' ends and other types of assistance begin, especially in situations of prolonged vulnerability.
>
> *(GHA 2012: 87)*

The traditional distinction based on point of time and duration of an intervention has lost significance in recent years, as activities of humanitarian organizations increasingly stretch beyond the immediate aftermath of natural disasters or violent conflicts. One reason for this development is the increasingly difficult distinction of 'natural' extreme events or violence as triggers of humanitarian crises from already existing vulnerabilities, political and economic inequalities within societies, and structural forms of marginalization. Those multifaceted crisis situations have been described as 'complex emergencies' (see, e.g., Goodhand and Hulme 1999). In the early 1990s, these changes initiated a debate within the humanitarian and development communities on the manifold intersections between different phases of humanitarian aid on the one hand, and peace building or development activities on the other hand. The debate identified a tendency towards a 'relief-development-continuum' (see Buchanan-Smith and Maxwell 1994; Duffield 1994), affecting both the work of humanitarian and development actors. Reflecting respective changes in the agenda of humanitarian organizations, Barnett (2009) differentiates 'emergency' from 'alchemic' organizations: while emergency organizations are described as being mainly engaged in alleviating human suffering in the face of acute emergency situations, alchemic organizations are characterized by the coexistence of various aims and differing operational contexts. After the Cold War, alchemic organizations have increasingly gained influence within the humanitarian community (Barnett 2009: 627).

Another criterion for delimiting humanitarian organizations from others is the basic principles of humanitarian aid, to which most actors in the field relate in one form or another. The humanitarian principles build upon the principles of the International Committee of the Red Cross (ICRC) that, with regard to its huge importance for the evolvement of modern humanitarianism, has been called the "high priest of humanitarianism" (Barnett and Weiss 2011: 9). In its statutes, the ICRC refers to the principles of humanity, impartiality, neutrality, and independence as the core pillars of humanitarian aid (ICRC 2007: 5). Those principles are often perceived as the quintessence of humanitarianism that defines the ethical basis of humanitarian action: "In sum, humanitarianism is defined as the desire to provide life-saving relief while honoring the principles of humanity, neutrality, impartiality, and independence" (Barnett and Weiss 2011: 10). Anyhow, the humanitarian principles are at the core of a debate that has been subsumed under the buzz-phrase of a 'crisis of humanitarianism' in the early 2000s. David Rieff (Rieff 2002), who coined the phrase, accused humanitarianism of falling behind its self-imposed moral codex, paradoxically induced by a turn towards human-rights questions in the 1990s. After the end of the Cold War,[1] humanitarian organizations had gradually shifted their focus, increasingly justifying interventions with human rights violations, and thereby weakening the strong moral stance of neutrality, impartiality, and, ultimately, independence from third parties. The move towards a human rights-driven agenda, as Rieff concluded, had opened humanitarianism for far-reaching political agendas of national and international actors and undermined the traditional neutral appearance of those organizations. Chandler (2001) similarly argues that by picking up human issues, humanitarian actors have made themselves connectable to political agendas also pursued with military means. The notion of 'humanitarian interventions' subsumes the increasing convergence of both fields, that led to a relatively open cooptation of humanitarian actors in the military interventions in Iraq and Afghanistan (Donini et al. 2004; Shannon 2009), especially since 9/11 and within the context of the US 'Global War on Terror.' Those dramatic changes have also been made possible by underlying discursive shifts, e.g., the revision of strong, state-centered security concepts through individual-based conceptions of 'human security,' often uncritically shared and supported by humanitarian and development actors (see Duffield 2006; Chandler 2008).

Consequently, the apolitical character of humanitarian aid has been relativized in recent years. Humanitarian actions do have political impacts and are perceived as being political on the ground (Barnett and Weiss 2011: 4). This politicization of humanitarianism has been explained with different developments and circumstances, such as the growing importance of states for funding and defining missions, as well as places of engagement. Besides, new funding mechanisms, changing characteristics of political conflicts, and a stronger focus on human rights issues have contributed to the adoption of new practices by humanitarian organizations that would have been rejected as too political some years before (Barnett 2009: 653). This at least partial renunciation of the principle of neutrality is particularly evident in the documentation and accusation of human rights violations in conflict situations, as practiced for example by the 'Médecins Sans Frontières' with reference to the 'witnessing' paradigm of the organization.

The humanitarian field is characterized by an abundance of different actor and organization types. The Global Humanitarian Assistance differentiates actors of the public sector (i.e., governmental agencies on different levels), multilateral organizations (especially UN agencies and other institutions like the World Bank), NGOs, and civil society organizations (CSOs), and the different institutions of the Red Cross and Red Crescent movement. To analyze the behavior of multilateral organizations bears some difficulties, as they intrinsically depend on the international community and powerful states within the international system. They have multiple roles and functions in the international political architecture, are both fora and actors of global governance processes, are at the same time influential and subservient, and they are part of negotiation processes regarding their own mandate (see Ivanova 2003: 20). The processes of change described in the following sections have to be understood and evaluated against this background and the multiple and contradictory roles of humanitarian organizations in international politics.

3 A new world order? Humanitarian organizations and global climate change

To determine the relevance of the climate change debate for the controversial discourses on principles and aims of humanitarianism is an ambiguous task. It seems clear that new challenges posed on humanitarian actors through climate change will not only change funding sources for aid missions, but also impact normative and operational positions within the humanitarian community. The German Committee for Disaster Reduction (Deutsches Komitee Katastrophenvorsorge, DKKV) states:

> [M]odifications in norms and structures are needed in terms of better bridging between short and immediate humanitarian aid and medium recovery and long-term resilience strategies in development. Consequently, humanitarian aid has to change its primarily shockdriven approach towards including a more medium- and long-term perspective for reducing vulnerability, especially in the light of climate change.
>
> *(DKKV 2011: 7)*

Those far-reaching claims are at least partially supported by the relevant international humanitarian organizations. In the forerun to the climate negotiations in Copenhagen, the Inter-Agency Standing Committee (IASC), a network of the big humanitarian agencies, suggested the need for a new humanitarian 'business model' to be able to react to climatic shocks and to focus more intensively on preventive measures (IASC 2009: 2).

Based on official documents as well as interviews led with representatives of humanitarian organizations in Geneva, Switzerland and Bonn, Germany in 2010, the following section describes the position of selected humanitarian organizations within the debates on the humanitarian consequences of climate change.

3.1 UNHCR and climate change

The United Nations High Commissioner for Refugees (UNHCR) has a narrow mandate defined through international law, and traditionally concentrates on the provision of shelter and emergency aid in the context of humanitarian, especially refugee crises.[2] As special agency of the United Nations, the UNHCR is accountable to the UN General Assembly and the Economic and Social Council (ECOSOC); the program and budget is approved by an Executive Committee. At least formally, programmatic changes and decision-making within the organization depend on the UN member states.

The internal climate change debates of the organization center around the question of potential impacts of environmental and climatic changes on migration and displacement that have been the subject of academic and humanitarian discussions alike. Since the early 2000s, a growing body of literature has been concerned with the question of whether and how environmental factors impact migratory movements and displacement, with different, often antithetical answers. One threat within the discourse considers environmental change as an important factor that directly impacts migration, concluding that climate change will produce massive displacement. Other, more skeptical views underline the complexity of migration decisions, deriving the need for contextual knowledge and the consideration of broader, political, economic, and social circumstances to be able to evaluate the extent to which the environment impacts migration. For a long time, UNHCR has clearly followed this second strand and, even more clearly, has doubted the legitimacy of establishing connections between environmental factors and displacement at all. By delimiting itself from establishing a causal relation between the two areas, UNHCR has also disclaimed a potential responsibility for possible new forms of refugees. For instance in 2001, UNHCR published a study (Black 2001) that heavily criticized the notion of environmental refugees as nonconstructive and politically questionable. The following quote underlines the ambivalent attitude of UNHCR towards the environment-displacement-nexus, in which the growing relevance of climate change for displacement is confirmed, but changes in the statutes of UNHCR and the operations are rejected: "Environment does and will play an increasing role in global displacement, but it would be wrong to squeeze environmental refugees into the existing legal frameworks. We would be opening the flood gates" (interview with representative of the UNHCR-headquarter, March 2, 2010 in Geneva).

Similarly, a strategy paper from 2009 argues that although there are potentially multiple impacts of climate change on displacement events, UNHCR is only able to engage in those situations under very specific conditions. For example, UNHCR operations are held possible when conflicts arise in the context of environmental change or when complete national territories are lost due to sea level rise. The main argument is, anyhow, that UNHCR strictly opposes a renegotiation of the existing refugee definitions as a basis of the UNHCR mandate:

> UNHCR considers that any initiative to modify this definition would risk a renegotiation of the 1951 Refugee Convention.... [I]n the current political environment, it could result in a lowering of protection standards for refugees and even undermine the international refugee protection regime altogether.
>
> *(UNHCR 2009: 9)*

This position has not fundamentally changed in recent years. Nevertheless, UNHCR nowadays engages on different levels in the debates on humanitarian consequences of climate change. For instance, UNHCR has sponsored empirical studies on the connections between environmental change and displacement, for example in Eastern Africa (see Afifi et al. 2012). Furthermore, the relevance of climate change for UNHCR operations has been discussed during the meeting of the UNHCR executive committee in 2011 (see UNHCR 2011) and has been set on the agenda of the ministerial conference in December 2011. The outcome of this process was the foundation of the Nansen Initiative in October 2012. This Swiss and Norwegian-led process tries to identify protection gaps in the international law and establish a protective status for persons displaced by disasters and extreme weather events. So far, the UNHCR mandate is not touched by those developments.

3.2 The Red Cross and Red Crescent Movement and climate change

Traditionally, the international Red Cross and Red Crescent Movement conceives itself as nonpolitical and mainly focused on the alleviation of human suffering in humanitarian crises. In the past decades, large parts of the movement have extended their engagement and are nowadays also concerned with long-term programs and measures that are independent from single disaster events. In principal the Red Cross/Red Crescent Movement is independent from the influence of nation states, although one of the central institutions, the International Committee of the Red Cross (ICRC) is provided with an official mandate by the international community of states (Seiters 2012) and integrated intensively in different fora of the international political system. Given its highly diversified structure[3] and the strong regional subdivision, it is difficult to identify a general policy change within the movement. Nevertheless, single sub-institutions can be taken as an indicator in how far the introduction of new issues is translated into policies and operational changes. With regard to the climate change debate, the International Federation of the Red Cross and Red Crescent Societies (IFRC) is an interesting example.

The IFRC describes itself as one of the first nonenvironmental organizations to realize the serious threats climate change is likely to pose on vulnerable populations in the near future (Red Cross/Red Crescent Climate Center 2006). As early as 2002, the IFRC established a Climate Center in The Hague with the clear aim to gather information and strategies developed throughout the movement and to provide the national societies of the Red Cross/Red Crescent with this information. The Center describes its tasks as follows:

> Our strategy to address these rising risks is to integrate climate risk management into existing Red Cross and Red Crescent priorities and programs, rather than to develop stand-alone climate change activities.... The Climate Centre facilitates access to climate-change related channels of funding ... and advocates for support to the most vulnerable people in the international climate policy arena.
>
> *(Red Cross/Red Crescent Climate Centre n.d.)*

Here, the Center's proximity to the Red Cross operations is obvious: besides an engagement in international policy fora and the proactive quest for new funding sources, the main focus is on the cooperation with the project level and the programs of the national societies (Red Cross/Red Crescent 2010).

Changes that have taken place in the head office of the IFRC in Geneva are more related to the Movement's attempts to set climate change and its humanitarian consequences on the agenda of the international community. Within the head office, working positions were established in the run-up to the Conference of Parties (COP) of the United Nations Framework Convention on Climate Change (UNFCCC) in Copenhagen 2009. Together with representatives of other humanitarian organizations, those positions were built to communicate the humanitarian position on climate change to the participants of the negotiations. A Task Force on Climate Change that had been created within the agency network IASC was coordinated by IFRC staff members. An IFRC representative describes the Task Force's main assignment: "And we have a very clear role. We have been requested by the leaders of the organizations to undertake strong advocacy on climate change adaptation and humanitarian action" (interview with IFRC representative,

March 3, 2010 in Geneva). Besides this clear concentration on advocacy issues, the IFRC staff showed a strong commitment to build the IFRC position and the information communicated in advocacy measures on tangible examples from the project level.

> A huge part of our activities as a Task Force were directed towards Copenhagen, but we did also have regional consultations, looking at more operational issues. And obviously that also needed to feed into the Copenhagen Conference and for the principals of the organizations to be able to say that they are working very much on, and also building up capacities on the ground.
>
> *(Interview with IFRC representative, March 3, 2010 in Geneva)*

This shows an interesting connection between the operational level and the activities during the international climate negotiations. A strong negotiation position of humanitarian organizations derives from their experiences and the link to the project level. After the COP in Copenhagen, the climate change activities of the IFRC headquarters have slowed down. Since the mandate of the IASC Task Force ended in 2011, no relevant statements and publications have been issued from here.

3.3 The Welthungerhilfe and climate change

The German NGO 'Welthungerhilfe' (WHH) was founded in the 1960s as an organization exclusively focusing on emergency aid. Today, the organization has a much broader mandate and pursues the objective to improve food security through projects all over the world. Besides emergency aid in famine situations, the WHH is engaged in rural development projects, as well as education programs. The WHH operates internationally and describes itself as being independent from governmental, political, or confessional influences (WHH n.d.). With a substantial share of its funding being raised through charitable donations (WHH 2012), the WHH only partially depends on funding from governmental sources.

Climate change plays an increasing role in the organization's concepts and activities. In 2010, the issue has been defined as one of the major future challenges to be addressed through the organization's activities. In a 2011 strategy paper, climate change is described as a cross-cutting issue with far-reaching consequences for all WHH operations. Long-term planning in regional and country programs will have to account for climate change in future, and climate change adaptation projects will be supported in all partner countries (WHH 2011: 9). A 'climate adaptation assessment' has been developed by WHH to support all staff members involved in the design of regional and country programs to "efficiently assess climate change impacts on measures and partner regions and adequately account for them" (WHH 2011: 10).

Following a WHH representative, the turn towards climate change reflects the long-standing experiences and observations made by WHH staff in the partner countries. The introduction of the issue in the policy debates of the organization is therefore considered as a reaction to locally observable changes:

> You can see that rain falls have changed, that the frequency and intensity of precipitation has shifted and that traditional cultivation techniques don't work as they used to. Cultivation practices have not been hampered by population growth, progressing wealth etc., but by changes in the climatic conditions, that's quite obvious.
>
> *(Interview with WHH representative, November 3, 2010 in Bonn)*

Nevertheless, it is stated that the preoccupation with environmental and climate change issues is not necessarily new for WHH. The focus on food security has been traditionally associated with questions of vulnerability and sustainability, which have always been part of WHH operations.

> For subsistence farmers, you need local pathways. You have to pick them up where they are. An increasing industrialization is not the right way, especially in the face of global climate change. The report 'Agriculture on a Crossroad' and the IPCC-reports have been valuable sources to support those lines of reasoning.
>
> *(Interview with WHH representative, November 3, 2010 in Bonn)*

In this context, the climate change debate is understood as an opportunity to strengthen the positions and support policy approaches and practices that have traditionally determined the organization's work.

4 Trends and characteristics of the climate change debate in the humanitarian community

The inner- and trans-organizational discourses described above underline some characteristics of how climate change is discussed within the humanitarian community. Humanitarian organizations generally share the belief that climate change is increasingly relevant for humanitarian crises and impacting their own work, as crisis situations are exacerbated or created. It is feared that the rising number and intensity of natural disasters might translate into growing numbers of people affected by crises. The degree to which humanitarian organizations have incorporated climate change into their programs and operations as well as their public relations, anyhow, is highly diverging. Differences can be explained by the general orientation of respective organizations, especially whether they follow a narrow humanitarian mandate or are also engaged in broader 'development' agendas. The example of the UNHCR shows that the organization type also has an impact on how quickly an organization is able to adopt a new issue: the accountability to the member states is apparently among the reasons that have made it difficult to broadly incorporate climate change into statutes and practices of the organization. In contrast, for a smaller, more independent NGO like WHH, it has been easier to develop new policies and project approaches in the context of the climate change debate.

The big humanitarian organizations devote a certain share of their activities to advocate for their position in different policy arenas. The IASC activities, for example, have clearly concentrated on attempts to influence the international climate negotiations, without being crowned with extensive success, though. One important objective pursued by humanitarian organizations in this and other policy contexts has been the integration of 'disaster risk reduction' (DRR) measures into international and national adaptation frameworks (Mitchell et al. 2010: 17; Kellett and Sparks 2012). The arguments and notions, to which respective claims refer (for instance the call for a convergence of emergency aid and long-term resilience strategies) show a surprisingly clear renunciation of classic humanitarian agendas. In this line of reasoning, humanitarian organizations would further extend the scope of their operations and would be provided with a broad justification for interventions in diverse situations. This would enable them to operate even more independently from the occurrence of disaster situations. At the same time, buzz phrases like 'resilience,' 'prevention,' and 'vulnerability' indicate a close orientation towards hegemonic discourses in the development community and a far-reaching authorization to intervene in various (geographic) contexts. How far the perspectives of new funding sources have influenced the position of humanitarian organizations can only be hypothesized. OCHA, for instance, has pointed to the high relevance of the Adaptation Funds of the UNFCCC for future funding of the humanitarian community, as DDR measures are explicitly supported in this context (OCHA 2010: 5).

It is obvious that the humanitarian community focuses almost exclusively on the question of adaptation to climate change; the second important pillar of action against climate change – mitigation – is widely marginalized in the debates. This reduction continues on the project level: apart from single attempts to reduce their own greenhouse gas emissions and thereby minimize the impacts on climate change (here, UNHCR is one of the few examples), changes in the design and implementation of projects mainly relate to the adaptation discourse. There is a strong emphasis on integrating adaptation and DRR into project activities and on an increasing consideration of existing information on weather and climate forecasts (Hellmuth et al. 2011). The IFRC, for example, strongly advocates for a better utilization of data and forecasts from local meteorological services to be able to prepare for extreme weather events before they occur. Experiences from the West African floods of 2008 triggered the IFRC's calls for including 'user-oriented climate services,' especially forecasts of extreme weather, into humanitarian operations (Braman et al. 2013).

We have seen that the climate change debate does have major impacts on the behavior of humanitarian organizations. A withdrawal from the rather *reactive* agendas of 'classical' humanitarian aid also suggests more *political* approaches of respective programs: reducing vulnerabilities to environmental and climatic changes could also spotlight economic and political inequalities, as well as local or global power structures. Whether or not climate change leads to a politicized humanitarian aid on a local level will have to be evaluated in the future. By now, the rather technical nature of discourses in the advocacy campaigns of the big humanitarian agencies, but also the changes on the operational level, do not allow a 'politicized humanitarianism' to be identified in the context of climate change. Instead, surprisingly enough, humanitarian actors seem to target a reduced exposition towards hazards without systematically addressing the inherent political aspects of the vulnerability concept. The climate change discourse in itself is highly political, as it bears the potential to raise questions of structural inequalities in North–South relations. Thereby, the historical responsibility of the OECD world for the emission of greenhouse gases is at the center of respective global environmental justice debates (Parks and Roberts 2007). If applied consequently, this would entail a degree of politicization far beyond the current position of humanitarian organizations, as it would also shed light on global contradictions and power imbalances. The analysis shows that so far, this has been hampered by the strong adaptation focus and the rather technical strategies that are derived from the debate.

An even more dubious view is discussed under the notion of a suspected 'securitization' of global climate change. It draws upon the observation that climate change is discussed as a security issue in diverse political contexts, such as the European Union, the Security Council of the United Nations, and different US political institutions (Herbeck and Flitner 2010; Trombetta 2012). Being more than just a rhetoric sparring, respective debates could also promote restrictive answers to climate change, such as border surveillance practices to prepare for potentially increasing migratory flows. In this context, the described convergence between security agendas and humanitarian aid, as well as alliances between respective actors, is of special interest. It seems highly unlikely that humanitarian organizations will support respective threat scenarios and security solutions rather than framing climate change as an issue of global justice. Anyhow, the observable trend to dramatize climate change effects and the uncritical use of ambiguous and politically flawed concepts, such as the 'human security' lens, could unwillingly support such undesirable discourses and practices.

5 Humanitarian organizations as actors of a global climate movement?

Humanitarian organizations participate intensively in the debates on social consequences of global climate change. The direct connections between the humanitarian community and the climate movement are surprisingly weak: although representatives of both groups are active in the same policy arenas, official partnerships are rare. Some of the bigger humanitarian agencies like Action Aid, Care International, Oxfam, or World Vision are part of the Climate Action Network CAN, but there are also significant gaps. Neither Red Cross/Red Crescent institutions, nor UN organizations are found among the CAN associates (CAN n.d.).

Discourses on the humanitarian impacts of climate change are shaped by a small number of organizations, coming predominantly from the Global North. This underlines the hegemonic, Western form of knowledge production in humanitarian organizations that is often associated with colonial and gender stereotypes in the visual representation of problems like poverty, vulnerability, and hunger:

> The combination of discursive framings of spatial and social marginalisations ... and the recurrent application of stereotypes in mapping vulnerability on a global scale seems to be quite typical of the campaigns of international human aid organisations. The images of these campaigns are not new, but expound a classical colonial perspective.
>
> *(Strüver 2007: 698)*

Such inconsiderate representations could support a populist framing of climate change that has been identified as an important element of an increasing "post-politicization" of the public sphere: "[T]he climate change argument and how it is publicly staged ... has been and continues to be one of the markers through which post-politicization is wrought" (Swyngedouw 2010: 216). This does not necessarily obstruct the possibility of a politicized and democratic climate change debate, but raises doubts regarding the adequacy of including humanitarian organizations as driving forces in such attempts.

Nevertheless, humanitarian organizations are important actors in the international political system, whose neutral and apolitical image could represent a valuable resource for the climate movement. Their joined appearance at the international climate conferences could be an interesting connecting factor that could help members of the climate movement to carry their positions into the international debates and to reach important decision-makers. The moral credibility of the humanitarian organizations could favor such attempts.

We have seen that there are shared concerns between humanitarian organizations and the climate movement, such as the advocacy for the protection of vulnerable people. Shared networks and fora of the international climate politics have established first contacts between the two spheres that could be extended in the near future. Humanitarian action could also benefit from cooperation with climate groups, as mutual adaptation strategies and local climate protection initiatives could be developed and supported. For joint action, a clearer political positioning of the humanitarian community, especially the unambiguous naming of responsibilities for global climate change, is a precondition.

Notes

1 Some NGOs like the 'Médecins Sans Frontières' (MSF) started to turn towards human rights as early as the 1960s (Heins 2005; Barnett 2009).
2 Since the 1990s, anyhow, also the UNHCR has gradually turned towards human rights questions and prevention policies (Forsythe 2001; Barnett 2009).
3 The Red Cross and Red Crescent Movement consists of the ICRC with its clear humanitarian focus, and of the national Red Cross and Red Crescent societies with their umbrella organization IFRC. The latter ones follow a much broader agenda and are also engaged in health programs, education projects, etc. (ICRC 2012).

References

Afifi, T., Govil, R., Sakdapolrak, P., and Warner, K. (2012) *Climate Change, Vulnerability and Human Mobility: Perspectives of Refugees from the East and Horn of Africa*, Report No. 1. Partnership between UNU and UNHCR, Bonn: United Nations University Institute for Environment and Human Security (UNU-EHS).

Barnett, M. (2009) Evolution Without Progress? Humanitarianism in a World of Hurt, *International Organization*, 63(4): 621–663.

Barnett, M. and Weiss, T. G. (2011) *Humanitarianism Contested. Where Angels Fear to Tread*, London and New York: Routledge.

Black, Richard (2001) *Environmental Refugees – Myth or Reality?*, New Issues in Refugee Research Series Working Paper 34, Geneva: Office of the United Nation High Commissioner on Refugees.

Braman, L., van Aalst, M., Mason, S., Suarez, P., Ait-Chellouche, Y., and Tall, A. (2013) Climate Forecasts in Disaster Management: Red Cross Flood Operations in West Africa, 2008, *Disasters*, 37(1): 144–164.

Buchanan-Smith, M. and Maxwell, S. (1994) Linking Relief and Development: An Introduction and Overview, *IDS Bulletin*, 25(4): 2–16.

CAN (Climate Action Network) (n.d.) *CAN Member Organizations*. Accessed 12/28/2012 from www.climatenetwork.org/about/members.

Chandler, D. (2001) The Road to Military Humanitarianism: How the Human Rights NGOs Shaped a New Humanitarian Agenda, *Human Rights Quarterly*, 23(2001): 678–700.

Chandler, D. (2008) Human Security: The Dog That Didn't Bark, *Security Dialogue*, 39(4): 678–700.

DKKV (Deutsches Komitee Katastrophenvorsorge e.V.) (ed.) (2011) *Refining the Agenda? Humanitarian Assistance in Times of Climate Change*, DKKV Publications Series 44, Bonn: DKKV.

Donini, A. (2008) Through a Glass, Darkly. Humanitarianism and Empire, in Gunewardena, N. and Schuller, M. (eds.) *Capitalizing on Catastrophe. Neoliberal Strategies in Disaster Reconstruction*, Plymouth, UK: AltaMira Press.

Donini, A., Minear, L., and Walker, P. (2004) Between Cooptation and Irrelevance: Humanitarian Action after Iraq, *Journal of Refugee Studies*, 17(3): 260–272.

Duffield, M. (1994) Complex Emergencies and the Crisis of Developmentalism, *IDS Bulletin*, 25(4): 37–45.

Duffield, M. (2006) Human Security: Linking Development and Security in an Age of Terror, in Klingebiel, S. (ed.) *New Interfaces Between Security and Development: Changing Concepts and Approaches*, Bonn: Deutsches Institut für Entwicklungspolitik.

Ecker-Ehrhardt, M. (2007) *Neue Autoritäten? Ein kommunikationstheoretischer Blick auf die Deutungsmacht inter- und transnationaler Akteure in der Darfurkrise*, Discussion Paper SP IV 2007–303, Berlin: Wissenschaftszentrum Berlin für Sozialforschung.

Fink, G. and Redaelli, S. (2011) Determinants of International Emergency Aid – Humanitarian Need Only?, *World Development*, 39(5): 741–757.

Forsythe, D. (2001) *UNHCR's Mandate: The Politics of Being Non-Political*, New Issues in Refugee Research Series Working Paper 33, Geneva: Office of the United Nations High Commissioner for Refugees.

GHA (Global Humanitarian Assistance) (2012) *GHA Report 2012*. Accessed 12/28/2012 from www.globalhumanitarianassistance.org/wp-content/uploads/2012/07/GHA_Report_2012-Websingle.pdf.

Goodhand, J. and Hulme, D. (1999) From Wars to Complex Political Emergencies: Understanding Conflict and Peace-building in the New World Disorder, *Third World Quarterly*, 20(1): 13–26.

Heins, V. (2005) Democratic States, Aid Agencies and World Society: What's the Name of the Game? *Global Society*, 19(4): 361–384.

Hellmuth, M., Mason, S., Vaughan, C., van Aalst, M., and Choularton, R. (eds.) (2011) *A Better Climate for Disaster Risk Management*, New York: International Research Institute for Climate and Society (IRI).

Herbeck, J. and Flitner, M. (2010) 'A New Enemy Out There'? Der Klimawandel als Sicherheitsproblem, *Geographica Helvetica*, 65(3): 198–206.

IASC (Inter Agency Standing Committee) (2009) Aid Agencies Call for Strong Agreement to Address 'Humanitarian Shocks' of Climate Change, press release, June 8. Accessed 08/17/2012 from www.icva.ch/doc00003615.doc.

ICRC (International Committee of the Red Cross) (2007) *Statutes of the International Red Cross and Red Crescent Movement*. Accessed 08/14/2012 from www.icrc.org/eng/assets/files/other/statutes-en-a5.pdf.

ICRC (2012) *Components and Bodies of the International Movement of the Red Cross and Red Crescent*. Accessed 08/20/2012 from www.icrc.org/eng/resources/documents/misc/components-movement.htm.

IDMC (Internally Displaced Monitoring Center) (2012) *People Displaced by Natural Hazard-induced Disaster, Global Estimates 2011*, Geneva: IDMC.

Ivanova, M. (2003) Partnership, International Organizations, and Global Environmental Governance, in Benner, T., Streck, C., and Witte, J. (eds.) *Progress or Peril? Networks and Partnerships in Global Environmental Governance. The Post-Johannesburg Agenda*, Berlin: Global Public Policy Institute.

Kellett, J. and Sparks, D. (2012) *Disaster Risk Reduction. Spending Where it Should Count*, Somerset, UK: Global Humanitarian Assistance.

Mitchell, T., van Aalst, M., and Silva Villanueva, P. (2010) *Assessing Progress on Integrating Disaster Risk Reduction and Climate Change Adaptation in Development Processes*, Strengthening Climate Resilience Discussion Paper 2, Brighton, UK: Institute of Development Studies.

OCHA (Office for the Coordination of Humanitarian Affairs) (2010) *Potential New Climate Change Adaptation Funding Sources for Disaster Preparedness Activities*, Background Paper. Accessed 11/05/2012 from www.preventionweb.net/files/14454_OCHAFeb2010.pdf.

OCHA (n.d.) Climate Change – Humanitarian Impact. Accessed 08/14/2012 from www.unocha.org/what-we-do/advocacy/thematic-campaigns/climate-change/humanitarian-impact.

Parks, B. J. and Roberts, J. T. (2007) Fueling Injustice: Globalization, Ecologically Unequal Exchange and Climate Change, *Globalizations*, 4(2): 193–210.

Red Cross/Red Crescent Climate Centre (2006) *Annual Report 2005. Preparing for Climate Change*, The Hague: Board of the Red Cross/Red Crescent Climate Centre.

Red Cross/Red Crescent Climate Centre (2010) *What the RC/RC Climate Centre Can Do for the RC/RC Partners*. Accessed 08/21/2012 from www.climatecentre.org/downloads/File/climate_centre/Technical%20support%20provided%20by%20the%20RCRC%20Climate%20Centre_final_.pdf.

Red Cross/Red Crescent Climate Centre (n.d.) *Mission & Strategy*. Accessed 08/21/2012 from www.climatecentre.org/site/mission.

Rieff, D. (2002) Humanitarianism in Crisis, *Foreign Affairs*, 81(6): 111–121.

Rodella-Boitreaud, A.-S. and Wagnery, N. (2011) *'Natural' Disaster, Conflict and Aid Allocation*, Graduate Institute of International and Development Studies Working Paper No. 09/2011, Geneva: Graduate Institute of International and Development Studies.

Seiters, R. (2012) Die internationale Rotkreuz- und Rothalbmond-Bewegung als globaler Akteur, *Zeitschrift für Außen- und Sicherheitspolitik*, 2012(5): 1–9.

Shannon, R. (2009) Playing with Principles in an Era of Securitized Aid. Negotiating Humanitarian Space in Post-9/11 Afghanistan, *Progress in Development Studies*, 9(1): 15–36.

Strömberg, D. (2007) Natural Disasters, Economic Development, and Humanitarian Aid, *Journal of Economic Perspectives*, 21(3): 199–222.

Strüver, A. (2007) The Production of Geopolitical and Gendered Images Through Global Aid Organisations, *Geopolitics*, 12(4): 680–703.

Swyngedouw, E. (2010) Apocalypse Forever? Post-political Populism and the Spectre of Climate Change, *Theory, Culture & Society*, 27(2–3): 213–232.

Trombetta, M. J. (2012) Climate Change and the Environmental Conflict Discourse, *Hexagon Series on Human and Environmental Security and Peace*, 8: 151–164.

UNHCR (United Nations High Commissioner for Refugees) (2009) *Climate Change, Natural Disasters and Human Displacement: A UNHCR Perspective*, Geneva: UNHCR.

UNHCR (2011) High Commissioner's Opening Statement to the 62nd Session of ExCom, Palais des Nations, Geneva, October 3. Accessed 10/11/2012 from www.unhcr.org/4e89a67b9.html.
UNHCR (2012) *Global Trends 2011. A Year of Crises*, Geneva: UNHCR.
WHH (Welthungerhilfe) (2011) *Klimaanpassungsprüfung. Ein Instrument zur Berücksichtigung des Klimawandels und seiner Auswirkungen auf die Programme und Projekte der Welthungerhilfe*, Bonn: Welthungerhilfe.
WHH (2012) *Jahresbericht 2012*, Bonn: Welthungerhilfe.
WHH (n.d.) *Welthungerhilfe Profil*. Accessed 08/21/2012 from www.welthungerhilfe.de/welthungerhilfe-profil.html.

38

CONCLUSION

Results of the handbook – state, influence, and future of the climate movement

Matthias Dietz

The handbook has dealt with the young climate movement that has not yet been researched systematically. The chapters came to several conclusions which fit together regarding many aspects and create the following picture of the climate movement:

1 Characterization of the climate movement

The climate movement is a vivid and plural collective whose members are active in several countries. There are strong and established national climate movements in Australia, the US, and Brazil. In Germany, China, and India, climate movements are currently emerging and growing. In contrast, the climate movement in Great Britain is in a deep crisis although climate activism has its origin in this country. A certain decline of the activities of climate groups can also be observed in the US, but this decline is less intensive than in Great Britain. Apart from these national submovements, the international climate movement consists of different organizations and activists particularly focusing on international climate politics. The members of the movement are very heterogeneous. Indigenous people, migrant workers, young people, and scientists participate in the movement as well as environmental organizations, church groups, and citizens' initiatives. The international visibility of the climate movement is dominated by huge networks such as CJN! and CAN. In addition, there are global campaigns and days of action organized by 350.org or La Via Campesina. Each year, many members of the movement travel to the United Nations Climate Change Conferences. They organize protests and critical side events and in part also participate in the official negotiations. The presence of the climate groups during these conferences has declined during the last few years.[1] In addition, some big players of the movement such as Climate Justice Action (CJA) and 1 Sky have become inactive. Nevertheless, the climate movement is still very active. There are climate protests all around the world and intensive processes of communication among the members of the movement (see for example the chapters by Tokar and Dietz in this book). In order to communicate the climate groups particularly use websites, blogs, mailing lists, and newsletters. Besides, reports on and discussions of the climate movement are published in magazines and newspapers such as the *Shift* magazine, *Turbulence*, the *Guardian*, and the *Rolling Stone* magazine. Furthermore, apart from the UN Climate Change Conferences, the members of the climate movement meet at events such as the Powershift conferences in the US or in Australia. The climate groups exchange information on protests and strategies across borders, mobilize for their campaigns (see for example the 'International Campaign of Struggles' by La Via Campesina in the context of the Rio+20 summit), and discuss the state and future of the climate movement (see for example McKibben 2012 and Petermann 2012). They show intensive processes of reflection and a strong self-consciousness as social movement. Due to these characteristics and a certain skepticism about technology and capitalism, the climate movement creates the impression of being a critical and intellectual actor.

2 Development of the climate movement

According to the chapters on Australia, Brazil, the US, and Great Britain, the roots of the climate movement often lie in the environmental or the environmental justice movement. At the beginning of the climate movement in the late 1990s and the early years of the new millennium, only a few civil society groups existed that focused on climate politics. They acted in a cooperative way and supported the climate negotiations of the UN (see chapters by Bedall and Görg and by Brunnengräber in this book). From the year 2005 on, when the Kyoto Protocol became active, also the popularity of the topic of climate change grew. In consequence, several climate groups were founded which set their hopes on the climate regime of the UN. After the peak of mobilization before and during the conference of Copenhagen in 2009, difficult and challenging years began for the movement. The failure of the summit led to a declining support and to an inner crisis of the climate movement. Following North (2011: 1587) one reason for this crisis was the Climate Gate scandal. On the basis of hacked and published emails, climate scientists were accused of manipulating their data and prognoses. Climate change skeptics have also been supported by the cold winters in northern countries since 2009 which seem to be in contradiction to the predictions about global warming. Finally, the financial crisis led to a change in public and political attention. Economic topics became more important and urgent than the question of climate change. This shift in the public agenda also had the consequence that some climate activists left the movement in order to engage in the protests against the financial crisis and the political reactions to it. These developments resulted in a decline of the climate movement in Great Britain. For example, at the beginning of 2011 the organizers of the British climate camps declared that they were to dissolve their organization (see chapter by Graham-Leigh in this book). Apart from these negative developments, there were continuously also positive developments in the climate movement. In 2011 and 2012 climate groups in Germany organized more climate camps than ever before. The national climate movement in Germany grew as the British climate movement slipped into a crisis. In Germany several citizens' initiatives and other organizations were founded that are engaged against the extraction and use of coal, the sequestration and capture of carbon, or the enlargement of the airport in Munich (see the chapter by Kössler in this book). The climate movements in Australia, Brazil, or China don't seem to have been influenced by the failure of the climate summit in Copenhagen (see chapters by Baer, Segebart and König, and Schroeder in this book).

Apart from its huge geographical scope, the climate movement profits from a core of consistently engaged organizations such as Rising Tide, 350.org, Climate Justice Now!, Stop Climate Chaos, and the Global Campaign for Climate Action. These organizations remained active despite the declining public attention on climate change. Besides, there are activists such as Bill McKibben, Tim DeChristopher, or Naomi Klein who are popular leading figures of the movement. They haven't shown discouragement, consistently work against climate change, and have motivated the members of the movement to keep engaged. Other reasons for the robustness of the climate movement are its reflection and discussion skills. After Copenhagen the climate activists held intensive debates on a new strategic orientation of the movement. The majority of the climate groups decided for a *local and more radical approach* of acting. Instead of cooperative action at the UN level, many climate groups now focus on confrontative protests against the fossil fuel industry at the local level. Furthermore, they are engaged in local projects to protect the climate and to reduce carbon emissions. As a consequence of this strategic shift, in 2011 more than a thousand climate activists were arrested in Washington, D.C. because they used forms of civil disobedience to protest against the construction of the Keystone XL pipeline (see chapter by Tokar in this book). Members of the US group Peaceful Uprising advertise such confrontative, but peaceful forms of protest. In their opinion, effective climate protests can mean it is necessary to accept disadvantages such as arrests, trials, and prison fines.

The authors of the handbook were able to explain these developments in the climate movement convincingly by the use of different theories such as democracy, critical, justice, and framing theory. Melanie Müller and Heike Walk stated democratic and participatory deficits of the UN negotiation system which in parts could have caused the disappointment of the climate groups and their growing avoidance of the UN level. Achim Brunnengräber observed a general trend of social movements to distance themselves from international institutions and to refocus on the national or even local level. The changing strategy of the climate movement could be part of this general trend in the movement sector. According to Brunnengräber, this renationalization is also caused by the selective participatory offers of international institutions which have become more restrictive in the last few years. This analysis fits with the observations of Müller and Walk. Besides, different economic, ecological, and political crises would currently undermine the trust in the problem-solving ability of international politics. Philip Bedall and Christoph Görg state a politicization of the climate regime. After the Kyoto Protocol was passed, it was supported by many climate groups. During the last few years this situation has changed and the protocol has been increasingly criticized. One reason for this was the approach of the protocol to treat climate change as a technical problem and not as a societal problem, which is related to questions of justice, distribution, and capitalism. Furthermore, criticism came from people being negatively influenced by the market-based instruments integrated in the protocol.

The described radicalization of the climate movement was accompanied by a rising popularity of the concept of climate justice. Since the concept was developed at end of the end 1990s, more and more climate groups support the demand that the industrial countries should take on more responsibility for climate change. For Donatella della Porta and Louisa Parks, the concept of climate justice is a frame and therefore a special interpretation of the problem of global warming. This frame has been diffused in the climate movement. Processes such as the diffusion of ideas and interpretations also take place in other social movements. From the perspective of climate justice, global warming should be seen in a broader, more normative and societal context: Many climate activists point at the guilt of the industrial countries for the high concentration of carbon in the atmosphere. They underline the special vulnerability of poor countries concerning the consequences of climate change as well as the resulting necessity that rich countries reduce their emissions, pay compensations, and transfer technology in poorer countries. As shown by Silja Klepp in her chapter on the Pacific island of Kiribati, by Vito De Lucia in his critical analysis of the concept of climate justice, and by Johannes Kruse in his investigation of the Cochabamba Conference, demands for climate justice are already articulated by state-run actors and politicians. For example, representatives of Bolivia or Venezuela as well as politicians of Kiribati demand the implementation of climate justice. Scotland has started a fund for climate justice projects (see chapter by Tokar in this book). In the view of Johannes Kruse, the conference of Cochabamba and its final declaration particularly strengthened the diagnostic and prognostic dimension of the frame of climate justice. On the whole, authors of the handbook convincingly used framing approaches to explain the attitudes and development of the climate movement.

The diffusion of the climate justice frame caused effects that were judged in different ways: A spreading of the concept could result in stronger relations among the members of the movement and increase its attraction for outsiders (see chapter by della Porta and Parks in this book). Other authors of the book described cases in which the concept of climate justice was used for general accusations and inappropriate financial demands (see chapters by Klepp and Kruse in this book).[2] For organizations such as the Indigenous Environmental Network and NTAC as well as for authors such as De Lucia (in this book) or Larry Lohmann (2006), technology transfers to southern countries do not help or support these countries because hegemonic, capitalistic, and Western dominated structures and addictions would be strengthened.

To judge the relevance of the idea of climate justice the chapter by the philosopher Andreas Niederberger is helpful. In his opinion, the concept in its basic form is not fully convincing because it does not include the perspective of future generations in a sufficient way. Meanwhile a concept like climate justice is very important for climate politics because it could help to find just solutions for the distribution of emission rights or the amount of compensation payments. Furthermore, the concept is attractive because its focus on justice corresponds to a normative trend in society. Demands for justice are articulated in different policy fields and by several social movements. Possibly a certain theoretical revision of the concept of climate justice could prevent its abuse for selfish demands and improve its positive and practical effects for dealing with climate change.

3 First conclusions

On the whole, the development of the climate movement currently seems to have taken a slightly positive turn. After the disappointing climate summit of Copenhagen, there was a crisis and reorientation of the movement. Some groups dissolved, others oriented towards a local, radical approach of acting, often including ideas and demands of climate justice. Before Copenhagen only a minority of the climate groups had followed this approach. While the climate summits lost their relevance as focal points of the movement, the rising popularity of the concept of climate justice perhaps made it to a new focal point of the movement which facilitates orientation and identificational support. This assumption also gets confirmed by documents such as the 'Durban Declaration on Carbon Trading' from 2004 and the declaration of the Cochabamba Conference from 2010. These documents refer to the concept of climate justice and are very popular in the climate movement. Hundreds of organizations and individuals have signed them. On the Internet it is possible to sign some of them up to today (see for example Durban Group 2013). Many members of the climate movement interpret climate justice and the local approach of acting more moderately than when they were described and implemented in 2009 and previously (see chapters by Brunnengräber and De Lucia in this book). Today, neither critics of capitalism and technology, nor fundamental opposition to economic and state-run actors are automatically or always related to these concepts and approaches. One could conclude that there has been not only a reframing of the problem of climate change, but also a reframing of climate justice. This new and moderate interpretation of climate justice, as shown by De Lucia's analysis based on critical theory, results in a loss of the counter-hegemonic character of the concept. Radical climate groups disapprove of this development and have started debates and conflicts. According to della Porta and Parks, such processes of demarcation, often initiated by radical actors, are typical of social movements. Nevertheless, there are no signs that intensive and antagonistic conflicts will arise in consequence of the shifting meaning of climate justice, as there were on the approach of acting of the movement around 2009 and 2010 (see for example Dietz 2010; RTNA 2010; Tokar 2010). Generally, the level of conflict in the climate movement is considerable. As shown in different chapters of the handbook, the members of the movement particularly fight over the use and harm of capitalism and of economic instruments for climate protection.

4 Strengths and restrictions of the climate movement

Despite its limited scope and popularity, the climate movement has a certain influence on politics and society. Although the movement was not able to contribute to a positive result of the UN Climate Change Conferences with its lobbying and protest activity, it has several times positively influenced climate politics. On the one hand, this has happened on a discursive level, in the form of words, ideas, and symbols, on the other hand in the form of small physical successes. For example, the climate movement is responsible for the creation and rising popularity of the concept of climate justice. Topically, the concept has an impact on the negotiations about compensatory payments for climate change and the transfer of technology. It also helped to change perspectives and to strengthen the position of the global south. The campaigns of the US organization 350.org have achieved the outcome that almost all states have accepted the aim to lower the concentration of carbon in the atmosphere to a level of 350 ppm. Complex questions on how to stop climate change and the judgment of the current situation can be answered by this figure in a concrete and measurable way: In 2012 the concentration of carbon in the atmosphere was at 392 ppm, so that there is no more buffer and an immediate reduction is necessary to achieve a sustainable level.[3] The aim of 350 ppm reduces the complexity of climate change and increases the pressure on international actors, because they committed themselves (even if just in a soft way) to engage in advanced climate protection by accepting the importance of this figure.

Furthermore, members of the climate movement have close relations to states such as Bolivia, Venezuela, Chile, Brazil, the Philippines, Kiribati, and to a certain degree also to China. By consulting these states, they are able to influence national and international climate politics (see chapters by Kruse, Klepp, Müller and Walk and Schroeder in this book). Besides, tens of thousands of people participate in the worldwide days of actions and in carbon reduction campaigns of the climate groups such as 10:10. These events probably have changed the attitudes of many people towards a more climate and environmental friendly lifestyle, for example regarding consumption or mobility behavior. Many people have been sensitized to the problem of climate change. This can also be considered as a success and shows the positive influence of the climate movement, even if such changes in individual behavior and attitudes are difficult to measure. Nevertheless, this development could also have negative effects because directing the responsibility for climate change to the individual level could lower the pressure on politics to act on this problem.

Small physical successes of the climate movement include the prevention of projects with a negative impact on the climate. Climate groups for example stopped the planned enlargement of the airports in Munich and London Heathrow (see chapters by Kössler and Graham Leigh in this book), the construction of a further coal plant in Kingsnorth (Stewart 2009), or the Keystone XL pipeline in the US which was intended to transport huge amounts of oil (see chapter by Tokar in this book). These successes have a symbolic and motivational effect, because they prove that the activity of the climate groups can have relevant and visible results.[4]

Apart from these successes, the rising popularity of the concept of climate justice, the reorientation of many climate groups, and the robustness of the climate movement, there are even more signals for a positive development of the climate movement. The chapters of the handbook have shown that the movement seems to be able to attract huge masses and has the potential to grow. According to Hans Baer, even syndicates, church groups, and parties are engaged in the Australian climate movement. The chapters on Brazil, Kiribati, the Cochabamba Conference, and the democratization potential of the climate negotiations mentioned cooperation between politicians and state-run organizations and members of the climate movement. Melanie Müller and Heike Walk described the integration of climate and environmental groups in the UN delegations of Chile and the Philippines. Johannes Herbeck analyzed in his chapter the role of humanitarian organizations. Due to their high credibility, positive public image, and rising engagement in projects related with climate change, they would be valuable partners for the climate movement. At least some relations to the movement already exist because Oxfam and Care International are members of the network CAN. Perhaps the new local focus of the climate groups could facilitate cooperation with humanitarian organizations. Whether in future the relations of climate groups to actors who are outside the field of climate politics will intensify is difficult to predict and probably depends on the general importance of and public attention on climate change. After the topic had lost attention because of the financial crisis, prognosis about a faster and more intensive process of climate change (see for example Ismar and Donnebaum 2012) could return it to the top of the political and public agenda.

Looking at the potentials of the climate movement, the positive judgment of the Cochabamba Conference by many authors of the book is striking. As Johannes Kruse described, actors of civil society participated in huge numbers in this conference and had full voting rights. Furthermore, a declaration including radical demands was signed. Bolivia brought this declaration into the UN negotiation process. For della Porta and Parks, the conference enhanced the communication between the radical and the moderate wing of the climate movement. Besides, it supported the spreading of the climate justice frame. According to Bedall and Görg, the conference strengthened the communication and networks of the climate movement. Another alternative and international conference like Cochabamba could possibly enhance the cohesion of the movement. It could concentrate its demands and create further pressure on political and economic actors. In particular due to the involvement of representatives of states, such conferences could get higher attention by international politics than events solely organized and attended by civil society. Nevertheless, it is difficult to find out whether the climate groups would join such an alternative summit again, because many of them have lost interest in international negotiations or simply cannot afford the traveling costs to such conferences anymore (lack of resources; see above). A huge part of the climate movement now focuses on actions and projects on the local level.

Many authors of the handbook also interpret the coalition attempts of the climate movement in a positive way. For Elaine Graham-Leigh, the cooperation between the British climate movement and the British anti-public-cuts movement is a big chance to realize common aims such as green jobs and to create a powerful mass movement. Della Porta and Parks as well as Tokar observed relations between the climate movement on the one side and the global justice movement, the antiwar, and the Occupy movements on the other side. In their eyes, such relations make it possible for the different radical and moderate wings of the movements to come into contact with their counterparts in other movements. This could strengthen these subgroups. According to della Porta and Parks, such coalitions are also possibly an indicator for the appearance of a multi-issue movement that is focused on questions of justice and equality. From the perspective of Vito De Lucia, it is also useful to connect the climate topic with further questions of global justice. Consequently, coalitions and cooperation with other movements are seen as possible strengthening of the climate movement.

Despite these potentials, the restrictions of the climate movement should not be ignored. First, the described alliances not only result in chances, but also in risks. Coalitions and unifications with other movements could weaken the profile, independence, and aims of the climate movement. It is uncertain which specific progress and success in climate politics could be achieved by a unified global movement aiming at justice. Due to negative experiences of the German climate movement in cooperation with the antinuclear movement, Georg Kössler argues against too many and too broad coalitions because these could absorb energy and make the participants forget the original aims of the movement. An important question in this context is how the Occupy movement as a potential partner of the climate movement will develop in the future. After high growth rates at the beginning, only one year after the appearance of the movement, it is again declining significantly (Vieregge 2012). Due to the dynamic and fluctuating development of social movements, it is difficult to build up a consistent partnership among them.

In addition, Graham-Leigh, Kössler, and Dietz observed negative consequences of the internal debates and reflections of the climate movement. The British movement suffered from severe internal problems because popular activists criticized the antinuclear position of the movement. Furthermore, they made the movement *itself* responsible for its crisis. Kössler reported that some of the debates of the German climate movement are very intensive, complex, and slow down the activity of the movement. In the chapter on debates and conflicts in the climate movement a negative and careful perspective, that a part of the climate groups has, was ascertained. The climate groups intensively search for possible disadvantages of existing instruments for climate protection. In consequence, these instruments are often rejected and regarded as false solutions. Besides, these groups demand precise strategic considerations before establishing cooperation and starting actions. It is possible to wonder whether this careful perspective does not have a negative effect on the ability of the movement to act and to cooperate. A related question is how such a skeptical approach could contribute to a fast and effective reaction to the problem of climate change. The observed disadvantageous effects of the debates and reflections of the climate movement are of importance because they have been confirmed by primary sources, both activists participating in the handbook.

Another restriction of the climate movement became clear in the chapters on peripheral actors. Municipal networks (see organization profile Climate Alliance), humanitarian organizations (see chapter by Herbeck in this book), syndicates, parties, and church groups often have a quite weak relation to the climate movement. Although these peripheral actors are engaged in climate politics and sometimes have even built up special networks such as the Climate Alliance and transition towns, they keep a certain distance from the core of the climate movement. Perhaps they feel threatened by the radical demands and forms of protest that parts of the movement favor and therefore prevent contact in order not to lose their positive public or neutral political image. Another reason could be that the organizational forms are too diverse. On the one side there are professional, highly structured, and established organizations such as syndicates, parties, or the administrations of communities. On the other side there are self-organized climate groups with decentralized and informal structures. A certain tolerance, more urgent problems resulting from climate change such as in Australia, as well as the readiness for political statements and commitments on the part of the peripheral actors could enable a closer cooperation. But again the question arises how strong the protest and acting ability of a broad climate movement, in which such moderate and established actors take part, would be.

A general problem of international active social movements is the availability and mobilization of resources (McCarthy and Zald 1977). Networking, protest, and lobbying activity at the international level is expensive and consistently only possible for actors with a sufficient financial budget. Many members of the climate movement have these resources. US organizations such as 1 Sky and 350.org are supported by several foundations. Brazilian environmental and climate NGOs are often financed by international donors (see chapter by Segebart and König in this book). Nevertheless, grassroots groups often just dispose of very small budgets. This clearly limits their opportunities for acting and participation.

5 Future of the climate movement

Due to the different and interconnected strengths and restrictions of the climate movement, it is difficult to make predictions about the future of the movement. A first prognosis is (i) that the climate movement will consistently live on in the next few years. The rising scope and speed of climate change will create duties and possible public attention for the movement. Furthermore, the movement has proven to be able to cope with periods of crises, so that possible problems in the future probably will not endanger its existence. This robustness of the movement is based on the broad regional spread of the climate groups as well as on the existence of a core of permanently engaged groups. In addition, uniting, strengthening, and identity spending effects are created by focal points such as the concept of climate justice and several basic documents of the movement. The two activists participating in the handbook (Graham-Leigh and Kössler) showed high motivation and stated that they will keep engaged in the climate movement despite difficulties and problems. A second prognosis (ii) is that the size of the movement in the future will be similar to its current size. A development towards a mass movement does not seem to be probable. The potential for such a growth exists because the mobilization before and during the Climate Conference of Copenhagen was impressive and there are many groups and people who have weak relations to the movement and could be activated. Nevertheless, climate change currently is not public and political topic number one. For example, the number of media reports about the UN climate summits is declining. The news factors 'sensation' and 'novelty' currently do not seem to be fulfilled by climate change. Perhaps people have heard enough threatening news about global warming and therefore are no longer interested in the topic. Despite prognosis on a faster and more dangerous climate change (Ismar and Donnebaum 2012), there probably will not be an increase in the value of the topic the next time. This assumption is also based on the role of politics which creates the impression that it is coping with climate change successfully by enlarging the period of validity of the Kyoto Protocol. Finally, the financial and economic crisis is not yet over and will determine media headlines and political agendas for more time to come. Despite all these points, (iii) a growth of the climate movement is at least possible, even if it is not probable. This could happen if disasters took place, which are related to climate change or if other windows of opportunities appeared. Such 'possibilities' could for example be the withdrawal of further states from the Kyoto Protocol (Canada left the treaty in 2011; see Pötter 2011) or the early missing of its aims. As a result, public attention on the climate movement and its positions could rise if people got the impression that official climate politics is no longer working. In single countries the climate movement could grow due to substantial results and successes of their work. This could raise the interest of the media and motivate people to join the movement. Consequently, the future of the climate movement depends on several external factors, in particular the public relevance of climate change.

As a further prognosis about the future of the movement (iv), the local and radical approach of acting could spread. First, this prognosis is based on the assumption that nothing of relevance or interest will happen at the UN negotiations in the next years because the enlargement of the Kyoto Protocol relieved the pressure of developing a post-Kyoto treaty. An intensification of civil disobedience and other radical forms of protest is possible (see chapter by Tokar in this book). Because fossil resources are becoming rare, risk related forms of extraction such as fracking and offshore drilling will be used more often. Besides, the prosperous situation of the fossil fuel industry, rising emissions, and the increasing threats of an accelerated global warming could motivate the climate movement to intensify and radicalize its protests. When the climate groups recognized in the aftermath of Copenhagen that their cooperative approach of acting wasn't working, they shifted towards local and confrontative forms of protest. If they recognize that their current strategy still is not having the intended effects, another radicalization could be the result. A good example for the possible development of the climate movement is the following message of Bill McKibben who for a long time held quite moderate positions: "the fossil fuel industry ... is Public Enemy Number One" (McKibben 2012). Anger and frustration of the climate activists could rise and have effects on their activity. Furthermore (v), the question of possible coalitions could lose importance because possible partners are currently losing their attraction. The Occupy movement is declining (Vieregge 2012), some authors are writing about the end of the global justice movement (Banse and Habermann 2012), and the antiwar movement is in a rather inactive period. Finally (vi), due to its high attractiveness, one could expect a further growth of the popularity of the concept of climate justice. This could also result in new declarations and documents on the topic. Because of the weaknesses of the concept (see chapter by Niederberger in this book) and its different interpretations by the actors, debates on and distinctions of the concept are probable. Perhaps in the future there will be different forms of original and new, moderate and radical as well as of 'false' and 'correct' climate justice. Among others, while developing the concept of climate justice a topic to deal with could be the implementation because politics is recognizing and using the concept more and more. A rising popularity and use of the concept could result in a stronger position of southern actors in the climate movement as well as in climate politics. Nevertheless, this prediction is not very probable because many Western NGOs are part of the climate movement and international politics is traditionally dominated by Western states.

6 Further research

Future research on the climate movement first could focus on the national level. A question of research could be which factors are responsible for the different state of the national climate movements. In Australia, for example, the climate movement is quite successful and deeply integrated into society, whereas in other countries it is smaller, less established, and less influential. Are the causal factors responsible for the success of a national climate movement transferable to other countries and can the movement influence them by itself? Furthermore, it would be important to analyze young and rising movements such as in India in order to learn about the development of social movements in general and the climate movement in particular. Besides, relatively unknown national climate movements in Eastern Europe, Africa, or Asia could be searched for and investigated (as an introduction, see for example Bates 2009; Neuner 2011; Rigg 2012). As a starting point of such research, the climate organizations of young people, which were looked at in this book only in part, could be used. For example, there are organizations and events such as the Asian Youth Climate Movement, South Asian Youth Summits on Climate Change, the Arab Youth Climate Movement, or the Russian Youth Climate Movement.

Research would be also useful regarding the role and influence of single actors and subgroups of the climate movement. Apart from the youth climate organizations, indigenous actors are very interesting in this context. Indigenous climate NGOs seem to be characterized by an advocacy approach and a consistently high orientation on the Climate Change Conferences of the UN (see, for example, IEN 2012a; La Via Campesina 2012a, 2012b). This approach requires explanation, because the climate summits have not produced relevant decisions and successes for a long time. Furthermore, the indigenous groups define themselves as a grassroots movement.

They are engaged in local conflicts and fight for self-determination (see for example IEN 2012b). The mixture of local and international as well as of radical and diplomatic acting seems to be contradictive. The questions are raised, why do they keep being oriented at the international level, how politicians react to them, and whether this orientation influences the relation to the other members of the climate movement who reject the climate regime. A first possible explanation could be that the indigenous actors are following their own strategy and generally have special relations with other members of the climate movement.

Furthermore, research could concentrate on the consequences of the shifted strategy in the climate movement. What are the internal and external effects of the local approach of acting? Is the communication of the movement and its cohesion influenced by the decentralized actions? Do politics and media recognize the local actions and if so, do they perceive them in context with the local protests of other climate groups?

Researchers of social movements could use the case of the climate movement to analyze the role, influence, and consequences of the participation of established and state-run actors in a social movement. Does the cooperation with communities, parties, or syndicates increase or lower the impact, legitimacy, and coherence of a movement? How is the identity of a movement affected if a part of its members act in a cooperative way (for example by participating in carbon offset projects) and therefore provoke the protest of another part of the movement with which they share basic values and general aims?

As regards the relations of the climate movement to other movements and the resulting exchange of activists, another interesting question for movement researchers occurs: Which effects do the exchanges of activists have? On the one hand positive effects could result because such activists transport experience and knowledge, for example about how to mobilize effectively for protests. On the other hand, as observed by Dietz and Kössler, former activists of the global justice movement now engaged in the climate movement provoke conflicts and blockades because they persist in their old positions and aims. It could be researched whether and with which consequences attempts of actors exist to overtake or to use social movements for their own purposes.

As methods for dealing with these questions, network analytical approaches could be useful. Following della Porta and Parks, these methods could visualize the relations between the climate movement and other movements. Furthermore, it would be possible to identify partnerships and dividing lines between social movements. Network analysis could be also used to analyze the weak relations between periphery and core of the climate movement.

Finally, the characteristics of the people taking part in the climate movement should be researched. While literature often relates protest behavior with middle class people due to the resources necessary for engagement, the question occurs about the situation in countries where no or only a weak middle class exists. Who participates in the climate movement in India or in China? The motivation of the climate activists in these countries could be analyzed. What motivates them to take part in the protest activity, although they have only few resources and the visibility of global warming in their region is still quite low? Which explanational power unfolds the concept of 'moral panic'? Does it, as argued by Lipschutz and McKendry (2011: 373), particularly explain protest behavior in Western democracies?

For further research on the climate movement, the use of theories can be recommended. As shown by the authors of the handbook, framing, justice, democracy, and critical theories help to explain the development and self-concept of the climate movement. The theories also delivered hints at the strengths and weaknesses of the climate movement.

7 Final comment

This handbook was intended to strengthen the research on the climate movement and to create a concrete, systematic, and complete picture of it as far as possible. Hopefully these aims have been achieved. In a final and condensed description, the climate movement can be characterized as a heterogeneous collective that is active in different countries as well as on the international level. It is somewhat small and unknown, but quite robust and focuses more and more on justice and a locally oriented approach of confrontative action. Despite its limited scope, the movement has already proven to be able to change society and to influence politics.

Due to these insights, there are many opportunities and open questions for further research. The climate movement is and also in future will be a scientifically, but also politically and socially exciting object of investigation.

Notes

1 While the Copenhagen summit in 2009 was attended by ten thousands of activists, there were only some thousand of them in Durban 2011 (Schwikowski 2011) and some hundred in Doha 2012 (Wille 2012).
2 The final Declaration of the Conference of Cochabamba demands from the developed countries compensatory payments of six percent of their GDP (see chapter by Kruse in this book). This is an enormous and unrealistic sum because the developed countries until today have not achieved their aim of spending 0.7 percent of their GDP for foreign aid. In the chapter on Kiribati (see Klepp in this book), an inhabitant is cited with the following statement: "Australia should pay us a wastewater system; they are responsible for climate change and should now pay compensation." This statement reveals a material instrumentalization of climate change as well as a simplified and wrong understanding of global warming.
3 Nevertheless, the orientation on 'objective' facts and figures of this kind implicates a homogeneity – for example of regions – which does not exist.
4 Research has proven that soft factors such as ideas or consulting activity could bear political influence (see for example Williams 2004; Derichs 2006; Yeoh 2010).

References

Banse, H. and Habermann, F. (2012) Vom Ende der Globalisierungsbewegung – und dem, was kommt: ein Rück- und Ausblick, *Forschungsjournal Neue Soziale Bewegungen*, 25(1): 51–60.
Bates, W. (2009) Applications Have Begun for the South Asian Youth Summit on Climate Change, March 8. Accessed 12/16/2012 from www.350.org/de/node/5437.
Derichs, C. (2006) *The Power of Ideas: Intellectual Input and Political Change in East and Southeast Asia*, Copenhagen: NIAS.
Dietz, M. (2010) Gibt es eine internationale Klimabewegung? Ein skeptischer Befund, *Forschungsjournal Neue Soziale Bewegungen*, 23(4): 6–18.
Durban Group for Climate Justice (2013) Climate Justice Now! The Durban Declaration on Carbon Trading. Accessed 01/15/2013 from www.durbanclimatejustice.org/durban-declaration/english.html.
Indigenous Environmental Network (2012a) Indigenous Environmental Network – Road to Rio+20 and BEYOND. Accessed 12/17/2012 from http://indigenous4motherearthrioplus20.org/ien-rio-20-earth-summit-delegates/.
Indigenous Environmental Network (2012b) About Us. Accessed 12/17/2012 from www.ienearth.org/about.html.
Ismar, G. and Donnebaum, D. (2012) Die Erde wird wärmer und wärmer – und die Gegenwehr der Staaten bleibt aus, *Trierischer Volksfreund*, November 26.
La Via Campesina (2012a) Rio+20: Opening Statement of the Farmers Major Group, June 29. Accessed 12/16/2012 from http://viacampesina.org/en/index.php/actions-and-events-mainmenu-26/-climate-change-and-agrofuels-mainmenu-75/1269-rio-20-opening-statement-of-the-farmers-major-group.
La Via Campesina (2012b) Governments Produce Blank Pages in Doha for Planet's Future; La Via

Campesina Farmers Are Cooling Down the Planet, December 7, 2012. Accessed 12/16/2012 from http://viacampesina.org/en/index.php/actions-and-events-mainmenu-26/-climate-change-and-agrofuels-mainmenu-75/1350-governments-produce-blank-pages-in-doha-for-planet-s-future-la-via-campesina-farmers-are-cooling-the-planet.

Lipschutz, R. and McKendry, C. (2011) Social Movements and Global Civil Society, in Dryzek, J., Norgaard, R., and Schlosberg, D. (eds.) *Oxford Handbook of Climate Change and Society*, Oxford: Oxford University Press.

Lohmann, L. (2006) *Carbon Trading. A Critical Conversation on Climate Change, Privatisation and Power*, Uppsala: Dag Hammarskjöld Foundation.

McCarthy, J. and Zald, M. (1977) Resource Mobilisation and Social Movements: A Partial Theory, *American Journal of Sociology*, 82(6): 1212–1241.

McKibben, B. (2012) Global Warmings Terrifying New Math, *Rolling Stone*, August 2.

Neuner, M. (2011) Klimabewegungen in Russland. Chancen und Probleme ihrer transnationalen Vernetzung, in Brunnengräber, A. (ed.) *Zivilisierung des Klimaregimes. NGOs und soziale Bewegungen als Akteure der nationalen, europäischen und internationalen Politik*, Wiesbaden: VS Verlag.

North, P. (2011) The Politics of Climate Activism in the UK: A Social Movement Analysis, *Environment and Planning*, 43(7): 1581–1598.

Petermann, A. (2012) The System Will Not Be Reformed. Accessed 09/28/2012 from http://climate-connections.org/2012/07/24/three-responses-to-bill-mckibbens-new-article-global-warmings-terrifying-new-math/.

Pötter, B. (2011) Austritt aus dem Kioto-Protokoll. Klima-Outlaw Kanada, *taz*, December 13.

Rigg, K. (2012) The Arab Youth Climate Movement Should Give Us Hope, *Guardian Environmental Blog*, November 9. Accessed 12/16/2012 from www.guardian.co.uk/environment/blog/2012/nov/09/arab-youth-climate-movement.

RTNA (Rising Tide North America) (2010) The Climate Movement Is Dead. Long Live the Climate Movement. Accessed 03/11/2010 from www.risingtidenorthamerica.org/wordpress/wpcontent/uploads/2010/03/climatemovement_longlive.pdf.

Schwikowski, M. (2011) Klimagipfel in Durban. 'Hört nicht auf die Verschmutzer,' *taz*, December 4.

Stewart, B. (2009) Kingsnorth Climb-down is the British Climate Movement's Biggest Victory, *Guardian Environmental Blog*, October 8.

Tokar, B. (2010) Toward a Movement for Climate Change. Accessed 02/08/2010 from www.social-ecology.org/wp/wpcontent/uploads/2009/12/brian1.Toward_a_Movement.pdf.

Vieregge, T. (2012) Wenn der Protest sich totgelaufen hat, *Die Presse*, September 23.

Wille, J. (2012) Halbstaatlicher Aufmarsch, *Frankfurter Rundschau*, December 3.

Williams, M. (2004) Why Ideas Matter in International Relations: Hans Morgenthau, Classical Realism, and the Moral Construction of Power Politics, *International Organization*, 58(4): 633–666.

Yeoh, M. (2010) Considering the Council's Counsel: Advisor Influence and Political Change, *Public Choice*, 142(3): 293–296.

INDEX

For Product Safety Concerns and Information please contact our EU representative GPSR@taylorandfrancis.com
Taylor & Francis Verlag GmbH, Kaufingerstraße 24, 80331 München, Germany

›duct-compliance

.206

3 8 2 3 2 2 3 5 *